STRUCTURAL FAILURE

STRUCTURAL FAILURE

Edited by

TOMASZ WIERZBICKI

Department of Ocean Engineering
Massachusetts Institute of Technology
Cambridge, Massachusetts

and

NORMAN JONES

Department of Mechanical Engineering
University of Liverpool
Liverpool, England

WILEY

A WILEY-INTERSCIENCE PUBLICATION

JOHN WILEY & SONS

New York / Chichester / Brisbane / Toronto / Singapore

Library of Congress Cataloging in Publication Data:

Structural failure/edited by Tomasz Wierzbicki and Norman Jones.
    p. cm.
    Lectures from the Second International Symposium on Structural
Crashworthiness, held June 6–8th, 1988, at the Massachusetts
Institute of Technology.
    "A Wiley-Interscience publication."
    Includes bibliographies and index.
    ISBN 0-471-63733-5
    1. Structural failures—Congresses. 2. Structural stability—
Congresses. I. Wierzbicki, Tomasz. II. Jones, Norman.
III. International Symposium on Structural Crashworthiness (2nd:
1988: Massachusetts Institute of Technology)
TA656.S773 1988
624.1'71—dc 19   88-18571
                                    CIP

Contributors

W. Abramowicz
Department of Ocean Engineering, Massachusetts Institute of Technology,
   Cambridge, Massachusetts 02139

S. T. S. Al-Hassani
Department of Mechanical Engineering, University of Manchester, Institute
   of Science and Technology, P. O. Box 88, Manchester M60 1QD, England

J. Amdahl
Marintek, SINTEF, Norwegian Marine Technology Research Institute A/S,
   P. O. Box 4125 Valentinlyst, N7001 Trondheim, Norway

A. G. Atkins
Department of Engineering, University of Reading, Whiteknights, Reading
   RG6 2AY, England

T. A. Duffey
Aptek, Inc., P. O. Box 4404, Albuquerque, New Mexico 87196

N. Elfer
Martin Marietta Corp., New Orleans, Louisiana 70113

A. H. Fairfull
Matsel Systems Ltd., Cunard Building, Liverpool, England

D. E. Grady
Sandia National Laboratories, P. O. Box 5800, Albuquerque, New Mexico
   87185

K. Goudie
Department of Mechanical Engineering, University of Manchester Institute of Science and Technology, P. O. Box 88, Sackville Street, Manchester M60 1QD, England

E. Haug
Engineering System International S.A., 20 Rue Saarinen, Silic 270, 94578 Rungis—Cedex, France

D. Hull
Department of Materials Science and Metallurgy, University of Cambridge, Pembroke St., Cambridge CB2 3QZ, England

S. Iwai
Research Section, Earthquake-Resistant Plasto-Ductile Structures, Disaster Prevention Research Institute, Kyoto University, Gokasho, Uji City, Kyoto Prefecture 611, Japan

N. Jones
Department of Mechanical Engineering, University of Liverpool, P. O. Box 147, Liverpool L69 3BX, England

M. E. Kipp
Sandia National Laboratories, P. O. Box 5800, Albuquerque, New Mexico 87185

J. B. Martin
Faculty of Engineering, University of Cape Town, University Private Bag, Rondebosch 7700, Republic of South Africa

T. Moan
Division of Marine Structures, The Norwegian Institute of Technology, N-7034 Trondheim, Norway

T. Nonaka
Research Section, Earthquake-Resistant Plasto-Ductile Structures, Disaster Prevention Research Institute, Kyoto University, Gokasho, Uji City, Kyoto Prefecture 611, Japan

A. M. Rajendran
The University of Dayton Research Institute, Structural Integrity Division, 300 College Park, Dayton, Ohio 45469-0001

S. R. Reid
Department of Mechanical Engineering, University of Manchester Institute of Science and Technology, P. O. Box 88, Sackville St., Manchester M60 1QD, England

A. de Rouvray
Engineering Systems International S.A., 20 Rue Saarinen, Silic 270, 94578 Rungis—Cedex, France

C. S. Smith
Admiralty Research Establishment, St. Leonards Hill, Dunfermline Fife KY11 5PW, England

J. M. Thomas
Failure Analysis Associates, Engineering and Scientific Services, 2225 East Bayshore Road, P.O. Box 51470, Palo Alto, California 94303

T. Wierzbicki
Department of Ocean Engineering, Massachusetts Institute of Technology, Cambridge, Massachusetts 02139

To Our Parents

T. Wierzbicki
N. Jones

Contents

Preface

This book contains 15 chapters, which were presented as invited lectures at the *Second International Symposium on Structural Crashworthiness*, held at the Massachusetts Institute of Technology, Cambridge, Massachusetts, June 6–8, 1988.

Failure of engineering structures is a frequent occurrence in the present industrial complex when structures are sometimes pushed to the limits of their performance capabilities, or have to function in unforeseen and hostile environments. Everyday life brings us examples of the catastrophic collapse of buildings, bridges, and other civil engineering structures, damage suffered by aircraft, ships, and land-based vehicles, and accidents to industrial plants. It is the responsibility of scientists and engineers to determine the cause of failure, to describe the failure process, and to suggest measures of preventing the failure from recurring. It is also the responsibility of managers and planners to assume an "acceptable" operational risk of a structure. Failure is understood differently from the decision-making and the structural-analysis viewpoints. This book seeks to contribute to the latter aspect of the failure process by discussing how to predict, resist, and control failure.

Many industries and regulatory agencies are becoming increasingly aware of the importance of making allowances for safety in the design phase. This is partly the result of economic pressures, which dictate lighter structural designs, and partly legal and environmental concerns. However, this situation requires an increased understanding of structural failure since it is important to take account of the total life-cycle costs, and trade off any reductions in structural weight with increases in damage frequency and severity.

Simple and reliable engineering methods are thus needed to ensure safe storage and transportation of nuclear materials and other hazardous substances, to determine the amount of damage sustained by a supertanker caused by an internal gas explosion, to predict the rupture of containment vessels subject to an airplane crash, or to analyse the damage to a power-plant piping system caused by the loss of coolant. The loading conditions that could lead to the breakup and fragmentation of space vehicles are required. New techniques should be devised to improve the collision protection of land, sea, and air vehicles, or to determine the residual strength of partially damaged onshore or offshore structures. Finally, the failure pattern of structures made of composite materials should be better understood. This list may be extended, and every reader will probably have different personal experiences with structural failure.

Various aspects of structural and material failure have already grown into separate and well-defined disciplines, such as constitutive modeling, linear and nonlinear fracture mechanics, damage mechanics, corrosion, fatigue, buckling, etc. Advances in the respective disciplines have been documented in a number of books, monographs, and conference proceedings. Despite these achievements and the considerable progress made in each of the individual disciplines, many important problems remain unresolved.

Perhaps the best example of the lack of a sufficiently general predictive technique is the tragic flight 51L of the space shuttle *Challenger*. The orbiter broke up into several pieces, which were further shattered upon water impact. It is symptomatic of the situation that no detailed analysis of the failure process for this historic accident was reported in the open literature. A satisfactory description of structural failure requires a complex, multidisciplinary approach, and the analysis must cross boundaries between various disciplines.

This book has grown from our experience with the *First International Symposium of Structural Crashworthiness*, which organized by the present editors in 1983, at the University of Liverpool. The invited lectures were published in the book *Structural Crashworthiness* [1], and the contributed papers appeared in special issues of the *International Journal of Impact Engineering* [2] and the *International Journal of Mechanical Sciences* [3]. These publications contain a conspectus of the literature on the structural crashworthiness of aircraft, automobiles, buses, helicopters, ships, trains, and offshore platforms, as well as the structural characteristics of basic structural members and energy-absorbing systems.

It is clear from the proceedings of the *First International Symposium* that a considerable bulk of knowledge existed on the ductile behavior of a wide range of individual structural members and structural systems when subjected to quasi-static and impact loads. However, the difficult problem of the failure of structures due to material failure and interactive failure were not addressed in the 1983 meeting.

A designer requires information on the failure characteristics of a structure so that the safety factor and margins against failure can be assessed accurately. For example, it is important for many practical applications to have a capability for predicting the damage or permanent deformations of a ductile structural member when subjected to large dynamic loads. Indeed, this is an important type of failure when *excessive deformations* can impede the safe operation of a system or cause damage to other important components. However, in many other practical cases, particularly for rare events (impact damage, explosions), it is important to predict when the *material will fail* after excessive deformations. If a material fails in a small-scale structural model, can the results be used to predict the behavior of a full-scale prototype of the same structure using the geometrically similar scaling laws? It appears that the geometrically similar scaling laws were not obeyed in the drop tests of welded plated structures, which were conducted by Booth et al. [4] and reported in the *First International Symposium* and as discussed further in Reference 5. Material failure is also a vital element in the energy-absorbing properties of composite structures [6–8].

Linear elastic buckling leads to a form of structural failure even though the material remains linearly elastic. However, there is a broad class of practical problems, such as those associated with various accident scenarios, that produce material failure. Thus, in this book, we are focusing on *a type of structural failure that originates from a material failure.*

While planning the present book, we came to realize that there are important gaps in the predictive capabilities of a number of individual disciplines to describe the just-mentioned process of interactive failure. These gaps can be most easily visualized if we think of the failure process as occurring at three different levels. The first is failure on a micro-, or material, level. The second is an intermediate, or through-thickness failure of an individual element. Finally, there is a global structural failure. The present book focuses on what we believe is a weak point in the understanding of the failure phenomenon, i.e., on the interaction between the three aspects of failure. In the remainder of the preface, we outline some features of the subsequent stages of the failure process, and, on that basis, highlight the connection between the various chapter contributions and indicate some areas of needed research.

An interesting and appealing description of the sequence of events starting from the micro level, through the macro level, up to global failure is given by de Rouvray and Haug in Chapter 7. We quote verbatim several definitions, since they present the scenario for the rest of the book:

"Microflaws are natural or nucleated load-induced material microvoids or microcracks."
"Microflaws cause damage."

"Damage is an abstract measure of microflaw nucleation and growth."
"Damage precipitates fracture."

"Fracture is local material separation or rupture due to microflaw growth and coalescence after a critical damage state has been reached. Brittle fracture is plane strain fracture predicted by linear elastic fracture mechanics. Ductile fracture is fracture that is not predicted by linear elastic fracture mechanics."
"Fracture precipitates failure."

"Failure is the global disability of a component or specimen to perform due
to stable or unstable fracture growth; rupture means tensile failure."
"Failure precipitates catastrophy."

Studies of the microprocesses in solids have led to the development of
physically based, or phenomenological, material models. The constitutive
modeling of engineering materials and the description of damage mechanics
have grown into separate well-defined disciplines. In this book, the constitutive
equations are taken as tools for predicting local material softening, localization,
through-thickness failure, shear band formation, etc.

An important area of difficulty is that damage mechanics does not automatically set the stage for the next phase of failure on the propagation and
branching of cracks. The initial length of a so-called "starter crack" cannot be
calculated from continuum damage theory alone. This is a severe restriction on
the presently available analytical tools for studying the failure of uncracked
bodies. The practical aspect of damage mechanics, as applied to the problem of
symmetric dynamic fragmentation, has been illustrated by Grady and Kipp in
Chapter 1. One of the first attempts to apply the damage theory to structural
failure was made in the area of the low-cycle fatigue. An up-to-date survey of the
respective methods is presented by Nonaka in Chapter 12. Damage theory is
also used in Chapter 7 to predict material separation and fracture in composite
fiber-reinforced materials.

The difficulty in extending similar concepts to more complicated structural or
loading configurations is illustrated by Rajendran and Elfer in Chapter 2 for the
protection of a spacecraft against meteoroid or debris impact. These authors
emphasize the fact that reliable predictions for the bulging, local thinning, and
bursting of the so-called rear wall of a spacecraft must rely on damage-based
criteria. The development of such criteria are being pursued in several research
centers around the world.

Next in the chain of events, after material damage, is a through-thickness
failure that can be either ductile, which is called yielding, or brittle, which is
called fracture. We understand fracture to be the creation of a new surface on the
macroscale. Several typical failure mechanisms were identified in the literature,
such as spalling, scabbing, edge and corner fracture, flaking, delamination,
through-thickness cracking, tearing, wall rupture, bursting, fragmentation,
shattering, cratering, penetration, shear punching, petaling, and perforation.
The existence of a great variety of failure modes within the thickness of a
structural member illustrates the complexity of the failure process. Activation of
any particular failure mechanism depends on many factors, such as the relative
amount of brittleness and ductility of the material, the shape of the body, and the
loading history.

Various chapters address one or several scenarios leading to a through-
thickness failure in the general context of structural failure. Al-Hassani in
Chapter 3 deals with controlled fracturing, and describes the application of
elastic wave-propagation theory to induce various failure modes of brittle solids.

Criteria for ductile shear and tensile rupture of dynamically loaded beams and plates are discussed by Jones and Duffey, respectively, in Chapters 5 and 6. Typical fracture modes of composite structures are described and analyzed in Chapters 7 and 8, and further examples of local fracture, leading to structural failure, are given in Chapters 9, 12, 14, and 15.

Special consideration should be given to the purely plastic failure modes, which do not necessarily lead to fracture. A through-thickness yielding of structural components can again take a different form, depending on the type of loading. In tension, we talk about necking; the concept of plastic "hinges" describes the bending response, whereas plastic shear bands can be formed under the appropriate loading.

The interaction of plastic flow and fracture, especially in the presence of gross yielding, is an area of intense research. This is illustrated by Atkins by the example of the continuous splitting process of a metal tube (Chapter 4). The tube can be plastically inverted, inside out, without fracturing, but cannot fracture without inducing plastic flow. Typically, the tube wall will tear at several locations to achieve a delicate trade off between flow and fracture. Plastic effects leading to fracture and tearing are further illustrated in Chapters 9 and 12.

Through-thickness brittle or ductile failure opens a way for the study of overall structural failure. We understand structural failure to be the inability of a structure to perform its function, due to a catastrophic change of shape and/or a dramatic loss of load-carrying capacity. For example, a structure is considered to fail if excessive deflections are reached (Chapter 13), or the survival space is not maintained.

In almost all instances, fracture and yielding within the structural thickness influences the overall structural response. The interaction of local and global failure modes of thin-walled structures has been studied extensively for the buckling of elastic structures. However, parallel understanding of the interaction between the through-thickness and overall failure modes of structures subjected to dynamic loads is lacking. Through-thickness failure is controlled by strains that often originate from a far-field loading. This loading may, in turn, be effected by large geometry changes. At the same time, the continuously changing local strength and stiffness of members caused by either sectional collapse, fracture, or tearing changes the overall structural response.

Proper coverage of the interaction between local and global effects was considered by us to be an important factor in planning the scope of the present book. An interesting method of accounting for the through-thickness failure in structural analysis is based upon the concept of a global "softening hinge," as presented by Martin in Chapter 11. This concept leads, in a simple and elegant way, to the prediction of an accelerated or catastrophic overall collapse. Chapters 9 and 10 contribute further to an understanding of the structural and sectional collapse of thin-walled members from the theoretical and experimental points of view, respectively. The unconventional problem of progressive plastic collapse and residual strength of damaged structural members is studied extensively in Chapter 14 by Moan and Amdahl. The effect of repeated loads on the interaction of local and global failure modes of bar and frame structures is described by Nonaka and Iwai in Chapter 12. In Chapter 13, Smith describes experiments and numerical modeling of various failure modes of stiffened plates and panels subjected to blast loadings. Finally, the last chapter in this book, by Thomas, presents an interesting summary of the problem of structural failure from a practical point of view. He finds that the four factors of stress, environment, defects, and material properties dominate structural-failure investigations.

In our opinion, the basic theoretical background for treating structural failure has been largely developed. Several failure criteria are in common use, including critical stress or strain criteria, critical stress-intensity factor K_{IC}, or its generalization for inelastic fracture, the resistance curve. Also, crack-displacement opening criterion (CDOC) is in widespread use. More advanced failure criteria, such as critical damage accumulation or the so-called "separation curve" for the process zone with softening, are becoming increasingly attractive for studying more complex cases of fracture.

What remains to be done is to link all those criteria with the overall structural response to form an interactive model of structural failure. It is hoped that this book will help to elucidate these important and complex interactive processes. In addition, the book presents a state-of-the-art review of predictive techniques for assessing the extent of catastrophic damage to engineering structures.

It is our intention, by revealing many of the challenges of the various chapters of this book, to stimulate scientists and engineers to contribute further to the understanding of failure processes of engineering structures. These new developments in structural failure will improve the protection of the public [9, 10] and of equipment from a wide range of potential hazards, and ensure safe and economical exploration of resources on our globe and beyond.

T. WIERZBICKI

Cambridge, Massachusetts

NORMAN JONES

Liverpool, England

References

1. N. Jones and T. Wierzbicki, Eds., *Structural Crashworthiness*, Butterworths, London, 1983.
2. *International Journal of Impact Engineering*, Vol. 1, No. 3 (1983).
3. *International Journal of Mechanical Sciences*, Vol. 25, Nos. 9–10 (1983).
4. E. Booth, D. Collier, and J. Miles, "Impact Scalability of Plated Steel Structures," in N. Jones and T. Wierzbicki, Eds., *Structural Crashworthiness*, Butterworths, London, 1983, pp. 136–174.
5. N. Jones, "Scaling of Inelastic Structures Loaded Dynamically," in G. A. O. Davies, Ed., *Structural Impact and Crashworthiness*, Vol. 1, Elsevier Applied Science Publishers, London and New York, 1984, pp. 46–74.
6. D. Hull, "Axial Crushing of Fibre-Reinforced Composite Tubes," in N. Jones and T. Wierzbicki, Eds., *Structural Crashworthiness*, Butterworths, London, 1983, pp. 118–135.
7. G. Dorey, "Impact and Crashworthiness of Composite Structures," in G. A. O. Davies, Ed., *Structural Impact and Crashworthiness*, Vol. 1, Elsevier Applied Science Publishers, London and New York, 1984, pp. 155–192.
8. J. D. Grundy, J. Blears, and B. C. Sneddon, "Assessment of Crash-Worthy Car Materials," *Chart. Mech. Eng.* **32**(4), 31–35 (1985).
9. G. A. Scott, "The Development of a Theoretical Technique for Rail Vehicle Structural Crashworthiness," *Proc.—Inst. Mech. Eng.*, **201**(D2), 123–128 (1987).
10. A. Scholes, "Railway Passenger Vehicle Design Loads and Structural Crashworthiness," *Proc.—Inst. Mech. Eng.*, **201**(D3), 201–207 (1987).

CHAPTER 1

Fragmentation of Solids Under Dynamic Loading

Dennis E. Grady and Marlin E. Kipp

Sandia National Laboratories
Albuquerque, New Mexico

ABSTRACT

Intense impulsive loading can lead to the catastrophic failure of solid structures. Issues of concern include the transient strength of such structures and the particle size and distribution of fragment ejecta created in these events. Recent theoretical progress and related experimental studies focused on dynamic fragmentation are reviewed. Directions of current research, including computational analysis of transient fracture and fragmentation, are discussed.

1. INTRODUCTION

The dynamic fragmentation of a solid body or structure can result from the application of an intense impulsive load. The scale of such events ranges from shaped-charge jet breakup and rock blasting to astrophysical impacts and creation of planetary debris. In rock blasting, for example, specific information on ejecta velocities and fragment-size distributions is sought, and methods to control resulting fragment sizes by proper placement and type of explosives are of interest [1]. In stretching shaped-charge jets, fragmentation characteristics, such as time-to-breakup and particle size, are intimately tied to performance [2]. Ejecta from planetary and meteoric impact provide information on the evolution and dynamics of the solar system [3]. The applications in which solids or structures are subjected to intense dynamic loading and when breakup must be mitigated or controlled are numerous and varied. The need to understand the dynamic-fracture mechanisms for such applications has provided the impetus for research in this rich area, and the field is currently quite active.

The response of a single crack or void, within a solid body, to both static and impulsive loading has received considerable attention over the past several decades and is reasonably well understood [4–6]. The mechanics of a system of cracks or voids under impulsive or stress-wave loading, and how the cooperative response of such a system relates to the transient strength and ultimate failure and fragmentation of a solid body is less well understood, and has been a subject of study over the past decade [1, 7–10]. Experimental studies of fracture under high-rate loading have revealed unusual features associated with the phenomenon, such as enhanced material strength and failure-stress dependence on loading conditions. Although such observations have led to the postulation of rate-dependent material properties, most of the features can be understood through fundamental fracture concepts when considered in terms of a system of interacting cracks or voids.

In events involving dynamic failure of solids or structures, there are a number of features for which a predictive capability is needed. Perhaps the first, and most fundamental, is the transient strength, or ability to support an impulsive load, either without sustaining fracture damage, or sustaining fracture damage within some tolerable level without permitting total failure. In partially fractured bodies, the density and location of cracks or voids can be important, along with the void volume and extent of intersection, which relate to the permeability of the crack system. In completely failed bodies, the degree of fragmentation is of interest in many applications. The size and velocity of ejected fragments is of concern, as is the distribution in fragment sizes and how this relates to the conditions of loading.

This chapter focuses primarily on processes of fragmentation when a body or structure is impulsively loaded well beyond the point of failure. Three fairly broad topics related to dynamic fragmentation are addressed. First, the intensity of fragmentation, or average fragment size resulting from an impulsive failure event and how this relates to material properties and specific loading conditions, is treated. Second, the mechanical and statistical conditions that determine the distributions in fragment size resulting from a catastrophic fracture event are considered. This is a diverse and complex topic that is not yet well understood. Finally, some of the various approaches to continuum and computational modeling of dynamic failure and fragmentation currently under consideration are discussed. Such modeling represents a necessary final goal in that the complexities of stress loading, geometry, and the interaction of stress and relief waves necessitate the use of wave-propagation codes to address realistic problems in the dynamic fragmentation of solid structures.

2. FRAGMENT-SIZE PREDICTIONS IN DYNAMIC FRAGMENTATION

In this section, attention focuses on the size of fragments created in a violent fragmentation event. The objective is to explore some theoretical ideas that are important to the dynamic-fragmentation process. The two underlying phenomena that have dominated theoretical efforts in this area of dynamic-fracture mechanics are the presence of an inherent-flaw structure and energy balance in the fracture process.

On one hand, inherent flaws or perturbations in a fracturing body, which are the sites of internal fracture nucleation, have been recognized as important in determining characteristic fracture spacing and, consequently, the nominal fragment size in a fracture event. Theoretical work based on a physical description of these material imperfections has been actively pursued [7, 11].

On the other hand, it has also been recognized that the energy of a newly created fracture surface must come from the work done on the body by the loading forces, and theories of fragment size have been developed based on energy-balance ideas, quite independent of inherent-flaw concepts. This connection was suggested early on by von Rittinger [12] and can be inferred from later studies [13, 14]. An explicit energy-balance approach to the prediction of fragment size has been proposed by Grady [15].

Each approach emphasizes different physical features observed in the fracture process. Either approach, under certain sets of conditions, can provide a satisfactory theory of fragmentation, although neither is apparently complete.

An acceptable reconciliation of inherent-flaw and fracture-energy concepts has not been achieved and provides an area of current study. The two theoretical concepts are discussed, and several applications in fragment-size prediction are described. We compare the two fragmentation approaches and attempt to identify some conditions that determine when one or the other method applies.

2.1.   Inherent Flaws and Fragment Size

The importance of inherent flaws as sites of weakness for the nucleation of internal fracture seems almost intuitive. There is no need to dwell on theories of the strength of solids to recognize that material tensile strengths are orders of magnitude below theoretical limits. The Griffith theory of fracture in brittle material [16] is now a well-accepted part of linear–elastic fracture mechanics, and these concepts are readily extended to other material–response laws.

In many materials, the inherent flaws are easily recognized. Brittle polycrystalline materials, for example, contain microcracks, voids, and other imperfections that can be identified in micrographs, and are expected to provide sites for internal-fracture activation. Artificial flaws introduced into a hollow metal shell by uniform scoring can be expected, under rapid expansion, to fracture the shell along the paths of scoring.

In other cases, the inherent flaws or perturbations responsible for fracture are less easily recognized. The internal spalling of glass or the cavitation of a rapidly expanding liquid are examples; although even here, some form of imperfection such as impurities, dislocations, or thermal fluctuations are expected to play an important role in the fracture process.

There is considerable literature on material imperfections and their relation to the failure process. Typically, these theories are material-dependent; flaws are idealized as penny-shaped cracks, spherical pores, or other regular geometries, and their distribution in size, orientation, and spatial extent is specified. The tensile stress at which fracture initiates at a flaw depends on material properties and geometry of the flaw, and scales with the size of the flaw [7, 17–19]. In thermally activated fracture processes, one or more specific mechanisms are considered, and the fracture-activation rate at a specified tensile-stress level follows from the stress dependence of the Boltzmann factor [20].

An eminently practical, if less physical, approach to inherent-flaw-dependent fracture was proposed by Weibull [21], in which specific characteristics of the flaws were left unspecified. Fractures activate at flaws distributed randomly throughout the body according to a Poisson point process, and the statistical mean number n of active flaws in a unit volume is assumed to increase with tensile stress σ through some empirical relations such as a two-parameter power law:

$$n = k\sigma^m \tag{1}$$

In the study of dynamic fracture and fragmentation, it has been assumed that there is a correspondence between the number of fractures that grow to completion and the number of fragments created in the breakage process [22]. Characterization of the inherent-flaw distribution then becomes an important material property in determining the number of fractures and the nominal fragment size in a fragmentation event. For example, considering the Weibull relation previously described, if a body were rapidly loaded to a tensile-fracture stress σ_c, then the fragment size d would be expected to be related to a critical fragment number N_c [11]:

$$d \sim N_c^{-1/3} = k^{-1/3}\sigma_c^{-m/3} \tag{2}$$

In many applications and experimental configurations leading to dynamic fracture and fragmentation of a body, it is convenient to characterize the motion of the event through a single strain-rate parameter $\dot{\epsilon}$. When the response prior to failure is elastic, the tensile stress increases linearly and

$$\sigma_c = E\dot{\epsilon}t_f \tag{3}$$

where E is the elastic modulus, and t_f is the time to spall fracture. If fracture proceeds with a nominal fracture velocity c_g, then we expect, approximately,

$$t_f \sim d/c_g \tag{4}$$

where d is a fracture-propagation distance, on the order of the fragment size, at crack coalescence. Equations (2)–(4) combine to provide a strain-rate-dependent fragment-size criterion:

$$d = f(k, m, c_g, E)\dot{\epsilon}^{-m/(m+3)} \tag{5}$$

Similarly, the same equations combine to provide a strain-rate-dependent fracture-stress criterion:

$$\sigma_c = g(k, m, c_g, E)\dot{\epsilon}^{3/(m+3)} \tag{6}$$

Therefore, at least for the case of brittle fracture, the picture is qualitatively clear. At relatively slow loading rates, the increase in tensile stress is low and relatively few fracture flaws activate. Fracture-propagation distances are large and the resulting fragments are correspondingly large. In contrast, at rapid loading rates, the stresses achieved are higher and many more flaws are activated. Consequently, fracture-propagation distances are shorter, the time to failure is less, and fragment sizes are smaller. In addition, the apparent fracture stress is not a unique property of the material, but is coupled to the flaws present and the rate of tensile loading.

2.2. Application to Rock Fragmentation

Applications in the field of rock blasting and fragmentation are appropriate to the theoretical concepts posed in the previous subsection. Laboratory and field testing have been performed to provide material property data important to the dynamic fracture and fragmentation behavior of oil shale. Properties such as the fracture stress and fragment size, and their dependence on the rate and intensity of loading, have been measured through the use of several experimental methods [11]. Data on the fracture stress and mean fragment size resulting from dynamic-fragmentation experiments are shown in Figure 1.

The data shown in Figure 1 are for an 80-ml/kg grade oil shale obtained from a mine near central Colorado. Oil shale grades from this region vary from about 40 to 320 ml/kg. Properties such as fracture toughness and elastic constants are found to depend on oil shale grade. For the oil shale in Figure 1, a fracture toughness of $K_C \simeq 0.9 \text{ MN/m}^{3/2}$, a density of $\rho \simeq 2000 \text{ kg/m}^2$, and an elastic wave speed of $c \simeq 3000 \text{ m/sec}$ are representative.

The fracture-stress and fragment-size data in Figure 1 are observed to depend on the loading strain rate over the three decades of strain rate covered in the experiments. This behavior is consistent with the power-law strain-rate-sensitive fracture properties predicted by the expressions in Equations (5) and (6). The strain-rate range of the data covers a significant portion of the range encountered in rock-blasting situations.

The theoretical fracture parameters in Equations (5) and (6), based on a model assuming an inherent power-law fracture-flaw distribution and a constant fracture-growth velocity, can be determined with the strain-rate-dependent fracture data in Figure 1 [11]. Using the fracture data for oil shale provides a value of $m = 8$ and a fracture-stress dependence on strain rate of approximately $\sigma_c \sim \dot{\epsilon}^{0.27}$. Similarly, a fracture velocity of $c_g \simeq 1300 \text{ m/s}$ is obtained, which is about 0.4 times the longitudinal wave velocity for oil shale, a reasonable upper limit for the fracture velocity.

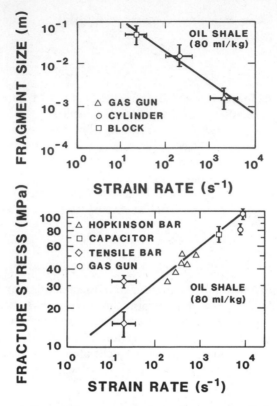

Figure 1. Fragment-size and fracture-stress dependence on tensile-loading strain rate for oil shale.

Further studies on rock and rock-like materials have tended to support the general modeling concept outlined in Equations (1) to (6) [1, 11, 23, 24].

2.3. Energy Effects and Fragment Size

Although accurate characterization of the inherent-flaw structure appears to be important, there are indications that such characterization alone may not be sufficient to explain all of the features observed in dynamic fragmentation. First, theories based on inherent-flaw concepts, although certainly necessary in some applications, are not fundamentally satisfying. At best, specification of the flaw structure is difficult; more frequently, it is impossible and curve fitting of indirect data is required to infer the inherent-flaw parameters. Second, theoretical models using inherent flaws have tended to ignore the energy consumed in fracture growth through the production of new surface area, plastic work, or viscous dissipation. Such energy must come from the applied load in the dynamic-fracture event, and it is reasonable to expect that some energy-balance principles can limit the amount of new fracture surface area that can be produced in a particular fracture event.

There is another way to model the dynamic-fragmentation process, which leads to an energy balance governing the nominal fragment size [15]. In this approach, kinetic energy introduced into the body by the dynamic load is considered to be the important energy fueling the fracture process. The model is thought to be most reliable in catastrophic fragmentation events; however, in application, it has been found quite useful over a fairly broad range of loading rates.

As an idealization, consider a body that has previously been compressed and is currently in a state of rapid expansion. The instantaneous kinematic state is determined by the density ρ and the density rate $\dot{\rho}$, which will be assumed uniform over a sufficiently large region encompassing the point of interest. The kinetic energy associated with the outward expansion is responsible for the fracturing forces, and surface tensions associated with the newly created fractures resist the fracture process. It is intuitive that, after fragmentation, particles will continue to fly apart at high velocities, as illustrated in Figure 2. Thus, the production of fragment surface cannot simply be governed by a balance of total kinetic energy and surface energy, since a large portion of the kinetic energy and surface energy remains after fragmentation. In contrast, in slower-rate crushing processes, a much larger percentage of the input energy is consumed in the fragmentation process.

An expression for the kinetic energy available for fragmentation can be determined by considering an element of the expanding body with a volume of the order of the fragment size expected from the event. With reference to a specific coordinate system, the kinetic energy of this element can be decomposed

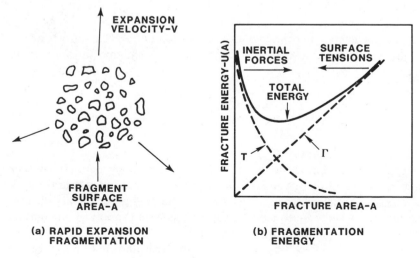

Figure 2.   Energy balance in catastrophic fragmentation.

into a center-of-mass kinetic energy E_k' and the kinetic energy relative to a coordinate system referenced to the center of mass of the element. Assuming average response, forces acting on the element during fragmentation should, due to symmetry, exert no net impulse and, consequently, both the center-of-mass velocity and kinetic energy of the element should remain invariant through the fragmentation event. Therefore, in the decomposition, the center-of-mass kinetic energy is conserved during fragmentation and only E_k' is available to drive the breakage process. This latter energy can be regarded as a local kinetic energy that is available for fragmentation without violating momentum conservation.

An explicit expression for the local kinetic energy can be obtained by considering a spherical mass element of radius r, expanding uniformly at a density rate $\dot{\rho}$. The kinetic energy about the center of mass for this single sphere is

$$E_k' = \frac{2\pi}{45} \frac{\dot{\rho}^2}{\rho} r^5 \tag{7}$$

Dividing by the volume of the spherical element and expressing in terms of the fragment-surface-area-to-volume ratio, $A = 3/r$, a measure of the local kinetic energy density, in terms of the surface-area density created by fragmentation, is obtained:

$$E_k = \frac{3}{10} \frac{\dot{\rho}^2}{\rho A^2} \tag{8}$$

The new fragment surface-energy density is simply

$$\Gamma = \gamma A \tag{9}$$

where γ is the energy associated with the creation of fragment surface area and depends on the specific material response during fragmentation. The total energy is given by

$$U = \frac{3}{10} \frac{\dot{\rho}^2}{\rho A^2} + \gamma A \tag{10}$$

as illustrated in Figure 2.

We assume that during the catastrophic fragmentation process, forces seek to minimize the energy density in Equation (10) with respect to the fracture surface-area density. This approach assumes that coordinate A is free to vary during the fracture process and, consequently, requires a sufficient supply of inherent flaws or sites of fracture initiation, although precise requirements are not yet clearly understood. This issue, which relates both inherent-flaw and energy effects, is considered further in a later section.

At energy minimum, $dU/dA = 0$, and the equilibrium fracture surface-area density is

$$A = (3\dot{\rho}^2/5\rho\gamma)^{1/3} \tag{11}$$

Equation (11) provides a quantitative measure of the fracture surface-area density created in the fragmentation process in terms of fundamental thermodynamic and kinematic properties. Interpretation of the fracture surface energy depends on the material response at fracture and the geometry of the event. For example, γ has been shown to be related to the flow stress in the fragmentation of ductile solids and to the viscosity in highly viscous fluids [25].

2.4. Application to Explosively Loaded Metal Cylinders

The application of Equation (11) to dynamic fragmentation can be illustrated with several examples. Typically, a nominal fragment size, rather than surface area, is of interest. Assuming spherical or cubical fragments of equal size, the fragment diameter or size is related to the surface-area density through $d = 6/A$.

In brittle fracture, the material-fracture toughness K_C provides a measure of resistance to fracture. If we identify the fracture energy $\gamma = K_C^2/2\rho c^2$ and a strain rate $\dot{\epsilon} = \dot{\rho}/3\rho$, an expression for the nominal fragment size is obtained from Equation (11):

$$d = \left(\frac{20^{1/2}K_C}{\rho c \dot{\epsilon}}\right)^{2/3} \tag{12}$$

An application of brittle fracture appropriate to Equation (12) is provided by the explosive fragmentation experiments of Weimer and Rogers [26] on high-strength steel cylinders. In that work, four cylinders of FS-01 steel with fracture toughness values ranging from about 20 to 60 MN/m$^{3/2}$ were explosively fragmented with Composition B explosive. Fragments were collected and the Mott size and number parameters [27] were determined for the distribution. The tensile-hoop strain rate can be calculated with the Gurney equation [28] and is approximately $\dot{\epsilon} = 4 \times 10^4$/sec. The remaining properties for the steel are $\rho = 7840$ kg/m^3 and $c = 5000$ m/sec. The Mott-size parameter (approximately, the mean fragment size) is plotted against fracture toughness in Figure 3. The fragment size based on Equation (12) is also shown in Figure 3 and observed to be reasonably consistent with the data both in magnitude and the two-thirds power dependence on K_C.

An application not involving brittle fragmentation is provided by a series of three experiments [15, 29] in which cylindrical shells of depleted uranium were imploded by varying the amount of high explosive surrounding the shell. The particular loading geometry is such that uranium is first compressed to quite high pressures (~ 500 GPa) before expansion and fragmentation. Due to shock heating during the compression phase, the uranium is a hot liquid at fragmenta-

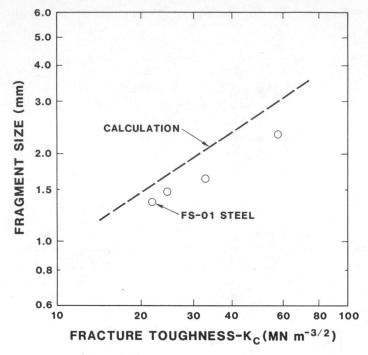

Figure 3. Comparison of calculation and experiment for dynamic fragmentation of steel cylinders.

tion. Fragment-size distributions in these experiments were determined by analyzing impact craters on lead witness plates. Data are shown in Figure 4. The expansion strain rates for each experiment were obtained from two-dimensional hydrodynamic simulations. The fracture-surface energy appropriate to the fragmentation process is provided by the surface tension of liquid uranium. This was estimated to be $\gamma = 1.75 \text{ J/m}^2$. The calculated fragment size based on Equation (11) is compared with the data in Figure 4.

In both applications just described, agreement is good, providing strong support for an energy-balance theory of fragmentation. Note in particular that no recourse to an inherent-flaw distribution was needed.

2.5. Momentum Balance and Fragmentation of Expanding Rings

Further insight into the role of energy balance in controlling the size of fragments produced under conditions of dynamic loading is provided through consideration of a special loading geometry and a particular constitutive behavior of the fragmenting material.

The one-dimensional geometry of a radially expanding ring is perhaps the simplest for considering fundamental aspects of the fracture and fragmentation

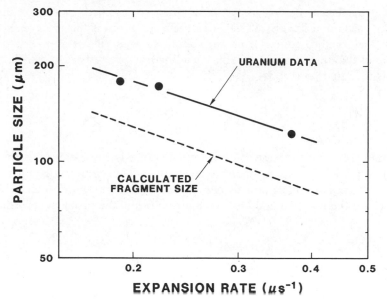

Figure 4. Comparison of calculation and experiment for explosive implosion fragmentation data on uranium cylindrical shells.

process. In a ductile metal ring, fracture proceeds through the multiple nucleation and growth of necking regions. Concepts that are found to govern fracture and fragmentation in expanding rings generalize to thin cylindrical and spherical shells.

In an incisive paper on the fragmentation of metal rings, Mott [30] outlined a solution procedure to a fracture problem relevant to the present application. The problem addressed by Mott is best visualized by a slender ductile rod flowing in tension under a constant force T_y. The introduction of an instantaneous break within the rod corresponds to fracture and, by assuming rigid-plastic behavior, he found through momentum considerations that stress unloading propagates away from the point of fracture according to the relation

$$x = \left(\frac{2T_y t}{\lambda \dot{\epsilon}}\right)^{1/2} \tag{13}$$

where x is the distance from the point of fracture to the unloading wavefront at time t, λ is the linear mass of the rod, and $\dot{\epsilon}$ is the initial rate of stretching. There is no fracture energy in Mott's model.

To model energy dissipation in the fracture process, Kipp and Grady [31] have extended the solution of Mott by including a crack-opening resistance. This is done by considering a crack-opening displacement concept in which the force in the crack (or necking region) is assumed to reduce linearly with crack-opening displacement. When the flow stress is specified, crack-opening resistance is characterized by a single parameter γ_r, which is the energy expended within the fracture zone during the failure process. The expression for the motion of the unloading wavefront, including fracture resistance, which corresponds to Equation (13), is

$$x = \frac{1}{12} \frac{T_y^2}{\lambda \gamma_r} t^2 \tag{14}$$

Equation (14) applies to the motion until fracture is complete. After this, the form of Equation (13) applies. Kipp and Grady [31] have shown that the time to fracture (time from inception to completion of failure at the fracture zone) depends on the material and kinematic properties of the problem according to

$$t_f = \left(\frac{72\lambda\gamma_r^2}{T_y^3 \dot{\epsilon}}\right)^{1/3} \tag{15}$$

Further, if one considers the distance the tensile-unloading wave can propagate over the time to fracture t_f from Equation (14), a lower-bound criterion for the fragment size can be established:

$$d = \left(\frac{24\gamma_r}{\lambda \dot{\epsilon}^2}\right)^{1/3} \tag{16}$$

Lastly, assuming that the rod fragments into pieces of nominal length given by Equation (16), a dynamic fracture-strain criterion $\epsilon_f = \dot{\epsilon} t_f$, depending on the problem parameters, can be determined:

$$\epsilon_f = \left(\frac{72\lambda\gamma_r^2 \dot{\epsilon}^2}{T_y^3}\right)^{1/3} \tag{17}$$

where ϵ_f is the dynamic strain accumulated after initiation of fracture regions.

Some qualifying comments should be made concerning the simplifications assumed in reaching the solutions provided by Equations (14)–(17). A ductile metal ring subjected to an outward radial impulse decelerates due to the circumferential tensile force. The magnitude of this force varies due to work hardening in the metal and geometric softening caused by thinning of the ring. Initially, hardening dominates, but, eventually, a maximum in the tensile force is achieved at which point current theories suggest a transition from stable uniform flow to unstable flow manifested by the onset of plastic necking and fragmentation. The present calculation is an attempt to describe the evolution of events from the onset of this transition, and the constant force T_y and the stretching rate $\dot{\epsilon}$ similarly correspond to this critical point. The law governing force relaxation in the necking zone in the present calculation is clearly a simplification of the complex stress-strain-strain-rate behavior actually occurring in the neck. Further aspects of this problem have been considered by Johnson [32], Fressengeas and Molinari [33], and Regazzoni et al. [34]. The intention of the present calculation, however, is not in achieving accurate constitutive descriptions of the material response, but in demonstrating the importance of the balance of fracture forces and inertial forces in governing fragment sizes, fracture times, and fracture strains in the dynamic-fragmentation process separate from other rate-sensitive effects.

2.6. Application to Magnetically Loaded Expanding Rings

An application of the results of the previous section to interpretation of the fracture energy in the energy-based fragment-size relation is provided by dynamic-fragmentation experiments on rapidly expanding ductile metal rings [35]. In that study, a range of expansion strain rates was achieved through a controlled magnetic loading technique. (Improvements on this technique have been described recently by Gourdin et al. [36].) Speciments were soft copper and aluminum rings 32 mm in diameter with a 1-mm-square cross section. Radial velocities were determined from displacement-time records obtained from streak camera photography of the events. The fragment number was determined by counting pieces after each test. Dynamic fracture occurred, through a process of ductile necking, followed by extension fracture near the end of the fracture process. A correlation of fragment number with the expansion velocity at fracture is shown in Figure 5.

The expected number of fragments N can be obtained from Equation (16):

$$N = 2\pi \left(\frac{\lambda R}{24\gamma_r} \right)^{1/3} V^{2/3} \tag{18}$$

where R is the ring diameter, λ is the linear density, and V is the radial expansion velocity at fracture. Again, γ_r is the fracture energy and, in this situation, corresponds to the plastic work dissipated within the neck region during fracture. A value of $\gamma_r = 0.02\,\mathrm{J}$ was determined for aluminum from static tensile tests. A comparison of Equation (18) with the data is provided in Figure 5.

2.7. Comparison of Inherent Flaws and Energy Concepts in Dynamic Fragmentation

Two theoretical approaches relating to the nominal fragment size achieved in dynamic-fracture events have been discussed. The first is based on the reasonable concept of existing fracture-producing flaws within the fragmenting body, where fragment size (or number) is necessarily correlated with the description of the flaw distribution. A second approach is based only on energy-balance concepts, where no recourse to a preexisting flaw distribution is made. Considering the diverse applications for which this latter approach provides

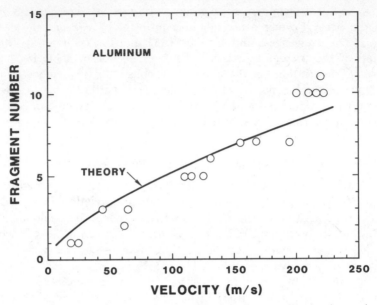

Figure 5.   Fragmentation data for rapidly expanding ductile aluminum rings.

reasonable fragment-size predictions, energy-balance ideas clearly play an important role in the fragmentation process. In outward appearance, these two theoretical descriptions do not seem compatible.

In the latter energy-balance approach, some tacit assumption of an adequate flaw distribution must be implied. Otherwise, tensile loading to the theoretical strength of the body is achieved before fragmentation could proceed. Consequently, the fragment surface-area density A, which determines the energy in Equation (10), should be viewed as a coordinate that seeks an equilibrium value through a minimum-energy principle, only if a sufficient fracture-producing flaw distribution exists. A less than sufficient distribution results in nonequilibrium values of A.

A reasonable estimate of the number of flaws needed to sustain an equilibrium value of A can be made based on the concepts introduced earlier in this section. To achieve a nominal fragment size of d, the number of activated flaws per unit volume should be of the order $d \sim N^{-1/3}$. From the equilibrium-energy theory, fragment size from Equation (11) or (12) is expected to depend on strain rate according to $d \sim \dot{\epsilon}^{-2/3}$. From this, we conclude that the number of flaws that activate an equilibrium fracture-surface-area density A (or fragment size d) should increase at least as rapidly as $N \sim \sigma^6$ [compare with Equation (1)].

In Figure 6, we illustrate this "just sufficient" distribution in comparison to a hypothetical flaw distribution for an actual material. In this example, we envision a solid of finite extent, which will have a single critical flaw that

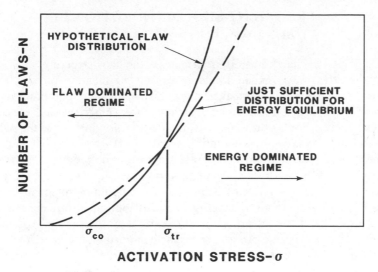

Figure 6. Energy- and flaw-dominated regimes in dynamic fragmentation.

activates at a minimum stress σ_{co}. With increasing tensile stress, the population of flaws that activate should increase rapidly, perhaps as illustrated in Figure 6. In contrast, a flaw distribution just sufficient to satisfy the energy-balance criterion increases smoothly as $N \sim \sigma^6$.

The illustration suggests that for sufficiently modest tensile-loading rates (the dynamic-fracture strength less than σ_{tr}), both the fracture stress and fragment size would be "flaw-dominated," since the flaw distribution within this range is not sufficient to allow equilibrium-energy behavior. In contrast, at higher loading rates, when the dynamic-fracture stress exceeds σ_{tr}. equilibrium-energy values of both fragment size and fracture stress are achieved since the flaw distribution is more than sufficient to allow equilibrium variation of the fragment surface-area parameter. Some computational studies of dynamic fragmentation discussed in the next section provide further support for this concept.

In other fracture processes, the ideas presented here would be couched somewhat differently. For instance, if fracture occurred through a rate-controlled thermally activated process, such as might apply in the dynamic spall of liquids, then a fracture-activation rate would be "sufficient" or "not sufficient", depending on whether an equilibrium A according to the energy-balance theory could, or could not, be achieved. In principle, the consequences would be similar to the results in Figure 6. Although speculative, the ideas of this section seem intuitively reasonable and provide a rationale for including both inherent-flaw and energy-balance concepts within a theoretical framework of dynamic fracture and fragmentation.

3. FRAGMENT-SIZE DISTRIBUTIONS IN DYNAMIC FRAGMENTATION

To this point, we have addressed primarily the flaw-structure and energy concepts in stress-wave-loaded solids governing the creation of new fracture surface area (or the mean fragment size) in catastrophic fragmentation events. In this section, we discuss in more detail what is frequently the end concern in impulsive-fracture applications, namely, the distribution in sizes of the particles produced in the dynamic-fragmentation event.

A theoretical understanding of dynamic fragmentation requires a satisfactory statistical, as well as physical, explanation. Even in situations of perfect symmetry, such as the rapid expansion of a circular ring or spherical shell, the process of breakup is random, leading to a distribution in fragment size and shape. This behavior seems almost intuitive and is borne out by experimental observation. Physical principles are also clearly important, however. The root of observed statistical distributions in fragment size is the random locations and activation levels of the flaws or fluctuations that are the sources of internal-fracture nucleation during stress-wave loading. These effects, when coupled with properties such as loading rate, material characteristics, and geometry, influence the dynamic-fragmentation process leading to a specific fragment-size distribution.

Because of the continuing practical interest in material fragmentation and fragment-size distributions, studies on various aspects of the problem can be traced through the literature for more than a century. The predominantly statistical nature of fragmentation was recognized early and stimulated efforts to identify size-distribution relations that described the resulting fragmentation. Various standard distributions such as Poisson [37, 38], binomial [39], log normal [40], and Weibull [41] have been used, and functions associated with the names of Rosin and Rammler [41], Schuhmann [42], Gaudin and Meloy [39], and Mott [27] have acquired popularity in certain applications.

Theoretical efforts a step beyond simply fitting standard statistical curves to fragment-size distribution data have involved applications of geometric statistical concepts; that is, the random partitioning of lines, areas, or volumes into the most probable distribution of sizes. The one-dimensional problem is reasonably straightforward and has been discussed by numerous authors [37, 39, 43, 44]. Extension to two and three dimensions apparently cannot be done without recourse to assumptions regarding the random placement of the partitioning boundaries. Agreement has not been achieved on this issue and, consequently, the applicability of geometric statistical fragmentation as a model for dynamic fragmentation cannot yet be determined.

It is clear, however, that a strictly statistical approach ignores the dynamics of the fragmentation event. In the actual situation, the partitioning boundaries represent growing, propagating, bifurcating, stress-relieving, energy-consuming, interacting cracks and fractures. These physical processes most probably influence the fragment-size statistics. A restricted set of those physical fracture features was considered in the studies of Mott [30], and examined more recently by Grady [45, 46] and Kipp and Grady [47]. In this approach, more significance is attributed to the dynamics of fracture activation and growth, including the nucleation process and the influence of material-deformation properties. Mott [30], in considering a restricted geometry, combined the spatial randomness of the fracture process with the growth of plastic tensile release waves and predicted fragment distributions dependent on both dynamic and material properties. More recent studies [22, 48, 49] have focused on developing physically founded laws governing the nucleation, growth, and coalescence of fracture during one- and two-dimensional stress-wave propagation. A different approach to the statistics of fragmentation has been proposed by Griffith [50], where particle-fracture surface energy is related to the distribution through a unique application of classical Boltzmann statistical concepts.

It is important to note that despite a fairly significant body of literature, an acceptable theoretical framework relating the processes of dynamic fragmentation to observed distribution is not yet available. In fact, in many cases the observed phenomena lack even a good qualitative explanation. Consequently, the topics discussed in the present section must be regarded as tentative. This area of study is, once again, becoming active and new ideas can be expected to emerge over the next few years. It is hoped that the topics in this section provide a partial background for the subject of dynamic statistical fragmentation.

3.1.   The Single-Fracture Process

In fragmentation phenomena, events in which abrupt, single, or initial one-time fragmentation occurs, as opposed to processes that involve continued reduction of already broken fragments, can be identified. For instance, a rapidly expanding spherical shell of material undergos abrupt fragmentation when the biaxial stress achieves the tensile strength of the material. Individual fragments continue to fly outward on roughly radial trajectories and no further breakage occurs. The distribution in fragment size is determined by the physics and statistics of the process at the time of failure. In contrast, when a brittle solid is deformed in shear, the initial fracture event is followed by continued comminution as fragments roll and tumble and repeatedly impact one another. As another example of continued comminution, if the spherical shell in the first example were a ductile solid or liquid, and if the initial fragments were ejected at a sufficiently high velocity into a finite atmosphere, further fragmentation of these particles would occur through aerodynamic breakup, ablation, or burning processes. The final size distribution of particles is observed to reflect this multiplicity of breakup processes. The present section focuses on theoretical concepts leading to the fragment-size distributions that result from the first example, namely, that of single fracture. Theories focused on predicting fragment-size distributions have typically started with the fundamental premise of a randomly cracked body. This immediately leads to concepts of Poisson statistics that were central to the developments of Bennett [38], Lienau [37], Mott and Linfoot [27], and Gilvarry [43].

Consequently, consider an infinite one-dimensional line or rod along which fractures occur randomly with an average frequency of N_0 per unit length, as illustrated in Figure 7. Randomly distributed points on an infinite line obey Poisson statistics and the probability of finding n fractures in a length l is given by

$$P(n, l) = (N_0 l)^n e^{-N_0 l}/n! \tag{19}$$

To determine the probability distribution in fragment lengths, first determine the probability of finding no fractures in length l:

$$P(0, l) = e^{-N_0 l} \tag{20}$$

and then the probability of finding one fracture in a length dl:

$$P(1, dl) = N_0 \, dl \tag{21}$$

Then the probability of finding one fracture whose length is within a range of l to $l + dl$ is given by

$$dP = P(0, l)P(1, dl) = N_0 e^{-N_0 l} \, dl \tag{22}$$

which provides an exponential fragment-size probability distribution in fragment length of

$$\frac{dP}{dl} = N_0 e^{-N_0 l} \tag{23}$$

The cumulative probability distribution of fragments larger than length l is obtained from

$$P(l) = \int_l^\infty \frac{dP}{dl} \, dl = e^{-N_0 l} \tag{24}$$

and the cumulative fragment number is given by

$$N(l) = N_0 e^{-N_0 l} \tag{25}$$

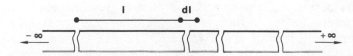

Figure 7.   Random one-dimensional fragmentation—a Poisson process.

3.2. Applications in One-Dimensional Statistical Fragmentation

Although a number of authors have discussed the one-dimensional statistical fragmentation problem, experimental efforts to test the model have been few. Lienau [37] discusses earlier work by Herrata [51], in which long thin strips of glass are broken by some random process, and states that agreement with the predicted exponential distribution appeared to be excellent. Lienau [37] performed similar fragmentation experiments on bundles of thin glass rods placed randomly into a number of envelopes and crushed in a press. He also found good agreement with an exponential distribution. The fragmentation process in both of these studies was quasi-static.

From the expanding-ring experiments discussed in Section 2.6, at the highest expansion velocities, rings were observed to fracture into 10 to 12 randomly sized fragments [35]. In Figure 8, a cumulative number distribution, obtained by combining the results of four tests performed on soft copper rings, in which exactly 10 fragments each were produced, is compared with an exponential fit to the data. The exponential prefactor was also adjusted rather than constrained by the total number of fragments.

Although an exponential curve provides a good description of 35 of the 40 fragments, there are obvious discrepancies at both the small- and large-fragment

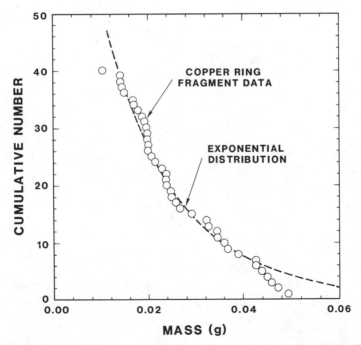

Figure 8. Cumulative number distribution of fragments from four expanding ring experiments (10 fragments each), and comparison with one-dimensional theoretical distribution based on Poisson statistics.

end of the distribution. The absence of small fragments results from the finite diameter of the wire and the energy dissipated in creating a fracture, both of which are ignored in a geometric statistical treatment of the fragmentation process. The behavior at the large end of the distribution is more difficult to understand. To account for the finite circumference of the ring, we compared the data with a binomial rather than with an exponential distribution, but this did not resolve the discrepancy. A finite rate of fracture activation is a possible explanation for the nature of the large end of the distribution.

The important observation from this data, however, is that it tends to support the basic premise suggested in the previous section: namely, geometric statistics appear to describe the gross trend of the distribution data, without alluding to specific material or kinematic properties. More specific details in the distribution require careful consideration of the actual processes involved.

3.3.   Computational Analysis of One-Dimensional Fragmentation

The statistical nature of dynamic fragmentation can be profitably explored in one dimension through further consideration of the ductile stretching-rod geometry or, equivalently, the expanding-ring geometry introduced previously. The growth and local stress-release behavior of ductile necking fracture under dynamic-loading conditions has been treated in detail both analytically and computationally [31]. The treatment of multiple fractures appears to preclude the use of analytic methods due to the complexity of wave interactions, and necessitates the use of numerical techniques [47].

In the numerical calculations [47], an elastic–perfectly plastic rod stretching at a uniform strain rate at $\dot{\epsilon} = 10^4$/sec was treated. A flow stress of 100 MPa and a density of 2700 kg/m^3 were assumed. A 1-mm-square cross section and a fracture energy of $\gamma_r = 0.02$ J were used. These properties are consistent with the measured behavior of soft aluminum in experimental expanding-ring studies of Grady and Benson [35]. Incipient fractures were introduced into the rod randomly in both position and time. Fractures grow in time, as described in Subsection 2.5, and subsequently interact due to local stress-wave release. Consequently, some fractures go to completion while others arrest at various stages of growth, depending on the relative spatial positioning and nucleation timing of the randomly introduced fracture sites.

In an initial calculation [47], fracture sites were introduced randomly in position but close enough together in time to be considered simultaneous for computational purposes. The fragment-size distribution resulting from the calculation was nearly exponential (or Poisson) in good agreement with the statistical theory described in the preceding section. An interesting complication noted was that only a fraction of the initial fracture sites introduced grew to completion and contributed to the final fragment distribution. This complexity is not addressed in the assumptions leading to a Poisson distribution in the preceding section.

In further calculations [47], a fracture-site nucleation rate uniform in time was used. Effectively, this implies that fracture sites are distributed randomly but with uniform density along the time axis as well as the position axis. Over the problem time, many fracture sites initiate. At a strain rate of 10^4/sec only 12 to 13 fractures grow to completion. This is consistent with the energy-limiting fragmentation discussed earlier and is also in agreement with experimental results on soft aluminum rings at comparable strain rates [35].

In a study where the uniform nucleation rate was varied over several values at a fixed strain rate of $\dot{\epsilon} = 10^4$/sec, the fragment number results shown in Figure 9 were obtained. At lower nucleation rates, nearly all fracture-initiation sites grow to completion and provide the reduced number of fragments shown. At higher nucleation rates, many of the initiation sites arrest before growth to failure is complete, resulting in the 12 to 13 fragments discussed in the previous paragraph. Thus, this calculation is consistent with the flaw-dominated and energy-dominated regimes of fragmentation discussed in Section 2.7.

At the conclusion of the calculation, a fragment-size distribution as well as a fragment number is provided. A cumulative number distribution is shown in Figure 10 and compared with aluminum-ring data acquired at $\dot{\epsilon} \simeq 10^4$/sec [35]. With the assumed fracture-site nucleation law, the calculated distribution appears to agree reasonably well with the data. The calculation better predicts the tails of the distribution that deviate from strict exponential behavior, as was noted in the previous section.

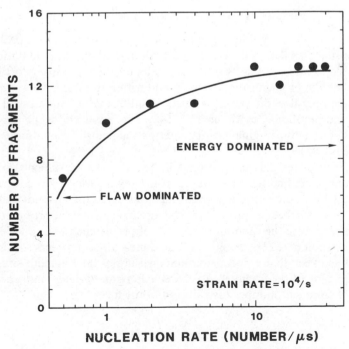

Figure 9. Numerical study of random fragmentation.

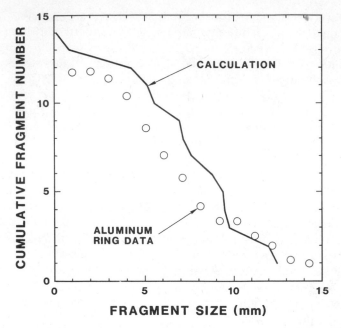

Figure 10. Comparison of numerical cumulative fragment-size data and aluminum expanding-ring data.

3.4. Statistical Fragmentation in Several Dimensions

The extension of the one-dimensional geometric fragmentation problem described in the previous sections to the two-dimensional fragmentation of an infinite sheet is an issue that is still being debated. Considerable effort has been expended on the problems that can be posed within this framework, not only in application but also on purely statistical mathematics questions [52] and random computational methods [53, 54]. The applications go far beyond problems of dynamic fragmentation, entering the fields of metallurgy, geology, and biology [55].

With the problem of dynamic fragmentation, the primary difficulty centers on constructing a method for randomly partitioning a surface. No consensus has been reached and the approaches that are considered here do not exhaust the methods that have been explored. Several methods have been proposed that do not actually involve the random placement of partitioning lines on the surface. Mott and Linfoot [27], using the one-dimensional problem as guidance, proposed, in effect, that a linear nominal measure of the fragment size, namely the square root of the fragment area a is a Poisson variable and arrived at a cumulative fragment number distribution of the form

$$N(a) = N_0 e^{-(2N_0 a)^{1/2}} \tag{26}$$

where N_0 is the number of fragments per unit area.

Although the distribution of Mott and Linfoot has had a long and relatively successful history of describing fragmentation data from rapidly expanding spherical- and cylindrical-shell experiments, there are reasons to suspect that Equation (26) may not be unique. An alternative assumption has been suggested [44] in which a scalar measure of the fragment area, a is considered as a Poisson variable. This leads directly to a cumulative fragment-number distribution:

$$N(a) = N_0 e^{-N_0 a} \tag{27}$$

This "linear exponential" distribution has also been used to represent dynamic-fragmentation data [44].

The more common approach is the actual positioning of random lines on a surface to create a statistical distribution of fragment sizes. One example of this, suggested by Mott and Linfoot [27], is a construction of randomly positioned and oriented infinite lines, as illustrated in Figure 11. If the random lines are restricted to horizontal or vertical orientation, an analytic solution can be obtained for the cumulative fragment number [27, 56]:

$$N(a) = 2N_0 (N_0 a)^{1/2} K_1 [2(N_0 a)^{1/2}] \tag{28}$$

where K_1 is a modified Bessel function. Mott and Linfoot [27] offered the close agreement between the Bessel distribution [Equation (28)] and the Mott distribution [Equation (26)] as partial justification for their proposed distribution.

Unfortunately, fragmentation by construction does not appear to be independent of the random construction algorithm. An alternative method of successive segmentation [44] of the surface, as illustrated in Figure 12, leads to a fragment distribution that agrees well with a linear exponential distribution [Equation (29)] and differs significantly from the Mott distribution [Equation (26)].

The random-area-fragmentation algorithm that has received by far the greatest attention in the literature is most frequently referred to as the Voronoi

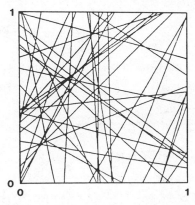

Figure 11. Geometric fragmentation with randomly positioned and oriented infinite lines.

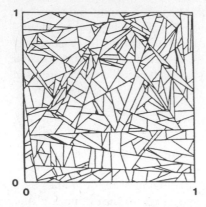

Figure 12. Geometric fragmentation with randomly oriented line segments.

construction. Points are first distributed as a random Poisson process on the surface, and perpendicular bisectors are then constructed that ultimately partition the area into a collection of randomly sized and shaped polygons. Kiang [57] has proposed that the Voronoi construction most closely describes a process of material fragmentation. The fragment distribution determined from the Voronoi construction differs sharply from both the Mott and the linear exponential distributions discussed earlier [44], as indicated in Figure 13.

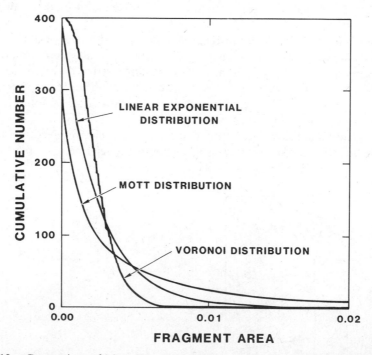

Figure 13. Comparison of Mott, Voronoi, and linear exponential cumulative fragment distributions.

It is still difficult to clearly discern the connection between the discipline of geometric statistical fragmentation and of dynamic fragmentation. At the very least, the geometric studies begin to provide a feel for how statistical processes enter into the random collection of particles generated in a catastrophic fracture event. It seems clear, however, that the subject is not yet sufficiently well developed to provide guidance for the statistics of dynamic fragmentation, although several important observations have emerged. It is now apparent that the Mott distribution has no stronger basis for acceptance than a number of other statistical assumptions for randomly fragmenting the region of interest. The authors favor the linear exponential distribution as a basis for single dynamic fragmentation.

It should be clear that most of the methods discussed in this section extend readily to three dimensions and the random geometric partitioning of a volume. For example, the linear exponential expression corresponding to Equation (27) is

$$N(v) = N_0 e^{-N_0 v} \tag{29}$$

where v is a measure of the fragment volume. A related expression for the Mott distribution is also apparent, and most of the area-fragmentation algorithms are readily extended.

3.5. Statistical Inhomogeneity and Fragment-size Distributions

In the geometric-fragmentation problems considered earlier, statistical homogeneity over the unit area was tacitly assumed. Namely, the average fragment size at a point within an element of area does not differ from any other point. In practical fragmentation events, a situation of uniform or homogeneous fragmentation is usually not achieved. Normally, due to the complexity of the dynamic-loading conditions leading to fragmentation, the intensity of fracture varies from point to point within the body, and consequently the average fragment size is also a function of position. The uniformly expanding ring discussed earlier or the uniform expansion of a spherical shell represents unique experimental geometries in which nearly homogenous fragmentation is realized. Most experimental fragmentation geometries lead to a situation of *statistically inhomogenous fragmentation*. This idea was also introduced by Lienau [37], but was not pursued.

To explore further the concept of statistically inhomogenous fragmentation, consider the geometric-fragmentation problem from the previous section that led to a linear exponential distribution of fragments,

$$N(a) = N_0 e^{-N_0 a} \tag{30}$$

for a statistically homogenous distribution. Then the statistically inhomogenous distribution,

$$N(a) = \sum_1^n g_i N_{i0} e^{-N_{i0} a} \tag{31}$$

would be obtained from a mixture of n statistically homogeneous distributions with corresponding fragment densities N_{i0} and fragment-area fractions g_i. Such a superposition approach, the method of Poisson mixtures [58], has been found useful in describing fragment-size distributions resulting from dynamic fracture events [44].

3.6. Application in High-Velocity Impact and Explosive Fragmentation

A descriptive example can be selected from fragmentation data resulting from high-velocity impact on metal plates [59]. This study is illustrated by the high-speed photograph of Figure 14 that shows the fragment-debris cloud behind a 6-mm-thick lead plate subjected to high-velocity ($\simeq 1500$ m/sec) normal impact by a lead cylinder of similar dimensions. Due to the high ductility of lead, the initial deformation is the rapid expansion of an intact shell or bubble of lead that thins and finally fragments due to plastic instabilities, leading to the behavior seen in Figure 14. The thickness of the bubble at the time of breakup is approximately 0.3 mm, so the process can be regarded as area fragmentation. Fragment-size distributions for these experiments were determined by photoimaging and computer processing methods.

One of these distributions is shown in Figure 15. Similar features were observed in the other distributions. A straight-line (linear exponential) behavior in the semilogarithmic plot was not observed due to the significant upturn at the small-particle end of the distribution. The data in Figure 15 are well represented by a statistically inhomogeneous bilinear distribution of the form

$$N(a) = g_1 N_{10} e^{-N_{10}a} + g_2 N_{20} e^{-N_{20}a} \tag{32}$$

The data could not be reasonably represented by a Mott distribution.

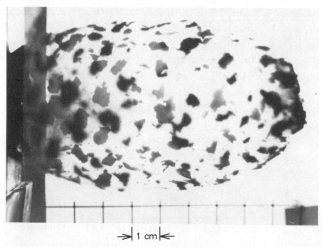

$\twoheadrightarrow| \ 1 \ \text{cm} \ |\twoheadleftarrow$

Figure 14. Fragment debris from high-velocity normal impact of lead projectile onto lead plate at approximately 1.5 km/sec. The high-speed photograph was taken approximately 200 μsec after impact. A 1-cm-interval grid indicates scale.

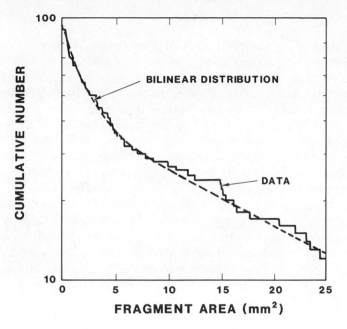

Figure 15. Cumulative number distribution data for lead impact experiment and comparison with bilinear exponential distribution.

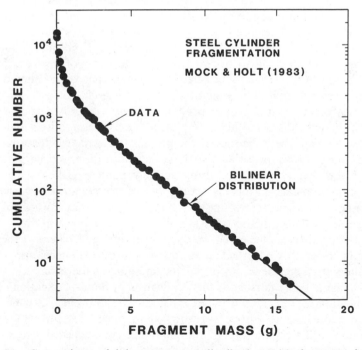

Figure 16. Comparison of inhomogeneous distribution with fragment data from explosively driven cylinders.

27

The comparisons with experimental distributions would be incomplete without examining some exploding munitions data. Data of this type have been published by Mock and Holt [60] in which explosive-filled cylinders of Armco iron and several heat-treated steels were detonated, and the fragments collected and analyzed. A cumulative number distribution from one of the heat-treated steel experiments is shown in Figure 16. The trend of the data in this example is typical of the six experiments performed.

Of the more than 10,000 fragments collected in this test, fewer than 80 retained segments of both inner and outer surface. So the fragment mass should be expected to scale approximately as the volume rather than the area and would suggest that a linear exponential distribution for volumetric fragmentation, Equation (29), would provide the basic theoretical curve for homogeneous fragmentation. As noted by Mock and Holt [60], the data is quite linear with the exception of the small-fragment end of the distribution. Again, the data can be represented with a bilinear exponential curve, although there are indications that the small-particle portion of the distribution is more complex than this.

3.7. Further Developments in Statistical Fragmentation

Perhaps the overriding conclusion that emerges from the discussions of fragment distributions in dynamic fragmentation is the difficulty of relating to physically realistic applications. The description of fragment-size distributions and the relation of these distributions to the different geometries, material properties, and loading conditions are interesting and complex problems that, at present, are poorly understood. Size distributions are observed to range from sharply centralized to broadly dispersed, and there are indications that these differences may be related to the mode and multiplicity of fracture, although material property effects cannot be ruled out. There is a need for better and more systematic experimental data to guide the theoretical efforts.

Fragmentation as a Poisson process, along with geometric statistical methods, has been discussed here. At best, however, observations and conclusions gleaned from these theories at the present time are very tentative.

Other theoretical approaches that have not been discussed here may have merit and, in several cases, are being actively pursued. Dynamic geometric statistical methods, which include active crack processes such as growing and intersecting cracks and interacting stress relief waves, may go a step beyond strictly static geometric methods [30, 44–46, 61]. Again, however, this is a study that is most easily implemented in one dimension and can be expected to increase in complexity when extended to two and three dimensions.

An alternative approach, not unrelated to some of the geometric statistics methods, that is being pursued is the application of bond-percolation concepts to fragmentation [62]. This is also a probabilistic approach in which lattice bonds in a preassigned lattice are allowed to open through some random prescription until a certain level of fracture intensity is achieved. Fragment-size distributions are assessed by determining the closed circuits obtained in the fractured lattice. In this approach, internal fracture damage not associated with fragment surfaces can occur. Statistical size distributions have been obtained by this method and compared with Mott's theoretical equation [62]. Work in this area is continuing in an attempt to demonstrate a relevance to fragmentation.

Recent suggestions have been made that the discipline of fractal geometry may have applicability to concepts of material fragmentation [63, 64]. In particular, the relation of the fracture surface area to fragment volume is perceived to be a fractal number rather than a simple integer, as is usually employed in the analysis. Considerable work remains to complete the coupling of the fractal analysis to the physical nature of the fragmentation process.

An intriguing approach to the problem of statistical fragmentation is offered by the maximum-entropy principle of statistical mechanics as recast in the more general terms of information theory [55, 65]. In the fracture of a body, physical forces that might bring about regularity in the fragment size or structure are assumed to be absent. Thus, fragmentation is totally random and the only constraints are that the process is space-filling and that certain observable features such as the average fragment size or total surface area are satisfied. Such constraints serve as undetermined multipliers in a variational expression of the maximum-entropy principle—the premise that the most probable configuration dominates. The theory has the potential for describing random fragmentation without the construction-dependent assumptions intrinsic to the geometric statistical methods. To date, however, only the underlying ideas and a few tentative examples have been explored [55, 64].

4. CONTINUUM MODELING OF DYNAMIC FRACTURE AND FRAGMENTATION

In practical applications, dynamic fracture and fragmentation usually result from loading sources that create stresses and stress rates that vary in both space and time. Problems of interest include explosive cratering or borehole blasting in rock, impact cratering and spall formation, projectile/armor interaction, shaped-charge jet stretching and breakup, and micromechanical behavior of grains in propellants, etc. These situations typically involve unique geometries with complex stress-wave propagation and degradation as the material fractures. In all but the most elementary of cases, numerical analysis is required to gain insight into the formation of regions of material failure.

The current availability of high-speed computers and shock-wave propagation codes makes possible the development of continuum models of fracture and fragmentation to include in these codes. Such numerical tools enable detailed analysis to be made of complex fracturing events. Davison and Stevens [66] have established the fundamental concepts necessary for a thermodynamically consistent continuum description of dynamic fracture. Subsequent development of these concepts have been extended to both brittle and ductile fracture formation. High-strain-rate ductile fracture of metals has been considered by many, including Cochran and Banner [67], Davison et al. [19], Curran et al. [7], and Davison and Graham [8]. Several groups have pursued models based on the activation, growth, and coalescence of inherent distributions of fracture-producing flaws to predict crack- and fragment-size spectra in brittle fracture [10, 11, 48, 49]. Others have preferred to apply well-developed concepts from plasticity to the problem of fracture [68–71], predicting regions and levels of damage, nonrecoverable void volume, and tensile or shear fracture.

4.1.   Current Directions in Continuum Modeling

This discussion of continuum modeling of dynamic fracture is not an exhaustive review. Rather, it points out the variety of approaches that have been, and are still being, pursued to provide methods for calculating dynamic-fracture phenomena. Such work is still quite active and considerable effort appears to remain before the best approaches emerge as mature computational models.

One of the earliest, and simplest, computational methods for establishing a state of dynamic fracture within a stress-loaded body was to specify a tensile-stress threshold (or spall threshold) at which point the corresponding elements were assumed to instantaneously fracture. The technique illustrated certain stress-wave propagation features characteristic of fracturing bodies; however, it is now recognized that a fracture threshold, independent of the size of the body and rate of loading, is not a physical property. In addition, instantaneous loss of strength is not consistent with particle-velocity records from impact experiments [19, 67, 72].

Davison and Stevens [66] studied the application of high-intensity and short-duration loads to brittle materials, which results in intense local fracturing or spalling of the body. They introduced a concept of fracture damage in terms of a vector field describing the size and orientation of distributed penny-shaped cracks throughout the region of fracture. Damage was allowed to occur gradually, according to some specified law of growth determined by the changing stress and current damage state at the point of fracture.

The general idea has been pursued by others. Curran and co-workers in a series of papers [7, 22, 73] have developed a theory of dynamic brittle fracture based on the nucleation and growth of penny-shaped cracks that evolve gradually to full coalescence of fragments. The inherent distribution of fracture-producing flaws is regarded as observable, and petrographic methods are described in their work for determining such distributions. Laws based on the current stress state are specified to drive fracture nucleation and growth. The model has been implemented in two-dimensional stress-wave codes and has been used extensively in various impact and explosive loading applications.

A model of continuum fracturing devoted primarily to explosive fragmentation of oil shale was developed by Grady and Kipp [11]. The general framework follows that of Davison and Stevens [66] in that fracture damage represented a scalar variable measure of crack-like defects that could grow under appropriate tensile-stress loading. Physics of the activation and growth process, however, was guided by a dynamic application of Weibull statistical concepts, which leads to the known size dependence of fracture stress observed in static testing [74]. This approach allowed the fracture-damage activation parameters to be determined directly from measured fracture stress and fragment-size dependence on strain rate [11, 72].

A bedded crack model of dynamic fracture for brittle and quasi-brittle materials has been developed [48, 49]. Fracture damage is based on a micro-physical description of fracture following the Griffith theory. Considerable care was taken to consistently relate damage and material integrity through an effective modulus theory. The model is amenable to computational implementation, and dynamic-fracture features, such as rate-dependent fracture strength, appear naturally within the workings of the model.

A somewhat different approach to stress-wave-induced fracture is represented in the work of Butkovich [68], developed to calculate underground explosive fracture and induced permeability in coal. The method is more akin to conventional elastic–plastic calculations in that stress-space surfaces of yield or failure are established to determine onset of fracture. Fracture due to shearing is explicitly treated, and two parameters are associated with fracture damage: a shearing related to distortional strain and a tensile-induced cracking or porosity that is related to permeability.

A similar plasticity model of dynamic fracture has been described by Johnson [70] and applied to explosive fracture in oil shale. A scalar fracture-damage parameter is related to the damage-induced reduction in the unconfined yield stress of the material, although the parameter is a mathematical concept rather than a measured property. Damage growth is related to the stress in excess of a pressure-dependent yield surface, with no damage growth above a brittle–ductile transition point on the yield surface. Computer simulations of explosives placed in boreholes provided successful descriptions of extent and regions of fracture damage and dependence on explosive energy and geometrical features.

A void-nucleation-and-growth-fracture model imbedded in a general visco–elastic–plastic material model is representative of approaches to ductile dynamic fracture [19, 75]. Other approaches include employing the plastic strain as a damage variable [71], so that both spall and large strain-to-failure can be treated.

4.2. An Application of Continuum Brittle-Fracture Modeling

Numerical simulation of a complex dynamic-fracture application can be illustrated by calculations of impact-induced damage in a ceramic cylinder. The computer model used was originally developed for oil shale explosive fragmentation [11] with various extended applications considered by Boade et al. [76] and Chen et al. [77]. In this model, the stresses σ_{ij} and strains ϵ_{ij} are related through

$$\sigma_{ij} = B_0(1 - D)\epsilon_{ii}\delta_{ij} + 2\mu_0(1 - D)(\epsilon_{ij} - \tfrac{1}{3}\epsilon_{ii}\delta_{ij}) \qquad (33)$$

where B_0 and μ_0 are the intrinsic moduli of the solid material, and the time-dependent effective moduli are determined by a scalar measure of the fracture damage D. The constitutive description in tension is augmented by two rate equations for the fracture damage and fracture-surface area:

$$\dot{D} = (36\pi)^{1/3}[n(\epsilon)]^{1/3}c_g D^{2/3}(1 - D) \tag{34}$$

$$\dot{A} = (48\pi^2)^{2/3}[n(\epsilon)]^{2/3}c_g D^{1/3}(1 - D) \tag{35}$$

where $\epsilon = \epsilon_{ii}$, and the specific forms are determined by physical considerations of internal cracks subjected to high-rate loading and the Weibull crack-statistics concepts noted earlier [11, 78]. The parameter c_g is a crack-propagation velocity and $n(\epsilon)$ is a crack-activation law driven by the bulk tensile strain ϵ and specified by the Weibull fracture theory:

$$n(\epsilon) = k\epsilon^m \tag{36}$$

The parameters for the model were originally evaluated for oil shale, a material for which substantial fracture-stress and fragment-size data dependent on strain rate were available (see Figure 1). In the case of a less well-characterized brittle material, the parameters can be inferred from the shear-wave velocity and a dynamic fracture or spall stress at a known strain rate. In particular, c_g is approximately one-third the shear-wave velocity, m has been shown to be about 6 for various brittle materials [79], and k can then be determined from a known dynamic-fracture stress using an analytic solution of Equations (33), (34) and (36) in one dimension for constant strain rate.

The complexity of the stress waves generated by explosive charges or projectile impact, and the appearance of relief waves that emanate from free surfaces or regions previously fractured, necessitate the use of wave-propagation codes to address realistic problems. The codes numerically integrate the conservation equations of mass, momentum, and energy, along with the constitutive equations for the material. The fracture model described was incorporated into the Lagrangian two-dimensional wave code, TOODY-IV [80].

A calculation of the lateral impact of a steel plate at approximately 700 m/sec onto a UO_2 fuel pellet clad with zirconium is illustrated in Figure 17, simulating experiments performed under similar conditions [81]. Based on previous computational simulations, a damage level of 0.2 was identified with the onset of catastrophic damage growth and selected as a criterion for fracture levels sufficient for complete fragmentation [11]. The numerical calculation established the damage contours (Figure 17a) and domains of fragment size (Figure 17b). We observe that a very small central region of fine fragments formed, surrounded by a gradation to larger fragments. There are also regions in which no fragmentation is expected to occur. Calculations were generally consistent with experimental results.

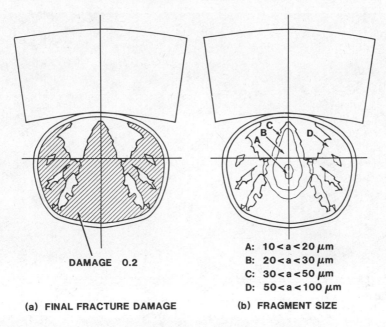

A: $10 < a < 20\,\mu m$
B: $20 < a < 30\,\mu m$
C: $30 < a < 50\,\mu m$
D: $50 < a < 100\,\mu m$

DAMAGE 0.2

(a) FINAL FRACTURE DAMAGE   (b) FRAGMENT SIZE

Figure 17. Calculation of impact between depleted UO_2 cylinder and steel pellet at 0.7 km/sec. (a) Damaged regions. (b) Fragment-size contours.

4.3. An Application of Continuum Ductile-Fracture Modeling

A common impact experiment to examine deformation in ductile metals is the Taylor impact test, in which a right circular cylinder of metal is impacted axially against a "rigid" flat target. A typical experiment results in a flared end of the cylinder at the impact interface, with decreased bulging toward the free end, and possibly internal ductile failure damage, as is illustrated by the sectioned Taylor impact specimen in Figure 18. In this case, the impact was symmetric at a relative velocity of 300 m/sec. A calculation was performed with the impact conditions and configuration corresponding to the 1100 aluminum cylinder of Figure 18 [82] using the void-nucleation-and-growth model of Davison et al. [19]. In this continuum model, the material degradation was patterned after Barbee et al. [83], as a scalar damage variable D, where the rate of damage growth is

$$\dot{D} = V_0 C_N (1 - D)^2 \exp\left[\frac{\sigma_{ii} - \sigma_N}{\sigma_1}\right] + 3D(1 - D)C_G \left[\frac{\sigma_{ii} - \sigma_G}{\sigma_1}\right] \qquad (37)$$

and V_0, C_N, σ_N, σ_1, σ_G, and C_G are material constants. The stresses σ_{ij} are related to strains ϵ_{ij} and temperature θ by

$$\sigma_{ij} = [(B - \tfrac{2}{3}\mu)\epsilon_{ii} - 3\alpha B(\theta - \theta_R)]\delta_{ij} + 2\mu\epsilon_{ij} \qquad (38)$$

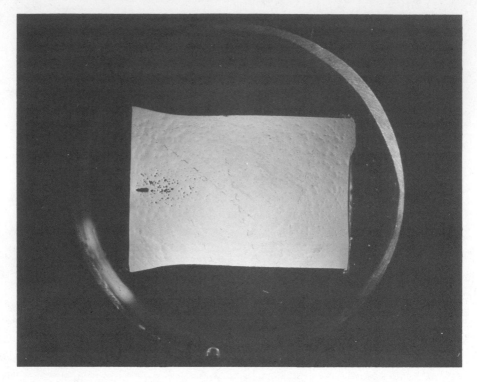

Figure 18. 1100-0 aluminum Taylor impact specimen (12.5 mm dia.) sectioned and polished illustrating void formation near impact interface.

The moduli B and μ are modified by the damage through

$$B = B_0 \left[1 - \frac{3(1 - v_0)}{2(1 - 2v_0)} D \right] \tag{39}$$

and

$$\mu = \mu_0 \left[1 - \frac{15(1 - v_0)}{(7 - 5v_0)} D \right] \tag{40}$$

Tensile stresses stimulate void growth, and subsequent loss in local strength of the material, hence simulating spall in ductile materials.

Figure 19 shows the evolution of the compressive shock wave as it propagates down the cylinder, moving away from the impact interface. Tensile stresses are observed to form along the axis near the impact interface as a result of interacting release waves emanating from the lateral free surfaces of the cylinder, with subsequent damage appearing there. The calculated distributed void damage that results from the impact is observed to agree reasonably well with the sectioned aluminum Taylor specimen shown in Figure 18.

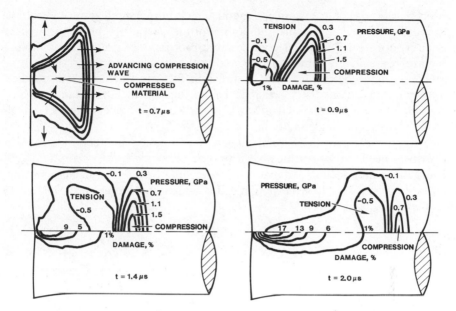

Figure 19. Calculation of Taylor impact of a 1100-0 aluminum cylinder including stress-wave evolution and damage formation. Symmetric impact at 0.15 km/sec.

4.4. Future Problems in Continuum Modeling

Although progress in continuum and computer modeling of dynamic fracture and fragmentation is encouraging, it is apparent that further advancements are needed. Many of the emerging physical and statistical concepts, some of which have been discussed in this chapter, are not yet included in these models. The brittle models based on analogy with theories of plasticity will probably not survive. The processes of dynamic plasticity and fracture are sufficiently different that it is unlikely that the theoretical framework formulated for the former will be appropriate for the latter.

The continuum models based on the statistical nucleation and growth of brittle and ductile fracture appear to be an attractive approach, especially within a framework that provides for some form of a continuum cumulative-damage description of the evolving fracture state. Despite the importance of inherent flaws in the dynamic fracture of solids, some of the concepts introduced in Section 2 clearly indicate the effectiveness of energy principles in determining dynamic fracture strengths and fragment sizes. In Section 2.7, qualitative arguments were advanced that suggested that both inherent-flaw and energy concepts play important roles in the dynamic fracture and fragmentation process. These ideas need to be pursued further and incorporated into continuum computational models.

Even in the loading regime in which inherent-flaw effects dominate the fracture process, further clarification of the fracture-activation and growth process is needed. For example, dynamic-crack branching leading to multiple fracturing is expected to constitute an important part of the breakage process. Such a cooperative and collective fracture process does not fit well within a picture of relatively simultaneous and isolated activation and growth of inherent-fracture sites. The lack of apparent flaws in ductile metals requires nucleation conditions to be specified before subsequent growth and material degradation can occur. Further work is needed on fragmentation of metals under extreme loading conditions.

Finally, the concepts of fragment size, and fracture number or frequency statistics, need to be included within the framework of continuum and computational modeling of dynamic fracture and fragmentation. This challenging area of research has the potential for addressing many needs related to dynamic fragmentation.

ACKNOWLEDGMENTS

This work performed at Sandia National Laboratories and was supported by the U.S. Department of Energy under contract number DE-AC04-76DP00789.

REFERENCES

1. D. E. Grady and M. E. Kipp, "Dynamic Rock Fragmentation," in B. K. Atkinson, Ed., *Fracture Mechanics of Rock*, Academic Press, London 1987, pp. 429–475.
2. P. C. Chou and J. Carleone, "The Stability of Shaped-Charge Jets," *J. Appl. Phys.*, **48**, 4187–4195 (1977).
3. H. J. Melosh, "Impact Ejection, Spallation and the Origin of Meteorites," *Icarus*, **59**, 234–260 (1984).
4. L. B. Freund, "Crack Propagation in an Elastic Solid Subject to General Loading. III. Stress Wave Loading," *J. Mech. Phys. Solids*, **21**, 47–61 (1973).
5. E. P. Chen and G. C. Sih, "Transient Response of Cracks to Impact Loads," *Elasto-Dynamic Crack Problems*, Vol. 4, Noordhoff, Gröningen, Leyden, The Netherlands, 1977.
6. M. E. Kipp, D. E. Grady, and E. P. Chen, "Strain-Rate Dependent Fracture Initiation," *Int. J. Frac.*, **16**, 471–478 (1980).
7. D. R. Curran, L. Seaman, and D. A. Shockey, "Dynamic Failure in Solids, "*Phys. Today*, **30**, 46–55 (1977).
8. L. Davison and R. A. Graham, "Shock Compression of Solids," *Phys. Rep.*, **55**, 255–379 (1979).
9. M. A. Meyer and C. T. Aimone, "Dynamic Fracture (Spalling) of Metals," *Prog. Mater. Sci.*, **28**, 1–96 (1983).
10. D. R. Curran, L. Seaman, and D. A. Shockey, "Dynamic Failure of Solids," *Phys. Rep.*, **147**, 253–388 (1987).
11. D. E. Grady and M. E. Kipp, "Continuum Modelling of Explosive Fracture in Oil Shale," *Int. J. Rock Mech. Min. Sci.*, **17**, 147–157 (1980).

12. P. R. von Rittinger, *Lehrbuch Der Aufbereitungskunde*, Ernst & Korn, Berlin, 1867, p. 19.

13. E. F. Poncelet, "Fracture and Comminution of Brittle Solids." *Trans. Am. Inst. Min. Metall. Pet. Eng.*, **169**, 37 (1946).

14. A. G. Ivanov and V. N. Mineev, "Scale Effects in Fracture," *Combust., Explos. Shock Waves* (*Engl. Transl.*), **15**, 617–638 (1980).

15. D. E. Grady, "Local Inertia Effects in Dynamic Fragmentation," *J. Appl. Phys.*, **53**, 322–325 (1982).

16. A. A. Griffith, "The Phenomena of Rupture and Flow in Solids," *Philos. Trans. R. Soc. London, Ser. A*, **34**, 137–154 (1920).

17. M. M. Carroll and A. C. Holt, "Suggested Modification of the P-α Model for Porous Materials," *J. Appl. Phys.*, **43**, 759–761 (1972).

18. M. M. Carroll and A. C. Holt, "Static and Dynamic Pore-Collapse Relations for Ductile Porous Materials," *J. Appl. Phys.*, **43**, 1626–1636 (1972).

19. L. Davison, A. L. Stevens, and M. E. Kipp, "Theory of Spall Damage Accumulation in Ductile Metals," *J. Mech. Phys. Solids*, **25**, 11–28 (1977).

20. N. A. Zlatin and B. S. Ioffe, "Time Dependence of the Resistance to Spalling," *Sov. Phys.—Tech. Phys.* (*Engl. Transl.*), **17**, 1390–1393 (1973).

21. W. Weibull, "A Statistical Theory of the Strength of Materials," *Ingretensk. Akad. Handl.*, 151 (1939).

22. D. A. Shockey, D. R. Curran, L. Seaman, J. T. Rosenberg, and C. F. Peterson, "Fragmentation of Rock Under Dynamic Loads," *Int. J. Rock Mech. Min. Sci.*, **11**, 303–317 (1974).

23. D. L. Birkimer, "A Possible Fracture Criterion for the Dynamic Tensile Strength of Rock," in G. B. Clark, Ed., *Dynamic Rock Mechanics*, 1971, pp. 573–590.

24. L. S. Costin and D. E. Grady, "Dynamic Fragmentation of Brittle Materials Using the Torsional Kolsky Bar," in J. Harding, Ed., *Mechanical Properties of Materials at High Rates of Strain*, Oxford Univ. Press, London and New York, 1984, pp. 321–328.

25. D. E. Grady, "Spall Strength of Condensed Matter," *J. Mech. Phys. Solids* (in press).

26. R. J. Weimer and H. C. Rogers, "Dynamic Fracture Phenomenon in High Strength Steels," *J. Appl. Phys.*, **50**, 8025–8030 (1979).

27. N. F. Mott and E. H. Linfoot, *A Theory of Fragmentation*, Extra-Mural Res. No. F72/80, Ministry of Supply, England, 1943.

28. G. E. Jones, J. E. Kennedy, and L. D. Bertholf, "Ballistics Calculations of R. W. Gurney," *Am. J. Phys.*, **48**, 264–269 (1980).

29. R. E. Luna, H. W. Church, R. M. Elrick, D. R. Parker, L. S. Nelson, and O. G. Raabe, "Combustion and Smoke Formation Following Exposure of Actinide Metals to Explosion," *Sandia Lab.* [*Tech. Rep.*] **SAND75-6246** (1976).

30. N. F. Mott, "Fragmentation of Shell Cases," *Proc. R. Soc. London, Ser. A*, **189**, 300–308 (1947).

31. M. E. Kipp and D. E. Grady, "Dynamic Fracture Growth and Interaction in One Dimension," *J. Mech. Phys. Solids*, **33**, 399–415 (1985).

32. J. N. Johnson, "Ductile Fracture of Rapidly Expanding Rings," *J. Appl. Mech.*, **50**, 593–600 (1983).

33. C. Fressengeas and A. Molinari, "Inertia and Thermal Effects on the Localization of Plastic Flow," *Acta Metall.*, **33**, 387–396 (1985).

34. G. Regazzoni, J. N. Johnson, and P. S. Follansbee, "Theoretical Study of the Dynamic Tensile Test," *J. Appl. Mech.*, **53**, 519–528 (1986).

35. D. E. Grady and D. A. Benson, "Fragmentation of Metal Rings by Electromagnetic Loading," *Exp. Mech.*, **23**, 393–400 (1983).

36. W. H. Gourdin, S. L. Weinland, and R. M. Boling, "Electromagnetic Ring Expansion as a High-Rate Test: Experimental Development," in S. C. Schmidt, Ed., *Shock Waves in Condensed Matter*, North-Holland Publ., Amsterdam, 1987.

37. C. C. Lienau, "Random Fracture of a Brittle Solid," *J. Franklin Inst.*, **221**, 485–494, 674–686, 769–783 (1936).

38. J. G. Bennett, "Broken Coal," *J. Inst. Fuel*, **10**, 22–39 (1936).

39. A. M. Gaudin and T. P. Meloy, "Model for Comminution Distribution Equation for Single Fracture," *Trans. Soc. Min. Eng. AIME*, **223**, 40–50 (1962).

40. A. N. Kolmogorov, "On the Log-Normal Law of Distribution of Fragment Dimensions in Crushing," *Dokl. Akad. Nauk SSSR*, **31**, 99 (1941).

41. P. Rosin and E. Rammler, "The Laws Governing the Fineness of Powdered Coal," *J. Inst. Fuel*, **7**, 29–36 (1933).

42. R. Schuhmann, "Principles of Comminution, in Size Distribution and Surface Calculations," *Min. Technol.*, Rep. AIME **1189**, 1–11 (1940).

43. J. J. Gilvarry, "Fracture of Brittle Solids. I. Distribution Function for Fragment Size in Single Fracture (Theoretical)," *J. Appl. Phys.*, **32**, 391–399 (1961).

44. D. E. Grady and M. E. Kipp, "Geometric Statistics and Dynamic Fragmentation," *J. Appl. Phys.*, **58**, 1210–1222 (1985).

45. D. E. Grady, "Fragmentation of Solids Under Impulsive Stress Loading," *J. Geophys. Res.*, **86**, 1047–1054 (1981).

46. D. E. Grady, "Application of Survival Statistics to the Impulsive Fragmentation of Ductile Rings," in M. A. Meyers and L. E. Murr, Eds., *Shock Waves and High-Strain-Rate Phenomena in Metals*, Plenum, New York, 1981, pp. 181–192.

47. M. E. Kipp and D. E. Grady, "Random Flaw Nucleation and Interaction in One Dimension," in L. E. Murr, K. P. Staudhammer, and M. A. Meyers, Eds., *Metallurgical Applications of Shock-Wave and High-Strain-Rate Phenomena*, Dekker, New York, 1986, pp. 781–791.

48. J. K. Dienes, "A Statistical Theory of Fragmentation," in Y. S. Kim, Ed., *Rock Mechanics*, Stateline, Nevada, 1978, pp. 51–55.

49. L. G. Margolin and T. F. Adams, "Numerical Simulation of Fracture," in R. E. Goodman and F. E. Heuze, Eds., *Issues in Rock Mechanics*, 1982, pp. 637–644.

50. L. A. Griffith, "A Theory of the Size Distribution of Particles in a Comminuted System," *Can. J. Res., Sect. A*, **21**, 57–64 (1943).

51. T. Herrata, "On Form and Growth of Cracks in Glass Plates," *Inst. Chem. Phys. Res., Tokyo*, **16**, 159 (1931).

52. H. Solomon, *Geometric Probability*, Soc. Ind. Appl. Math. Philadelphia, PA, 1978.

53. I. K. Crain, "The Monte-Carlo Generation of Random Polygons," *Comput. Geosci.*, **4**, 131–141 (1978).

54. I. K. Crain and R. E. Miles, "Monte Carlo Estimates of the Distributions of the Random Polygons Determined by Random Lines in a Plane," *J. Stat. Comput. Simul.*, **4**, 293–325 (1976).

55. D. Weaire and N. Rivier, "Soap, Cells and Statistics—Random Patterns in Two Dimensions," *Contemp. Phys.*, **25**, 59–99 (1984).

56. S. Goudsmit, "Random Distribution of Lines in a Plane," *Rev. Mod. Phys.*, **17**, 321–322 (1945).

57. T. Kiang, "Random Fragmentation in Two and Three Dimensions," *Z. Astrophys.*, **64**, 433–439 (1966).

58. P. S. Purie and C. M. Goldie, "Poisson Mixtures and Quasi-Infinite Divisibility of Distributions," *J. Appl. Prob.*, **16**, 138–153 (1979).

59. D. E. Grady, T. K. Bergstresser, and J. M. Taylor, "Impact Fragmentation of Lead and Uranium Plates," Sandia Lab. [Tech. Rep.] **SAND85-1545** (1985).

60. W. Mock, Jr. and W. H. Holt, "Fragmentation Behavior of Armco Iron and HF-1 Steel Explosive-Filled Cylinders," *J. Appl. Phys.*, **54**, 2344–2351 (1983).

61. E. N. Gilbert, in B. Noble, Ed., *Applications of Undergraduate Mathematics in Engineering*, Macmillan, New York, 1967, Chapter 16.

62. R. Englman, Z. Jaeger, and A. Levi, "Percolation Theoretical Treatment of Two-Dimensional Fragmentation in Solids," *Philos. Mag. B*, **50**, 307–315 (1984).

63. D. L. Turcotti, "Fractals and Fragmentation," *J. Geophys. Res.*, **91**, 1921–1926 (1986).

64. R. Englman, N. Rivier, and Z. Jaeger, "Information Theoretical Derivation of Fragment-size-Distribution in Disintegration," *Philos. Mag. B* (in press).

65. E. T. Jaynes, in R. D. Levine and M. Tribus, Eds., *The Maximum Entropy Formalism*, MIT Press, Cambridge, MA, 1979, p. 15.

66. L. Davison and A. L. Stevens, "Thermomechanical Constitution of Spalling Elastic Bodies," *J. Appl. Phys.*, **44**, 668–674 (1973).

67. S. Cochran and D. Banner, "Spall Studies in Uranium," *Lawrence Livermore Lab.* [*Rep.*] **UCRL-78716** (1976).

68. T. R. Butkovich, "Calculations of Fracture and Permeability Enhancement from Underground Explosions in Coal," *Proc. Am. Soc. Mech. Eng.*, pp. 19–23 (1976).

69. L. A. Glenn, "The Fracture of a Glass Half-Space by Projectile Impact," *J. Mech. Phys. Solids*, **24**, 93–106 (1976).

70. J. N. Johnson, "Explosive Produced Fracture of Oil Shale," *Los Alamos Sci. Lab.* [*Rep.*] **LA-7357-PR**, 38–45 (1978).

71. G. R. Johnson and W. H. Cook, "Fracture Characteristics of Three Metals Subjected to Various Strains, Strain Rates, Temperatures and Pressures," *Eng. Fract. Mech.*, **21**, 31–48 (1985).

72. D. E. Grady and M. E. Kipp, "The Micromechanics of Impact Fracture of Rock," *Int. J. Rock Mech. Min. Sci.*, **16**, 293–302 (1979).

73. L. Seaman, D. R. Curran, and D. A. Shockey, "Computational Models for Ductile and Brittle Fracture," *J. Appl. Phys.*, **47**, 4814–4826 (1976).

74. J. C. Jaeger and N. G. W. Cook, *Fundamentals of Rock Mechanics*, Chapman & Hall, New York, 1969.

75. M. E. Kipp and A. L. Stevens, "Numerical Integration of a Spall-Damage Viscoplastic Constitutive Model in a One-Dimensional Wave Propagation Code," *Sandia Lab.* [*Tech. Rep.*] **SAND76-0061** (1976).

76. R. R. Boade, M. E. Kipp, and D. E. Grady, "A Blasting Concept for Preparing

Vertical Modified in situ Oil Shale Retorts," *Sandia Lab.* [*Tech. Rep.*], **SAND81-1255** (1981).

77. E. P. Chen, M. E. Kipp, and D. E. Grady, "A Strain-Rate Sensitive Rock Fragmentation Model, in K. P. Chong and J. W. Smith, Eds., *Mechanics of Oil Shale*, Elsevier, London, pp. 423–456, 1983.

78. D. E. Grady and M. E. Kipp, "Oil Shale Fracture and Fragmentation at Higher Rates of Loading," *Proc. 20th U.S. Symp. Rock Mech.*, pp. 403–406 (1979).

79. D. E. Grady and J. Lipkin, "Criteria for Impulsive Rock Fracture," *Geophys. Res. Lett.*, **7**, 255–258 (1980).

80. J. W. Swegle, "TOODY IV-A Computer Program for Two-Dimensional Wave Propagation," *Sandia Lab.* [*Tech. Rep.*], **SAND78-0552** (1978).

81. J. L. Alvarez, R. C. Green, L. Isaacson, B. B. Kaiser, R. W. Marshall, Jr., and V. J. Novick, "Waste Forms Response Project Correlation Testing," *E. G. G. Tech. Rep.*, **EGG-PR-5590** (1982).

82. M. E. Kipp and L. Davison, "Analyses of Ductile Flow and Fracture in Two Dimensions," in W. J. Nellis, L. Seaman, and R. A. Graham, Eds., *Shock Waves in Condensed Matter–1981*, Am. Inst. Phys., New York, 1982, pp. 442–445.

83. T. W. Barbee, L. Seaman, R. Crewdson, and D. R. Curran, "Dynamic Fracture Criteria for Ductile and Brittle Metals," *J. Mater.*, **7**, 393–401 (1972).

CHAPTER 2

Debris-Impact Protection of Space Structures

A. M. Rajendran

University of Dayton
Research Institute
Dayton, Ohio

and

N. Elfer

Martin Marietta Corp.
New Orleans, Louisiana

ABSTRACT

Response of space structures to debris impact is modeled through a combined experimental and analytical approach. The laboratory tests include experiments with and without any intermediate layer of thermal insulation between the bumper shield and the rear wall. Different failure modes are observed in the rear wall. Spallation is the predominant mode in experiments without any intermediate layers. Bulge and burst-type failure seem to be the mode when thermal blankets are used. The modeling of the debris cloud requires a careful interpretation of the cloud characteristics. Various characteristics are explored in the loading-model formulations. The experimentally observed salient features of the pressure distribution on the rear wall are reproduced by the models that assume either mass-filled or hollow debris clouds. These models predict spallation-type dynamic failure in the rear wall. Modeling of the pressure distribution, when intermediate layers are used, requires major modifications to these cloud models. This leads to the use of simple impulse-velocity-based models for the analysis of the bulge and burst-type failure in the rear wall. A critical strain-based failure criterion is successfully applied in the numerical simulation of rear-wall deformation.

1. INTRODUCTION

The main objective of this chapter is to model the deformation and failure of space structures due to debris impact. Concepts related to spacecraft designs for meteoroid and space-debris protection are briefly described for completeness. The descriptions include definitions of *meteoroids*, *space debris*, etc., and also the safety requirements for meeting a specified probability of no penetration or failure in the design life. The structural-failure aspects due to space-debris impact is covered in detail. The failure analyses considered combined experimental and numerical methods. A simple critical strain-based engineering model was used successfully to describe *failure* or *no failure* of the thin sheet (rear wall) in the numerical simulations.

Spacecraft designs for meteoroid and space-debris protection must account for several different mechanisms of penetration to satisfy requirements on probability of no penetration or failure. At different impact velocities and obliquities, multiwall designs can be penetrated by projectile or bumper-shield fragments, spall, melting, or by a late-time momentum failure of the final wall. Projectile breakup is analyzed with a combination of experimental data, scaling, hydrocodes, and phenomenological models. Intermediate shields are useful in further breaking up projectile fragments and reducing spall in a rear wall. At very high velocities, the bumper and projectile melt or even vaporize. If the final wall is not penetrated by fragments, it must still absorb the momentum of the initial projectile. The failure of the rear wall from this blast loading can be very sensitive to the momentum distribution, momentum multiplication due to rebound or cratering, and damage to the surface from early time cratering. Results of several calculations are discussed. If penetration should occur, the next step is to calculate the distance cracks propagate, so that a traditional fracture-mechanics analysis can be applied. While workable approaches have been developed in each area, more rigorous analyses are still being developed.

Spacecraft designers have commonly included meteoroid protection in the basic structure. However, space debris, man-made orbiting objects, is becoming a bigger threat than meteoroids to spacecraft survivability [1–4]. Most space-debris impacts, and all meteoroid impacts, fall under the classification of hypervelocity. The most general definition for hypervelocity is that the material strength is negligible and the impact process is initially hydrodynamic. The impact velocity can be less than the speed of sound in the material, 5 km/sec in aluminum, but can also be much higher.

To protect spacecraft from meteoroids, Whipple [5] proposed a sacrificial shield, called a bumper, which would vaporize the meteoroid and distribute the momentum on a rear wall. The subject of hypervelocity impact flourished in the 1960s with the Apollo program; waned in the 1970s; and has received renewed interest in the 1980s with the Strategic Defense Initiative program and the advent of space debris as a design driver. Important references include recent

Hypervelocity Impact Symposia in 1986 [6], Comet Halley Micrometeoroid Hazard Workshop [7], and two texts: *High Velocity Impact Phenomena* [8], and *Impact Dynamics* [9].

The typical design requirement is to meet a specified probability of no penetration or failure in the design lifetime. While there are fewer large particles than small particles, large-area and long-duration spacecraft require substantial weight of protective shielding. Part of the problem is understanding the nature of the threat. Rather than having a unique threat size and velocity that must be defeated, there are spectra of sizes, velocities, and obliquities, and consequently a variety of penetration mechanisms that must be considered. These aspects of the problem are discussed first.

2. METEOROIDS

As a matter of definition, meteoroids are a threat to spacecraft. *Meteors* are visual phenomena produced by *meteoroids* entering the earth's atmosphere. *Meteorites* are typically iron–nickel objects that reach the earth's surface. They are the remnants of asteroidal meteoroids after they enter the atmosphere. However, the meteoroid environment beyond the atmosphere is dominated by cometary meteoroids, which consist of loosely packed ice with a density of $0.5\,g/cm^3$ [8]. While this is not very substantial, the impact velocity can range from 16 to 72 km/sec, with an average of 20 km/sec. There are many more small meteoroids than larger ones. The average flux F_{met} is defined as the number of meteoroids of a diameter d (cm) or larger that pass through one side of a randomly oriented 1 square meter area per year, and is described in terms of d:

$$\log F_{met} = -6.2 - 3.66 \log d_{met} \qquad (1)$$

For short-duration missions, a specific analysis for meteoroid *showers* may be more appropriate. Since the meteoroid environment is essentially omni-directional, the earth provides significant geometrical shielding. At an altitude of 500 km, the earth blocks 34.4 percent of the view angle. One small further correction in the meteoroid environment is a focusing factor due to the effect of the earth's gravity on attracting meteoroids. Deep-space fluxes are predicted to be 56 percent of the near-earth meteor flux. At 500 km, it is 97 percent.

For a spacecraft with a fixed orientation relative to the earth, the directional aspects of earth shielding, as well as the spacecraft velocity through a random environment can be used to calculate a distribution of impacts around the spacecraft. The forward-facing side of a spacecraft in low earth orbit (LEO) can have several times as many impacts as a rear-facing side due to the spacecraft's velocity in sweeping a volume of space [4]. As mentioned before, the space structure must not only be protected under meteoroid impact but also from the impact of space debris. Examples of space debris and their average flux F_{SD} is defined in the next section.

3. SPACE DEBRIS

Space debris includes nonfunctioning satellites, fragments from upper-stage rocket explosions, impact-generated debris, and satellites purposefully destroyed (including antisatellite tests), and ranges down in size to paint flakes or Al_2O_3 generated from solid rocket propellants. At this time, the best definition of the space-debris environment at 400 and 500 km is given by Kessler [3]. The average density of a debris particle is $2.8\,g/cm^3$, only slightly higher than aluminum alloys, $2.7\,g/cm^3$, which constitutes most debris. For sizes over 1 cm, the density begins to drop because it tends to be folded sheet metal. While there were 7000 objects approximately 10 cm or larger tracked by NORAD (North American Air Defense Command) in 1987, the number of smaller objects is orders of magnitude larger. Similar to meteoroid definitions, the flux is given in terms of number of impacts of a diameter d_{SD} (cm) or larger per square meter per year as a function of d.

$$\log F_{SD} = -5.46 - 2.52 \log d_{SD} \text{ for } d \leq 1 \text{ cm at } 500 \text{ km} \qquad (2)$$

$$\log F_{SD} = -5.82 - 2.42 \log d_{SD} \text{ for } d \leq 1 \text{ cm at } 400 \text{ km} \qquad (3)$$

This can be revised by a factor of 2 or more in flux as more information is available, and the reader should seek the most current definition if needed. Figure 1 compares the flux of meteoroids and space debris in low earth orbit. The space-debris flux has a pronounced altitude dependence. It is not only a function of the popularity of certain orbits, but at low earth orbits, atmospheric drag tends to remove debris, although it is continuously replenished from higher orbits [1].

Space debris is very directional. For low earth orbit debris can be described by the intersection of circular orbits. Besides the dominant polar, 60° and 28° inclinations, precession of orbits can result in high impact velocities from orbits with the same inclination. Figure 2 shows the fraction of the total flux coming from angles relative to the direction of flight in 5° increments [3]. The relative impact velocity for the intersection of 500-km orbits is also labeled on the plot. The relative impact velocity in LEO is determined by the orbital velocity and the intersection angle of the two orbits.

When the spacecraft attitude is fixed relative to the earth, each facet of the surface has its own distribution of probable obliquity angles and velocities. While no debris overtakes from the rear, a rear-facing facet can be impacted by debris that appears to come from the side (an orbit with a shallow intersection angle). Debris cannot intercept a spacecraft from more than 10° above or below

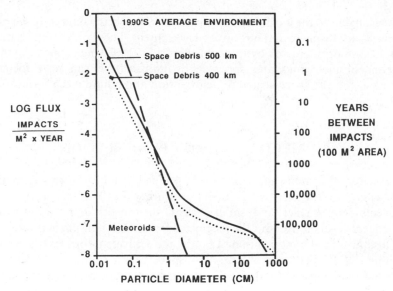

Figure 1. The flux of meteoroid and space-debris particles, equal or larger than the diameter indicated, which impact a randomly oriented area on a spacecraft is shown as a function of particle diameter [3]. The left axis shows the average time between impacts on a 100-m² surface area.

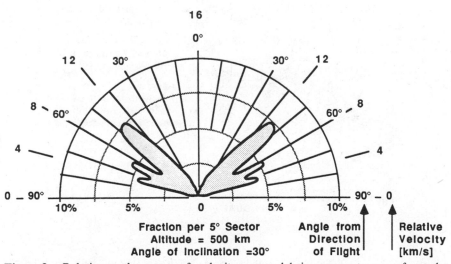

Figure 2. Relative to the spacecraft velocity, space debris appears to come from the sides [3], at an average of 51° from the direction of flight. This is due to the intersection angle of different orbital inclinations. The relative impact velocities are also shown for a 500-km circular orbit.

45

a plane tangent to the local earth normal, since the debris would otherwise enter the earth's atmosphere and be removed as a threat. The typical angle is only 3° for a 500-km orbit crossed by trackable debris [10]. Having defined the threats in terms of meteoroid and space-debris impacts, we can next focus our discussions on the space-structure deformation and failure mechanisms due to such impact.

4. MECHANISMS OF PENETRATION

The reaction of a multiwall shield to a projectile impact is shown schematically in Figure 3. With increasing shock pressures, the original projectile first fragments into smaller and smaller sizes, then melts and even vaporizes. Subsequent layers must stop the fragments from the initial impact, resist spall, and also absorb the momentum. Tests at the University of Dayton Research Institute (UDRI [11] and at Marshall Space Flight Center (NASA–MSFC) [12] have demonstrated that an intermediate layer of thermal-insulation material can be very effective in preventing cratering and spallation of the rear wall, but the momentum of the initial impact must still be absorbed by the rear

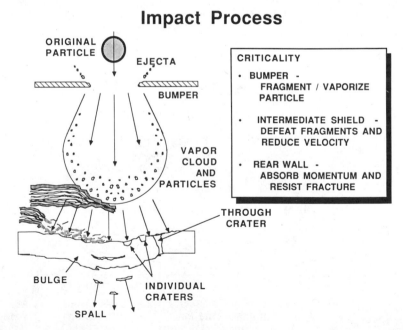

Figure 3. The bumper ideally fragments or vaporizes the incident projectile. If fragments are still present, they can be further fragmented or slowed by intermediate layers. The rear wall must then resist craters, pinhole penetration, spall, and finally rupture from the projectile momentum.

wall. The multilayer insulation typically consists of 0.0005-in. sheets of aluminized mylar or polyimide film (KAPTON) and a protective cloth. There is a possibility of rear-wall penetration even at very high obliquity angles. Even if the original particle is fragmented and ricochets from the bumper, bumper fragments can still penetrate the rear wall. Figure 4 shows typical rear-wall damage that can result from hypervelocity impact.

Testing of shield concepts is typically done using light-gas guns. An explosive charge is used to accelerate a piston and pump hydrogen (or helium gas) that then bursts a diaphragm and accelerates the projectile and sabot. The maximum velocity of a light-gas gun is theoretically limited to about 11 km/sec, but in practice it is typically on the order of 6 to 8 km/sec, depending on the size of the projectile. While electric rail guns are being designed for velocities of up to 50 km/sec, they are limited to about 6 km/sec for routine testing. A shaped charge can be used to launch a discrete projectile at 10 to 12 km/sec, but the projectile is severely deformed by the launching process, and the target can be destroyed by subsequent debris. Therefore, shaped-charge tests require several flash X-rays to measure projectile size, velocity, and target damage [13].

Most of the experimental observations discussed here are from light-gas gun tests of aluminum-on-aluminum impacts at 6–7 km/sec. Several empirical relationships and guidelines based on experimental observations are available for design in this range [14]. Craters are approximately three times as deep as the initial projectile diameter, and because of spall, a single sheet five times the projectile diameter can be penetrated. For a two-wall system, the optimum ratio of bumper thickness t_b to projectile diameter d_p is from 0.15 to 0.25. For a two-wall design, the standoff S between the bumper and the rear wall should be $30d_p$. Larger standoffs do not prevent the fragment penetration that can still occur, and smaller standoffs can lead to penetration by multiple impacts in the same location, or to momentum overload. The total thickness of bumper and rear wall to prevent penetration is 0.6 to 0.7 times d_p. At a lower velocity, such as 3 km/sec, almost twice that thickness, or 1 to 1.2 times d_p, is required to prevent penetration, although the spacing is no longer crucial. However, these observations are only adequate for cursory feasibility studies and apply only for certain range of parameters. In practice, the shield materials, thicknesses, and spacings are also influenced by many other design requirements, and the threat material, velocity, shape, and obliquity can range over wide spectra.

Figure 5a illustrates an example ballistic-limit surface for a multiwall shield. A projectile diameter below the surface does not penetrate, whereas a larger projectile does. The surface contours are dictated by the operative penetration mechanism. For example, a change in spacing strongly influences momentum failure, has a mild influence on fragment penetration, and almost has no effect on single-particle or richochet penetration. It is interesting to note that in the valley, or *bucket*, of the surface, an oblique impact can be more penetrating than a normal impact because the projectile is not as effectively fragmented. Figure 5(b) shows the distribution of penetrations when a geometry, the flux from Figures 1 and 2, and penetration resistance in Figure 5(a) are all combined.

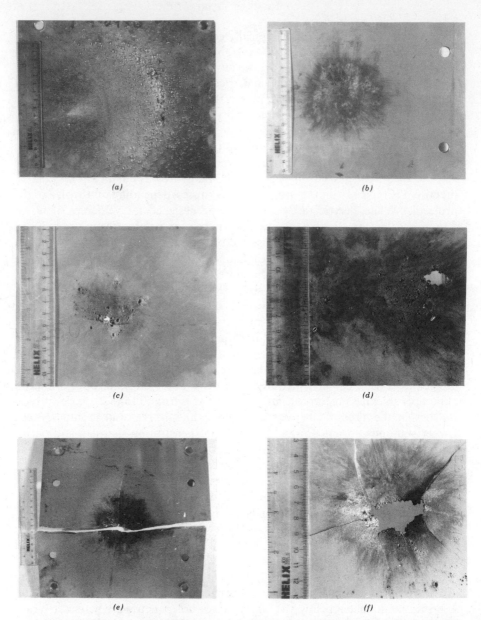

Figure 4. Typical rear-wall damage is shown from hypervelocity impacts. The basic test configuration was a 2-mm 6061 Al bumper, thermal insulation, and a 3.2-mm 2219 Al rear wall with 114-mm spacing, impacted by a 1-g Al cylinder ($L/D = 1$) at 6.4 ± 0.1 km/sec, with differences as noted. (a) 4.4-mm wall, 305-mm spacing, no thermal layer; penetrated. (b) Basic configuration; slight bulge. (c) 3.2-km/sec impact; penetrated. (d) 45° obliquity; penetrated. (e) 1.25-g projectile; momentum rupture. (f) 1.6-mm bumper; both craters and tears.

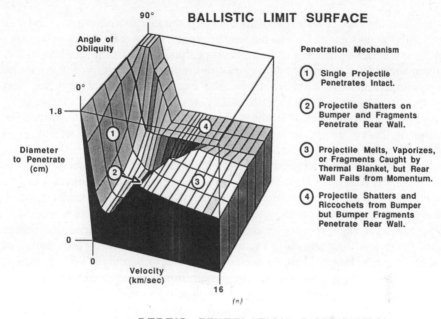

BALLISTIC LIMIT SURFACE

90°

Angle of
Obliquity

0°

1.8

Diameter
to Penetrate
(cm)

0

0

Velocity
(km/sec)

16

(a)

Penetration Mechanism

① Single Projectile
   Penetrates Intact.

② Projectile Shatters on
   Bumper and Fragments
   Penetrate Rear Wall.

③ Projectile Melts, Vaporizes,
   or Fragments Caught by
   Thermal Blanket, but Rear
   Wall Fails from Momentum.

④ Projectile Shatters and
   Riccochets from Bumper
   but Bumper Fragments
   Penetrate Rear Wall.

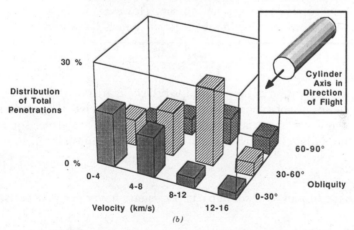

DEBRIS PENETRATION DISTRIBUTION

30 %

Distribution
of Total
Penetrations

Cylinder
Axis in
Direction
of Flight

0 %

0-4

4-8

8-12

12-16

Velocity (km/s)

60-90°

30-60°   Obliquity

0-30°

(b)

Figure 5. (a) A ballistic limit surface from the basic configuration in Figure 4 separates penetration, above the surface, from no penetration, below the surface. The surface is defined by different penetration mechanisms, each with a unique shape. (b)

5. PROBABILITY ANALYSIS

The traditional analysis is based on the Poisson distribution for a probability P of zero impacts:

$$P = (F \times A \times T)^n \exp(-F \times A \times T)/n! \qquad (4)$$

where F is the flux, A is the exposure area, n is the number of impacts, and T is time, all in consistent units. The simplest approach is to use the total surface area of the spacecraft or the area of a simple shape that envelopes the spacecraft [15]. Given the area, the design lifetime, and the required reliability, a flux can be determined and from that a particle size can be determined that has that probability of no impact. If this size can be stopped by the basic structure at all velocities and obliquities, then the requirement has been met. If, however, the design is driven by the penetration resistance, then a proper weighting of probable velocities and obliquities must be done.

For a spacecraft with a fixed orientation relative to the earth, a more accurate calculation is possible [11, 12]. The space-debris and meteoroid threats can be broken down into streams of flux from different directions at specific velocities, as shown in Figure 2. The diameter to penetrate every area on the spacecraft is determined for each stream element. The flux for each diameter is determined from the relationship shown in Figure 1 and then multiplied by the fraction of the flux in each stream. Then, for each element, the probability of no penetration is the product of the probability of no penetration from each stream, based on the flux that penetrates, the projected area normal to the flux, and the exposure time. However, it must be noted that the definition of flux used previously is the flux F on a randomly rotating object. This means that the area spends half its time with no exposure to a given stream. However, when a stream from a particular direction is used with the projected area in that direction, then flux J is given by

$$J = 4F \tag{5}$$

The simplest derivation of this is that the projected area of a sphere, πr^2, is one-fourth of the surface of a sphere, $4\pi r^2$. Whereas each receives the same number of impacts from a stream perpendicular to the disk, the impacts per square meter are four times higher for the disk. An alternative definition is

$$j = F/\pi = J/4\pi \tag{6}$$

where j is the flux in terms of impacts per square meter per year per steradian.

6.   CRATERING

A single wall can be very vulnerable to hypervelocity impact. At low velocities, the depth of the crater is controlled by erosion of the projectile. However, as velocity increases, the crater becomes hemispherical and the volume of material removed is roughly proportional to the energy of the impacting particle. Empirical fits to the data yield slightly different results. The equation used in [16] is

$$P_d = K_s m^{0.352} \rho_p^{0.167} V^{0.667} \tag{7}$$

where P_d is the depth of penetration, cm; K_s is a material constant for semiinfinite materials, 0.42 for Al alloys and 0.25 for 304 and 316 stainless steel; ρ_p is projectile density, g/cm^3; and V is projectile velocity (km/sec). Tests indicate oblique impacts should use only the normal component of the impact velocity to calculate energy and crater volume [17]. As the obliquity angle increases, greater than 45°, the crater is elongated, and at greater than 65° from the normal, the projectile primarily ricochets from the surface.

In practical applications, there are very many subtle nuances in understanding the effects of variable densities, strengths, and thermodynamic properties, and there have been numerous publications on the phenomena. The applications range from prehistoric cratering of the moon, earth, and the planets by meteoroids, to cratering of runways by bombs, and microscopic cratering of spacecraft by debris. The initial impact is hydrodynamic; the stresses are well above the strength of the materials involved. However, the final size of the crater is controlled by the strength of the material. For exact curve-fitting to test data, it is necessary to include an effect due to momentum as well as energy [18].

For a finite-thickness sheet, the threshold of penetration is reached when the bottom of the crater links with spall from the back of the sheet. Spall occurs when the compressive shock wave accelerates the rear of the sheet and reflects as a tensile wave with an amplitude exceeding the spall strength of the material. Spall strength is given by the maximum longitudinal tensile stress that the material experiences in a one-dimensional strain experiment (plate impact-test configuration) [19] just before spalling. Empirical equations for critical sheet thickness to resist penetration take a similar form to the depth of penetration equations:

$$h_{\mathrm{cr}} = K_1 m^{0.352} \rho_p^{0.167} V^{0.875} \tag{8}$$

where h_{cr} is the critical thickness on the verge of penetration, cm, and K_1 is a material constant for single-sheet penetration. Typical constants for penetration resistance are 0.57 for aluminum alloys 2024, 7075, and 6061; 0.32 for 304 and 316 stainless steel; 0.38 for annealed 17-4PH stainless; and 0.34 for Cb-1 Zr. It is interesting to note that for the same area density, aluminum can stop larger projectiles than much stronger steels. If the same projectile material and velocity impacts equal-areal density steel and aluminum sheets, the ratio of the projectile sizes to cause penetration in each sheet is given by

$$(m_{\mathrm{al}}/m_{\mathrm{ss}}) = \{(K_{1\mathrm{ss}} \times \rho_{\mathrm{ss}})/(K_{1\mathrm{al}} \times \rho_{\mathrm{al}})\}^{1/0.352} = 3.82 \tag{9}$$

Even though the volume of material removed is much larger in aluminum, the lower density is an advantage in preventing performance. If equal weights of steel and aluminum are equally effective bumpers for a large-design projectile, then aluminum is the better choice for a bumper for spacecraft applications, because thermal-insulation layers beneath the bumper suffer less degradation from micrometeoroids that certainly impact the spacecraft.

7. SINGLE-PARTICLE PENETRATION

The penetration of a single particle through multiple sheets is a standard type of ballistic analysis. Even this is complicated by a variety of penetration mechanisms such as shearing a plug versus tearing a petalled hole (see Figure 4). However, for the purposes of a space-debris analysis, it is sufficient to use an empirical approach such as the Thor equations [20, 21]. The Thor equations have empirical constants for a variety of target materials that describe the critical velocity for a specified projectile to penetrate a single sheet, and if penetration occurs, different equations and constants can describe both the residual velocity as well as the residual mass for calculation on subsequent layers.

8. FRAGMENT PENETRATION

As velocity increases, the shock pressure also increases and the release wave causes fragmentation. This was discussed by Grady and Kipp in Chapter 1. What this means to a multiwall design is that as velocity increases, the projectile breaks into finer and finer fragments that become less penetrating to subsequent layers. At high velocities, the projectile melts and even vaporizes.

Parametric experiments show that the minimum total shield weight, bumper plus rear wall, is achieved when the bumper thickness is between 0.15 and 0.25 times the projectile diameter [12]. This is for bumpers and projectiles of the same materials. When different materials are used, the areal density of the bumper should still be about 20 percent of the projectile density times the diameter. However, it was shown by Swift [22] that if the bumper has a significantly lower density, then the projectile does not fragment as well. Materials with higher density than the projectiles were all equally effective bumpers for the same areal density, and differences between different materials were due to whether bumper fragments remained intact and penetrated the rear wall.

There have been several phenomenological approaches to calculate initial fragment sizes from impact velocities and pressures. These fragments are then given a spatial distribution, and the penetration of subsequent walls is calculated [12, 23–26]. The concept of a double bumper to first fragment a projectile and then further fragment the fragments was proposed by Richardson and Sanders [27].

9. HYDROCODE MODELS

Hydrocodes have been used extensively to model high-velocity impact. The codes track material mass, momentum, and energy, and include relationships for hydrodynamic, elastic–plastic, and failure behavior [28, 29]. The applicability of hydrocodes is limited by the material properties that are available and also by the expense of running a large model with a fine mesh over a long time period. While most impacts have both yaw and obliquity, these are three-dimensional calculations, and with a large standoff, the computational time and expense become too high to do more than selected cases.

The large deformations in hypervelocity impact are well suited to Eulerian models, where mass is transported through a stationary grid. Lagrangian codes can use automatic schemes to rezone the grid to handle large deformations. Lagrangian codes are much better than Eulerian codes in handling late-time small deformations where strength is most important. Rather than trying to include early time fragment penetration and late-time momentum failure in the same calculation, the simpler approach is to perform two separate calculations. The pressures and velocities in the initial impact can be used to drive a much larger but simpler blast-loading calculation.

Hydrocodes provide the best means of extrapolating beyond test capabilities, but modeling of impacts is hardly a turn-key operation. Adequate equation-of-state (EOS) properties are only available for selected materials, and even when they are adequately defined, the ability of a hydrocode to implement them is questionable. Hydrocodes do not produce the shape of debris clouds observed in X-ray radiographs of hypervelocity impacts [30, 31]. The material-failure models in most hydrocodes are crude and simple. Due to the nonavailability of sophisticated realistic failure models in these codes, it is not, at present, possible to reproduce accurately the debris cloud shapes. Simpler models of the blast loading is taken for the remainder of this chapter to review momentum failure of rear walls.

10. MODELING OF IMPACT

As mentioned earlier, several approaches [32–38] were considered for modeling the penetration of a hypervelocity particle onto the bumper shield and the rear wall. Upon impact, the particle and a portion of the bumper shield vaporize to form a debris cloud. Due to the complex nature of the penetration process, the model formulations generally include several assumptions and approximations. The main objective in modeling the debris cloud is to estimate the spatial and temporal variation of pressure or velocity that is applied by the debris cloud onto the rear wall. The estimation is directly influenced by the interpretation of the cloud characteristics from the radiographs of the experiments. The interpretation of debris cloud has always been a controversial subject in the field of particle impact on the bumper-shield problem. Recently, several papers appeared on this subject in the proceedings of a hypervelocity impact symposium [39]. Most of them addressed the debris cloud formation in terms of hydrocode calculations. A spherical projectile was mostly used in the calculations.

Typical shapes of the debris clouds due to spherical and cylindrical projectiles are shown in Figure 6. The fundamental question is whether these clouds are filled with materials or are empty. The radiographs can not distinguish between the two conditions. Based on the finite-difference simulations [40] and a few experiments conducted by Swift [41], it appears that the cloud is empty. Examinations of several radiographs from the recently conducted hypervelocity experiments [31] with copper and aluminum projectiles, either spheres or cylinders, indicated that the cloud may be partially filled with particles from the bumper shield and the projectile. However, the problem is still far from being resolved. Several experiments [32–38] modeled the cloud using a relatively

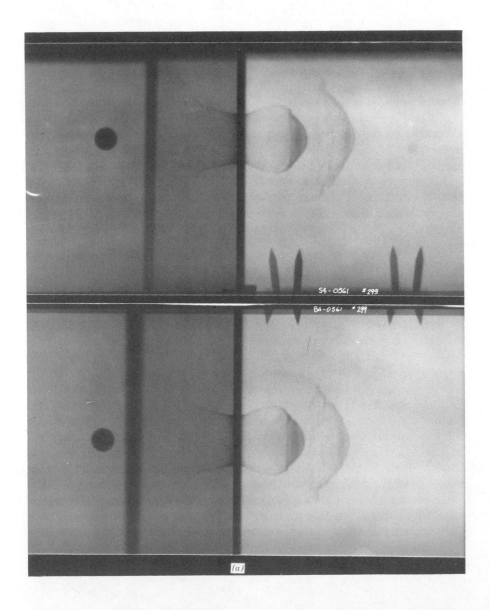

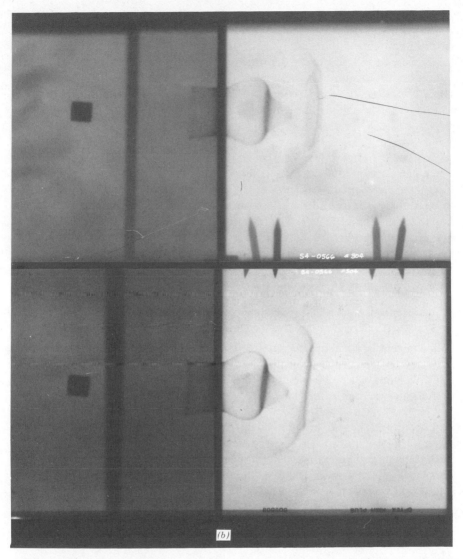

Figure 6. Radiographs of debris clouds due to (a) spherical projectile and (b) cylindrical projectile [31].

simplistic approach. In this approach, the time duration of the momentum transfer from the cloud to the rear wall is assumed to be instantaneous. Due to this assumption, a circular portion of the rear wall receives a sudden velocity. The problem is then reduced to the estimation of the velocity distribution that is imparted to the wall. Various empirical formulas were suggested for the velocity distribution [32–35].

Improved loading models consider the shape, areal density, and velocity distribution of the particle, and the pulse duration of the cloud. Since deformation and failure analyses of the rear wall depend on the dynamic-loading conditions due to the debris cloud, it is essential to critically evaluate the validity and sensitivity of the models that describe the cloud effects. Different cloud characteristics are explored and the corresponding loading conditions are evaluated in the following section.

10.1.   Impulse-Velocity-Based Models

Since the debris-cloud formation and the mass and velocity distribution of fragmented or melted particles within the cloud are extremely complex, one way to circumvent the difficulties in modeling the cloud is to assume that the cloud imparts a sudden velocity to the rear wall. In that case, the details of the cloud are not important. It is possible to estimate the impulse velocity based on conservation of momentum. However, the amount of momentum transfer between the debris cloud and the rear wall is still a debatable issue. Depending on the hypervelocity regime and the projectile and bumper materials, the cloud impact on the rear wall can be an elastic collision, an inelastic collision, or somewhere in between. If the projectile is completely vaporized, the momentum of the cloud must be doubled for an elastic collision. In the case of an inelastic collision, that is, when the molten debris sticks to the rear wall, the momentum is the same as that of the projectile $(m_p V_p)$. If cratering occurs, then momentum multiplication by a factor of 5 is possible. When the impact is neither fully elastic or inelastic, the momentum may have to be multiplied by a factor between 1 and 2.

The projectile and bumper-shield interaction is schematically shown in Figure 7. Many authors [32–35] considered the spray angle of the cloud as an important parameter in the determination of the velocity distribution that is imparted to the rear wall. The first model, used by Wilkinson [35], suggested the following expression for the initial velocity of the rear wall:

$$V_i(r) = \xi \exp\left(-r^2/2\Delta^2\right) \tag{10}$$

where

$$\xi = \frac{m_p V_p}{4 m_r S^2 \tan^2 \overline{\theta}} \tag{11}$$

and

$$\Delta = (2/\pi)^{1/2} S \tan \overline{\theta} \tag{12}$$

m_p and V_p are the mass and velocity of the projectile, respectively. S is the spacing between the bumper and the rear wall, m_r is the mass per unit area of the

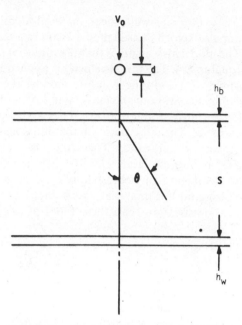

Figure 7. Schematic of a projectile, bumper shield, and rear wall.

rear wall, and r is the radial distance from the center of the circular portion on the rear wall through which the momentum of the debris cloud is transferred. In this model, the parameters m_p, V_p, S, and m_r are known based on the experimental configuration. The only parameter that has to be determined from the experiments is the spray angle. Wilkinson presented an empirical expression for $\tan \bar{\theta}$ as

$$\tan \bar{\theta} = \begin{cases} \beta \left(\dfrac{m_b}{\rho d} \right)^{1/2} & \dfrac{m_b}{\rho d} < 1 \\ \beta & \dfrac{m_b}{\rho d} \geqq 1 \end{cases} \tag{13}$$

where m_b is the mass per unit area of the bumper sheet, ρ is the density, d is the diameter of the projectile, and β is an empirical parameter. Based on the various hypervelocity experiments with lead and cadmium and the hydrodynamic computer code calculations of vaporized aluminum, Wilkinson selected a value of 0.6 for β.

The second model, suggested by McMillan [33], assumes an initial velocity V_i imparted to a circular portion of the rear wall as

$$V_i = \frac{32 m_p V_p}{\pi S^2 \rho_w h_w} \tag{14}$$

where ρ_w and h_w are the rear-wall sheet material density and thickness, respectively. This impulse-velocity based model assumes a momentum multiplication factor of 2. The model also assumes that the radius of the circular portion of the rear wall is equal to $S/4$. The initial velocity is applied uniformly over this central circular portion, whereas in the Wilkinson model, the initial velocity was based on a Gaussian distribution of the mass as it strikes the rear wall.

These two models are fairly straightforward. They avoid modeling the details of the cloud characteristics. The pulse duration that is associated with the debris cloud is assumed to be instantaneous. These models can be reasonable for velocities > 15 km/sec and for materials with low melting temperatures, such as cadmium, lead, etc. Modeling of debris clouds, under low-velocity impact (between $6-15$ km/sec) and for materials with high melting temperatures, require much more sophistication than these types of impulse-velocity-based models.

10.2. Mass-Filled Cloud Model

A typical radiograph of a debris cloud due to an impact of a right circular solid aluminum cylinder on an aluminum bumper shield is shown in Figure 6. The cloud at different instances during its flight toward the rear wall is idealized in Figure 8. In this model, it is assumed that the diameter of the plug from the bumper that fills the front portion of the debris cloud is the same as that of the projectile. Large fragments of the bumper shield and projectile are assumed to be at the leading and trailing ends of the cloud, respectively. Two approaches can be considered in the interpretation of the radiographs.

The first approach, as suggested by Piekutowski [42], assumes that the cloud is filled with material particles and that the density varies along the length of the cloud (X-direction in Figure 8). By conserving the mass within the cloud, the areal density along the cloud can be estimated. The axial velocities of the leading and the trailing cross sections AB and EF, respectively, can be estimated from the radiographs taken at three different locations during the flight of the cloud toward the rear wall. The average axial velocity of the interface section in several experiments is equal to 80 percent of the impact velocity V_p. From the radiographs, it appears that the leading and the trailing portions of the cloud apparently do not have any appreciable radial velocities. Therefore, the area of cross sections at AB and EF do not change throughout the flight of the cloud toward the rear wall. However, the radiographs indicate a radial expansion of the cloud's central section (CD in Figure 8). In general, the loading model can be formulated based on these radiograph measurements. By knowing the axial and radial velocity distribution, the duration of the loading pulse can be determined.

The debris cloud shown in Figure 8 is due to impact velocities in the range of $6-7$ km/sec. Typically, in this range, it is reasonable to believe that the debris cloud consists of molten aluminum. Therefore, the momentum can be transferred to the rear wall assuming an inelastic collision. The pressure history that the rear wall experiences due to the impact of a mass-filled debris cloud (from an

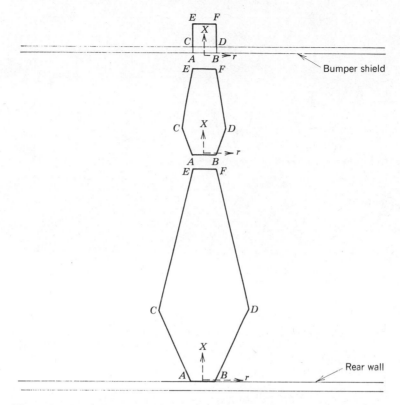

Figure 8. Modeling of an expanding debris cloud (cylindrical projectile).

impact of a 1-g mass 0.304-in. diameter aluminum projectile onto a 0.08-in. thick aluminum bumper) is shown in Figure 9. The impact of the leading face of the cloud is at $t = 0$, and the trailing face is at $t = 23.6$ microseconds. The higher pressures at these two times are the direct consequence of the assumption that the cloud is denser at the leading and trailing faces. The kink in the pressure at around 4 microseconds is due to a discontinuity in the axial velocity gradient across the interface (*CD* in Figure 8). The momentum transfer due to this pressure history is

$$m_p V_p = \int_0^t P(t)A(t)\, dt \tag{15}$$

where

$$P(t) = \rho(x, t)V^2(x, t) \tag{16}$$

Note that $\rho(x, t)$ and $V(x, t)$ are calculated from the cloud information. The mass-filled cloud model is much more complex than the impulse-velocity-based

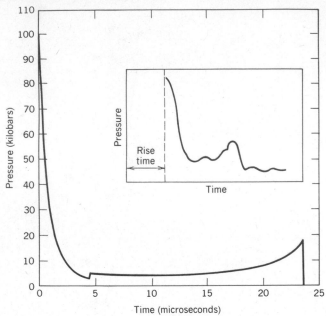

Figure 9. Pressure history based on a mass-filled cloud model. The inset is the pressure record for lexan on lead impact, with 15.3-cm (6-in.) spacing, and an 7.5-km/sec impact velocity [43].

models. The cloud characteristics as read from the radiograph are incorporated into the model formulation. The model-calculated pressure history contained all the salient features of a typical experimentally obtained pressure history in Reference 43, as shown in the inset of Figure 9.

10.3. Hollow-Cloud Model

The loading model in the second approach is based on a hollow cloud. The cloud is assumed to contain mass distributed only along its periphery, as shown in the inset of Figure 10. The axial and radial velocity distributions can be estimated from the radiographs as in the mass-filled cloud model. The area of load application of the cloud on the rear wall is circular for the leading and trailing edges. However, the area is annular for any other section of the cloud. The area of load application computed from the hollow-cloud model is compared with that of the mass-filled cloud in Figure 11. The pressure history based on the hollow-cloud characteristics is shown in Figure 10. It can be seen that the experimentally observed final-pressure peaks (at $t = 0$ and at $t = 25$) appear in the hollow-cloud model also. Again, this is because of the presence of large fragments at the leading and trailing edges of the cloud.

The two impulse-velocity models, the mass-filled cloud model and the hollow-cloud model, can be used to evaluate the response of the rear wall. For this purpose, one of the experiments conducted at the University of Dayton

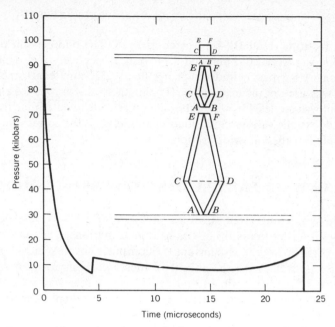

Figure 10. Pressure history based on a hollow-cloud model. The inset shows the schematic of a typical hollow-cloud expansion.

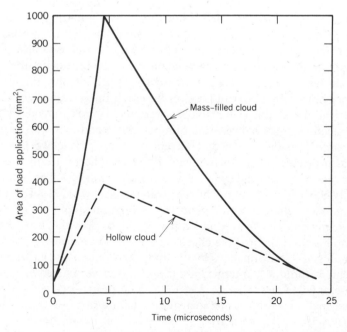

Figure 11. Comparison of area of load application with respect to time between mass-filled and hollow-cloud models.

Research Institute (UDRI) [42] was considered. The bumper-shield and rear-wall thicknesses were 2.032 mm and 3.175 mm, respectively. The spacing S was 109.093 mm. A 1-g mass cylindrical projectile of 7.722-mm length and 7.722-mm diameter was used in the experiment. The impact velocity was 6.5 km/sec. This experiment was modeled using a finite-difference code to determine the rear-wall response due to the various cloud-loading models and the corresponding results are presented in the following section.

11. REAR-WALL FAILURE DUE TO DEBRIS-CLOUD IMPACT

During the sixties and seventies, several investigators [32–38] modeled the rear-wall response using thin-strip or beam approximations. They modeled only a portion of the rear wall to circumvent the computer cost associated in modeling the entire rear wall. With the advent of increased computer capabilities and availabilities, it is now feasible to model the entire rear wall with appropriate boundary conditions. Several state-of-the-art finite-difference/element codes are available to calculate the rear-wall response to debris-cloud impact. Use of a three-dimensional code is more appropriate than a two-dimensional code. However, due to an almost axisymmetric nature of the geometry and the response, it can be approximated reasonably well as a two-dimensional problem. In our present study, we selected a two-dimensional finite-difference code, called STEALTH, for calculating the rear-wall response. This well-documented code was developed by Hoffman [44]. The accuracy and validity of this code have been demonstrated exhaustively by comparing the code results with already existing, either analytical or numerical, results from other sources. STEALTH solves the conservation laws using a well-established central-difference finite-difference numerical scheme. A few hypervelocity-impact experiments that were conducted by Piekutowski [42], at the Impact Physics Laboratory of the University of Dayton Research Institute, were considered in the numerical simulation. Two kinds of experiments were conducted: (1) without any intermediate layers between the bumper shield and rear wall, and (2) with thermal blankets as the intermediate layers. The impact velocity of the 2024-T3 aluminum projectile ranged between 6–7 km/sec. The projectile was either a sphere or right circular solid cylinder with an L/D ratio of 1. The bumper shield was mostly 6061-T6 aluminum. The rear-wall material was 2219-T87 aluminum. The diagnosis included X-ray pictures (radiographs) of the debris cloud. Measurements on the deformed/failed rear wall also provided some indications of the response and the loading characteristics on the rear-wall surface.

Depending on the impact velocity and shock intensity, the rear wall in experiments without any intermediate layers was either bulged locally without any failure or failed by spallation. Use of thermal-insulation intermediate layers eliminated spall-type failure, and instead, the failure was due to tensile necking of the rear wall leading to tearing. Similar type of failure occurred when the projectile velocity was high enough to cause complete vaporization of the projectile upon impacting the bumper shield. Under this condition, the response of the rear wall even without any intermediate layer was observed to be of bulge and burst-type failure [33–35].

The experiments at UDRI were conducted with a thicker rear wall compared to the earlier reported experiments of Wilkinson [35], Gehring [34], and McMillan [33]. In UDRI experiments, as mentioned earlier, the aluminum projectiles did not vaporize because of the relatively low impact velocity (6–7 km/sec). In the following section, the response of the rear wall with and without the intermediate layers is reported. Both spall and tearing failures are considered. The STEALTH finite-difference code is employed in the response simulation. The rear wall is modeled as a two-dimensional axisymmetric geometry.

11.1.   Spallation

Due to the inherent nature of the impulse-velocity-based models, they cannot predict or produce spall-type failure in the rear wall. In these models, a portion of rear-wall mass with a circular surface area is given a sudden initial velocity. Since all the material particles within this mass move together with some velocity, high-amplitude shock waves are not produced. As a result, there are no wave interactions that can produce high tensile stresses to cause spall. However, the mass-filled and hollow-cloud models can produce such tensile stresses, and therefore, it is possible to model spall failure. In both these models, the load transfer occurs over a finite pulse duration and momentum is transferred to the rear wall through shock pressures. As mentioned earlier, the pressure versus time and the area of load application on the rear wall versus time can be estimated from the loading models. The pressure and loading-area histories were employed as the loading conditions in the rear-wall response calculations using finite-element/difference codes.

The experimental configuration described earlier was modeled. In this experiment, there was no intermediate layer and the rear wall failed by spallation. The debris cloud was modeled from radiographs. Assuming a mass-filled cloud, the corresponding time histories of pressure and loading area are shown in Figures 9 and 11, respectively. The peak pressure at impact is around 100 kbars. The response of the 0.125-in. (3.175-mm) rear wall due to this pressure history was calculated using the STEALTH code. The momentum transferred to the wall was around 6.5 g-km/sec. Solutions were carried out for 10 microseconds. The undeformed and the excessively deformed configurations of the rear wall are shown in Figure 12.

It can be seen that the central portion has thinned down immediately upon the impact of the leading face of the cloud. To investigate spall-type failure in the rear wall, the stress history of the rear wall was obtained. Figure 13 shows the stress history near the central portion *AB* (see Figure 8). The initial compressive stress reaches a peak value of 6000 MPa (point *C*) and the reflected tensile stress from the back free surface of the rear wall reached a value of 2000 MPa (point *D*).

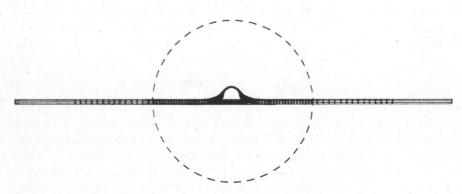

Figure 12. Undeformed and excessively deformed rear-wall grid due to a mass-filled cloud as simulated by STEALTH.

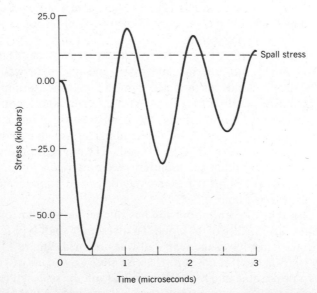

Figure 13. Stress history in a rear wall due to a mass-filled cloud.

For aluminum, the tensile stresses to cause spall are typically between 800–1200 MPa. This clearly predicts spallation over a radius of 5–10 mm in the rear wall. Further loading would eventually lead to a catastrophic failure of the rear wall due to spall fragmentation.

As an additional exercise, the hollow-cloud model was used in the rear-wall response calculations. The time histories of pressure and loading area due to a hollow cloud were shown earlier in Figures 10 and 11. Using these loading histories, the STEALTH simulation of the rear-wall response was carried out. The rear wall deformed excessively. The stress history also indicated spall-type failure in the rear wall. Thus, the mass-filled and hollow-cloud models can predict spallation failure in the rear wall. It appears that the models that follow the cloud characteristics as closely as possible can predict the response of a rear wall more realistically than the ad hoc models.

11.2. Tearing Failure

The next step is to model the experiments in which thermal blankets as intermediate layers were used. These experiments, conducted at the impact facility of UDRI, revealed bulging and bursting- (tearing-) type failure in the rear wall. The projectile, bumper shield, and rear wall were all made of aluminum. The velocity range was 6–7 km/sec. Between two otherwise identical tests, the one with a layer failed by tearing and the other without a layer failed by cratering and spallation. It is fairly well established, based on the numerous tests with aluminum [33–35], that the debris cloud consisted of partially molten fragmented particles. The radiographs indicated denser material along the leading and trailing edges of the cloud. Without the protection of the intermediate layers, these denser parts invariably produced crater and spall in the rear wall. The intermediate layer changed the cloud characteristics, either by fragmenting the bigger fragments into finer particles or by redistributing the

**TABLE 1 Description of the Bumper Shield:
Rear-Wall Experimental Configuration**

Case No.	Mass (g)	Velocity (km/sec)	Bumper Thickness (mm)	Spacing (mm)	Rear-Wall Thickness (mm)	Comments
4-0502	1.0028	6.49	2.032	114.3	3.175	Bulged; no failure
4-0506	2.0232	6.36	3.175	228.6	3.175	Insignificant bulge
4-0510	1.0038	6.96	2.032	114.3	3.175	Tearing
4-0528	1.0020	6.16	2.032	114.3	3.175	Bulged; no failure
4-0529	1.0004	6.17	1.600	114.3	3.175	Tearing

mass of the cloud. Also, the layer could have slowed the debris cloud. At present, there are not many experimental measurements that would indicate or provide any solid evidence regarding the effects of thermal blankets on cloud characteristics. The mechanics or physics associated with debris-cloud kinetics after it passes through the layers are mostly speculative. Elimination of spall indicates that the intermediate layers suppress the high peak in the pressure history.

A few experiments were selected for modeling bulge and tear-type failure. Table 1 describes these experiments. The projectile was a right circular aluminum cylinder with a diameter of 0.304 in. (7.7 mm). The bumper was 6061-T6 aluminum and the rear wall was 2219-T87 aluminum. The objective was to model the tests in Table 1 using the various loading models described earlier.

11.3. Modified Mass-Filled and Hollow-Cloud Models

The main effect of the intermediate layer was to eliminate spallation-type failure in the rear wall. In an ad hoc manner, the mass-filled and hollow-cloud models were modified so that the high-frequency pressure due to the denser material at the leading face of the cloud was eliminated. The assumptions were (1) the momentum was conserved, (2) the cloud characteristics between the leading edge and the interface (front portion) were altered due to the layer; (3) a Gaussian pressure distribution was assumed for the pulse duration of the front portion, and (4) the rear portion between the interface and the rear end of the cloud was unchanged. In Figure 14, comparison is made between the pressure histories due to the original and modified mass-filled clouds. It can be seen that the initial peak-pressure value has been considerably reduced. The STEALTH simulation of test 502 was carried out with the modified pressure history. The stress versus time plot of the rear wall showed stresses higher than the stress to cause spallation. Also, the deformations were excessive and the rear wall stretched to well above 100-percent strain. These results do not agree with the experiment in which the rear wall did not fail. It must be noted that the typical failure strains in aluminum are far below the values obtained in the simulation. The modifications introduced into the mass-filled cloud model are inadequate to represent the actual cloud dynamics.

To further investigate the loading models, the hollow-cloud model was considered. Modifications were also introduced into the hollow-cloud model to suppress high peak pressure. The corresponding pressure history was applied to the rear wall. The results were similar to those of the modified mass-filled cloud model. The excessive bulging and higher stresses indicate that the pressure and pulse duration have been overestimated.

Instead of modifying the original cloud model, other types of simpler models were considered. One such model is based on a uniform pressure applied over a circular portion of the rear wall for a finite-time duration. The area of application at any given instant for the mass-filled cloud and simple uniform-pressure models are compared in Figure 15. The momentum $(= \int_0^t P(t)A(t)\,dt)$ is conserved in both models. To evaluate the effect of uniform-pressure loading on

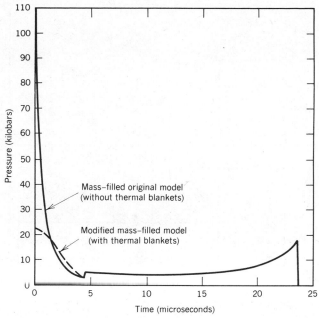

Figure 14. Comparison of pressure histories between original and modified mass-filled cloud models.

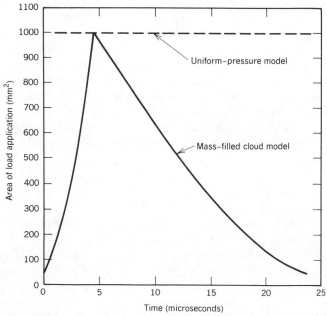

Figure 15. Comparison of area of load application with respect to time between mass-filled and uniform-pressure models.

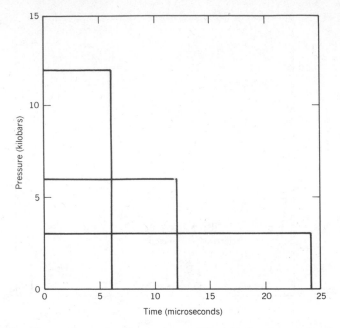

Figure 16. Comparison of uniform-pressure histories for three pulse durations.

the rear-wall response, three cases were considered, as shown in Figure 16. In these cases, conservation of momentum required a different pulse duration for each pressure level. The rear-wall response to these three pressure histories were calculated through the STEALTH code. The maximum strains in the rear wall were excessive; the wall necked down beyond 100-percent strain. In the experiment, the wall had bulged without failure. More refined models are needed to incorporate the effects of spacing, bumper thickness, and realistic pulse durations. For these reasons, the impulse-velocity-based models of McMillan [2] and Wilkinson [4] were considered in the following section to model bulge and tear failure in experiments with intermediate layers.

11.4.   Impulse-Velocity-Based Models

The response of the rear wall of test 0528 was calculated using the two impulse-velocity models discussed earlier. In these calculations, one-to-one momentum was assumed. In the numerical simulation, the value of β that appears in the Wilkinson model, as given earlier in Equation (13), was assumed to be 0.392. This value was chosen based on the Wilkinson model ability to reproduce the test 0528. The velocity distributions that were imparted to the wall in these models are shown in Figure 17. The responses of the wall for the two models were calculated using the STEALTH code. The strain and deflection histories at the center of the rear wall are compared in Figure 18. They differed significantly.

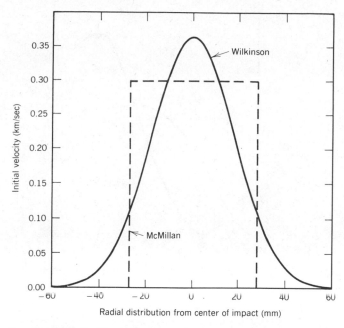

Figure 17. Impulse-velocity distribution on the rear wall of test 0528 as calculated from the Wilkinson and McMillan models.

Since the McMillan model imparts the velocity uniformly over an area of radius $S/4$, and beyond this radius, the velocity is abruptly dropped to zero, it leads to excessive deformation. The McMillan model predicted much more bulging than that was observed in the experiment. However, the results from the Wilkinson model substantiated the experimental observation. To further evaluate and validate the use of these two models, a test 0529, in which the rear wall failed, was considered.

In order to predict failure or no failure, we require a criterion. A strain-based criterion was used. Based on the split Hopkinson bar test at a strain rate of 1200/sec performed on 2219-T87 aluminum, the maximum principle failure strain was around 20 percent, which was much higher than in a quasi-static tensile test. The strain history for test 0529 at the central portion of the rear wall using the Wilkinson model is shown in Figure 19. It can be seen that the maximum strain in the rear wall exceeded the strain at failure.

As far as the McMillan model is concerned, the simulation of tests 0528 and 0529 are identical. This is because the only difference between the two tests is in the bumper-shield thickness (see Table 1). In test 0529, the thickness is about 25 percent lower than in test 0528. Since the impulse-velocity calculation [see Equation (14)] is independent of bumper-shield thickness in the McMillan model, the estimated histories on the rear wall are identical. This clearly shows that the loading model, to be precise, requires much more sophistication than

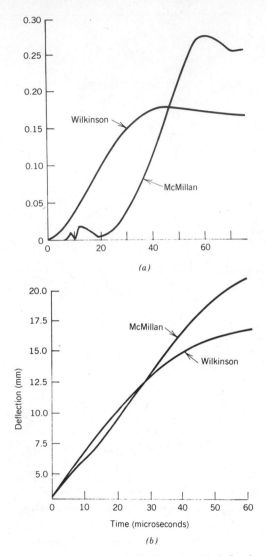

Figure 18. Comparisons of (a) strain and (b) rear-wall deflection of test 0528 was calculated from the Wilkinson and McMillan models.

the McMillan model. This model may be suitable when the cloud consists of completely vaporized particles. Depending on the projectile and bumper materials, and also the velocity regime, the loading model due to the debris cloud can be independent of bumper thickness. In the velocity regime where the debris consists of fragmented molten particles and where intermediate thermal-insulation layers were used between the shield and the rear wall, Wilkinson-type impulse-velocity models appear to be more suitable.

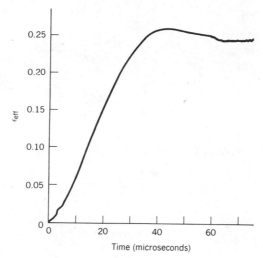

Figure 19.   Simulated strain history of test 0529 using the Wilkinson model.

The Wilkinson model has several advantages: (1) pulse duration is not required; (2) there is only one model parameter, β; and (3) the model is sensitive to bumper and projectile materials, their thicknesses, the spacing between the bumper and rear wall, and the impact velocity. As it was shown, test 0529 was successfully modeled using the Wilkinson model because the model formulation included the effect of bumper thickness. The area of load or velocity application onto the rear wall is reduced because the spray-angle parameter, $\tan \bar{\theta}$, is smaller. [Note that the radius of the circular portion, which is imparted an initial velocity, is equal to 3Δ, where Δ is calculated from Equation (12).] The physical interpretation is that when the bumper-shield thickness is not large enough to fragment the projectile into finer particles at any given velocity, the larger fragments tend to spread at reduced spray angles. The larger fragments are more detrimental to the rear wall. Because of the presence of the intermediate layer, these fragments could not produce spall. This could be due to further fragmentation of the large particles upon impacting the intermediate layers. However, the rear wall was necked and torn into several petals. Using the Wilkinson model, the other tests in Table 1 were simulated and the correspond-ing strain histories are shown in Figure 20. It can be seen that the response calculations agreed with all of these tests.

Considering the complex loading nature of the problem, especially with an intermediate layer, the simplistic approach based on the Wilkinson model seems to provide a reasonable response calculation for the rear wall. In design and engineering analyses, the simple critical strain-based failure maps give a first-order estimate of the hypervelocity-impact response of the rear wall. However, accurate prediction of the onset of necking, which leads to tearing in the rear wall, requires a better understanding of biaxial stretching of thin shells or sheets

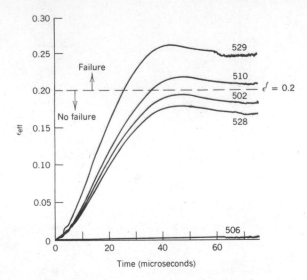

Figure 20. Strain histories of the tests in Table 1 as calculated by STEALTH using the Wilkinson model.

under dynamic loading conditions. Not only the inertia effects but also the strain-rate effects play important roles on the onset of necking of thin sheets. These various aspects are covered in other chapters and are also discussed in the following section.

11.5. Analysis of Thin-Sheet Failure by Tearing

Ductile failure as a phenomenon associated with plastic instability has been addressed by several investigators. Marciniak and Kuczynski [45] considered this phenomenon in metal sheets under biaxial tension for quasi-static loading conditions. They developed a mathematical model (M–K model) to construct forming-limit diagrams. This model is based on an initial geometric imperfection, which is a localized thickness reduction in the sheet lying perpendicular to the principal strain direction. The thickness is assumed to be uniform everywhere else. By this method, they predicted that the strain in the uniform region reaches a limiting value at which point the local strain in the defective region becomes unstable. Chu and Needleman [46] used the M–K model approach for constructing forming-limit diagrams for biaxially stretched sheets under quasi-static loading conditions. They considered nucleation and growth of microvoids in the local site and predicted necking.

In the high-strain-rate loading regime, where the quasi-static loading assumption no longer applies, considerable evidence exists, Wilson [47], that an increase in strain rate can in fact result in an increase in the ductility of most

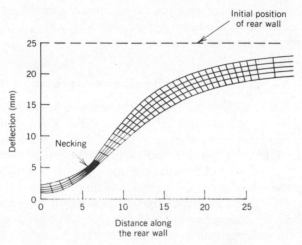

Figure 21.   A deformed rear-wall configuration exhibiting strain gradients.

metals. In a series of experiments at a high strain rate, Fyfe and Rajendran [48] confirmed the increase in ductility that occurs in this regime. To model failure by necking under dynamic-loading conditions, Rajendran and Fyfe [49] considered a failure mechanism in which microvoids nucleate, grow, and coalesce. Their combined analytical and experimental approach confirmed the increase in ductility due to inertia. Rajendran [50] showed that the high-strain-rate effects on ductility are less pronounced under dynamic-loading conditions.

When inertia effects were included into the analyses, Rajendran and Fyfe [49] found that even though local strain grows faster than the uniform strain, no plastic instability (necking) conditions occur before the material has actually failed. They proposed and validated a critical-void-content-based failure criterion under dynamic-loading conditions. To predict or model necking in the rear wall, it is possible to incorporate the void-growth-based failure model. In a finite-element/difference simulation of the rear-wall response to debris-cloud loading, the assumption of an initial geometric imperfection (M–K model) is not necessary, because of the strain gradients that arise from the nature of loading on the rear wall, as can be seen from the deformed configuration in Figure 21. These strain gradients lead to the spatially nonuniform void nucleation and growth. Thus, the basic concepts that describe the necking process can be described through a constitutive model in the code that can describe the dilatancy due to the presence of voids. Chu [51] modeled the localized necking in out-of-plane punch stretching using a finite-element method. Later Chu and Needleman [46] used the void-growth-model approach in their punch-stretching analysis. A similar approach can be considered to predict failure under biaxial stretching in the rear wall due to bulging. By using a finite-difference/element code, the void nucleation and growth process can be modeled. Since the inertial loads are internally handled by the code, the only addition that is required in the code calculations for modeling the rear-wall failure is the constitutive model for a porous material. At this time, the results using this approach are not available.

11.6.   Fracture-Mechanics Approach

Depending on the severity of the impact or momentum intensity, rear walls that failed usually developed multiple cracks. These cracks can dynamically run, leading to the complete fracture or separation of the rear wall. The experiments conducted at UDRI, with the intermediate thermal blanket, revealed this type of failure in the rear wall. The strong elastic waves reflected back from the boundaries of the rear wall can drive the cracks that were created by the debris-cloud impact to complete failure. Dynamic-crack-propagation analysis is quite complex. Most fracture-mechanics-based dynamic-crack-propagation analyses considered [52] uniform loading conditions applied at regions away from the crack. The problem was mathematically tractable. However, in rear-wall fracture, the dynamic load was applied directly on the crack face, such as in a wedge crack problem. The boundary condition is dictated by a nonuniform out-of-plane velocity distribution. Estimates of this nonuniform velocity distribution depend on the loading conditions due to debris-cloud impact. At present, due to lack of any direct experimental measurements on the rear wall, modeling of the dynamic-crack propagation is not possible.

In general, the following steps are necessary for the fracture-mechanics-based crack analysis: (1) estimation of the applied in-plane stretch velocity at the bulging regions where cracks initially appear, (2) determination of dynamic-stress intensity factors, (3) simulation of dynamic-crack propagation using finite-element/difference codes, and (4) determination of dynamic-fracture toughness and arrest toughness for the rear-wall material.

12.   SUMMARY

The modeling of the debris cloud requires a careful interpretation of the cloud characteristics. The mass-filled and hollow-cloud models could model spallation-type failure in experiments without any intermediate layers, whereas the impulse-velocity-based models are not suitable for predicting spall. The experimentally observed salient features of the pressure distribution on the rear wall were reproduced by the mass-filled and hollow-cloud models. The use of thermal blankets as intermediate layers introduces major modifications into these cloud models. This led to the simple impulse-velocity-based models as more suitable to predict bulging and necking-type failure modes. We also showed that the Gaussian-distribution-based Wilkinson model predicted the rear-wall response in several experiments reasonably well. At present, it appears that for design and engineering calculations, the Wilkinson model may provide a reasonable solution. As far as the biaxial in-plane stretching of the rear wall, it can be modeled using a finite-difference code. The prediction of necking regimes based on a simple critical-failure strain seems to match with the experiments. However, in the numerical simulation of rear-wall deformation, localized necking was not present at $\epsilon = \epsilon^f$. The void-growth-based failure models are more appropriate to model the process of necking. An outline has been provided for the void-growth-model calculations through a finite-difference/element method. Biaxial stretching can be modeled with both inertia and strain-rate effects included in the analysis. The final failure is due to dynamic-crack propagation that leads to the complete split of the rear wall. A fracture-mechanics-based approach to model the dynamic-crack propagation and the possible crack-arrest analysis is essential.

12.1. Recommendations

The following recommendations are made to improve the modeling efforts: (1) conduct instrumented experiments to provide the fine details of the debris-cloud interaction with thermal insulation; (2) develop new uniquely designed experiments to understand the momentum transfer to the rear wall; (3) obtain high-speed photographs of the rear wall, measurements of rear-wall deflections with respect to time, and pressure distribution on the rear wall due to debris-cloud impact; (4) develop improved debris-cloud models using these direct measurements; (5) model necking in the rear wall using either void-growth-based constitutive models or dynamic forming-limit diagrams; and (6) perform an out-of-plane motion fracture-mechanics analysis to model dynamic-crack propagation and arrest.

ACKNOWLEDGMENTS

The authors wish to acknowledge the support of the Martin Marietta Independent Research and Development program, as well as internal funding by the University of Dayton Research Institute. The authors are grateful to the following individuals for their dedicated analysis: D. Grove, D. Jurik, B. Roberts, and G. Kovacevic. The authors also wish to acknowledge many stimulating discussions with Mssrs. D. Kessler, B. Cour-Palais, and Dr. R. Becker. Particular thanks go to Mr. A. Piekutowski, who directed all of the testing discussed here, and proposed new mass distributions in the debris cloud.

REFERENCES

1. D. J. Kessler and S. Y. Su, "Orbital Debris," *NASA Conf. Publ.* **2360** (1985).
2. D. J. Kessler, "Orbital Debris Issues," *Adv. Space Res.* **5**(2), 3–10 (1985).
3. D. J. Kessler, "Orbital Debris Environment for Space Station," *NASA*, **NASA JSC-20001** (1984).

4. D. J. Kessler, "A Guide to Using Meteoroid-Environment Models for Experiment and Spacecraft Design Applications," *NASA Tech. Note*, **NASA TN D-6596** (1972).

5. F. Whipple, *Meteoric Phenomena and Meteorites. The Physics and Medicine of the Upper Atmosphere*, Univ. of New Mexico Press, Albuquerque, 1952, pp. 137–170.

6. C. E. Anderson (Guest Ed.), "Hypervelocity Impact—Proceedings of the 1986 Symposium," *Int. J. Impact Eng.* **5** (1987).

7. Comet Halley Micrometeoroid Hazard Workshop, *Eur. Space Agency [Spec. Publ.] ESA SP*, **ESA SP-153** (1979).

8. R. Kinslow (Ed.), *High Velocity Impact Phenomena*, Academic Press, New York, 1970.

9. J. A. Zukas (Ed.), *Impact Dynamics*, Wiley, New York, 1982.

10. R. E. Smith and G. S. West, "Space and Planetary Environment Criteria Guidelines for Use in Space Vehicle Development," *NASA Tech. Memo.*, **NASA TM-82478** (1982).

11. N. C. Elfer et al., Martin Marietta IR&D M-015, unpublished research (1987).

12. A. R. Coronado et al., *Space Station Integrated Wall Design and Penetration Damage Control*, Contract NAS 8-36426, NASA, Marshall Space Flight Center, 1987.

13. A. Wenzel, "A Review of Explosive Accelerators for Hypervelocity Impact," *Int. J. Impact Eng.*, **5**, 681–692 (1987).

14. B. G. Cour-Palais, "Hypervelocity Impact in Metals, Glass and Composites." *Int. J. Impact Eng.*, **5**, 221–238 (1987).

15. N. C. Elfer and G. Kovacevic, (1985). "Design for Space Debris Protection," *AIAA-GNOS Symp. Pap.*, **85-009** (1985).

16. V. C. Frost, "Meteoroid Damage Assessment," *NASA [Spec. Publ.] SP*, **NASA SP-8042** (1970).

17. D. R. Christman, J. W. Gehring, C. J. Maiden, and A. B. Wenzel, "Study of the Phenomena of Hypervelocity Impact," *Gen. Mot. Def. Lab. Rep.*, **TR 63-216** (1963).

18. K. A. Holsapple, "The Scaling of Impact Phenomena," *Int. J. Impact Eng.*, **5**, 343–356 (1987).

19. J. R. Asay and G. I. Kerley, "The Response of Materials to Dynamic Loading," *Int. J. Impact Eng.*, **5**, 69–99 (1987).

20. Johns Hopkins University, *The Resistance of Various Metallic Materials to Perforation by Steel Fragments; Empirical Relationships for Fragment Residual Velocity and Residual Weight*, Ballistic Res. Lab. Proj. Thor TR No. 47, Johns Hopkins University, Baltimore, MD, 1961.

21. Johns Hopkins University, *The Resistance of Various Non-Metallic Materials to Perforation by Steel Fragments; Empirical Relationships for Fragment Residual Velocity and Residual Weight*, Ballistic Res. Lab. Proj. Thor TR No. 47, Johns Hopkins University, Baltimore, MD, 1961.

22. H. F. Swift and A. K. Hopkins, "The Effects of Bumper Material Properties on the Operation of Spaced Hypervelocity Particle Shields," *AIAA Hypervelocity Impact Conf.*, Cincinnati, OH, April 1970.

23. A. J. Richardson, "Theoretical Penetration Mechanics of Multisheet Structures Based on Discrete Particle Modeling," *J. Spacecr. Rockets*, **7**, 486–489 (1970).

24. B. J. Henderson and A. B. Zimmerschied, "Very High Velocity Penetration Model," **NSWC TR 83-189** (1983).

25. J. D. Yatteau, "High Velocity Penetration Model," **NSWC TR 82-123** (1982).

26. D. L. Dickinson, J. D. Yatteau, and R. F. Recht, "Fragment Breakup," *Int. J. Impact Eng.*, **5**, 249–260 (1987).

27. A. J. Richardson and J. P. Sanders, "Development of Dual Bumper Wall Construction for Advanced Spacecraft," *J. Spacecr. Rockets*, **9**, 448–451 (1972).

28. C. E. Anderson, Jr., "An Overview of the Theory of Hydrocodes," *Int. J. Impact Eng.*, **5**, 33–60 (1987).

29. W. E. Johnson and C. E. Anderson, Jr., "History and Application of Hydrocodes in Hypervelocity Impact," *Int. J. Impact Eng.*, **5**, 423–440 (1987).

30. K. S. Holian and M. W. Burkett, "Sensitivity of Hypervelocity Impact Simulations to Equation of State," *Int. J. Impact Eng.*, **5**, 333–342 (1987).

31. A. J. Piekutowski, "Debris Clouds Generated by Hypervelocity Impact of Cylindrical Projectiles with Thin Aluminum Plates," *Int. J. Impact Eng.*, **5**, 509–518 (1987).

32. R. Madden, "Ballistic Limit of Double-Walled Meteoroid Bumper Systems," *NASA Tech. Note*, **NASA TN D-3916** (1967).

33. A. R. McMillan, "Experimental Investigations of Simulated Meteoroid Damage to Various Spacecraft Structures," *NASA* [*Contract. Rep.*] *CR*, **NASA-CR-915**, TR 66-67 (1966).

34. J. Gehring, Jr., "Theory of Impact on Thin Targets and Shields and Correlation with Experiment," in R. Kinslow, Ed., *High Velocity Impact Phenomena*, Academic Press, New York, 1970, Chapter VV.

35. J. P. D. Wilkinson, "A Penetration Criterion for Double-Walled Structures Subject to Meteoroid Impact," *AIAA J.* **7**(10), 1937–1943 (1969).

36. J. F. Lundeberg, P. H. Stern, and R. J. Bartow, "Meteoroid Protection for Spacecraft Structure," *NASA* [*Contract. Rep.*] *CR*, **NASA-CR-54201**, D2-24056 (1965).

37. T. D. Riney and E. J. Halda, "Effectiveness of Meteoroid Bumpers Composed of Two Layers of Distinct Materials," *AIAA J.* **6**(2), 338–344 (1968).

38. C. J. Maiden, *Meteoroid Impact. Space Exploration*, McGraw-Hill, New York, 1964, pp. 236–284.

39. W. Johnson, N. Jones, S. R. Ried, and C. E. Anderson, Jr. (Eds.), *Hypervelocity Impact—Proceeding of the 1986 Symposium*, 1986; also *Int. J. Impact Eng.*, **5** (1–4) (1987).

40. N. Elfer, unpublished works.

41. H. Swift, R. Bamford, and R. Chinn, "Designing Dual Plate Meteoroid Shields—A New Analysis," *JPL Publ.* **82-39** (1982).

42. A. Piekutowski, private communication—unpublished works.

43. W. H. Friend, C. L. Murphy, and I. Shanfield, "Review of Meteoroid Bumper Interaction Studies at McGill University," *NASA* [*Contract. Rep.*] *CR*, **NASA-CR-54857** (1966).

44. R. Hoffman, "STEALTH, Lagrange Explicit Finite-Difference Code for Solids, Structures and Thermohydraulic Analysis," **EPRI NP-2080** (1981).

45. A. Marciniak and K. Kuczynski, "Limit Strain in the Process of Stretch Forming Sheet-Metal." *Int. J. Mech. Sci.*, **9**, 609 (1967).

46. C. C. Chu and A. Needleman, "Void Nucleation Effects in Biaxially Stretched Sheets," *J. Eng. Mater. Technol.*, **102**, 249–256 (1980).

47. F. W. Wilson, *High Velocity Forming of Metals*, Am. Soc. Tool Manuf. Eng., New York, 1964.

48. I. M. Fyfe and A. M. Rajendran, "Dynamic Pre-Strain and Inertia Effects on the Fracture of Metals," *J. Mech. Phys. Solids*, **28**, 17 (1980).

49. A. M. Rajendran and I. M. Fyfe, "Inertia Effects on the Ductile Failure of Thin Rings," *J. Appl. Mech.*, **49**, 31–36 (1982).

50. A. M. Rajendran, "A Critical Void Growth Failure Criterion Developed for Dynamic and Static Loading Conditions," Ph.D. Dissertation, Dept. Aeronaut. & Astronaut., University of Washington, Seattle, 1980.

51. C. C. Chu, "An Analysis of Localized Necking in Punch Stretching," *Int. J. Solids Struct.* (1981).

52. M. F. Kanninen and C. H. Popelar, "Advanced Fracture Mechanics," *Oxford Eng. Sci. Ser.*, **15** (1985).

CHAPTER 3

Controlled Fracturing of Structures by Shockwave Interaction and Focusing

Salim T. S. Al-Hassani

Department of Mechanical Engineering
University of Manchester Institute of Science and Technology
Manchester, England

ABSTRACT

The development of cracks and fracture damage in bars, plates, and solids of simple shapes subjected to explosive loading is described and the controlling mechanisms are considered. High-speed photography of model Perspex and plaster specimens was employed to observe the fractural behavior and the growth of the internal fissures. The results are used to verify the predictions of simple ray theory. The location and the rate of growth of the cracks are found to depend mainly on the geometry of the surface of the specimens. Straight, curved, and angled bars, triangular and trapezoidal plates, as well as solids of spherical, cylindrical, paraboloidal, hyperboloidal, and elliptical shapes, are considered. Reference is made to cutting of steel plate and pipe by controlled shocks generated by a surface explosive charge.

1. INTRODUCTION

This chapter describes attempts to constructively utilize the long-known, but inadequately understood, phenomenon of fracture due to stress waves. The characteristic features that typify the phenomenon are that (1) the high-speed loading is accomplished in a few microseconds, (2) the applied stress pulse is of a transient nature, (3) the stress waves reflect and interact destructively, and (4) the fracture is generally located at a point remote from the application of the load. The various pieces that fracture from the parent body are called *scabs*. The term *spalling* applies specifically to bars.

Figure 1 shows spalling in curved perspex bars. The straight bar was

79

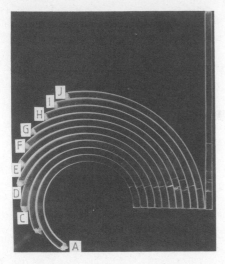

Figure 1. Spalling in curved Perspex bars of 3/8 in. sq. section and 12 in. circumferential length, exploded with ICI. No. 6 detonators. Mean radii of curvature are (A) 2.75, (B) 3.19, (C) 3.63, (D) 4.07, (E) 4.51, (F) 4.95, (G) 5.39, (H) 5.83, (I) 6.27, and (J) 6.71 in. [13].

subjected to explosive loading at the upper end, resulting in multiple spalling at the free end.

Spallation is the result of tensile reflection from the free end of a bar of the transient compressive pulse induced by the explosion. The number of spalls depends on the ratio of the compressive stress to the tensile strength. The lengths and escape velocities of the spalls depend on the pulse shape.

1.1. Origins of the Problem and Related Investigations

The first observations on fracture due to stress waves were made by John Hopkinson [1] about a century ago in his experiments "On the rupture of iron-wire by a blow." His work was followed up by his son, Bertram [2], and a number of others [3] at the turn of the century. The impetus for further work in this field came during World War II when aerial bombardment and field artillery were sometimes directed to both military and civilian targets.

Spalling, scabbing, cracking, and fracture damage are often encountered when geological and engineering structures are subjected to intense impulse loading. An understanding of the development of such damage can aid in the design of structures and missiles to resist failures and in devising novel methods for demolition purposes.

Military engineers [4, 5] are primarily concerned with the destruction caused to buildings, underground shelters, bunkers, and other structures, and this has led to the study of penetration, scabbing, perforation, and cavity formation. It was observed, for instance, that a missile that could penetrate a semifinite block of concrete could also perforate a wall of the same material of thickness twice the penetration depth. On examination, this was attributed to scabbing from the rear free surface, which assisted perforation.

Another vitally important phenomenon discovered was the scabbing of the inner face of walls of tanks and armored vehicles when subjected to attack by shells, bombs, and mines in contact with the outer surface. A more recent development is the study of seismic effects and the cracking of subterranean rock due to underground nuclear explosions [6]. Industrial interest in the phenomenon mainly centers around comminution of rock [7] and fracture during explosive- and shock-hardening of metals [8].

An interesting investigation reported by Miklowitz [9] concerns the tensile testing of brittle materials, in which it was noted that fractures were produced at two different cross sections of the specimen when the rupture load was reached. Experiments were conducted on Plexiglas at a loading rate of 300 lb/min. The situation was analyzed by considering two types of waves emanating from the initial rupturing section. One is a longitudinal compression wave that is, in fact, an unloading wave associated with the decrease in load in the fracture process. The other is a group of flexural strain waves produced by the moment that develops as the crack propagates across the bar from its origin at a surface of discontinuity. It was found that the second fracture usually occurred at a gauge cross section adjacent to the specimen head farthest from the initial fracture surface. It was shown in the analysis that this second failure is due to the superposition of the longitudinal strain from the unloading wave that, through reflections, becomes tensile and the resulting flexural strain, which together exceed the original tensile-fracture strain.

Many other related investigations are reported in the literature on stress waves, e.g., that of Davies [10], Rinehart [11–13], Kolsky [14, 15], and Johnson [16].

The effect of subjecting model brittle engineering structures to point explosive loads has recently received considerable attention. The geometries investigated include long, thin, straight, and curved rods [16, 17], spheres [18–23], thick-walled spherical shells [24], bars with transverse discontinuities [25], tapered bars and plates [26], cones [27], paraboloids and hyperboloids of revolution [28], ellipsoids of revolution [29], hollow cylinders packed with explosive [30], and hemispherically ended rods loaded at their plane end [31] and at their curved end [32]. The last paper includes a discussion of the formation of cardioid-shaped fractures stemming from the growth of *Hertzian* cone cracks emanating from the impact zone.

Particularly interesting features found in axisymmetric solids, explosively loaded on the surface, are fractures resulting from the focusing of tensile waves reflected from the curved boundaries. These features can be understood by reference to *ray theory*, commonly used in optics and acoustics, which enables the stress wavefronts propagating through the solid to be traced. The theory for the transmission and reflection of stress discontinuities in a solid of revolution is outlined in References 23, 28, and 33.

1.2. Recent Applications of Explosive Loading

Apart from their conventional use in the military field, explosives are being used in engineering applications. They have been used in welding dissimilar plates and cylinders, forming and sizing of a variety of components, compacting and hardening of powder, extensively in cutting ropes and bolts, and in ejecting aircraft canopies to allow pilot escape. They are now becoming increasingly used in the offshore environment. In particular, they are used for perforating well casings to start the flow of oil in a production well, cutting stuck drill strings, severing wellheads for seabed clearance, cutting wellhead guideposts, cutting piles and also cross members and legs of offshore platforms that are to be abandoned.

1.3.   The Present Chapter

This chapter presents a review of some experiments in which solids and structural members undergo fracturing and cracking due to explosive loading. In view of their practical importance, the following cases are considered:

1. Bars (straight, curved, and L-shaped)
2. Plates (triangular, trapezoidal, and rectangular)
3. Rectangular blocks
4. Solids of revolution (spherical, cylindrical, paraboloidal, hyperboloidal, and ellipsoidal)

2.   FRACTURING BARS

Bars of $\frac{3}{8}$ in. square section of various shapes were cut from Perspex (PMMA) sheets such that they all had equal 12 in. lengths. Each was subjected to a single explosive blow from an ICI No. 6 detonator placed at one end. The specimens were freely supported and the distal end of each was free.

2.1.   Straight Bars

A straight bar spalled at a section about 1.0 in. from the free end. Other incipient spalls farther away also occurred. The fractures took place at sections perpendicular to the axis of the bar. When shallow grooves of rectangular and V shapes were introduced at various sections along similar bars, the spalls occurred at the reduced sections. Therefore, a method was found for controlling the position of fracture along straight bars. Similar effects were found to occur in bars of circular section with steel wire wrapped tightly once around various sections along the bar. Explanation of the phenomenon of a stress wave and an effective Young's modulus is given in Reference 31.

2.2.   Curved Bars

Figure 1 shows photographs of the fractured curved bars. The specimens have been closely packed to facilitate visual comparison of the modes of fracture.

Spalling occurred in each specimen and the spalls were completely separated. It can be clearly seen that all fractures are strictly radial and compare well with a straight bar. The spalls are nearly equal with a mean arc length of 1.3 in.; the mean variation is about $\pm 3\%$.

2.3. Bent Bars

Fractured U-shaped specimens with bends of mean radii 3.0, 2.0, 1.5, 1.0, and 0.5 in. are shown in Figure 2. It is observed that two spalls occurred in the 3 in. radius specimen and a single spall in the case of the 2 in. radius of bend. It seems that in both cases, the stress pulse was transferred wholly and smoothly around the bend. In the specimens with bend radii of 1.5 and 1.0 in., radial fractures occurred in the middle of the bends, almost along the axis of symmetry; spalling in the loading arm can be seen in the 1.0 in. radius case. In the case of the 0.5 in. radius bend, four fractures occurred within the relatively small region of bend along with a fracture in the loading arm, which can be counted as spalling judging from its appearance.

In the case of L-shaped bars, spalling was also observed, but the length of spall reduced gradually as the radius of bend decreased. The fracture surfaces of the spalls are very nearly parallel to the free-end surface. Multiple spalls broke off from the 2 and 1.5 in. radius bend bars. For 1, 0.5, and 0.25 in. mean radii of bend, secondary fractures near the ends were observed. These secondary fractures tend to make spalled angles with the axis of the arm for diminishing radii of curvature. The fracture arm decreased until it made an angle of nearly 45° with the axis for a 0.25 in. radius bend. Multiple spalling was observed in the loading arm for the 0.25 in. mean radius, possibly because the radius of curvature was too short, with tensile reflections taking place from the side of the second arm.

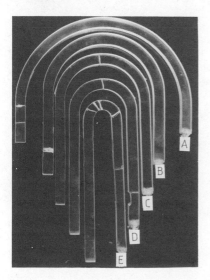

Figure 2. Fractures in U-shaped bars of 3/8 in. sq. section and 12 in. overall length, exploded with ICI No. 6 detonators. Mean radii of bend are (A) 3.0, (B) 2.0, (C) 1.5, (D) 1.0, and (E) 0.5 in [13].

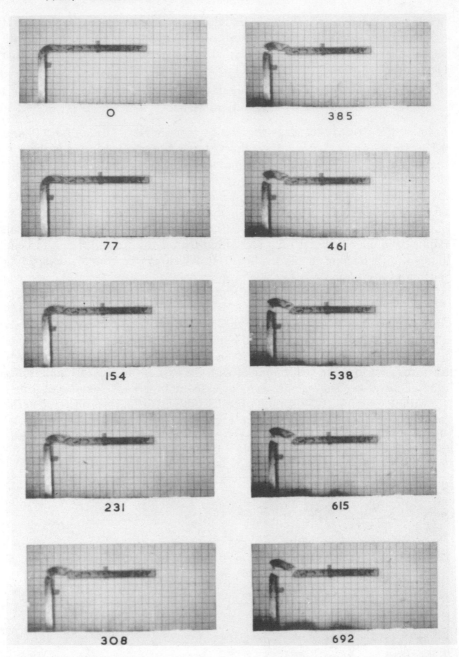

Figure 3. High-speed photographs of an L-shaped Perspex bar of 3/8 in. sq. section, 12 in. total length, and $1\frac{1}{2}$ in. radius of bend exploded with an ICI No. 6 detonator placed at the end of the vertical leg (not shown) [13].

84

High-speed photographs of the loading of an L-shaped bar with 0.5 in. radius of curvature were taken, as shown in Figure 3. The framing speed of the Hy Cam camera used was 13,000 frames/sec. At such a speed, a pulse would advance nearly 4.7 in. between successive frames. The first frame shows the specimen intact in the initial position. In the next frame, 0.077 msec after the first, all the fractures are completed; spalling almost occurred with a crack across most of the section in the free arm, but inclined fractures in the free and loading arms are seen. In successive frames, it is seen that the fragments, constituting the bend, moves along the axis of the loading arm, while rotating clockwise about a vertical axis and remaining in the horizontal plane.

3. RAY THEORY AS APPLIED TO REFLECTION OF STRESS WAVES FROM BOUNDARIES IN PLATES AND SOLIDS

Before examining the fracture behavior of plates and solids, it is necessary to consider aspects of ray theory as applied to reflection from free surfaces. An appreciation of the general concepts of the propagation and reflection of stress wavefronts in solids can be gained by considering Figure 4. When a homogeneous isotropic elastic axisymmetric solid is subjected to a localized pressure at point O on the surface, two spherical wavefronts emanate from the point of loading, that is, a dilatational wave (P wave) and a slower moving shear wave (S wave). Only one of the incident waves and only one of its reflected waves are shown in Figure 4. On reflection, a P wave produces a P wave, P_p, and an S wave, S_p. An incident S wave also produces a P wave, P_s, and an S wave, S_s.

Rayleigh surface waves also propagate over the surface. A detailed study of their behavior as they travel over the surface of a sphere subjected to impact is given by Silva-Gomes [34]. If the surface is flat, another type of shear wave is generated due to the grazing incidence of the P wave. It is usually referred to as a *head wave*, or H wave, and is discussed by Pao and Mow [35].

The jump in stress or particle velocity V across the wavefront at any time is related to its initial magnitude V_0 by

$$V = V_0 \left(\frac{R_0 S_0}{RS} \right)^{1/2} \tag{1}$$

where R and S are the instantaneous principal radii of curvature of the wavefront, and R_0 and S_0 are values corresponding to $V = V_0$.

In the case of *plates*, we have a two-dimensional situation, where $S = S_0 \to \infty$, and Equation (1) gives $V = V_0 (R_0/R)^{1/2}$. For a spherically symmetric wavefront, as in the case of three-dimensional solids, $R = S$, and we have $V = V_0 R_0/R$. On encountering the free curved boundary of the body, the wavefronts are reflected back into the body, resulting in a change in direction of propagation as well as in the radii of curvature of the wavefronts. If the surface is flat, there are no changes of curvature. The angles of the incident and reflected

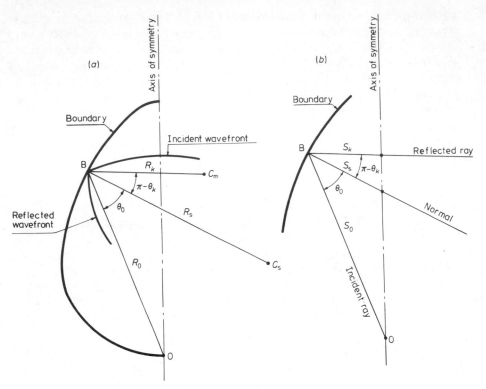

Figure 4.   (a) Wavefront geometry, where C_s is the center of curvature of the surface at B, and C_m is the center of curvature of the reflected wavefront [29]. (b) Tangential radii of curvature [29].

rays are related through Snell's law:

$$\frac{\sin \theta_0}{c_0} = \frac{\sin \theta_k}{c_k} \tag{2}$$

for $k = 1, 2$. The value of $k = 1$ refers to the reflected dilatational wave, and $k = 2$ refers to the reflected shear wave.

   In an axisymmetric situation, the reflected wavefronts are all surfaces of revolution about the axis of symmetry. At the boundary, the radius of meridional curvature R_k of the reflected wavefront is given by [23]

$$R_k = -\left[\frac{\tan \theta_k}{\tan \theta_0}\left(\frac{\cos \theta_0}{R_0} - \frac{1}{R_s}\right) + \frac{1}{R_s}\right]^{-1} \cos \theta_k \qquad k = 1, 2 \tag{3}$$

where R_s is the meridional radius of curvature of the boundary at the point of incidence, and R_0 is the radius of curvature of the incident wavefront at the same point. Equation (3) gives the position of the center of curvature C_m of the reflected wavefront at point B on the boundary surface [Figure 4(a)]. This center of curvature lies on the reflected ray at an algebraic distance R_k, measured from B toward the direction of propagation of the reflected wavefront. The center of tangential curvature of the reflected wavefront lies on the axis of symmetry [Figure 4(b)] at a distance S_k from point B given by

$$S_k = -S_2 \left[\cos \theta_k - \sin \theta_k \frac{S_0 \cos (\theta_k - \theta_0) - S_s \cos \theta_k}{S_0 \sin (\theta_k - \theta_0) - S_s \sin \theta_k} \right] \tag{4}$$

where $k = 1, 2$, and S_s and S_0 are the tangential radii of curvature of the boundary surface and incident wavefront at point B, respectively.

3.1. Stress-Wave Focusing

In wavefront-analysis nomenclature, a *caustic* is an envelope of a set of converging rays. In general, in a three-dimensional situation, a caustic consists of two distinct sheets and the reflected wavefront has an edge of regression where it touches either sheet of the caustic. When there is axial symmetry, one of the sheets of the caustic degenerates into the axis of symmetry and, instead of an edge of regression, there is a singular point on the axis of the solid. For this case, the only sheet of the caustic is given by Equation (3), which defines a surface of revolution about the axis of symmetry. The amplitude of the discontinuity at the wavefront becomes larger and larger as the caustic is approached. This would imply, of course, that the linear theory breaks down before the caustic is reached. It is, however, reasonable to infer that as the reflected rays converge (focus) onto the caustic, the amplitude of the stress (whether tensile or shear) increases dramatically, such that a fracture can result close to the caustic.

Another form of stress-wave focusing is developed in situations when a plane stress wave propagates along a bar with a tapering cross section. In the case of a cone [27], for example, the apex breaks off due to a tensile tail developing in a compression pulse as it approaches the tip.

4. CRACKS IN PLATES

Multiple tip fractures, similar to those in tapered bars [27], also develop in triangular plates. Figure 5 shows plaster triangular plates subjected to uniformly distributed explosive loading at the base [26]. It was noticed, through high-speed photography, that the cracks occurred progressively, with those nearest to the base occurring first. This is in reverse order to that of bars undergoing multiple spalling. This is due to the compressive-stress pulse developing a tensile tail as it propagates toward the tip. When a fracture occurs behind the wavefront, the pulse continues toward the tip, undergoing further amplification and causing more cracks to develop.

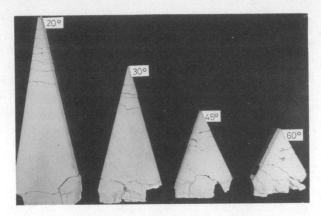

Figure 5. Fractures in 5.66 in. base plaster triangular plates due to uniformly distributed explosive loading with Cordtex [26].

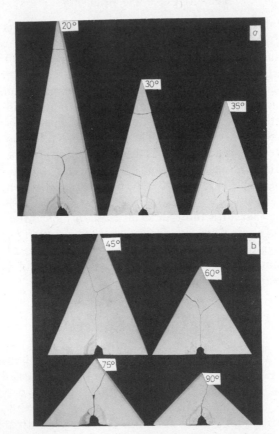

Figure 6. Fractures in 1 in. thick plaster of Paris triangles loaded with ICI No. 6 detonators [26].

If the plaster plates were subjected to single concentrated explosive loads (by a No. 6 detonator), additional axial cracks develop, as shown in Figure 6. Axial cracks are thought to be due to the reflections of the cylindrical compressive wave as tensile waves from the slanted sides of the plate. The two tensile wavefronts meet at the axis of the triangle and generate an axial crack. For further details on attenuation and on a method of using the ray theory to predict the positions of cracks and on their rate of growth, see Reference 26.

In another set of experiments, axial cracks were caused by applying explosive shock on the slanted sides of a triangular plate. The principle here is to generate plane shock fronts propagating toward the base and reflecting back as tension waves. When they meet at the axis, a central crack develops [36], as shown in Figure 7. In this figure, the Perspex plate ($\frac{1}{4}$ in. thick and with a 6 in. base) was subjected to Cordtex explosive initiated simultaneously at the lower corners of the slanted edges. The detonation wave travels toward the apex, causing a plane oblique shock front to travel toward the free base, where it is reflected according to Equation (2). This phenomenon of fracturing provided the basis of developing a new stress waveguide for use as an explosive cutting tape. The principle involves the use of a trapezoidal section waveguide (rubber or steel) to direct the shock waves, generated by the explosive on its sides, into the target to be cut such that the reflected waves from the base meet at the intended line of cut. The obliquity of the shock is controlled by the angle of the slanted edges as well as by the ratio of the velocity of detonation of the plastic explosive to the dilatational wave speed of the waveguide material. Further changes of the angle and

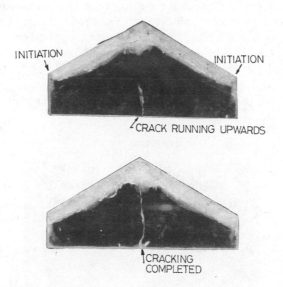

Figure 7.   Front view of a Perspex specimen plate with a 1/4 in. thick 6 in. base subjected to explosive loading on the slanted sides. The central crack started from the free base and continued to the apex [36].

amplitude of the stress wave would occur at the interface between the waveguide and the target plate.

4.1. Transverse Cutting of Plates by a Flat Charge

Oblique shock waves can be entered into a flat plate by means of a flat charge placed on the surface. By initiating the charge at both ends simultaneously, two oblique shock waves are generated and propagate across the thickness of the plate. When they reflect back into the plate from the free bottom surface, they meet at the central axis, engendering a high-intensity tensile stress. The component of the reflected stress parallel to the plate surface causes a transverse crack to develop, which starts at the bottom surface and runs upwards to the top surface as the wavefronts pass each other. The vertical stress components, however, are responsible for the usual scabbing phenomenon. Careful design of the explosive charge can generate complete transverse fracture of plates without any distortion of the rest of the body of the plate, as shown in Figure 8. The figure shows the top of a 12 × 6 in. mild-steel 1.5 in. thick plate cut into halves by a surface charge whose geometry is revealed by the imprint it left on the surface. The line marks are traces of the detonation fronts initiated at the extreme corners of the diamond shape. It is interesting to note that the paths followed by the detonation fronts are controlled to make them arrive simultaneously at the line to be cut, a technique referred to as *plane-wave generation*.

A flexible tape or belt encased in rubber that contains a series of such diamond charges can be made for cutting large plates or tubular shells. It was found that using this technique for cutting tube can save about 50% of the weight of the charge in comparison with the conventional method of linear shaped charge. As is well known, a linear shaped charge employs the energy of the explosive to accelerate a thin metal copper liner in order to focus the collapsing material at the collision point to generate an ultrahigh-speed metal jet that cuts through the target, as shown in Figure 9.

A circular form of the shaped charge is used in cutting tubular members.

INITIATION AT E

Figure 8. Top view of a mild steel plate, 6 × 12 × 1.5 in. thick, transversely cut through by a diamond-shaped flat surface charge the imprint of which is seen on the surface [36].

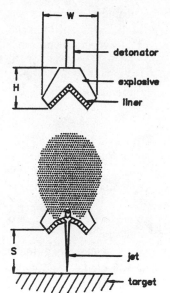

Figure 9. Linear shaped charge

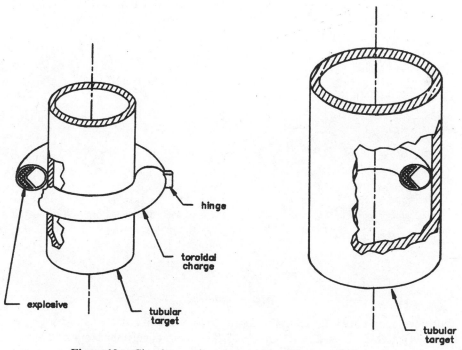

Figure 10. Circular cutting charges for offshore applications.

When used underwater, hermetically sealed toroidal casings to contain the charge are employed, as shown in Figure 10. Unlike the linear shaped charge, the flexible flat tape does not require an air environment. However, the presence of water or concrete backing can affect the efficiency and performance of the oblique shock waves. For further details on the application of oblique shock collision in cutting plate and tubular structures, see Reference 36.

5. FRACTURING SOLIDS

We now describe the development of cracks and fissures in Perspex models subjected to concentrated loading on the surface by means of a No. 6 detonator.

5.1. Rectangular Blocks

Figure 11 shows the cracks developed in a 3 × 3 × 1.5 in. Perspex block due to an explosion at the center of the top square face. The reflected tensile waves cause an incipient scab near the bottom surface. Surface stress waves reflected from the four sides of the square cause edge fractures while the converging wavefronts toward the corners cause corner fractures. Depending on the dimensions of the block, it may be possible to cause a central axial crack due to tensile stress tensile stress waves reflecting from the sides and meeting at the axis.

Examples of rocks of various shapes fractured by stress waves are given in Reference 7.

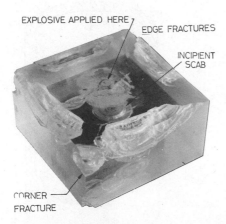

EXPLOSIVE APPLIED HERE
EDGE FRACTURES
INCIPIENT SCAB
CORNER FRACTURE

Figure 11. Cracks in a 3 × 3 × 1½ in. Perspex block due to detonation of an ICI No. 6 detonator at the center of the top square face [26].

5.2. Spheres

The development of internal fractures due to focusing of reflected stress waves in spheres is fully discussed in Reference 23. The main experiment reported is that of a Perspex sphere subjected to a surface explosion at one pole. Three distinct

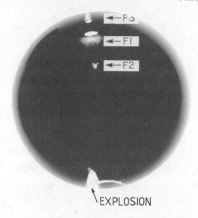

EXPLOSION

Figure 12. Side view of the Perspex sphere showing internal fractures F1, Fw, and F3 [23].

regions of fracture (F1, F2, and F3) were noticed inside the sphere, as shown in Figure 12, and one shallow and concentrated region (F4) on the surface at the antipole, not shown in the figure. Each region appeared at a different time and was associated with a different type of stress wave. The Perspex sphere was 73.7 mm in diameter and was subjected to an explosion at the lower pole. The times at which the three internal fracture regions (F1–F3) occur, their positions in relation to the point of loading, and their instantaneous shapes have been shown [23] to agree remarkably well with the predictions of simple ray theory [viz Equations (1)–(4)]. F1 is thought to be initiated by the reflected P initiated by the incident P wave, but later reinforced by the reflected S_s wave. Its position matches well with the upper regions of the caustics of the P_p and S_s waves. F2 is associated with the reflected shear wave generated by the incident P wave, i.e., S_p, and fracture F3 with the caustic of the reflected P wave originating from the incident S wave, i.e., P_s wave.

5.3. Surface Rayleigh Waves

The antipodal surface cracking F4 is thought to be due to focusing of the Rayleigh waves emanating from the loaded pole.

It is interesting to note that when a segment of the sphere is cut and the load is applied to the center of the flat surface of the remaining hemisphere, the antipodal surface cracks disappear. Numerous photographs of specimens loaded in this way are given in References 19 and 20. The reason for the disappearance of surface cracks at the antipole is believed mainly due to the effect of the sharp corners between the flat and curved surfaces that the Rayleigh waves encounter. A large proportion of their energy is reflected back toward the loading location.

Surface waves are almost eliminated in specimens having the geometry of a rod with a hemispherical end. Reference 26 reported the occurrence of only fracture types F1–F3 when such specimens were loaded at the plane end. However, internal cracks were eliminated, whereas the surface antipodal crack was exaggerated when a flat circular disk was loaded at the edge in the same way. Reference 18 shows how a radial crack is generated at a point diametrically opposite the point of loading. In a plate, the surface area of the flat edge is too small to produce high-amplitude reflected waves. Yet the circular edge allows the Rayleigh waves to travel around and meet again at the antipole.

5.4. Cylinders

Especially interesting features were found [32] in the neighborhood of the loading point when hemispherically ended rods were loaded at their hemispherical end. Figure 13 shows a typical set of fracture patterns found in such specimens. Of particular interest is an inward-curving truncated-cone fracture that develops toward the axis to produce a cardioid surface. This is explained in terms of propagating Hertzian cone cracks, normally found in rocks and glasses [37, 38], which change direction due to incident compressive P waves reflected from the flat surface as a tensile P wave (or S waves reflected as S waves). The

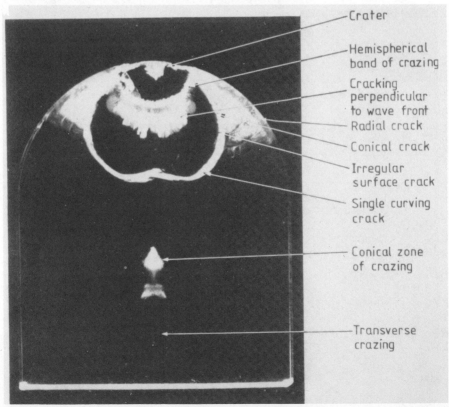

Figure 13. Section of a perspex specimen loaded at the hemispherical end and showing fracture features, including cardioid fracture [32].

next zone of crazing, the conical zone, is thought to correspond to the location of the caustic of the reflected S wave derived from the incident P wave. The other fracture features at the zone of impact are discussed in detail in Reference 32.

A most interesting and more frequently occurring situation is that of a thick-walled cylinder packed with an explosive charge along its axis. The mechanics of deformation and fragmentation of ductile cylinders is fully discussed by the author in Reference 39. When detonation is initiated at one end, a conical fracture normally occurs at the other end. This is especially noticeable in brittle materials, where a conical crack can be seen as shown in Figure 14(a). Such a phenomenon was used by Nash and Cullis [40] to study the dynamic fracture of materials. In order to explain this fracture geometry, it is envisaged that a detonation wave runs from the point-of-initiation end to the free end. Because of the movement of the point of loading along the axis of the cylinder, the wavefront is oblique, as shown in Figure 14(b). In this case, it is represented by lines parallel to AB. In a cylindrical situation, the wavefront is a conical surface moving along rays GB. As the P wave is reflected back from the cylindrical surface, it undergoes a change of sign, so that when the P wave (AF) reflected from the flat base meets with that reflected from the curved surface (BE), a region of high tensile stress is formed and a conical fracture occurs.

Another interesting situation occurs when the explosive is initiated simultaneously at both ends of the cylinder. The cylinder is cut into two equal lengths, with the fracture section af perpendicular to the axis, as shown in Figure 15(a). The occurrence of disk crack af is thought to be due to the interaction of the two

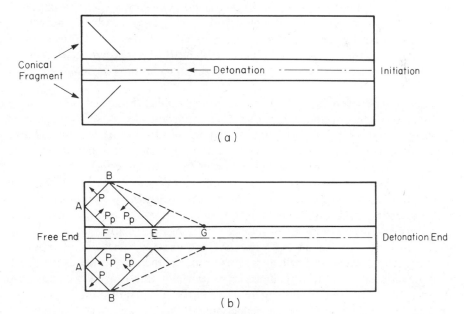

Figure 14. (a) Conical fragment about to be formed at the distal end. (b) Stress wavepoints just after complete detonation of the explosive.

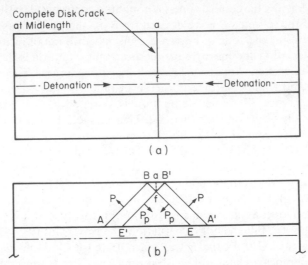

Figure 15. (a) Disk crack completely separates the two halves of the cylinder. (b) Stress wavepoints after the arrival of the two detonation fronts at the middle of the cylinder [30].

reflected P_p waves BE and $B'E'$ as the P wavefronts AB and $A'B'$ pass each other at the middle of the cylinder, as shown in Figure 15(b). The triangular region $BB'f$ at this instant is subject to two tensile waves. The stresses can be resolved into two components, parallel and perpendicular to af. A charge can be designed with a suitable detonation speed such that the obliquity of the wavefronts can be optimized to maximize the tensile stress perpendicular to af and thus cause a brittle fracture along af. As the P_p waves BE and $B'E'$ pass each other, point f propagates toward the inner surface of the cylinder until complete fracture is accomplished.

This phenomenon is finding wide-ranging application in the demolition industry and, more recently, in the cutting of entrapped drill strings in the oil production industry. The particular advantage of this feature is that a charge can be designed that cuts a pipe with no excessive diametral expansion. This reduces the pull-out forces on the portion of the pipe to be recovered.

The collision of detonation waves has been usefully employed in the design of efficient explosive charges for wellhead severance offshore. A cylindrical charge that can be initiated simultaneously at both ends is contained in a thin-walled container and lowered down whole to about 5–10 m below the sea bed, as shown in Figure 16. In order to reduce the damaging effect of the explosive pressure on the wellhead itself, as well as to reduce the magnitude of the shock emerging from the well mouth, an energy-absorbing plug is usually placed atop the charge. By careful matching of the charge shape and weight with the deformation behavior of the plug, it is possible to design an efficient cutting system that cuts through the multicasings with least damage to nearby structures.

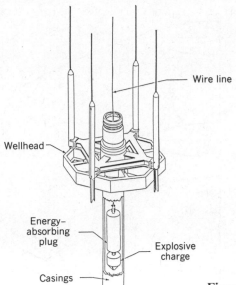

Wire line

Wellhead

Energy–
absorbing
plug

Explosive
charge

Casings

Figure 16. Explosive charge inside a wellhead.

6. PARABOLOIDS AND HYPERBOLOIDS

Full discussion of the subject of fracture generation in paraboloids and hyperboloids by explosive loading at the focus is given in Reference 28. The main findings, however, can be summarized by referring to Figures 17–19. The manner in which P, S, and H wavefronts propagate and reflect from the free boundary in paraboloids and hyperboloids loaded at focus 0 (i.e., at the center of the base whose radius is a) is indicated in Figure 17. The reflected P_p wavefront in the paraboloid is a right cylindrical surface converging, and intensifying, onto axis OV. It arrives at OV at $t = 2a/C_L$, which is equal to 36.3 microseconds for Perspex with $2a = 73.7$ mm. In the case of a hyperboloid, however, the reflected P_p wavefront has a curved cylindrical surface that reaches the axis first at the focal point 0, the point of loading. This focal point moves toward the apex along the axis with a finite speed governed by the meridional radius of curvature of the reflected wavefront and the dilatational wave speed of the material.

High-speed photographs for a paraboloid revealed that an axial fracture F1 appeared instantaneously at 39 microseconds after detonation, whereas at the same framing speed (10^6 frames/sec), the hyperboloid, see Figure 18, showed an axial fracture emanating from the base center 0 and propagating toward the apex with a high but finite speed. Very close agreement was found between the instantaneous position of the head of fracture F1 and the focal point of the P_p wavefront along the axis.

Another type of fracture having the shape of a disc parallel to the base occurs both in the hyperboloid and paraboloid. Figure 19 shows this type of disc

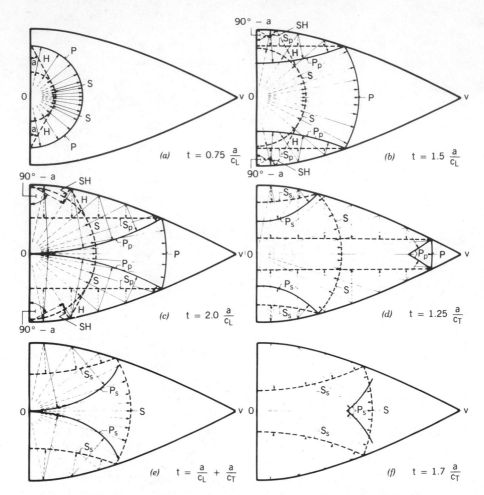

Figure 17. Wavefronts in a hyperboloid loaded at the center of the base.

fracture, F2, in a paraboloid. The high-speed photographs revealed that F2 appeared at $t = 68$ microseconds. This matches well with the instant of arrival at the axis of symmetry of P_s, the reflected dilatational wave from the incident shear S wavefront. On the other hand, its position corresponds to reflected rays resulting from waves incident at angles greater than 26° to the normal to the boundary. This, according to ray theory [23], produces high-intensity reflected waves. A P_s ray reflected from a ray incident at 26° reaches the focal point on the axis of symmetry at a time approximately 65 microseconds after detonation. Therefore, it seems reasonable to associate this second type of fracture with the reflected P_s wavefront.

Another possible cause for F2 is proposed in Reference 28 to be associated with the focusing of the reflected head wave SH, as shown in Figure 17. The fact

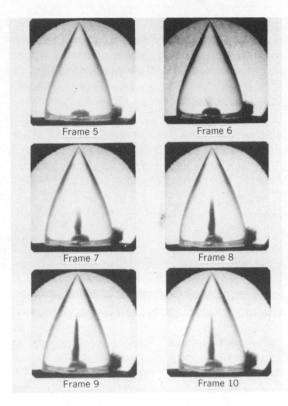

Figure 18. High-speed photographs (10^6 frames/sec) of internal fractures in a 73.7 mm base diameter Perspex hyperboloid of revolution due to the explosion of an ICI No. 6 detonator at its base.

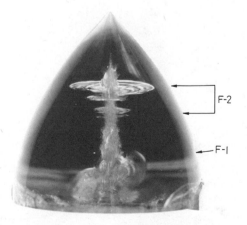

Figure 19. Internal fractures in a Perspex paraboloid loaded at its base (73.7 mm diameter). Axial fracture is shown at F1 and disc fractures at F2.

that *SH* originates from the *H* wave, which has less curvature than *P* and *S* wavefronts, could result in a relatively large amplitude SH wave. The focus of the *SH* wave, however, does coincide with the position of the upper disc fracture.

6.1. Ellipsoids of Revolution

When an ellipsoid of revolution is impulsively loaded at one focus F_1, two spherical stress waves, as well as a headwave, emanate, which upon reflection at

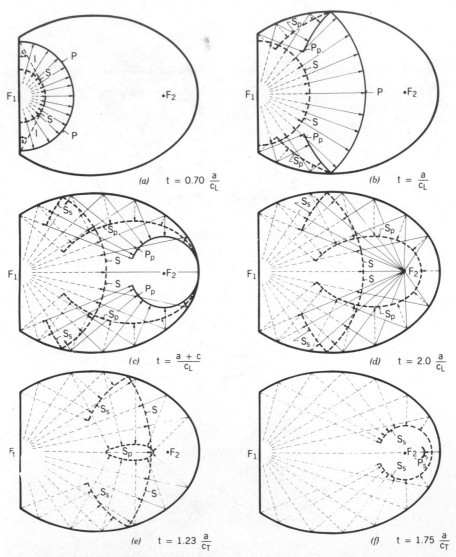

Figure 20. Wavefronts in an ellipsoid loaded at focus F_1. The major axis is $2a$ and $F_1F_2 = 2c$.

the boundary produce a total of four distinct wavefronts, as shown in Figure 20. The dilatational wave P_p resulting from the incident P wave is a spherical wavefront converging onto F_2, which is its only focal point. The reflected S wavefront S_p, originated by the incident P wave, first reaches a focal point on the axis of symmetry and later a second focal point at the end of the major axis. These two focal points move toward each other. There is a caustic $C_s^{(P)}$ associated with the reflected wavefront S_p, as shown in Figure 21. It is seen that each reflected ray touches the caustic on the other side of the axis of symmetry after having passed through a first focal point on the axis of symmetry. The S_s wavefront is geometrically similar to the P_p wave, converging onto F_2. The reflected P_s wavefront originates at the far end of the ellipsoid from shear wavefronts incident at angles below $28.7°$. This has a caustic $C_p^{(S)}$, as shown in Figure 21.

Experiments are reported in Reference 29 on Perspex ellipsoid models with a major semiaxis equal to $a = 50.8$ mm and eccentricity $e = 0.677$. The final observation of an exploded specimen shows that apart from the fracture in the immediate vicinity of the loaded region, three other distinct regions of internal fractures can be identified, as shown in Figure 22. An upper fracture F1 is

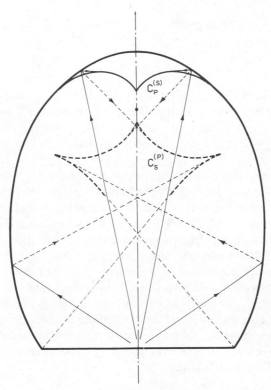

Figure 21.   Caustic of reflected wavefronts in an ellipsoid loaded at a lower focus.

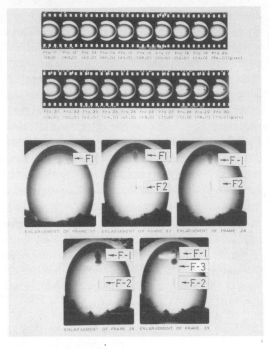

Figure 22. High-speed photographs (5×10^5 frames/sec) of internal fractures in an ellipsoid loaded at a lower focus ($a = 5$ cm, $b = 3.7$ cm).

concentrated around the axis of symmetry in the region near the focus. The fractured zone extends over 15 mm. The lower fracture F2 is also a narrow elongated type about 12 mm long and located in the central region of the specimen. The middle fracture F3 is very small and appears just below fracture F1.

Figure 22 shows selected photographs taken at a rate of 500,000 frames per second (2.0 μsec between frames). The full film shows that fracture F1 initiates in frame 22 (60 μsec after detonation) and by frame 24 (64 μsec after detonation), it has developed completely. Then by frame 28 (72 μsec after detonation), it is observed that fracture F1 extends farther toward the center of the ellipsoid. Finally, in frame 29 (74 μsec after detonation), the third fracture F3 occurs.

The formation of these regions of internal fractures corresponds to the arrival of the different types of stress waves, which were discussed.

Perhaps one of the most important applications of ellipsoidal stress waveguides is the development of an electrohydraulic discharge machine for destroying kidney stones. This was recently reported in Reference 41, where stress impulses are initiated at one focus of the ellipsoid and are collected at the other focus, in the patient's kidney.

7. RECENT OFFSHORE APPLICATIONS

With the advent of North Sea oil exploration, numerous structures have been installed. Many of these, in particular, deep sea platforms, will have to be removed soon. The number of installations that have to be removed to leave a clean seabed is beginning to increase rapidly. Of the four cutting methods for use in a deep sea environment, the explosive cutting technique is available. Explosive tools are becoming safer and adapted to a deep sea environment [42, 43]. A wide variety of cutters have been developed for cutting large-diameter tubulars, chain and wire ropes, gas and oil pipeliners, drill collars and stuck drill pipes, piles, wellhead guideposts, and wellhead casings, for joint jumping, and for debris clearance. Most of these cutters utilize in one or another form some of the principles discussed in this chapter. Some of these methods, and their effects on nearby structures, and service methods are reviewed in References 42–44.

8. CONCLUSION

This chapter reviews a variety of fracture features occurring in solids and simple structural members. These features are mainly governed by the manner in which stress waves interact to form regions of tensile stress of high magnitude that cause failure. The location of the onset fracture and the extent and direction of the fracture zone were found to be predictable using simple ray theory. Stress-wave reflection from a flat boundary plays a major role in determining the spalling behavior of straight, curved and bent bars and plates. Stress-wave focusing seems to control the fracture behavior of brittle solids with convex surfaces that are concave when viewed from inside the solid and that reflect the incident wavefronts back into the body of the solids with different radii of curvature and increasing amplitudes. Such phenomena are used in developing effective cutting methods in the demolition industry and more recently in the offshore environment.

ACKNOWLEDGMENTS

The author wishes to acknowledge the United Kingdom Science and Engineering Research Council for their support under Contract No. GR/C/7136.1, which funded the work concerned with offshore applications referred to in this chapter.

REFERENCES

1. J. Hopkinson, *Original Papers*, Vol. II, Cambridge Univ. Press, London and New York, 1901, pp. 316–320.

2. B. Hopkinson, *Scientific Papers*, Cambridge Univ. Press, London and New York, 1921, pp. 423–474.
3. J. W. Landon and H. Quinney, "Experiments with the Hopkinson Pressure Bar," *Proc. R. Soc. London, Ser. A*, **103**, 622 (1923).
4. W. N. Thomas, "The Effects of Impulsive Forces on Materials and Structures. The Civil Engineer at War," *Proc.–Inst. Civ. Eng*, **3** (1948).
5. H. E. Wessman and W. A. Rose, *Aerial Bombardment Protection*, Wiley, New York, 1962.
6. M. L. Baron and R. Chech, "Elastic Raleigh Wave Motion due to Nuclear Blast," *Proc. Am. Soc. Civ. Eng. EMI, Eng. Mech. Div.*, Feb., p. 57 (1963).
7. J. S. Rinehart, "The Role of Stress Waves in the Comminution of Brittle Rocklike Materials," in N. Davids, Ed., *Stress Wave Propagation in Materials*, Wiley (Interscience), New York, 1968.
8. J. P. Mykanen et al., "A New Method for Explosive Strengthening of Metals with Application to Produce Type Parts," *Proc.—Int. Conf. Cent. High Energy Form.*, *2nd*, Vol. 1 (1969).
9. J. Miklowitz, "Elastic Waves Created During Tensile Fracture," *J. Appl. Mech.*, p. 122, March (1953).
10. R. M. Davies, "A Critical Study of the Hopkinson Pressure Bar," *Trans. R. Soc. London, Ser. A*, **240**, 375 (1948).
11. J. S. Rinehart, "Some Quantitative Data Bearing on the Scabbing of Metals under Explosive Attack," *J. Appl. Phys.*, **22**, 555–560 (1962).
12. J. S. Rinehart and J. Pearson, *Explosive Working of Metals*, Pergamon, Oxford, 1963, p. 47.
13. J. S. Rinehart and J. Pearson, *Behaviour of Metals Under Impulsive Loading*, Dover, New York, 1965.
14. H. Kolsky, "Fracture Produced by Stress Waves," in B. L. Averbach et al., Eds., *Fracture*, MIT Press, Cambridge, MA, 1959.
15. H. Kolsky, *Stress Waves in Solids*, Dover, New York, 1963.
16. W. Johnson, *Impact Strength of Materials*, Edward Arnold, London, 1973.
17. M. Nasim, S. T. S. Al-Hassani, and W. Johnson, "Stress Wave Propagation and Fracture in Thin Curved Bars," *Int. J. Mech. Sci.*, **13**, 599 (1971).
18. E. Lovell, S. T. S. Al-Hassani, and W. Johnson, "Fracture of Spheres and Circular Discs due to Explosive Pressure," *Int. J. Mech. Sci.*, **16**, 193 (1974).
19. J. F. Silva-Gomes, S. T. S. Al-Hassani, and W. Johnson, "A Note on Times to Fracture in Solid Perspex Spheres due to Point Explosive Loading," *Int. J. Mech. Sci.*, **18**, 543 (1976).
20. W. Johnson and A. G. Mamalis, "Fracture Development in Solid Perspex Spheres with Short Cylindrical Projections (Bosses) Due to Point Explosive Loading," *Int. J. Mech. Sci.*, **19**, 309 (1977).
21. H. Rumpf and K. Schonert, "Die Brucherscheinungen in Kugeln bei elastischen sowie plastischen Verformungen durch Druckbean-sprunchun. *DECHEMA-Monogr.*, **69**, 51 (1972).
22. H. Rumpf, "Fracture Physics in Comminution," *Int. Tag. Bruch, 3rd*, Plate IX, p. 142 (1973).
23. J. F. Silva-Gomes and S. T. S. Al-Hassani, "Internal Fractures in Spheres due to Stress Wave Focussing," *Int. J. Solids Struct.*, **13**, 1007–1017 (1977).

24. J. D. Stackwi and O. H. Burnside, "Acrylic Plastic Spherical Shell Windows under Point Impact Load," *J. Eng. Ind.*, **98**, 563 (1976).

25. W. Johnson and A. G. Mamalis, "The Fracture in Some Explosively End-loaded Bars of Plaster of Paris and Perspex Containing Transverse Holes or Changes in Section," *Int. J. Mech. Sci.*, **19**, 169 (1977).

26. S. T. S. Al-Hassani, W. Johnson, and M. Nasim, "Fracture of Triangular Plates due to Contact Explosive Pressure," *J. Mech. Eng. Sci.*, **14**(3), 173–183 (1972).

27. M. Kolsky and A. C. Shearman, "Investigation of Fractures Produced by Transient Stress Waves," *Research*, **2**, 383 (1972).

28. S. T. S. Al-Hassani and J. F. Silva-Gomes, "Internal Fracture Paraboloids of Revolution due to Stress Wave Focussing," *Conf. Ser.—Inst. Phys.*, **47**, 187–196 (1979).

29. S. T. S. Al-Hassani and J. F. Silva-Gomes, "Internal Fractures in Solids of Revolution due to Stress Wave Focussing," in M. A. Meyers and L. E. Murr, Eds., *Shock Waves and High-Strain-Rate Phenomena in Metals*, Plenum, New York, Chapter 10, pp. 169–180.

30. S. T. S. Al-Hassani, "Fracturing of Explosively Loaded Solids," in S. R. Reid, Ed., *Metal Forming and Impact Mechanics*, Pergamon, Oxford, 1986, Chapter 18.

31. W. Johnson and A. G. Mamalis, "High Velocity Deformation of Solids," in K. Kawata and J. Shioiri, Eds., *IUTAM Symposium, Tokyo, Japan*, 1977, pp. 228–246.

32. D. J. Williams, B. J. Walters, and W. Johnson, "Crack Patterns in Cylinders Explosively Loaded at an Hemispherical End," *Int. J. Fract.*, **23**, 271–279 (1983).

33. S. T. S. Al-Hassani and W. Johnson, "Stress Wave Fracturing of a Bar," in F. Koenigsberger and S. Tobias, Eds., Proc. 12th Int. Conf. on Machine Tool Design and Research, Macmillan, New York, 1972, pp. 129–139.

34. J. F. Silva-Gomes, Ph.D. Thesis, University of Manchester, U.K., 1978.

35. Y. H. Pao and C. C. Mow, *Diffraction of Elastic Waves and Dynamic Stress Concentrations*, Adam Hilger, Bristol, U.K., 1973.

36. L. Davis, "Explosive Cutting of Plates by Stress Wave Interaction," M.Sc. Thesis, University of Manchester, U.K., March 1984.

37. M. V. Swain and B. R. Lawn, "Indentation Fracture in Brittle Rocks and Glasses," *Int. J. Rock Mech. Min. Sci., Geomech. Abstr.*, **13**, 311–319 (1976).

38. M. V. Knight, M. V. Swain, and M. H. Chaudri, "Impact of Small Steel Spheres on Glass Surfaces," *J. Mater. Sci.*, **12**, 1573–1587 (1977).

39. S. T. S. Al-Hassani and W. Johnson, "The Dynamics of the Fragmentation Process for Spherical Shells Containing Explosives," *Int. J. Mech. Sci.*, **11**, 811 (1969).

40. M. A. Nash and I. G. Cullis, "Numerical Modelling of Fracture," *Conf. Ser.—Int. Phys.*, **70**, 307–314 (1984).

41. "Shock Treatment for Stones," *Newsweek*, June 25, p. 54 (1984).

42. S. T. S. Al-Hassani, *Explosive Requirements and Structural Safety Aspects in Offshore Decommissioning Applications*, Offshore Decomm. Conf. ODC86, Heathrow Penta., November 1986.

43. S. T. S. Al-Hassani, "Underwater Cutting," *IBC Conf. Decomm. Removal North Sea Struct.*, London, Marriott Hotel, April 1987.

44. S. T. S. Al-Hassani, "Concepts of Explosive Cutting for Platform Removal," *Int. Conf., Offshore Mech. Arct. Eng. (OMAE)*, 7th, ASME, Houston, TX, February 1988.

CHAPTER 4

Tearing of Thin Metal Sheets

A. G. Atkins
Department of Engineering
University of Reading
Reading, England

ABSTRACT

The energy-based equations governing crack initiation and propagation in elastic, elastoplastic, and plastic deformation fields are briefly reviewed, and the important parameters concerned with material properties and body geometry are identified. The mechanics appropriate to ductile tearing fractures in thin metal sheets are developed for the cases of (1) when there are starter cracks present (propagation), and (2) when the sheets are supposedly flawless (initiation and propagation). The role of microstructure in the case (2) is highlighted, particularly for commercial metals in which void growth and coalescence control ductile fracture. It is shown that the rigid-plastic approximations to tearing problems, in place of full elastoplastic solutions, often adequately represent observed behavior when fracture is accompanied by, or preceded by, extensive plastic flow. Methods of measuring the specific essential work of crack propagation (R) in the presence of large amounts of plastic flow are described for crack-opening modes I, II, and III (i.e., tensile, shear, and twisting fractures, respectively) along with experimental results for metal sheets in different states of work-hardening. Applications to tearing problems are given. Traditional empirical criteria for predicting crack initiation in ductile fracture are reviewed and most are shown to be equivalent to the McClintock equations for void growth and coalesence. How sheet-metal-forming-limit diagrams (which give the locus of necking strains for in-plane biaxial loading) can be extended to include fracture strains is discussed. A possible connection between initiation strain pairs and the propagation specific work of fracture for the same material is explored, and applied to the fracture of shells, along with the implications for a supposed one-parameter characterization of ductile fracture. The conditions governing the transition from plasticity alone preceding fracture to plastic flow combined with fracture during ductile crack propagation is investigated. The case of energy-absorbing tubes, which split after some ductile expansion, is used as an illustration. It is shown how the number of splits can be predicted from the size of the tube and the toughness-to-strength ratio of the material from which it is made.

1. INTRODUCTION

The tearing of thin sheets of ductile metals is a process of combined flow and fracture. There is plastic work both preceding the onset of crack propagation and also accompanying fracture (i.e., tearing). Familiar examples are the opening of metal food or drink cans of various types [1], and controlled tearing accompanied by plastic flow has been suggested for energy-absorbing devices [2]. An understanding of the mechanics of crack initiation and propagation in these situations is important for the assessment of shell structures, safety screens, and so on under dynamic loading.

Deformation models that give predictions of the forces required, displacements, and energy involved have components of plastic work increment and fracture work increment; and friction work increment, too, when the device involves sliding of metals over tooling. In ideal cases, the elastic work increment component should also be included, but the solution of fully elastoplastic fracture problems is formidable. It is important to appreciate that the fracture work component can be as significant as the plastic work component in these sorts of problems [1] and the relative magnitude of flow and fracture components is an important element in determining the transition from plastic flow preceding fracture to crack initiation and propagation.

While the mechanics of elastic fracture are well-founded, the mechanics of elastoplastic fracture are still being developed. Progress on criteria for cracking has been made where the level of strains in plastic flow fields is not much greater than yield strain levels, but the mechanics of elastoplastic fracture when the remote strains are very great remains a lacuna. However, the author and colleagues [1, 3, 4] have shown that an acceptable line of attack to this sort of problem is available using rigid-plastic fracture mechanics. Tearing is tackled in this way in this chapter. It is appropriate first to review the general mechanics of fracture in order to make clear the approach and relate it to well-known (but often inapplicable) criteria for fracture, such as the critical stress-intensity factor K_c, the J integral, the J_R resistance curves, critical crack-opening displacements δ_c, and so on.

1.1. The Mechanics of Fracture

Quasi-static elastic fracture is governed by equations of the form [5]

$$X\,du = d\Lambda + R\,dA \tag{1}$$

where X is the load, u is the load-point displacement, Λ is the elastic strain energy, R is the specific work of fracture (the fracture toughness), and A is the crack area. The body behaves in a globally elastic manner with all the irreversibilities associated with R restricted to boundary layers contiguous to the crack faces. In linear elastic fracture, the strain energy $\Lambda = \frac{1}{2}Xu$; in nonlinear elastic fracture, $\Lambda = \int X\,du$ and the complementary strain energy $\Omega = \int u\,dX = Xu - \Lambda$ [6].

It follows that

$$R = -(\partial\Lambda/\partial A)_u \tag{2}$$

etc., and five other equivalent relations for $R(\Omega)$, which follow from the Jacobian $J(Xu/RA) = 1$ [6].

In linear elastic fracture, Equation (2) becomes

$$R = \frac{1}{2}X^2\frac{d}{dA}\left(\frac{u}{X}\right) \tag{3}$$

which with G_c replacing R is the well-known Irwin compliance calibration equation for linear elastic fracture [7]. G_c is the critical strain-energy release rate of the cracked body at which fracture occurs, and, in quasi-static fracture, matches the specific energy required by the material (i.e., it is equal to fracture toughness R). In unstable elastic fracture, whereas G_c matches R at the onset of fracture, thereafter $\partial G_c/\partial A > \partial R/\partial A$, which is why instability occurs [1, 8]. A discussion of crack stability is beyond the scope of this book, but the reader should note that there is always interplay between a material-dependent term and a cracked-body-dependent term [9]. The reader should also note that the term strain-energy *release* rate is a misnomer, since many fractures occur when the elastic energy of the body under load *after* propagation is *greater* than before, owing to the importance of external work done [1].

In place of an energy-based approach to linear elastic fracture, the direct approach can be employed and conditions for fracture obtained from consideration of the complicated fields of stress and strain around crack tips. This leads to well-known critical stress-intensity factor K_c formulas for fracture, viz,

$$K_c = \sigma(\pi a)^{1/2}Y \tag{4}$$

where σ is the fracture stress, a is the crack size, and Y is a nondimensional factor dependent on the size and shape of the body, type of loading, and orientation of crack. K_c is often written K_{Ic}, K_{IIc}, or K_{IIIc}, indicating the mode of crack-tip propagation, i.e., tensile, shear, or twist, respectively [10].

The energy-based and direct approaches (which give, of course, the same answer, but couched in different terms) are related in the same way that the Castigliano and the theory of elasticity approaches are related.

K_c is related to G_c or R by

$$K_c^2 = ER \tag{5}$$

where E is Young's modulus, and we see that the rate of change of the compliance term $d/dA(u/X)$ in the Irwin equation is implicit in the K formulas. Since $K_c = (ER)^{1/2}$, we see the origin of the peculiar units of K_c, e.g., Pam$^{1/2}$, N/m$^{3/2}$, ksi (in.)$^{1/2}$, etc. We also see that Equation (4) can be recast as

$$\sigma = \frac{1}{Y}\left(\frac{ER}{\pi a}\right)^{1/2} \tag{6}$$

which, with $Y = 1$, is the seminal Griffith expression [11].

K_c is a parameter of great utility, combining fracture stress and crack length. While linear elastic fracture mechanics (LEFM) is the vehicle by which fracture-stress analysis is most often taught and written about (and design codes are usually based on K_c rather than the energy approach using G_c or R), it must be emphasized that K_c applies *only* for *linear elastic* fracture: it is not an appropriate parameter for postyield fracture and is not even the correct parameter for nonlinear elastic fracture (i.e., reversible deformation with stiffness varying with displacement, such as produced when bending resistance is reinforced at high deflections by membrane stresses—in shell structures, for example—and the inverse situation—buckling—where resistance to deformation from membrane stresses is reinforced at large deflections by bending resistance [1, 6]).

An extension to Equation (1) to accommodate remote flow is

$$X\,du = d\Lambda + R\,dA + d\Gamma \tag{7}$$

where Γ represents the remote irreversibilities accompanying crack propagation. The form of Equations (1) and (7) presumes that the terms are uncoupled (i.e., the variables are separable) and, in particular, that independent values of fracture toughness R can be determined.† Clearly, the latter requirement can be met in adhesive fracture, where, by suitable choice of specimen geometry, elastic fracture experiments controlled by Equation (1) can be performed, and R for the glue can be measured at various crack speeds, temperatures, and in various environments. Then, in other situations, where there is plastic flow in the adherand, appropriate values for R are available for use in Equation (7). A closed-form solution to the problem of an elastoplastic beam glued to a massive

---

†An alternative formulation for elastoplastic flow and fracture that is multiplicative rather than additive is due to Andrews and Bhatty [12] and gives $\mathscr{F} = \mathscr{F}_0\Phi$, where $\mathscr{F}$ is identified with surface free energy, and Φ is the loss function by which the actual specific work required for fracture (owing to local and remote irreversibilities) is greater than the thermodynamically least work. Φ is $\Phi(\dot{a}, T, \Gamma, \ldots)$.

substrate is available [13] and is of interest pedagogically in explaining the different interchanges of energy during fracture, the transition from flow to fracture, and so on.

Application of Equation (7) to the cohesive fracture of monolithic bodies, although intuitively correct, is contentious for a number of reasons.

(a) Independent measurements of R in monolithic materials undergoing combined elastoplastic flow and fracture are in some doubt, since, for example, it has not been clear how to partition work areas after some propagation. In addition to flow and fracture components, whenever there has been prior nonuniform plastic flow, there will be a component of residual strain energy Λ_r and Equation (7) is better written

$$X \, du = d\Lambda + R \, dA + d\Gamma + d\Lambda_r \tag{8}$$

Λ_r should be accounted for in recovered work areas.

(b) There can be extensive remote flow in bodies before crack propagation and such an initiation phase is controlled by equations of the form

$$X \, du = d\Lambda + d\Gamma \tag{9}$$

with an appropriate criterion for the initiation of concern, for example, void growth and coalescence in metals or craze formation in polymers.

(c) Initiation in precracked bodies can be expressed in terms of a critical value of the Eshelby–Rice J integral J_c, which either concerns the same work areas represented by Equation (9) operated on by Turner's η factor [14], or, alternatively, concerns the rate of change of those work areas with respect to the crack area.

(d) Propagation can be expressed in terms of J_R, which, in essence, corresponds with $R + d\Gamma/dA$ in Equation (7) and makes no attempt to separate flow and fracture components during propagation. While the J_R approach may be acceptable for limited elastoplastic flow (and certain conditions are laid down for J_R-controlled growth [15]), the J integral is unable to cope with large amounts of stable crack growth accompanied by large-scale plasticity. As explained by, for example, Kanninen et al. [16], there is pronounced specimen-dependency in J after some small amount of crack growth (say 10% of the remaining ligament—although some finite-element calculations seem to indicate that J_R-controlled growth can be greater than this). If remote flow and fracture processes are truly uncoupled, $d\Gamma/dA$ can be varied independently of R, and it should not be surprising that J_R has been found to be geometry-dependent. This is easily seen if we consider the mode-III tearing of soft ductile metal sheet: the width of tear can be varied (and hence the plastic bending work represented by $d\Gamma/dA$ changed) even though the crack area generated and the specific work of fracture remain constant.

(e) It has been found that cracks, whether surrounded by elastic or plastic fields, tend to propagate with their own same natural bluntness that has led to the Wells–Cottrell concept of a critical value of crack-opening displacement (COD) δ_c as a criterion of fracture [17]. The COD design curve, which is a combination of theoretical predictions and experimental observations in situations where plastic strains are not much greater than yield strain levels, has had success in elastoplastic structural design [18]. R and δ_c are related by

$$R \approx m\sigma_y\delta_c \tag{10}$$

where σ_y is the flow stress, and $1 \gtrsim m \gtrsim 3$ represents particular constraints (i.e., biaxiality) to crack-tip flow.

Application of the variables-separable energy-rate equations, Equations (7) and (8), to cohesive elastoplastic fracture is formidable, both owing to the lack of independent data for R and the complicating factor of Λ_r. Again, the J_R approach has present currency in this area, even though it must now be realized that J_R is likely to be geometry-dependent. In the case of extensive plastic flow, however, where it may be permissible to neglect $d\Lambda$, application of the rigid-plastic version of Equation (7), i.e.,

$$X\,du = R\,dA + d\Gamma \tag{11}$$

has found success in Cotterell and Mai's studies of fracture in cohesive modes I, II, and III in ductile sheets [19] and in metal-forming fracture such as cropping and can-opening [20, 21]. In these analyses, the equation is used for algebraic prediction of loads and deflections during cracking, to interpret work areas given by load-deflection diagrams both under load and off load, and to explain why fracture supervenes plastic flow alone. It has been shown that geometry-independent R values do characterize the fracture behavior and that the plastic flow and fracture terms in Equation (11) can be separated. Of course, R can be a function of rate, temperature, thickness, environment, and so on, as found in elastic fracture studies; whether R has the same values in ductile fracture as in elastic fracture is the subject of current research.

The analyses presented thus far are quasi-static and a kinetic-energy incremental term dH has not been included. We should write

$$X\,du = d\Lambda + R\,dA + d\Gamma + d\Lambda_r + dH \tag{12}$$

where certain terms can be omitted, as appropriate, as was explained. H incorporates inertia forces in the equation of motion of the cracked body, stress-wave interactions with the boundaries of the body during propagation, and so on. Insofar as we may be interested in the dynamic tearing of ductile metal sheets, we should consider the conditions where dynamic effects become important. It happens [6] that the limiting crack speeds consistent with a quasi-static approximation are usually some 1–10 m/sec: for most materials and cracked geometries, equating the term in kinetic energy to zero in Equation (12) has negligible effect on the quantities determined by experiment or used for design purposes. Thus, we are able to use Equation (11) in many dynamic tearing cases of interest.

The problems associated with obtaining elastic, elastoplastic, or plastic fracture in the same materials (by changing specimen size, degree of constraint, rate, etc.) are part of the wide problem of deformation transitions discussed by Atkins and Mai [22].

2. MEASUREMENT OF THE SPECIFIC WORK OF FRACTURE IN THE PRESENCE OF EXTENSIVE REMOTE PLASTIC FLOW

Consider a tensile specimen of a ductile metal sheet into which symmetrical edge cracks have been cut, as shown in Figure 1(*a*). On monotonic loading, plastic

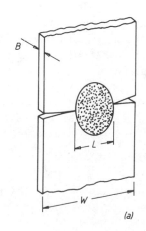

(a)

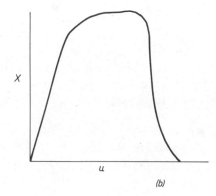

(b)

Figure 1. (*a*) The deep double-edge notched (DEN) tensile specimen of metal showing contained plastic flow in a circular patch on the ligament as diameter. (*b*) Typical load-deflection (*Xu*) diagram for DEN test piece.

flow occurs in an approximately circular region with the ligament between crack tips as diameter. Eventually, necking occurs in the line between the crack tips along the ligament followed by crack propagation from the two crack tips to give separation in the middle. A typical load-deflection diagram is shown in Figure 1(b); the size and extent of such Xu diagrams depends, of course, on the ligament length L for the same sheet material and the size of the test piece.

The situation is nonsteady for X and u, and whereas an incremental analysis has been presented [23] for this problem using Equation (11), experiments show that it is adequate to consider the integrated Equation (11) and apply it to the work areas ($\int X \, du$) under complete Xu plots through to final fracture. That is,

$$\int X \, du = \Gamma + RA \qquad (13)$$

The plastic work is given by

$$\Gamma = W^P V \qquad (14)$$

The plastic work/volume $W^P = \int \bar{\sigma} \, d\bar{\varepsilon}$, with $\bar{\sigma}$ the von Mises effective stress, and $\bar{\varepsilon}$ the von Mises effective strain, which is some average value representative of the nonuniform plastic strain field in the circular patch of deforming metal contained between the notches. The plastic volume $V = (\pi L^2/4)B$, where L is the ligament length, and B is the sheet thickness. Insofar as it is impossible to produce fracture under these circumstances without prior necking, it is legitimate to consider the whole of the necked region as the fracture-process zone, and we use the original sheet thickness and not its necked-down value for the crack area, which is, therefore, given by LB. The height of the necked-down zone is of the order of the sheet thickness [24]. Thus,

$$\int X \, du = \left(\int \bar{\sigma} \, d\bar{\varepsilon} \right)(\pi L^2/4)B + RLB$$

or

$$\left(\int X \, du \right) \bigg/ LB = \left(\int \bar{\sigma} \, d\bar{\varepsilon} \right) L + R \qquad (15)$$

It is found that $\int \bar{\sigma} \, d\bar{\varepsilon}$ for completely developed circular patches is constant for all practical purposes. From this, it follows that the testing-machine autographic external-work-done areas $\int X \, du$ normalized by LB should plot linearly against ligament length L, with slope $(\int \bar{\sigma} \, d\bar{\varepsilon})(\pi/4)$ and ordinate intercept R. This result is due to Cotterell and Reddell [25].

Figure 2 shows representative results for low-carbon steel sheet at various prestrain levels. The method is seen to work very well and the simplicity of the testing procedure is obvious. Other results on various types of metal and polymer sheets can be found in Reference 1.

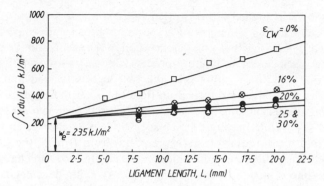

Figure 2. Plots of specific fracture work $\int X\,du/BL$ against ligament length L for a low-carbon steel at various prestrain levels.

There are refinements of the analysis that need not concern us here, such as the fact that a distinction sometimes has to be made between R for initiation and R for propagation in materials of low work-hardening index n in $\sigma = \sigma_0\varepsilon^n$ (this arises because, in such materials, crack propagation starts before the neck along the ligament between the crack tips has developed fully). The interested reader should consult Reference 1 and the original Cotterell–Mai papers [3].

Crack opening in the DEN test piece shown in Figure 1(a) occurs in tensile mode I. It is possible to obtain combined mode I/II (shear) fracture by employing staggered notches, as shown in Figure 3. The $\int X\,du$ work areas can be partitioned into the separate mode-I and mode-II components and values for R_{I} and R_{II} obtained [26].

Again, mode-III fractures (tearing fractures) can be analyzed in a similar way, which is highly relevant to our topic. Consider a sheet of ductile metal being

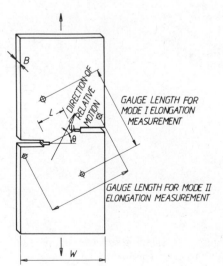

Figure 3. The staggered deep-edge notched tension specimen giving combined tensile and shear (modes I and II) crack propagation.

ripped as shown in Figure 4; this is often called a trousers tear test. The figure shows the deformation geometry of a specimen after steady-state tearing has been reached. There is virtually no deformation in the sheet until it reaches point B, where the metal enters the fracture-process zone BD, whose width is $2s$. There the metal that eventually flows into the two legs is twisted and eventually fractures. The torn edges are turned up (like burrs on the edges of guillotined sheets) owing to the shearing experienced in the process zone (see cross sections X-X and Y-Y). While tearing takes place, the legs of the specimen are bent plastically to a mean radius of curvature ρ; but from D to F, the legs undergo elastic unbending, with the bending moment reaching zero and changing sign at E along the line of action of the pulling load. At F, the legs begin to unbend plastically, which is complete at G. In a nonwork-hardening material, all the unbending takes place at G, where the unbending moment is a maximum. From G to H, the legs deform plastically, so that the load line can adjust itself to maintain the necessary bending moments.

The work of remote flow in this trousers geometry concerns the work of plastic bending and unbending of the legs of the specimen. The mean plastic strain, for a radius of curvature ρ, is $B/4\rho$ in a sheet of thickness B. Thus, the mean work/volume for bending in a material following $\sigma = \sigma_0 \varepsilon^n$ is

$$W^P_{\text{bend}} = \int \bar{\sigma} \, d\bar{\varepsilon} = \sigma_0 \varepsilon^{n+1}/(n+1)$$

$$= \sigma_0 B^{n+1} \qquad)4^{n+1}\rho^{n+1} \qquad\qquad (16)$$

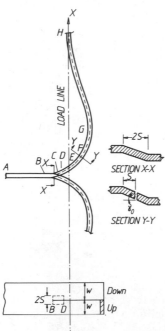

Figure 4. Steady-state tearing of thin-metal trousers specimen.

For unbending, where the mean unbending strain is equal in magnitude to the bending strain and has to be added to it,

$$W^P_{\text{unbend}} = \frac{\sigma_0}{n+1}\left[\left(\frac{B}{2\rho}\right)^{n+1} - \left(\frac{B}{4\rho}\right)^{n+1}\right] \tag{17}$$

Thus,

$$W^P_{\text{total}} = \sigma_0 B^{n+1}/2^{n+1}\rho^{n+1}(n+1) \tag{18}$$

The incremental volume dV^P passing through the process zone is

$$dV^P = 2wB\,du \tag{19}$$

where w is the width of each leg, and u is the displacement of the load as previously (elasticity is neglected). The corresponding incremental crack area generated is $dA = B\,du$, so in Equation (11),

$$2X\,du = \frac{\sigma_0 B^{n+1}}{2^{n+1}\rho^{n+1}(n+1)}2wB\,du + RB\,du$$

$$\frac{X}{B} = \frac{\sigma_0 B^{n+1}w}{2^{n+1}\rho^{n+1}(n+1)} + \frac{R}{2} \tag{20}$$

ρ is expected to be a weak function of the leg width w and it was assumed to be approximately constant for the range of w investigated by Mai and Cotterell in 1984 on 1.6 mm thick low-carbon steel and 2 mm thick 5251 aluminum alloy sheets [27]. According to Equation (20), a plot of X/B vs. w should be a straight line of intercept $R/2$. Figure 5 shows that the agreement is good with R values of 1040 kJ/m^2 (steel) and 600 kJ/m^2 (aluminum). The prediction of ρ from the slopes of the lines also agrees well with experimental values of about 30 mm.†For the steel, $\sigma = 893\varepsilon^{0.21}$ MPa; for the 5251 aluminum alloy, $\sigma = 182\varepsilon^{0.5}$ MPa.

Experiments on trousers tests with three or more legs, in which the bending/unbending plastic work can be altered independently of the fracture work, gives the same values for R. This is important for the application of Equation (11) and demonstrates that the flow and fracture work components can indeed be uncoupled, and the toughness so determined is a characteristic material property.

Typical values for R the specific work of fracture in the presence of large amounts of remote flow obtained in these different tests are given in Table 1.

---

†Such large ρ values are associated exclusively with plastic bending in the legs. For elastic bending of the legs, such as occurs in the tearing of materials with very large flexibilities, Gurney and Ngan [6] showed that $\rho = (EI/RB)^{1/2}$ and thus ρ is small.

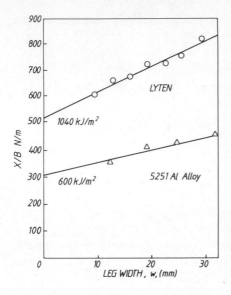

Figure 5. Tearing force X/B plotted against leg width w for 1.6 mm thick low-carbon steel and 2 mm thick 5251 aluminum alloy.

TABLE 1   Typical Plane-Stress Values of Toughness R (kJ/m^2)

Sheet Material	Mode I	Mode II	Mode III
all (1–2 mm thick)			
Low-carbon steel	250	200	1000
70/30 brass	200		
5251 Aluminum alloy	125		600
Ultrahigh-molecular weight polyethylene	50 (initiate) 80 (propagate)		
High-impact polystyrene	1 (plane strain) 18 (plane stress)		

The R values in the trousers tear test are very much larger than the corresponding values of the essential work of fracture obtained with double-edged notched (DEN) tensile tests. The mode of deformation is, of course, different, being ideally out-of-plane mode III in the trousers test, but practically involving some mode-I deformation, and there is no reason why the R values should coincide. The essential work of fracture R consists of two components: (1) the work necessary to produce out-of-plane fracture, and (2) the plastic work associated with the thin lip or burr of width s on the edges of the torn legs (which are inherent to the process if the metal is ductile). The shear zones are similar, but physically much larger, than the burrs produced on cut edges by guillotines with blunt blades [28]. It turns out that most of the essential work of fracture goes into deforming the burr zones rather than producing the final tear.

A shear strain γ_0 in the lip (Figure 4) corresponds with an effective strain $\gamma_0/3^{1/2}$, so the plastic shearing work/unit volume is $\sigma_0\gamma_0^{n+1}/(n+1)3^{(n+1)/2}$. The incremental lip volume generated on the edge of each leg in passage through the process zone is $sB\,du$, so the total incremental work in forming the two lines of burred edges is

$$2sB\sigma_0\gamma_0^{n+1}\,du/(n+1)3^{(n+1)/2} \tag{21}$$

When this work, which is local to the torn edges and which must be included in the essential work of fracture (i.e., in R) is normalized by the incremental crack area $B\,du$, its contribution to R is

$$w_{e(\text{burr})} = 2s\gamma_0^{n+1}/(n+1)3^{(n+1)/2} \tag{22}$$

It is difficult in practice to measure both s and γ_0. Roughly, $s \approx 2B$, and Mai and Cotterell found $0.5 < \gamma_0 < 0.9$ for both their steel and aluminum samples. Therefore, $380 < w_{e(\text{burr})} < 620\,\text{kJ/m}^2$ for the aluminum alloy and $980 < w_{e(\text{burr})} < 2520\,\text{kJ/m}^2$ for the low-carbon steel. Clearly, the formation of the burred edges is the major effect in setting the level of R.

3. PREDICTION OF TEARING FORCES

The analysis of the trousers tear test in Section 2 as a means of determining toughness provides an analysis that, knowing the toughness, can be employed in reverse to predict loads for tearing sheet metals.

Sometimes there are joints, glue lines, etc., along which the crack may run, in which case, the path is prescribed and a similar rigid-plastic approach gives opening loads [1]. A familiar example is the old-fashioned sardine can with soldered lid, as shown in Figure 6. The rolling up of the lid on to the key consumes plastic work, and the external incremental work done ($M\,d\theta$ rotating the key) has to provide that work in addition to the fracture work in the solder

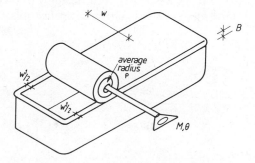

Figure 6. Opening a soldered can lid.

layer. Let us assume an average wound-up radius ρ; the size of the key limits ρ and the plastic strain, as plastic hinges are continuously collapsed onto it. Using $dA = w'\rho\, d\theta$ and $dV^P = wB\rho\, d\theta$, where w' is the width of the joint and w that of the whole can, Equation (11) gives

$$M\, d\theta = (\sigma_y B/4\rho)wB\rho\, d\theta + Rw'\rho\, d\theta$$

i.e.,

$$M = \sigma_y B^2 w/4 + Rw'\rho \tag{23}$$

The moment is greater as the thickness B of the metal increases and the wider the rolled-up tongue of metal becomes. Note also that M increases as ρ increases, although, in practice, σ_y falls during opening, as the plastic bending strain (proportional to $1/\rho$) is smaller.

Another, more complicated problem, is the elastoplastic peeling of strips from substrates (Figure 7). This problem has been solved in closed form [13] and is one of the few cracking problems for which a continuous solution is available ranging over elastic bending of the strip being peeled, elastoplastic bending coupled with fracture, or complete collapse of the strip with no fracture. The bending moment to cause combined flow and fracture is

$$M = wh^2\sigma_y[1 - \tfrac{3}{4}(1 - RE/\sigma_y^2 h)^2] \tag{24}$$

where $2h$ is the depth of the beam, and w is the width. This expression covers the whole range of possible behaviors. If ξ defines the proportion of the beam depth that is elastic, it is shown in Reference 13 that

$$h = \left(\frac{3RE}{\sigma_y^2}\right)\left(\frac{1}{3 - 2\xi}\right) \tag{25}$$

defines the transition from flow to fracture. With $\xi = 1$ (first plasticity on the outside of a beam), $h > 3RE/\sigma_y^2$ ensures elastic behavior in the beam during fracture; with $\xi = 0$, $h < RE/\sigma_y^2$ ensures that the beam will collapse and curl with no fracture at all. Intermediate h give fracture with elastoplastic deformation in the beam.

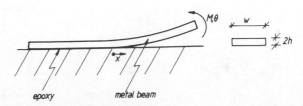

Figure 7.   Peeling an elastoplastic metal–epoxy joint by an end couple.

The interesting fact about this problem is that there are *two* energy sinks when cracking occurs (plastic flow in the beam and fracture work), in contrast to merely the *one* sink for continuing plastic flow without fracture. But at the transition from flow to flow plus fracture, the net work input is lower for the two sinks than the one because of changing plastic strains and changing plastic volumes as material is released alongside a propagating crack. Minimum-energy arguments of this sort have found success in cropping [4] and are used later in Section 7.

4. FRACTURE IN CRACK-FREE BODIES

As discussed, a proper mechanics of combined flow and fracture is only now being developed. In metal forming, limitations are set on many operations by the appearance of surface or interior cracks. Thus, to achieve maximum deformation and process yields, and avoid excessive scrap, crack nucleation must be prevented in the likes of forging, stamping, rolling, pressing, and extrusion. On the other hand, in processes that concern the *separation of parts*, crack nucleation, initiation, and propagation are desirable. Thus, cracking is an inherent part of can-opening, blanking, cropping, guillotining, piercing, nibbling, slitting, and machining. Problems in ballistics, penetration of armor, plugging of plates, and so on are all related to this general theme. In all these processes, cracks nucleate within regions of deformation fields that are highly strained—either simply by extensive overall flow (as in forging) or by localized flow produced by the external stress-raising effect of sharp tooling (as in cropping and armor piercing).

Thus, in some circumstances, crack initiation is the end of the process; in others, crack initiation is the beginning. It is rather like whether in strength of materials calculations, we call critical combinations of stresses a "failure" criterion or a "yield" criterion. Traditional elastic design supposedly keeps stresses below yield-point level: plastic flow cannot occur unless the yield level has been exceeded. Again, the propagation of initiated cracks is of little interest to those who want to avoid their formation, but propagation paths are important in all tearing and penetration problems.

In the absence of a full mechanics analysis for ductile cracking, one historical approach to establish a criterion for crack initiation in ductile fracture has been to build up a body of experimental data concerning plastic strains local to the fracture sites in monolithic pieces of metal or sheets and to construct a fracture locus. A review is given in Reference 1. In this way, as far as metal sheets are concerned, fracture-forming-limit diagrams, showing fracture strains as well as the necking strains (which often set limits in sheet metal forming) can be constructed [29].

Knowledge of the plastic strain history prior to fracture permits the corresponding stress components to be evaluated in the usual way using the theory of plasticity (e.g., Appendix 3 of Reference 1). Calculations are usually performed in terms of total strains, assuming linear strain paths rather than incremental strains along the curvilinear strain paths, although that situation is changing with developments in finite-element/finite-difference large-strain calculations [30]. Combined stress–strain calculations can then be performed to assess various empirical mechanics criteria of crack initiation—some of which are given in Table 2 [31–47]—and also to construct the corresponding fracture-strain loci that may be contrasted with experimental loci.

It is shown in Reference 1 that nearly all the integrated stress–strain criteria are versions of $\int \bar{\sigma}\, d\bar{\varepsilon}$, i.e., a criterion of a critical plastic work per unit volume in the "representative volume of microstructure" in which cracks initiate. This accords with McClintock-type calculations [40] for microvoid growth and coalescence as a criterion of fracture, which reduce to versions of $\int \sigma_H\, d\bar{\varepsilon}$, where σ_H is hydrostatic stress. Certainly, in most commercial metal alloys containing inclusions and/or hard second phases, crack initiation is usually associated with microvoiding.

For nonhardening materials, these criteria reduce to the attainment of critical

TABLE 2   Criteria for Ductile Fracture in Metal Forming

Criterion[a]	Reference
σ_{max} (maximum tensile stress)	Rankine [31]
ε_{max} (maximum tensile strain)	St. Venant [32]
$\int \sigma_{max}\, d\bar{\varepsilon}$	Cockcroft and Latham [33]
$\int \bar{\sigma}\, d\bar{\varepsilon}$	Gillemot [34]
$\int \sigma_H\, d\bar{\varepsilon}$	—
$(d\sigma_H/ds)_{max}$	Noble and Oxley [35]
$\int (1 + \sigma_H/c\bar{\sigma})\, d\bar{\varepsilon}$	Oyane [36]
$1/\mathscr{H} = 1/\mathscr{H}_0 - \int g(\sigma_H/\bar{\sigma} - \frac{2}{3})\, d\bar{\varepsilon}$	Marciniak and Kuczynski [37]
$\int d\bar{\varepsilon}/(1 - c\sigma_H)$	Norris et al. [38]
$\int (1 + \frac{1}{2}L)\, d\bar{\varepsilon}/(1 - c\sigma_H)$	Atkins [39]
Void growth	
$\ln(\ell/2r) \propto \int \sinh(\sigma_H/\bar{\sigma})\, d\bar{\varepsilon} \approx \int (\sigma_H/\bar{\sigma})\, d\bar{\varepsilon}$	McClintock et al. [40]
$(1 + 1/m)\sigma_1^2 = $ constant	Ghosh [41]
	Also see Thomason [42],
	Rice and Tracey [43],
	Palmar and Mellor [44],
	Dillamore and Maynard [45]
Void closure	Keife and Stahlberg [46]
Probablistic approach	
(size of distribution of voids)	Van Minh et al. [47]

[a]The integrated stress–strain criteria are supposed to attain critical values at fracture.

strains at fracture (whose magnitude *varies* with loading-strain history owing to the different σ_H history), but we should note the interesting example of crack formation in the ductile expansion of thick-walled rings made of hardening metals [48]: the shear strains are greatest at the bore, and the hydrostatic stress is greatest at the rim. Cracks initiate in *midannulus*, which is, perhaps, unexpected and which is certainly not predicted by criteria based simply on maximum stress or maximum strain.

The critical integrated value of damage at first cracking is related in McClintock-type models to inclusion spacing (ℓ)-to-size (r) ratios through a $\ln(\ell/2r)$ term. In practice, a quantitative link between microstructure and mechanics does not work too well, because (a) practical microstructures do not consist of a single array of uniformly spaced inclusions of uniform size, (b) the holes of the model are actually filled and cannot distort freely, (c) deformations become localized between holes (shear banding) so that failure occurs before the holes touch, which is the criterion of fracture in the original McClintock models. Even so, hole-growth models predict the correct functional dependence of variables in crack initiation.

What determines in practice the particular critical integrated value of work per unit volume and how deformations with different loading histories can be calibrated from tests with simple strain paths is the subject of current research. Usually, the critical value is obtained from simple uniaxial tests, measuring $\int \bar{\sigma} \, d\bar{\varepsilon}$ or whatever up to the first appearance of cracks. This method is not always as successful as might be imagined, and, recently, a plausible alternative method has been proposed [49] that employs propagation R values (determined on precracked test pieces as described in Section 2 from the same material). The essence of the method is the conversion of the specific work of fracture into a critical work per unit volume from knowledge of the size of the process zone in which cracking occurs. Details follow in Section 5. It is to be noted that this line of attack has implications as to whether ductile fracture mechanics can be characterized by one parameter like elastic fracture mechanics. A discussion of that point is given later in Section 6.

Given the attainment of a known critical value of plastic work/volume along a known loading path, the critical strains at fracture along other loading paths can be predicted. For example,

(a) the locus of surface strain pairs $(\varepsilon_1, \varepsilon_2)$ at fracture in upset cylinders of ductile metals that barrel to varying degrees, depending on platen friction and specimen aspect ratio

(b) the in-plane strains at fracture in sheet metals hydraulically bulged over orifices of different shapes

Figure 8 shows some of the fracture loci obtained in this way. Experiments over a wide range support certain of these predictions [50] and enable a discrimination to be made between the various empirical models of crack initiation. A review can be found in Reference 1.

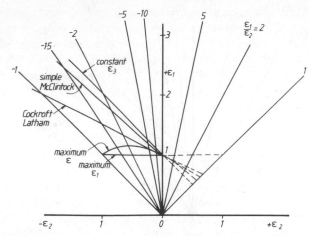

Figure 8. Empirical fracture criteria (from Table 2 for plane stress, $\sigma_3 = 0$) plotted in ε_1–ε_2 space.

This line of attack can be used in related problems such as the occurrence of first cracking caused by impact of projectiles on bodies, perforation of flawless sheets, and so on.

5. PREDICTION OF FRACTURE INITIATION IN CRACK-FREE SHEETS FROM PROPAGATION R DATA

The local stress and strain conditions of material ahead of a propagating crack must be, in a general sense (making allowance for change in constraint to flow), similar to those that exist in loaded crack-free material prior to initiation. That is, crack propagation can be viewed as a process of continuous reinitiation along the path of cracking. This seems a particularly reasonable hypothesis for materials that crack by void growth and coalesence. That is, there should be a connection between the specific essential work of fracture R and the work required locally to nucleate a crack.

The tensile fracture of thin ductile sheets in which starter cracks have been cut is characterized by necking down in the line of the crack [25]. Crack nucleation in flaw-free sheets also takes place within localized necks [51]. The changes in geometry of sheets at first fracture in, for example, biaxially bulged domes or even plain tensile sheet test pieces (where the plain strain necks are inclined to the axis of pulling) are similar to those at the tips of cracks that have propagated from long preexisting cracks.

The specific work of fracture can be converted to a local work/unit volume at the crack tip by dividing by the height h of the necked-down region, along which the cracks propagate, since the whole of the necked region forms the fracture-process zone in which all crack-tip microstructural events take place. That is, the incremental work $R\,dA$ is dissipated within a volume $h\,dA$ at the crack tip, which, therefore, corresponds with a critical plastic work done/unit volume of some $R\,dA/h\,dA = R/h$. If the material of the sheet work hardens according to $\bar{\sigma} = \sigma_0\bar{\varepsilon}^n$, the plastic work done/unit volume up to the crack-nucleation effective strain $\bar{\varepsilon}_f$ is $\sigma_0\bar{\varepsilon}_f^{n+1}/(n + 1)$. If crack-tip events are indeed the same, then

$$\sigma_0\bar{\varepsilon}_f^{n+1}/(n + 1) = R/h \tag{26}$$

and the effective strain at, say, the pole of a bulged sheet when cracks are just initiated is

$$\bar{\varepsilon}_f = [R(n + 1)/(\sigma_0 h)]^{1/(n+1)} \tag{27}$$

In order to test Equation (27) and compare the predicted $\bar{\varepsilon}_f$ with the fracture strain pairs actually measured in biaxially stressed crack-free sheets, R should be that measured under the same loading conditions along and perpendicular to the plane-strain neck region in which cracks form. That is, for nucleation under biaxial tension with stress ratio M, the necessary R should be determined *not* from simple uniaxial Cotterell–Mai tests (Section 2), but rather from special tests in which stresses in the ratio M are applied along the perpendicular to the line between the starter cracks (i.e., the direction of neck formation) from the outset of the test. Such data are not yet available. The basic Cotterell–Mai test with *remote* uniaxial loads corresponds with a *local* plastic stress ratio of 2:1 in the region of the neck if the crack tips do not approach one another before necking [24]. Consequently, a basis for comparison between predicted $\bar{\varepsilon}_f$ and experimental data relates to bulging over an elliptical orifice that produces a 2:1 stress ratio.

Caddell et al. [52] studied the fracture of hydraulically bulged diaphragms of annealed and cold-rolled aluminum and brass sheets under a variety of biaxial in-plane strain ratios. Their data for annealed commercially pure aluminum following $\bar{\sigma} = 155\bar{\varepsilon}^{0.28}$ MPa give a necking-limit and fracture-limit diagram, Figure 9, with the following crack-nucleation strains in 0.79 mm thick sheet, viz: $\varepsilon_{1f} = 0.97$ and $\varepsilon_{2f} = 0.36$ for balanced biaxial tension (circular diaphragm) in which a plane-strain neck formed at $\varepsilon_{1n} = \varepsilon_{2n} = 0.36$; and $\varepsilon_{1n} = 0.34, \varepsilon_{1f} = 1.07$, $\varepsilon_{2n} = \varepsilon_{2f} = 0.14$ for 2:1 bulging over an elliptical orifice. These ε_{1f} and ε_{2f} combinations (and similar strain pairs for brass) can be converted into effective strains $\bar{\varepsilon}_f$ (taking account of the change in strain ratios when the plane neck sets in) and then compared with the predictions from Equation (27). Table 3 compares the predicted and experimental effective strains at fracture not only for annealed material, but also for aluminum and brass sheets prestrained by cold-rolling.

The comparisons are promising and the work is being extended to establish, in particular, the possible influence of biaxial loading on experimentally determined R values.

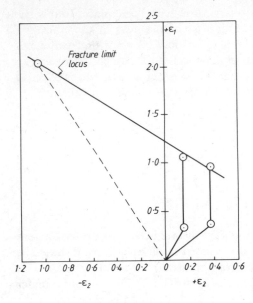

Figure 9. Fracture-forming-limit diagram for an annealed aluminum alloy.

TABLE 3   **Comparison of Predicted and Experimental Fracture Strains** $\bar{\varepsilon}_f$ **for Bulging Over a 2:1 Orifice**

Material	$\bar{\varepsilon}_{cw}$	Experimental $\bar{\varepsilon}_f$	Predicted $\bar{\varepsilon}_f$
Aluminum	0	1.33	1.49
	0.12	1.23	1.28
	0.42	1.06	1.17
70/30 α-brass	0	0.85	0.79
	0.14	0.81	0.66
	0.46	0.70	0.33

6.   IMPLICATIONS FOR A SUPPOSED ONE-PARAMETER CHARACTERIZATION OF DUCTILE FRACTURE TOUGHNESS

The previous section compared predicted and experimental fracture-strain pairs in sheet metals for one particular loading path. It is clear, however, that Equation (27) can be used to predict the complete ε_{1f} and ε_{2f} fracture loci for a range of loading histories (cf. Figure 8, using the criteria of Table 2). In the absence of contrary evidence at present, it is assumed that R and h are path-independent in such calculations. Examples can be found in Reference 1.

However, since the (ε_{1f} and ε_{2f}) fracture loci of such *flawless* sheets can be expressed in terms of R measured in experiments on sheets *containing starter cracks*, and since the (ε_{1f} and ε_{2f}) fracture loci of flawless sheets can be written in terms of McClintock-type analyses, there must be an R-to-void-growth-model connection, too. The simplest version of the full McClintock criterion is [40]

$$\ln(\ell/2r) = \int (\sigma_H/\bar{\sigma}) \, d\bar{\varepsilon} \tag{28}$$

It has been shown [1] that this relation is path-independent, so

$$\ln(\ell/2r) = (\sigma_H/\bar{\sigma})\bar{\varepsilon}_f$$

$$= (L+1)\bar{\varepsilon}_f/[3(L^2+L+1)]^{1/2} \tag{29}$$

where $L = \varepsilon_1/\varepsilon_2$, the in-plane strain ratio.

But $\bar{\varepsilon}_f$ is also given by Equation (27). Thus, for an annealed metal,

$$\ln(\ell/2r) = \{(L+1)/[3(L^2+L+1)]^{1/2}\}[R^{n+1}/\sigma_0 h]^{1/(n+1)} \tag{30}$$

and, if σ_y is constant

$$\ln(\ell/2r) = \{(L+1)/[3(L^2+L+1)]^{1/2}\}[R/\sigma_y h] \tag{31}$$

A similar relation is available for prestrained work-hardening metals.

Again, Equation (30) can be recast in the form

$$R = \frac{\sigma_0 h}{n+1} [(\bar{\sigma}/\sigma_H) \ln(\ell/2r)]^{n+1} \tag{32}$$

and, if σ_y is constant,

$$R = \left\{ \frac{h[3(L^2+L+1)]^{1/2}}{L+1} \ln(\ell/2r) \right\} \sigma_y \tag{33}$$

$$= (\sigma_y/\sigma_H)[h \ln(\ell/2r)]\sigma_y$$

Now, since $\ell/2r$ has a *fixed* value, characteristic of the microstructure, it follows from Equation (30) that R/h must be *path-dependent* via the L function, i.e., different for different applied stress or strain biaxialities. It could be that R and h both change with biaxiality or that only one of them does so; experiments are under way to investigate this important prediction. The assertion is highly relevant, of course, as to whether a one-parameter characterization of fracture is possible.

Equation (33) has the same form as the relation between COD δ_c and R, i.e.,

$$R = m\sigma_y \delta_c \tag{10}$$

There is correspondence between m and σ_y/σ_H and δ_c and $h \ln(\ell/2r)$.

The factor m represents the constraint to plastic flow that, in plane stress, is sensibly represented by σ_y/σ_H, which has a value of 3 for simple tension, $3^{1/2}$ for plane strain, and $\frac{2}{3}$ for equibiaxial tension (prior to any necking). The connection between δ_c and microstructure is sensible in that δ_c/h is some measure of crack-tip strain given by $\ln(\ell/2r)$ at fracture [53]. Clearly, if h changes with biaxiality, δ_c will do so as well. Experiments are in progress to look at this aspect, which, again, relates to whether a one-parameter characterization of fracture is possible.

Little experimental information is available, even for quasi-static fracture. It is suggested that experiments at high rates should also be performed.

7.   TRANSITION FROM FLOW TO FRACTURE: NUMBER OF CRACKS IN TUBE-SPLITTING

The transition from flow to the onset of crack propagation in deforming bodies occurs when the next work increment for continuing plastic flow is greater than the work increment for the alternative mode of deformation that involves fracture. Usually, the alternative mode involves both fracture *and* some continuing plastic flow, so that there are *two* energy sinks and it therefore seems odd at first sight that the associated work increment can be less than that for flow alone (cf. Section 3). But an increment of plastic flow $d\Gamma$ can be written

$$d\Gamma = d(WV) = W\,dV + V\,dW \tag{34}$$

where W is the work/volume, and V is the volume being plastically deformed. We see that if W and/or V, and their increments, change when crack propagation ensues, it is possible for combined flow and fracture to require a smaller work increment than flow alone. Examples in which the effect is discussed fully relate to cropping and guillotining of ductile metals [4, 20] and to elastoplastic adhesive fracture [13]; as explained in Section 3, the latter has an additional pedagogical interest in that the problem of peeling a strip from a substrate is unusual in that it has a closed-form algebraic solution.

An example of particular relevance concerns the axial splitting of expanded ductile metal tubes. Depending on geometry and material properties, circular metal tubes axially compressed on to curved dies can invert completely (either inside or outside by plastic flow) or, alternatively, cracks can initiate and propagate as the end of the tube is flared out in outside inversion (Figure 10). In the latter circumstance, extremely interesting behavior is observed because the number n of propagating cracks always falls into a characteristic range: if starter saw cuts are employed, they will bifurcate to bring the number of cracks up into the characteristic range if their number is smaller than n; alternatively, if there are more than n to start with, some starter cuts never propagate so as to bring the number of propagating cracks into the characteristic range. For the 50 mm

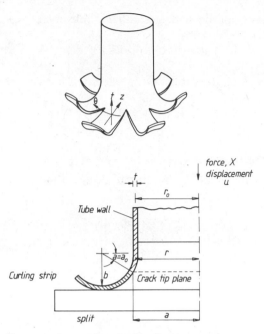

Figure 10.   Shear cracks produced when forcing a tube on to a mandrel.

diameter 1.6 mm thick mild-steel and aluminum tubes studied by Reddy and Reid [2], the characteristic number of cracks fell in the range between about 8 and 12 for both as-received and annealed tubes.

Prior to fracture, Equation (34) governs the external work required and applies with incremental changes dV and $d\Gamma$ since there is new material coming over the mandrel from the tube and the leading edge of the flare experiences larger plastic strains than before as it advances toward radius a. Increments of external work done are, therefore,

$$X \, du = W \, dV + V \, dW \tag{35}$$

During fracture dW is absent in Equation (34) since we assume no further plastic flow in the leading edge of the flare after cracks begin to propagate and the plastic-deformation field becomes steady. However, additional work $d\Gamma^c$ is required to propagate the crack(s), the magnitude of which is given for n cracks by

$$d\Gamma^c = nR \, dA \tag{36}$$

where dA is the increment of crack area. The incremental external work required in this case is

$$X \, du = W \, dV + nR \, dA \tag{37}$$

A minimum-energy approach suggests that stable cracks will propagate when

$$W \, dV + nR \, dA = W \, dV + V \, dW \tag{38}$$

over the same external displacement or velocity. That is, when

$$nR \, dA = V dW$$

or

$$n = (V/R)(dW/dA) \tag{39}$$

which gives an estimate for the number of propagating splits.

It can be shown that [54]

$$n = (8\pi/3^{1/2})(\bar{\sigma}/R)r_0\varepsilon_{\theta f} \tag{40}$$

where $\varepsilon_{\theta f} = \ln(a/r_0)$ is the hoop strain at radius a at which fracture occurs. The magnitude of $\varepsilon_{\theta f}$ can be related back, in principle, to microstructure, as discussed in Section 4. Alternatively [54],

$$n \approx (4\pi/3^{1/2})(\bar{\sigma}/R)(u_f^2/b) \tag{41}$$

where u_f is the axial displacement of the tube at which fractures start. Note, however, that in this form, u_f is not independent of b.

These formulas predict the number of cracks very well under both static and dynamic conditions. For example, $(\bar{\sigma}/R) \approx 10^3 \, \text{m}^{-1}$ for mild-steel and aluminum tubes, so, with $u_f \approx 4 \, \text{mm}$ and $b = 10 \, \text{mm}$ (from Reddy's and Reid's experiments [2]), $n \to 12$. However, Equations (40) and (41) are sensitive to the material properties $\bar{\sigma}$ and R, as well as to the geometrical parameters, which again points to the need for more experimental data—particularly at high rates.

We note that the relation for n [Equation (40)] involves the product of $\bar{\sigma}/R$ and $\varepsilon_{\theta f}$. In very ductile solids, $\bar{\sigma}/R$ is small, but the strain to fracture (however defined) is correspondingly high. The reverse is true in less-ductile solids, as $\bar{\sigma}/R$ increases, but the fracture strain decreases. Thus, according to this rigid-plastic analysis, the number of cracks cannot change much between the same size tubes in the annealed and work-hardened states. Results in Reference 2 support this conclusion.

8.   CONCLUSION

It has been shown how the rigid-plastic simplification of the full elastoplastic mechanics of fracture can be applied to a variety of sheet-tearing problems.

Successful application of this approach, and related lines of attack to the transition from flow to fracture, require knowledge of the new property R in particular cases, which is the specific work of fracture in the presence of large amounts of remote plastic flow. Clearly, this has to be measured as a function of rate, temperature, and so on. At the moment, high-rate mechanical-property measurements tend to be of traditional properties such as flow stress and strain to fracture. We are, at the University of Reading, now beginning to perform such tests at intermediate strain rates ($500 \sec^{-1}$) to provide R data.

REFERENCES

1. A. G. Atkins and Y.-W. Mai, *Elastic and Plastic Fracture*, Wiley, New York, 1985.
2. T. Y. Reddy and S. R. Reid, *Int. J. Mech. Sci.*, **28**, 111 (1986).
3. Y. W. Mai, B. Cotterell, and A. G. Atkins, in D. S. Mansell and G. H. Vasey, Eds., *Fracture*, Melbourne Univ. Press, Melbourne, Australia, 1979.
4. A. G. Atkins, *Philos. Mag. [Part] A*, **43**, 627 (1981).
5. C. Gurney and J. Hunt, *Proc. R. Soc. London, Ser. A*, **299**, 508 (1967).
6. C. Gurney and K. M. Ngan, *Proc. R. Soc. London, Ser. A*, **325**, 207 (1971).
7. G. R. Irwin and J. A. Kies, *Weld. J. Res., Suppl.* **17** (1952).
8. C. Gurney and Y.-W. Mai, *Eng. Fract. Mech.*, **4**, 853 (1972).
9. Y.-W. Mai and A. G. Atkins, *J. Strain Anal.*, **15**, 63 (1980).
10. J. F. Knott, *Fundamentals of Fracture Mechanics*, Butterworth, London, 1973.
11. A. A. Griffith, *Philos. Trans. R. Soc. London, Ser. A*, **221**, 163 (1921); see also *Proc. Int. Congr. Appl. Mech., 1st, 1924*.
12. E. H. Andrews and J. I. Bhatty, *Int. J. Fract.*, **20**, 65 (1982).
13. A. G. Atkins and Y.-W. Mai, *Int. J. Fract.*, **30**, 203 (1986).
14. C. E. Turner, in D. G. H. Latzo, Ed., *Post Yield Fracture Mechanics*, 2nd ed., Applied Science Publishers, Barking, U.K., 1984.
15. C. F. Shih and J. W. Hutchinson, *J. Eng. Mater. Technol.*, **98**, 289 (1976).
16. M. F. Kanninen, I. S. Abou-Sayed, C. W. Marshall, and D. Broek, Workshop Report NP-80-10-LD: WS 80-912, Electr. Power Res. Inst., Palo Alto, CA, 1980.
17. A. A. Wells, *Symposium on Crack Propagation*, Vol. 1, College of Aeronautics, Cranfield, 1961, p. 210; see also A. H. Cottrell, *ISI Spec. Rep.*, No. 69, p. 281 (1961).
18. 18. F. M. Burdekin, *Proc.—Inst. Mech. Eng.*, **195**, 73 (1981).
19. Y.-W. Mai and B. Cotterell, *J. Mater. Sci.*, **15**, 2296 (1980).
20. Z. G. Atkins, *Int. J. Mech. Sci.*, **22**, 215 (1980).
21. A. G. Atkins, unpublished research (1987).
22. A. G. Atkins and Y.-W. Mai, *J. Mater. Sci.*, **21**, 1093 (1986).
23. B. Cotterell and Y.-W. Mai (1982) *Adv. Fract. Res., Proc. Int. Conf. Fract., 5th, 1981.* Vol. 4, p. 1683 (1982).
24. R. Hill, *J. Mech. Phys. Solids*, **1**, 19 (1952).
25. B. Cotterell and J. K. Reddell, *Int. J. Fract.*, **13**, 267 (1977).
26. B. Cotterell, E. Lee, and Y.-W. Mai, *Int. J. Fract.*, **20**, 243 (1982).
27. B. Cotterell and Y.-W. Mai, *Int. J. Fract.*, **24**, 229 (1984).
28. A. G. Atkins and Y.-W. Mai, *J. Mater. Sci.*, **14**, 2747 (1979).

29. J. D. Embury and G. H. LeRoy, *Proc. Int. Conf. Fract.*, *4th*, Vol. 1, p. 15 (1977).

30. G. W. Rowe, C. E. N. Sturgess, P. Hartley, and I. Pillinger, *Finite Element Plasticity and Metalforming Analysis*, Cambridge Univ. Press, London and New York (to be published).

31. See H. Fromm, *Handb. Phys. Tech. Mech.*, **4**, 359 (1931).

32. See S. Timoshenko, *Strength of Materials*, 3rd ed., Vol. 2, Van Nostrand, Princeton, NJ, 1956.

33. M. G. Cockcroft and D. J. Latham, *J. Inst. Met.*, **96**, 33 (1968).

34. L. F. Gillemot, *Eng. Fract. Mech.*, **8**, 239 (1976).

35. C. F. Noble and P. Oxley, *Int. J. Prod. Res.*, **2**, 265 (1964).

36. M. Oyane, *Bull. JSME*, **15**, 1507 (1972).

37. Z. Marciniak and K. Kuczynski, *Int. J. Mech. Sci.*, **21**, 609 (1979).

38. D. M. Norris, J. E. Reaugh, B. Moran, and D. F. Quinones, *J. Eng. Mater. Technol.*, **100**, 279 (1978).

39. A. G. Atkins, *Met. Sci.*, **15**, 81 (1981).

40. F. A. McClintock, *J. Appl. Mech.*, **35**, 363 (1968); see also F. A. McClintock, S. K. Kaplan, and C. A. Berg, *Int. J. Fract. Mech.*, **2**, 614 (1966).

41. A. K. Ghosh, *Metall. Trans. A*, **7A**, 523 (1976).

42. P. F. Thomason, *Int. J. Mech. Sci.*, **11**, 187 (1968).

43. J. R. Rice and D. M. Tracey, *J. Mech. Phys. Solids*, **17**, 201 (1969).

44. A. Palmar and P. B. Mellor, *Int. J. Mech. Sci.*, **20**, 707 (1978); **22**, 133 (1980).

45. I. L. Dillamore and R. A. Maynard, *Br. Steel Corp. Rep.* **CDL/MT/9/74** (1974).

46. H. Keife and U. Stahlberg, *J. Mech. Work. Technol.*, **4**, 133 (1980).

47. H. Van Minh, R. Sowerby, and J. L. Duncan, *Int. J. Mech. Sci.*, **16**, 31 (1977).

48. B. Tomkins and A. G. Atkins, *Int. J. Mech. Sci.*, **23**, 395 (1981).

49. A. G. Atkins and Y.-W. Mai, *Eng. Fract. Mech.*, **27**, 291 (1987); see *Proc. Aust. Fract. Conf.*, *1986*, *1* (1986).

50. H. A. Kuhn, *ASTM Spec. Tech. Publ.*, **STP647**, 206 (1978).

51. W. F. Hosford and R. M. Caddell, *Metal Forming—Metallurgy and Mechanics*, Prentice-Hall, Englewood Cliffs, NJ, 1982.

52. R. M. Caddell, J. L. Duncan, and W. Johnson, *Proc. Conf. Dimens. Strength Calc.*, *3rd*, Budapest. p. 385 (1968).

53. W. A. Backofen, *Deformation Processing*, Addison-Wesley, Reading, MA 1972.

54. A. G. Atkins, *Int. J. Mech. Sci.*, **29**, 115 (1987).

CHAPTER 5
On the Dynamic Inelastic Failure of Beams

Norman Jones
Department of Mechanical Engineering
The University of Liverpool
Liverpool, England

ABSTRACT

This chapter examines the dynamic inelastic failure of beams that are subjected to either impulsive velocities or impact loadings.

It is observed that the strain-rate-insensitive aluminum alloy beams fail with the mode-I (large ductile deformation), Mode-II (tensile-tearing), and mode-III (transverse-shear) failure modes. These modes were defined by Menkes and Opat [4] for the impulsive-velocity loading of beams and remain valid for impact loadings, except that mode-III failure plays a minor role for impact loadings.

In contradistinction to the impact tests on the aluminum alloy beams, the failure of the strain-rate-sensitive mild-steel beams due to impact loads were dominated by shear failures. Moreover, these shear failures are more complicated than the mode-III response for impulsive-velocity loading, which developed only when the mass struck near a support.

The rigid-plastic methods of analysis can be modified to predict the dynamic failure of beams. These methods show promising agreement with experiments, but further study is required to examine the new shear-failure modes that were observed in beams subjected to impact loads.

1. INTRODUCTION

The behavior of structural members when subjected to large dynamic loads has been examined by many authors with the aid of the plastic methods of analysis. In general, the maximum displacement caused by an impact is sought when the material is assumed to have an unlimited ductility. A considerable body of theoretical and experimental information is now available on the design of various structural members subjected to dynamic loads [1–3].

In some practical applications, failure, or damage, is associated with excessive displacements of a structure. However, in other situations, a designer is often required to estimate the maximum possible impact load or maximum energy absorption of a structure or component. These quantities are limited by the ductility of the material, which, however, is assumed infinite in most theoretical and many numerical studies. In practice, a structure, or component, may tear and fail due to excessive local strains.

Menkes and Opat [4] conducted an experimental investigation into the dynamic plastic response and failure of fully clamped metal beams that were subjected to uniformly distributed velocities over the entire span, as shown in Figure 1. Menkes and Opat observed that the beams responded in a ductile manner and acquired permanently deformed profiles when subjected to velocities less than a certain value,† as shown in Figure 2(a). However, when the impulsive velocities were equal to this critical value, then the beams failed due to tearing of the beam material at the supports. As the impulsive velocities were further increased beyond this critical value, failure occurred and the plastic deformation of the beams became more localized near the supports (as shown in Figure 3 of Reference 4) until another critical velocity was reached that was associated with shear failure at the supports, as shown in Figure 2(c). On the basis of these experimental tests, Menkes and Opat [4] classified the three failure modes for impulsively-loaded fully clamped beams with rectangular cross sections as:

Mode I: Large inelastic deformation of the entire beam [Figure 2(a)]

Mode II: Tearing (tensile failure) of the beam material at the supports [Figure 2(b)]

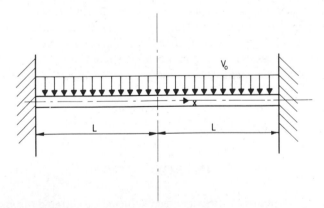

Figure 1. Fully clamped beam subjected to a uniformly distributed impulsive velocity V_0.

---

†The impulsive velocities in the tests were always larger than those necessary to produce a wholly elastic response.

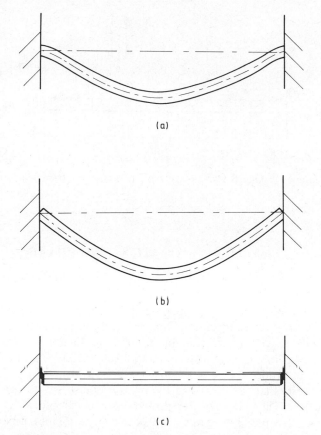

(a)

(b)

(c)

Figure 2. Permanent profiles of impulsively-loaded beams: (*a*) mode I, large ductile deformations, (*b*) mode II, tensile tearing at supports, and (*c*) mode III, transverse shear failure at supports.

Mode III: Transverse shear failure of the beam material at the supports [Figure 2(*c*)].

It is shown in Reference 5 that the theoretical rigid-plastic predictions for an impulsively-loaded fully clamped beam could be used to estimate the threshold velocities for tensile-tearing (mode II) and transverse shear failures (mode III).

More recently, the failure of beams when subjected to impact loads, as shown in Figure 3, has been examined in References 6–9.

This chapter contains a summary on the failure of beams when subjected to either impulsive velocities (Figure 1) or impact loadings (Figure 3). Particular attention is given to the areas where further studies are required in order to develop failure criteria that can be used with confidence for a wide range of practical problems.

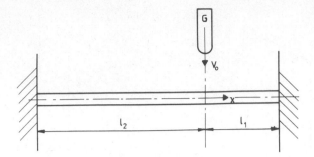

Figure 3. Clamped beam struck transversely by a mass G traveling with an initial velocity V_0.

2. IMPULSIVE LOADING OF BEAMS

2.1. Introduction

Consider the idealized dynamic-pressure pulse in Figure 4, with a time-independent magnitude p_0 for a time duration τ. If this pressure pulse acts on a beam, which has a static plastic collapse pressure p_c, then it is well known that for a given pulse, the maximum transverse displacement is insensitive to the dynamic-pressure ratio $\eta = p_0/p_c$ when $\eta > 20$, approximately [10]. This observation suggests that the details of the pressure pulse are not important for large values of η. Thus, a particular problem can be simplified by taking the loading as impulsive (i.e., $\eta \to \infty$ and $\tau \to 0$ with a finite impulse). In other words, a beam of unit breadth acquires instantaneously a uniform transverse velocity

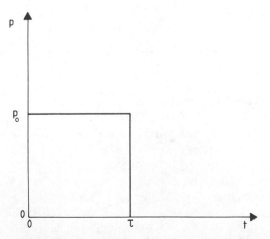

Figure 4. Rectangular-shaped pressure pulse.

V_0, as shown in Figure 1, where, to conserve linear momentum,

$$m2LV_0 = p_0 2L\tau$$

or

$$V_0 = p_0\tau/m \qquad (1)$$

where m and $2L$ are the mass per unit length and span of the beam, respectively.

The approximation of an impulsive loading usually simplifies a theoretical analysis and is an acceptable idealization for many practical problems.

As stated previously, any beams with a Mode I failure exhibited a ductile behavior throughout the entire response and acquired permanently deformed profiles.

It is evident from the theoretical analysis of Symonds and Mentel [11], and the more recent studies of other investigators, that the influence of finite deflections, or geometry changes, must be retained in the basic equations when the maximum permanent transverse displacements of axially restrained beams exceed the corresponding beam thickness, approximately. A theoretical procedure has been developed in Reference 12 to examine the dynamic response of arbitrarily shaped rigid-plastic plates and beams when the influence of finite deflections is retained in the governing equations. Equation (B4) of Reference 12 predicts that the permanent transverse displacement W_m at the midspan of a fully clamped impulsively-loaded rigid perfectly plastic beam is

$$W_m/H = [(1 + 3\lambda/4)^{1/2} - 1]/2 \qquad (2)$$

where

$$\lambda = 4\rho L^2 V_0^2/\sigma_0 H^2 \qquad (3)$$

is a nondimensional form of the initial kinetic energy, and $\rho_1 \sigma_0$ and H are the density, static uniaxial yield stress and beam thickness, respectively.

Equation (2) is developed for a square yield curve that circumscribes the exact maximum normal stress-yield curve [12]. Another yield curve that is 0.618 times as large would inscribe the exact yield curve. The theoretical predictions of Equation (2) and Equation (2) with σ_0 replaced by $0.618\sigma_0$ bound the theoretical predictions according to the exact yield curve, as well as bounding [5, 12] the experimental results on aluminum 6061 T6 beams, which are reported in References 4 and 13.

If a beam is made from a strain-rate-sensitive material (e.g., mild steel), then simple modifications can be made to Equation (2) in order to provide good engineering estimates of the maximum permanent transverse displacement [14, 15].

Equation (2) and its modified versions for strain-rate-sensitive materials and other dynamic loadings can be used to obtain the energy absorbed for a given transverse displacement. Thus, a designer can estimate the damage of a beam for a specified kinetic energy. This is known as a mode-I failure, which is shown in Figure 2(*a*), and has been studied extensively for many structural members and is, therefore, not examined further in this chapter.

2.2. Tensile Tearing

In some practical applications, failure, or damage, is associated with excessive transverse displacements of a structure, in which case the methods of the previous section for mode-I behavior are relevant. However, Equation (2) is developed for a beam that is made from a perfectly plastic material with unlimited ductility. In practice, a structure can tear and fail due to excessive local strains, as shown in Figure 2(*b*). A failure that is caused by excessive tensile strain was identified by Menkes and Opat [4] as a mode-II failure.

The principal assumption that a material has an unlimited ductility makes rigid-plastic analyses simple and attractive. However, one consequence of this idealization is that the strain magnitudes in these analyses are indeterminate because the plastic hinges have an infinitesimally short length. Thus, it is necessary to estimate the hinge lengths so that the engineering strains can be calculated.

If it is assumed that a beam deforms with the kinematically admissible displacement profile in Figure 5, then plastic hinges would develop at locations *a*, *b*, and *c*. The plastic hinges at the ends of the rigid members *ab* and *bc* are assumed to have a length ℓ and a total hinge length of 2ℓ at the center of the beam. Thus, the maximum total strain (ε_m) at the supports of a fully clamped beam is

$$\varepsilon_m = \varepsilon + H\kappa/2 \tag{4}$$

where

$$\varepsilon = \{(1 + W^2/L^2)^{1/2} - 1\}2L/4\ell \tag{5}$$

or

$$\varepsilon \cong W^2/4L\ell \tag{6}$$

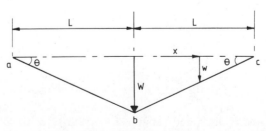

Figure 5. Transverse-displacement profile for a fully clamped beam that is subjected to a uniformly distributed impulsive velocity.

and

$$\kappa \cong W/L\ell \qquad (7)$$

when the angle θ in Figure 5 remains small.

The actual value of ℓ in Equations (5)–(7) is unknown but is estimated in Reference 5 with the aid of Nonaka's [16] suggestion that slip-line theory can be used to obtain the shape of a plastic hinge across the thickness of a beam. If $W_f/H \geq 1$, where W_f is the transverse displacement when tensile tearing occurs, then

$$\ell \cong 3H/2 \qquad (8)$$

when $0 \leq W/H \leq 1$. It is evident from Equations (6)–(8) that the maximum strain at the beam supports is

$$\varepsilon_{m1} = H/2L \qquad (9)$$

when $W/H - 1$.

A fully clamped beam commences to behave as a string once the membrane state $N = N_0$ is reached at $W/H = 1$, where N is the membrane force. Thus, it is assumed in Reference 5 that

$$\ell = L/2 \qquad (10)$$

for $1 \leq W/H \leq W_f/H$, and, therefore, Equation (6) gives

$$\varepsilon_m = W^2/2L^2 - H^2/2L^2 \qquad (11)$$

for the strain accumulated between $W/H = 1$ and W/H. In particular,

$$\varepsilon_{m_2} = [(W_f/H)^2 - 1]H^2/2L^2 \qquad (12)$$

when $W = W_f$.

The total tensile strain at the supports when $W = W_f$ is the sum of Equations (9) and (12), or

$$\varepsilon_m = 2[(W_f/H)^2 + L/H - 1](H/2L)^2 \qquad (13)$$

where

$$W_f/H = [(3\lambda)^{1/2}/2 - 1]/2 \qquad (14)$$

from Equation (2) when $3\lambda/4 \gg 1$. Substituting Equation (14) into Equation (13) gives

$$\lambda^{1/2} = 2\{1 + 2^{1/2}[2 + \varepsilon_m(2L/H)^2 - 2L/H]^{1/2}\}/3^{1/2} \qquad (15)$$

TABLE 1   Comparison of the Theoretical Threshold Impulses for a Mode-II Response from Equations (16) and (17) with the Corresponding Experimental Results of Menkes and Opat [4] on Aluminum 6061-T6 Impulsively-Loaded Beams

H (mm)	$2L$ (mm)	I Equations (16) and (17) (ktaps)	I Equations (16) and (17) with $0.618\sigma_0$ (ktaps)	I Experiments [4] (ktaps)
4.76	203.2	27.1	21.3	26
4.76	101.6	26.4	20.8	26
6.35	203.2	35.9	28.2	32
6.35	101.6	34.7	27.3	32
9.53	203.2	52.9	41.6	45
9.53	101.6	50.6	39.7	45

Thus, if $\varepsilon_m = \varepsilon_r$, where ε_r is the tensile strain at which a material ruptures in a uniaxial tensile test, then Equation (15) predicts the value of λ required for the onset of a mode-II response in a beam with a given ratio $2L/H$. Equation (15) can be rearranged to predict the initial impulse velocity:

$$V_0 = \{1 + 2^{1/2}[2 + \varepsilon_r(2L/H)^2 - 2L/H]^{1/2}\}(H/L)(\sigma_0/3\rho)^{1/2} \qquad (16)$$

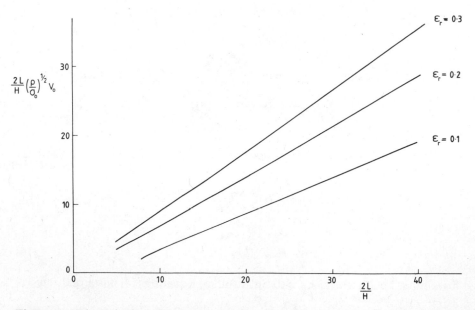

Figure 6.   Dimensionless threshold velocities according to Equation (15) for fully clamped beams with various tensile-failure strains ε_r.

and the initial impulse per unit area

$$I = \rho H V_0 \tag{17}$$

It is evident from Table 1 that the experimental results of Menkes and Opat [4] are bounded by the theoretical predictions of Equations (16) and (17) with square yield criteria that inscribe and circumscribe the exact yield surface.

The critical dimensionless energy according to Equation (15) is plotted in Figure 6 for several values of the tensile strain ε_r. The curves in this figure were obtained for a square yield curve that circumscribes the exact yield curve. However, these curves can also be used for an inscribing yield curve by simply replacing σ_0 in the parameter λ by $0.618\sigma_0$.

2.3. Transverse Shear Failure

Another type of failure mode is due primarily to the influence of large transverse shear forces and is identified as a mode-III failure by Menkes and Opat [4], as noted in Section 1. Several theoretical analyses have been developed for beams that are made from a rigid perfectly plastic material that yields from the combined influence of a bending moment and a transverse shear force [17, 18]. These theoretical analyses are simplified by using bending hinges and transverse shear slides. Bending hinges allow angular changes across a region of infinitesimal length, which is an idealization of a small region with a large curvature, and are familiar in the static plastic analysis of structures. In the same spirit, transverse shear slides are transverse-displacement discontinuities that are the idealization of rapid changes in transverse displacement across a short section of a beam.

The maximum transverse shear force occurs at the supports in the impulsively-loaded fully clamped beam in Figure 1. It was found in References 5 and 17 that transverse shear slides develop during the first phase of motion, as indicated in Figure 7. The transverse shear velocity at the supports of a beam of breadth B is

$$V_s = V_0 - Q_0^2 t / 3\rho BHM_0 \tag{18}$$

for $0 \leq t \leq t_s$, where

$$t_s = 3\rho HBM_0 V_0 / Q_0^2 \tag{19}$$

is the time when transverse shear sliding ceases (i.e., $V_s = 0$), and M_0 and Q_0 are the fully plastic capacity of a beam cross section when subjected to a pure bending moment and a transverse shear force without any bending, respectively. Thus, in order to avoid severance at the supports,

$$\int_0^{t_s} V_s \, dt \leq kH \tag{20}$$

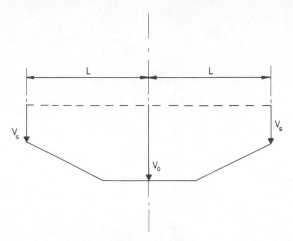

Figure 7. Transverse-velocity field for an impulsively-loaded fully clamped beam with shear sliding at the supports.

where k is a constant $(0 < k \leqq 1)$. Clearly, complete severance occurs in a beam with a thickness H when $k = 1$, but transverse shear failure is likely to develop for smaller values of k.

Equations (18)–(20) predict that the velocity just sufficient to cause severance of a beam with a rectangular cross section is

$$V_0 = 2(2k\sigma_0/\rho)^{1/2}/3 \tag{21}$$

or

$$\lambda = 32kL^2/9H^2 \tag{22}$$

TABLE 2   Comparison of the Theoretical Threshold Impulses for a Mode-III Response from Equations (17) and (21) ($k = 1$) with the Corresponding Experimental Results of Menkes and Opat [4] on Aluminum 6061-T6 Impulsively-Loaded Beams

H (mm)	$2L$ (mm)	I Equations (17) and (21) (ktaps)	I Equations (17) and (21) with $0.7245\sigma_0$ (ktaps)	I Experiments [4] (ktaps)
4.76	203.2, 101.6	39.1	33.3	40
6.35	203.2, 101.6	52.2	44.5	48
9.53	203.2, 101.6	78.3	66.7	65

where λ is defined by Equation (3), and $M_0 = \sigma_0 BH^2/4$, and $Q_0 = \sigma_0 BH/3^{1/2}$.

It is evident from Table 2 that the theoretical predictions of Equations (17) and (21) with circumscribing and inscribing yield criteria and $k = 1$ almost bound all the experimental results that Menkes and Opat [4] obtained on aluminum 6061-T6 beams.

2.4. Comments

The dimensionless critical energies for tensile-tearing (mode II) and transverse shear (mode III) failures according to Equations (15) and (22), respectively, are compared in Figure 8 for several values of the parameters ε_r and k. The dimensionless velocities for Mode II and Mode III failures in Figure 9 indicate that transverse shear forces govern failure for small values of k, whereas tensile tearing controls the failure when $k \geq 0.88$, approximately, and $\varepsilon_r = 0.3$ at least for $10 \leq 2L/H \leq 40$.

The theoretical predictions in Sections 2.2 and 2.3 were developed for a beam with a solid rectangular cross section. However, it is possible to use these methods for a beam having any cross-sectional shape that is symmetrical about the horizontal and vertical planes in Figure 1.

The relation $M_0 = N_0 H/4$ was used in the derivation of Equation (2), where $M_0 = \sigma_0 H^2/4$, and $N_0 = \sigma_0 H$ are the fully plastic bending moment and axial force of a beam with a solid rectangular cross section of unit width, respectively. It is straightforward to derive the values of M_0 and N_0 for a beam with another cross-sectional shape and to obtain the maximum permanent transverse displacement from the theoretical analysis in Reference 12 and, therefore, the corresponding modified form of Equation (2). Thus, the threshold velocity for ductile tearing can be obtained using the general theoretical procedure that is outlined in Section 2.2.

Several theoretical analyses, which retain the influence of the transverse shear force and bending moment in the yield condition, have been developed for beams with cross-sectional shapes that are characterized by the dimensionless ratio

$$v = Q_0 L/2M_0 \tag{23}$$

For example, an impulsively-loaded rigid perfectly plastic beam with simple supports was examined in Reference 18. The maximum transverse shear sliding at the supports is

$$W_s = \int_0^{t_s} V_s \, dt = 3mL^2 V_0^2/16M_0 v^2 \qquad v \geq 1.5 \tag{24}$$

which using the inequality of Equation (20) gives the threshold velocity

$$V_0^s = (16M_0 k H v^2/3mL^2)^{1/2} \qquad v \geq 1.5 \tag{25}$$

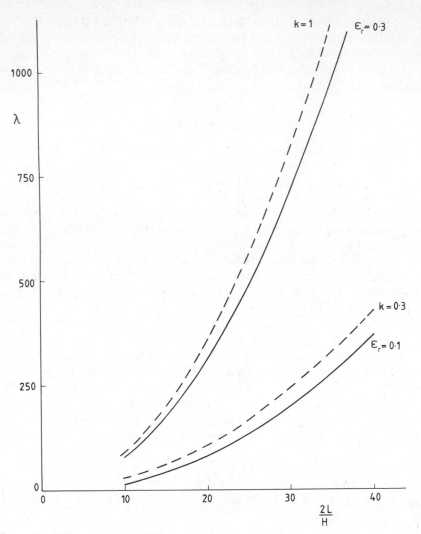

Figure 8. Dimensionless critical energies for tensile-tearing and transverse-shear failures of impulsively-loaded beams. Mode II is shown by the solid line [Equation (15) with $\varepsilon_m = \varepsilon_r$]. Mode III is shown by the dashed line [Equation (22)].

for a mode-III failure. Similarly,

$$V_0^s = (4M_0 kHv/mL^2)^{1/2} \qquad v \leq 1 \tag{26}$$

and

$$V_0^s = \{4M_0 kH(4v - 3)/mL^2\}^{1/2} \qquad 1 \leq v \leq 1.5 \tag{27}$$

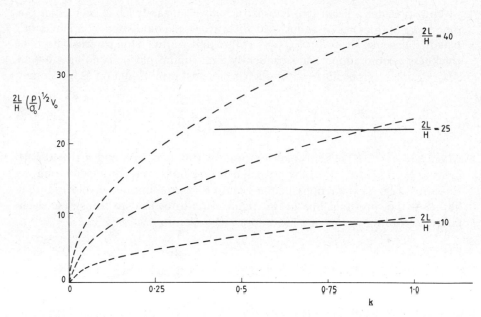

Figure 9. Dimensionless velocities for tensile-tearing and transverse-shear failures of impulsively-loaded beams. Mode II is shown by the solid line [Equation (15) with $\varepsilon_m = \varepsilon_r = 0.3$]. Mode III is shown by the dashed line [Equation (22)].

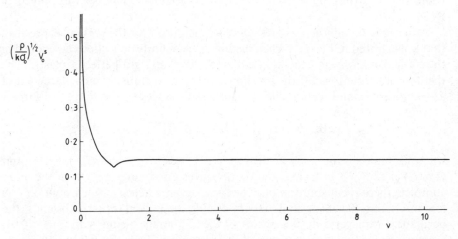

Figure 10. Variation of the dimensionless threshold impulsive velocities for a simply supported beam with a sandwich cross section, having $\sigma_0/2\tau_0 = 8$ and $h/H = 0.714$.

145

Now consider a beam that is constructed with a sandwich cross section. An inner core of thickness h and a yield shear stress τ_0 could support a maximum plastic transverse shear force $Q_0 = \tau_0 h$ (per unit width). Thin exterior sheets of thickness t could support independently a maximum plastic. bending moment $M_0 = \sigma_0 t(h + t)$ (per unit width). In this circumstance, Equation (23) becomes

$$v = \frac{\tau_0}{\sigma_0/2} \frac{h/H}{1 - (h/H)^2} \frac{L}{H} \qquad (28)$$

when $H = h + 2t$. Thus, a sandwich beam with $\sigma_0/2\tau_0 = 8$ and $h/H = 0.714†$ gives $v = 0.182 L/H$, and the associated threshold velocities according to Equations (25)–(27) are presented in Figure 10. Initial impulsive velocities that lie below the threshold line in this figure would not lead to transverse shear failure according to the simplified approach taken in Section 2.3.

3. IMPACT LOADING OF BEAMS

3.1. Introduction

The theoretical predictions in Section 2 were developed for beams that were subjected to an initial impulsive velocity. It was found that the theoretical estimates for mode-II and mode-III failures gave encouraging agreement with the corresponding experimental results of Menkes and Opat [4], which were conducted on impulsively-loaded aluminum 6061 T6 beams. However, the impact loads in many practical problems can arise from a dropped object with an impact velocity that may not be idealized as impulsive. Thus, the failure of a beam that is struck by a mass, as shown in Figure 3, has been examined in References 6–9.

A theoretical analysis was presented in Reference 7 for the particular problem that is illustrated in Figure 3 when the influence of finite deflections is retained in the basic equations and the material is idealized as rigid perfectly plastic. The theoretical predictions for the maximum permanent transverse displacement of a beam with a solid rectangular cross section reduces to the simplified form

$$W_m/H = 0.5\{[1 + 2\Omega/(1 + r)]^{1/2} - 1\} \qquad (29)$$

when the mass ratio g is small ($g/r^2 \ll 1$), where $g = ml_1/G$, $r = l_1/l_2$ and $\Omega = GV_0^2 l_1/2M_0 H$. In fact, Equation (29) gives good agreement with the more complete theoretical solution in Reference 7 even when $g = 0.1$. Equation (29) for inscribing ($0.618\sigma_0$) and circumscribing (σ_0) yield criteria bounds the permanent transverse displacements of the aluminum beams with $g/r^2 < 0.01$, approximately, which were reported in Reference 8. Thus, Equation (29) can be

---

†For example, a 12.7 mm thick core with 2.54 mm sheets gives $h/H = 0.714$.

used to predict the mode-I behavior of beams that are impacted by a mass at any point on the span except that it overpredicts the permanent deformations for impacts very close to a support.

3.2. Experimental Results

Experimental tests have been reported in Reference 8 for fully clamped aluminum alloy beams that are subjected to impact loads, as shown in Figure 3. The beams had a span of 4 in. (101.6 mm) and a width of 0.4 in. (10.16 mm) with four different thicknesses, 0.15 in. (3.81 mm), 0.2 in. (5.08 mm), 0.25 in. (6.35 mm), and 0.3 in. (7.62 mm). The mechanical properties for the beams, which had either flat or enlarged ends, are listed in Table 3. The striker mass is 5 kg and the impact velocities lie within the range 1.78 to 11.65 m/sec and strike the beam at the dimensionless impact locations, $r = 0.067, 0.143, 0.333, 0.6$, and 1. Various other details of the experimental tests are reported in Reference 8.

In the initial phase of the experimental investigation, the Mode I response was obtained for the lower range of impact velocities. The impact velocities were then increased further until a beam cracked or broke at the supports or at the impact point.

It transpires that the flat aluminum beams failed generally due to tensile tearing. The cracking was initiated by the maximum tensile strain at the supports on the upper surface (impacted surface) of a beam, or underneath the impact point on the lower surface. The location of a tensile-tearing failure changed from the impact point to a support with a reduction in the value of l_1. Thus, the tensile-tearing failure occurred at the impact point when l_1 was 2 in. (50.8 mm), 1.5 in. (38.1 mm), and 1.0 in. (25.4 mm), and a tensile-tearing failure might occur at the support or at the impact point when $l_1 = 0.5$ in. (12.7 mm).

TABLE 3　Mechanical Properties of the Aluminum Alloy Beams that Were Subjected to Impact Loads in Reference 8. (The Quoted Values of σ_0, σ_u, and ε_r Are the Average Values from Several Static Uniaxial Tensile Tests.)

Type of supports	H (in.)	H (mm)	σ_0 (N/mm^2)	σ_u (N/mm^2)	ε_r (%)
Enlarged ends	0.15	3.81	182	318	19
	0.20	5.08	182	318	19
	0.25	6.35	182	318	19
	0.30	7.62	182	318	19
Flat ends	0.15	3.81	354.5	475.5	19
	0.20	5.08	354.5	475.5	19
	0.25	6.35	354.5	475.5	19
	0.30	7.62	412	553	15

Not many failures were obtained in the present test program when $l_1 = 0.25$ in. (6.35 mm), but all the cracked specimens failed due to tensile tearing at the supports with the exception of one specimen that failed due to excessive transverse shear forces at the impact point. The values of l_1, or r, are nominal values and the actual values of l_1 were measured after each test. It turns out that in this particular test, which was conducted on the thickest beam (7.62 mm), the actual value of l_1 was 4 mm, giving $r \cong 0.04$ instead of the nominal value $r = 0.067$. Incidentally, a tensile-tearing failure did develop in the thickest beam when $r = 0.064$ (actual value of $l_1 = 6.1$ mm, which was fairly close to the nominal value of $l_1 = 6.35$ mm). Thus, it appears that the flat aluminum beams always suffer a tensile-tearing failure (mode II) for sufficiently large impact loads except when struck very close to a support ($r < 0.067$, approximately), which then produces a transverse shear (mode III) failure.

It is evident that large concentrated stresses would develop at the interface between the section of a beam and an enlarged end, which can cause plastic flow and lead eventually to the failure of a beam in this region. Indeed, it was found that all of the aluminum beams with enlarged ends cracked or broke at a support due to tensile tearing (Mode II).

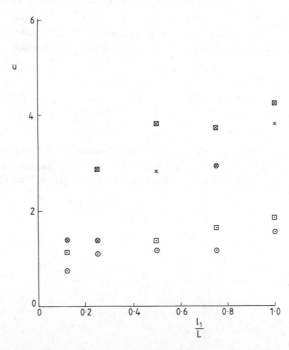

Figure 11. Variation of the dimensionless initial kinetic energy u that was absorbed by the flat aluminum alloy beams tested in Reference 8 with the dimensionless impact position l_1/L. x: $H = 0.15$ in. (3.81 mm); ∵: $H = 0.3$ in. (7.62 mm); □: broken beams; ○: cracked beams; otherwise no failure.

Several formability curves for aluminum alloys are presented by Duffey in Figure 6 of Chapter 6. It is evident that the magnitude of the major failure strain, which is required to produce localized necking in a fully clamped thin plate, is influenced by the value of the associated minor strain. Thus, it is possible that the different strain distributions in the vicinity of the supports of beams with flat and enlarged ends would influence the type of failure.

Some of the experimental results that are reported in Reference 8 are presented in Figures 11 and 12. The maximum dimensionless dynamic energy, $(u = GV_0^2/2M_0)$, which was absorbed by the beams through large ductile deformations without failure, is also shown for two cases. It is evident that there is a tendency for the dimensionless initial kinetic energy, which is required for failure, to decrease as the impact point approaches a support. Figures 11 and 12 also indicate that less dimensionless energy is required for the failure of the thicker beams. However, it should be noted that the dimensionless parameter u is inversely proportional to H^2, since the plastic limit moment is $M_0 = \sigma_0 BH^2/4$ for a beam with a rectangular cross section of width B.

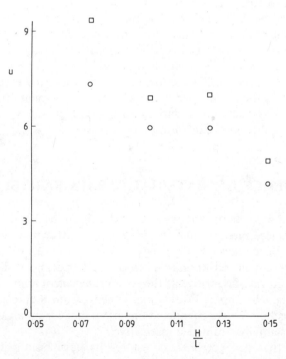

Figure 12. Variation of the dimensionless initial kinetic energy u that was absorbed by the aluminum alloy beams tested in Reference 8 with the dimensionless ratio H/L. The beams have enlarged ends and were struck at the midspan. □: broken beams; and ○: cracked beams.

3.3.   Theoretical Considerations

It is evident from Section 3.2 and the experimental results in Reference 8 that the mode-II tensile-tearing failure mode, which was introduced by Menkes and Opat [4] for an initial impulsive velocity loading, governs the failure of strain-rate-insensitive beams when subjected to impact loads. Thus, the general approach, which was outlined in Section 2.2 for the tensile tearing of impulsively-loaded beams, was employed in Reference 9 to examine the impact-loading case. Again, geometrical arguments were used to evaluate the strains, and Equation (29) is employed to predict the maximum transverse displacement that occurs at the onset of tensile tearing.

This theoretical procedure was used in Reference 9 to predict the onset of tensile-tearing failures (mode II) for the aluminum alloy beams that were tested in Reference 8. The theoretical predictions and experimental results are presented in Tables 4 and 5 for beams with enlarged and flat ends, respectively. It is evident from Table 4 that the theory provides acceptable bounds on the experimental results for the minimum values of λ with failure, particularly when considering the various difficulties inherent in dynamic tests of this nature. However, the comparisons between the theoretical predictions and the experimental results are not as satisfactory in Table 5 for the beams with flat ends. It should be noted from Table 3 that the material properties of these beams are significantly different from those for the specimens with enlarged ends. Moreover, the beams with flat ends have less ideal supports.† These factors may have contributed to make less satisfactory agreement between the theoretical predictions and the experimental results. Nevertheless, the theoretical predictions do show the same general trend as the experimental results and the theoretical method does predict satisfactory bounds in some cases and would provide an adequate approximation for several others.

4.   INFLUENCE OF MATERIAL STRAIN-RATE SENSITIVITY

The theoretical predictions that are discussed in Sections 2 and 3 are developed for beams that are made from a strain-rate-insensitive material. The experimental results of Menkes and Opat [4] on the impulsive loading of beams in Section 2 were conducted on aluminum 6061 T6 specimens. This material is essentially strain-rate-insensitive at the strain rates encountered in the tests on the beams [14]. The impact-loading experiments from Reference 8, which are discussed in Section 3, were conducted on beams made from an aluminum alloy that is also essentially strain-rate-insensitive.

It appears that no experimental test results have been published on the dynamic failure of beams that are made from a strain-rate-sensitive material and

---

†All failures in aluminum alloy beams with enlarged ends occurred at a support, whereas about two-thirds of the aluminum alloy beams with flat ends failed at the impact point.

TABLE 4 Comparison Between the Experimental Results and Theoretical Predictions for the Tensile-Tearing Failure of the Aluminum Alloy Beams with Enlarged Ends That Were Subjected to Impact Loads and Reported in Reference 8

H (in.)	H (mm)	l_1 (in.)	l_1 (mm)	ε_r (%)	Experimental Results		Theoretical results	
					Maximum λ without failure	Minimum λ with failure	λ circumscribing yield	λ inscribing yield
0.15	3.81	2	50.8	19	16.3	24.2 C[a]	32.2	19.9
0.2	5.08	0.5	12.7	19	0.55	1.6 B[b]	1.1	0.69
		2	50.8	19	10.2	14.1 C	19.1	11.8
0.25	6.35	2	50.8	19		8.2 C	12.9	7.9
0.3	7.62	2	50.8	19	5.0	6.6 C	9.4	5.8

[a]C: cracked beams
[b]B: broken beams

TABLE 5 Comparison Between the Experimental Results and Theoretical Predictions for the Tensile-Tearing Failure of the Aluminum Alloy Beams with Flat Ends That Were Subjected to Impact Loads and Reported in Reference 8

H (in.)	H (mm)	I_1 (in.)	I_1 (mm)	ε_r (%)	Experimental results		Theoretical results	
					Maximum λ without failure	Minimum λ with failure	λ circumscribing yield	λ inscribing yield
0.15	3.81	0.25	6.35	19	0.31	0.69 C[a]	0.59	0.36
		0.5	12.7		0.53	1.24 C	1.7	1.1
		1	25.4		4.5	6.16 B[b]	6.2	3.9
		1.5	38.1		5.8	7.48 C	15.3	9.4
		2	50.8		19.3	13.21 B	32.2	19.9
0.2	5.08	0.25	6.35	19	0.51	0.84 C	0.40	0.25
		0.5	12.7		0.53		1.1	0.69
		1	25.4		2.2	3.2 B	3.8	2.4
		1.5	38.1		6.1		9.2	5.7
		2	50.8		7.5	8.0 B	19.1	11.8
0.25	6.35	0.25	6.35	19	0.47		0.30	0.18
		0.5	12.7		1.4	1.0 C	0.81	0.50
		1	25.4		2.1	2.6 B	2.7	1.7
		1.5	38.1		4.3		6.3	3.9
		2	50.8		5.6	6.6 B	12.9	7.9
0.3	7.62	0.25	6.35	15	0.11	0.15 C	0.17	0.11
		0.5	12.7		0.37	0.53 C	0.45	0.28
		1	25.4		0.52	1.0 C	1.4	0.86
		1.5	38.1		0.78	1.5 C	3.1	1.9
		2	50.8		2.2	2.6 C	6.2	3.8

[a] C: cracked beams
[b] B: broken beams

subjected to an initial impulsive loading. Thus, the influence of material strain-rate sensitivity cannot be assessed properly for the failure of beams when impulsively loaded. However, experimental test results are reported in Reference 8 for the impact loading of beams that are made from mild steel, which is a highly strain-rate-sensitive material [14].

It transpires that for sufficiently large impact energies, the flat steel beams broke or cracked at the impact point with a shear mode, possibly because the rupture strain ε_r of steel is much larger than that of the aluminum alloy used in Reference 8. However, the angle of the broken section varied with the position of the impact point. When the mass struck close to a support, shear sliding occurred between two adjacent transverse cross sections, as shown in Figure 13(a). This behavior is similar to the mode-III response that Menkes and Opat [4] observed for impulsively-loaded beams, as discussed in Section 2.3 and shown in Figure 2(c). However, shear sliding occurred along planes that were inclined at about 45° to the beam axis when the impact point was near the midspan of a beam, as shown in Figures 13(b)–(d).

In some cases, the transverse-shear failure originated from the corner of the indentation caused by the striker on the upper surface of a beam. In other cases, the behavior shown in Figure 14 was identified as a transverse-shear failure even though the influence of membrane forces and bending moments may be important.

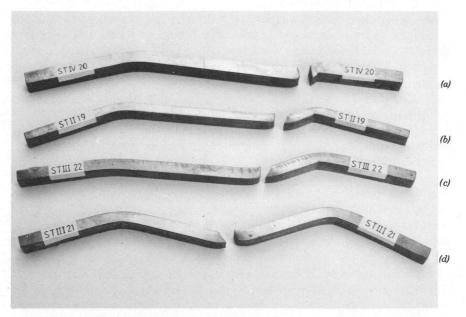

(a)

(b)

(c)

(d)

Figure 13. Broken mild-steel beams with flat ends. (a) Transverse-shear failure at the impact point. (b)–(d) Shear failures at the impact point when the broken section is inclined at about 45° to the beam axis.

Figure 14. Plastic deformation at the impact point and at the support when the mass strikes at $l_1 = 0.25$ in. (6.35 mm) on a 0.25 in. (6.35 mm) thick mild-steel beam with flat ends.

Fewer experimental tests were conducted on the steel beams with enlarged ends. It turns out that they cracked or broke at a support due to tensile tearing when $l_1 = 0.5$ in. (12.7 mm), whereas a shear failure occurred at the impact point when struck near the midspan.

An approximate method was developed in Reference 8 in order to modify the plastic-flow stress in Equation (29) to cater for material strain-rate-sensitive effects. Reasonable agreement was found between the approximate theoretical predictions for the maximum permanent transverse displacements (mode I) and the corresponding experimental results on mild-steel beams subjected to impact loads. Reference 15 contains a similar simplified theoretical procedure for an impulsively-loaded fully clamped beam.

It is evident from the foregoing summary of the experimental results on the mild-steel beams that failure is dominated by a shear mode. The theoretical predictions for the mode-III behavior assume that transverse shear sliding occurs across a transverse plane, as shown, for example, in Figures 2(c) and 13(a). Some theoretical predictions are developed with the aid of this idealization in Reference 7. However, this particular mode describes only one of the several shear-failure modes that were observed in References 8 and 9. The other shear failure modes are considerably more complex than the mode-III response and further experimental and theoretical work is required.

A comparison of the theoretical predictions for the mode-II and mode-III dynamic-failure modes of beams reveals that a beam is less resistant to a tensile-tearing failure when the rupture strain ε_r of the material is small, as expected, whereas a beam is likely to suffer a transverse shear failure when the impact point is close to a support (i.e., l_1 is small). Thus, the type of beam failure can change from tensile tearing to a transverse shear failure with an increase of ε_r or a decrease of l_1.

5. DISCUSSION

This chapter has focused on the dynamic failure of beams because of the paucity of experimental data on other inelastic structures. In fact, it is clear that insufficient information is available to understand fully the dynamic failure of beams.

A theoretical solution has been published in Reference 19 on the dynamic plastic behavior of a circular plate that is subjected to a uniformly distributed impulsive velocity. This particular analysis retains the influence of the transverse shear force as well as bending moments in the plastic yield condition. It was used in Reference 15 to study the perforation of a plate by a blunt projectile and the theoretical predictions of Recht and Ipson [20] were recovered when $k = 1/2$.

Liss, Goldsmith, and Kelly [21] have compared the residual velocities of various perforation theories with some experimental results on aluminum 2024-T3 target plates. It was shown in Reference 15 that the analysis of Reference 19 with $k = \frac{3}{4}$ gave fair agreement with the experimental results. However, a reasonable estimate of the numerical results was obtained with $k = 1$.

The theoretical predictions in Reference 8 have used an approximate method to examine the influence of material strain-rate sensitivity on the plastic-flow stress during a Mode I response. A considerable amount of experimental and theoretical work is required in order to remove the approximations made in Reference 8 for the Mode I response and to incorporate material strain-rate-sensitivity effects into the theoretical predictions for Mode II and Mode III failures. Moreover, the static uniaxial rupture strains have been used in the theoretical predictions for the tensile-tearing (Mode II) failures. It is likely that the rupture strain ε_r varies with strain rate $\dot{\varepsilon}$. Unfortunately, insufficient experimental data is available on this phenomenon [22–24].

The well-known Cowper–Symonds uniaxial constitutive equation relates the dynamic flow stress σ_0' to strain rate $\dot{\varepsilon}$ by the expression

$$\sigma_0'/\sigma_0 = 1 + (\dot{\varepsilon}/D)^{1/p} \tag{30}$$

where $D = 40.4\ \sec^{-1}$, and $p = 5$ for mild steel, and the values of the constants for some other materials are given in Reference 14. If the influence of material strain hardening is ignored, then the maximum inelastic energy absorbed per unit volume of material for static uniaxial behavior is

$$E_s = \sigma_0 \varepsilon_r \tag{31}$$

whereas the maximum energy per unit volume that can be absorbed in the dynamic uniaxial case is

$$E_d = \int_0^{\varepsilon_r'} \sigma_0'\, d\varepsilon \tag{32}$$

where ε_r' is the dynamic uniaxial rupture strain.

Now, substituting Equation (30) into Equation (32) gives

$$E_d = \int_0^{\varepsilon_r'} \sigma_0[1 + (\dot{\varepsilon}/D)^{1/p}] \, d\varepsilon \tag{33}$$

and, therefore,

$$E_d/E_s = \int_0^{\varepsilon_r'} [1 + (\dot{\varepsilon}/D)^{1/p}] \, d\varepsilon/\varepsilon_r \tag{34}$$

If it is assumed that $\dot{\varepsilon} = \dot{\varepsilon}_c$, where $\dot{\varepsilon}_c$ is a characteristic constant uniaxial strain rate, which can be obtained using the method discussed in References 15, 25, and 26 for a particular problem, then

$$E_d/E_s = [1 + (\dot{\varepsilon}_c/D)^{1/p}]\varepsilon_r'/\varepsilon_r \tag{35}$$

or

$$\varepsilon_r'/\varepsilon_r = [1 + (\dot{\varepsilon}_c/D)^{1/p}]^{-1}(E_d/E_s) \tag{36}$$

when neglecting any change in the coefficients D and p with strain [15]. The dimensionless energy E_d/E_s is a function of many parameters that control the static and dynamic rupture behavior of ductile materials. However, if it is assumed that the energy to fracture is invariant, i.e., $E_d/E_s = 1$, then the dynamic uniaxial rupture strain is

$$\varepsilon_r' = [1 + (\dot{\varepsilon}_c/D)^{1/p}]\varepsilon_r^{-1} \tag{37}$$

which could be evaluated for a given problem. For example, $\varepsilon_r' = \varepsilon_r/2$ when $\dot{\varepsilon}_c = D = 40.4 \, \text{sec}^{-1}$ for mild steel.†

It is evident that further experimental work is required on this topic before it is established whether equation (37) is a reasonable approximation for the influence of strain rate on the uniaxial rupture strain. However, the simple form of Equation (37) or another modified form of Equation (34) or (36) would prove attractive for design purposes.

This chapter has examined the threshold conditions required for a mode-II or a mode-III failure in a beam. Menkes and Opat [4] obtained both types of failure in impulsive-loaded beams. They encountered first the mode-II response and then increased the impulsive velocity imparted to fresh specimens until a mode-III failure was observed. The beams failed with a mixed tensile-tearing and transverse-shear mode for impulsive velocities that lie between the thres-

†It is interesting to observe that the necking strain in a dynamic uniaxial test on a material with $p = 5$ is 0.5 times the corresponding static necking strain when the dynamic-flow stress is twice the static value. This conclusion is predicted by Equations (33) and (34) of Campbell [22] for a material characterized by Equation (31) of Reference 22 with $n = 0.2$.

hold values for mode-II and mode-III responses. This aspect of the behavior was not explored in References 6 to 9, but the same trend was also found in several impact-loading tests.

The striker in Figure 3 has a mass of 5 kg for all of the impact tests reported in this chapter. However, the mode of failure may depend on the dimensionless mass ratio g.

6. CONCLUSIONS

This chapter has examined the dynamic failure of beams that are subjected to either impulsive velocities or impact loadings.

The strain-rate-insensitive aluminum 6061 T6 beams of Menkes and Opat [4] had large permanent ductile deformations (mode I) at low impulsive velocities, tensile tearing (mode II) at the supports for intermediate impulsive velocities, and transverse-shear failures (Mode III) at the highest impulsive velocities. The theoretical methods that are discussed in Section 2 are capable of predicting the three different failure modes with reasonable accuracy, but no experimental data are available for the dynamic failure of impulsively-loaded beams that are made from mild steel or any other strain-rate-sensitive material.

It appears that the dynamic failure of strain-rate-insensitive aluminum alloy beams, which are subjected to impact loads, is more complicated than for impulsive velocities. Again, at low impact velocities, the mode-I ductile response is obtained, as expected, and good agreement is found between the theoretical predictions of rigid-plastic methods and the corresponding experimental results. As the impact velocity is increased, all the beams with flat ends, except one, failed due to tensile tearing (mode II) either at the supports or underneath the striker. The location of the mode-II failure tended to change from the impact point to the support with a reduction in the value of I_1. However, a tensile-tearing (mode-II) failure occurred always at the support for the aluminum alloy beams with enlarged ends. Thus, it appears that the dynamic failure of aluminum alloy beams due to impact loads is governed largely by the mode-II failure mode that was introduced in Section 2.2 for impulsive-velocity loadings.

The impact experiments were repeated on beams made from mild steel, which is a highly strain-rate-sensitive material. It happens that for sufficiently large impact energies, the beams with flat ends failed in a shear mode at the impact point. However, this shear failure is more complex than the mode-III response for impulsive loading, which was found only when the mass struck close to a support. It turns out that the steel beams with enlarged ends failed due to tensile tearing (mode II) when struck near a support and in a shear mode when struck near midspan.

The results in this chapter emphasize the sensitivity of dynamic failure to the material properties and support conditions, show the value of simple rigid-plastic methods of analysis, and illustrate our poor understanding of this important practical topic.

ACKNOWLEDGMENTS

The author wishes to thank the Science and Engineering Research Council for their partial support of this study through Grant GR/B/89737. In addition, thanks are due to Dr. Jianhui Liu, who was supported partly by a University of Liverpool Studentship and partly by the British Council through the Academic Links with China Scheme between the Huazhong University of Science and Technology and the Department of Mechanical Engineering at the University of Liverpool.

The author is also indebted to the Impact Research Centre at the University of Liverpool and in particular to Dr. Jilin Yu for his comments and Mrs. M. White for her secretarial assistance. Mrs. A. Green's assistance with the preparation of the drawings is also gratefully acknowledged. Thanks are also due to Dr. T. A. Duffey and Professor T. Wierzbicki for their constructive comments on a draft version of this chapter.

Finally, the author wishes to thank Professor N. W. Murray and the Department of Civil Engineering at Monash University, Australia, for a visiting professorship during July and August 1987, which expedited the completion of this chapter.

REFERENCES

1. P. S. Symonds, *Survey of Methods of Analysis for Plastic Deformation of Structures under Dynamic Loading*, Div. Eng., Rep. BU/NSRDC/1–67, Brown University, Providence, RI, 1967.
2. W. Johnson, *Impact Strength of Materials*, Edward Arnold, London, 1972.
3. N. Jones, "A Literature Review on the Dynamic Plastic Response of Structures," *Shock Vibr. Dig.*, **7**(8), 89–105 (1975).

   "Recent Progress in the Dynamic Plastic Behaviour of Structures," *Shock Vibr. Dig.*: Part 1, **10**(9), 21–33 (1978); Part 2, **10**(10), 13–19 (1978); Part 3, **13**(10), 3–16 (1981); Part 4, **17**(2), 35–47 (1985).
4. S. B. Menkes and H. J. Opat, "Broken Beams," *Exp. Mech.*, **13**, 480–486 (1973).
5. N. Jones, "Plastic Failure of Ductile Beams Loaded Dynamically," *J. Eng. Ind.*, **98**, 131–136 (1976).
6. N. Jones and J. H. Liu, "Local Impact Loading of Beams," in *Intense Dynamic Loading and its Effects*, Science Press, Beijing, China, 1986, and Pergamon Press, 1988, pp. 444–449.
7. J. H. Liu and N. Jones, "Dynamic Response of a Rigid Plastic Clamped Beam struck by a Mass at any Point on the Span," *Int. J. Solids Struct.* **24**, 251–270 (1988).
8. J. H. Liu and N. Jones, "Experimental Investigation of Clamped Beams Struck Transversely by a Mass," *Int. J. Impact Eng.* **6**, 303–335 (1987).
9. J. H. Liu and N. Jones, *Plastic Failure of a Clamped Beam Struck Transversely by a Mass*, Rep. ES/31/87, University of Liverpool, Dep. Mech. Eng., Liverpool, 1987.

10. N. Jones, "Dynamic Plastic Behaviour of Beams," Chapter 3, in *Structural Impact*, Cambridge Univ. Press, London and New York (to be published).

11. P. S. Symonds and T. J. Mentel, "Impulsive Loading of Plastic Beams with Axial Constraints," *J. Mech. Phys. Solids*, **6**, 186–202 (1958).

12. N. Jones, "A Theoretical Study of the Dynamic Plastic Behaviour of Beams and Plates with Finite-Deflections," *Int. J. Solids Struct.*, **7**, 1007–1029 (1971).

13. N. Jones, R. N. Griffin, and R. E. Van Duzer, "An Experimental Study into the Dynamic Plastic Behaviour of Wide Beams and Rectangular Plates," *Int. J. Mech. Sci.*, **13**, 721–735 (1971).

14. N. Jones, "Response of Structures to Dynamic Loading," *Conf. Ser. No. 47—Inst. Phys.*, J. Harding, Ed., 254–276 (1979).

15. N. Jones, "Some Comments on the Dynamic Plastic Behaviour of Structures," in *Intense Dynamic Loading and its Effects*, Science Press, Beijing, China, 1986, and Pergamon Press, 1988, pp. 49–71.

16. T. Nonaka, "Some Interaction Effects in a Problem of Plastic Beam Dynamics, Parts 1–3," *J. Appl. Mech.*, **34**, 623–643 (1967).

17. P. S. Symonds, "Plastic Shear Deformations in Dynamic Load Problems," in J. Heyman and F. A. Leckie, Eds., *Engineering Plasticity*, Cambridge Univ. Press, London and New York, 1968, pp. 647–664.

18. N. Jones and J. G. Oliveira, "The Influence of Rotatory Inertia and Transverse Shear on the Dynamic Plastic Behaviour of Beams," *J. Appl. Mech.*, **46**, 303–310 (1979).

19. N. Jones and J. G. Oliveira, "Dynamic Plastic Response of Circular Plates with Transverse Shear and Rotatory Inertia," *J. Appl. Mech.*, **47**, 27–34 (1980).

20. R. F. Recht and T. W. Ipson, "Ballistic Perforation Dynamics," *J. Appl. Mech.*, **30**, 384–390 (1963).

21. J. Liss, W. Goldsmith, and J. M. Kelly, "A Phenomenological Penetration Model of Plates," *Int. J. Impact Eng.*, **1**, 321–341 (1983).

22. J. D. Campbell, "Plastic Instability in Rate-Dependent Materials," *J. Mech. Phys. Solids*, **15**, 359–370 (1967).

23. K. Kawata, S. Hashimoto, and K. Kurokawa, "Analyses of High Velocity Tension of Bars of Finite Length of BCC and FCC Metals with Their Own Constitutive Equations," in K. Kawata and J. Shioiri, Eds., *High Velocity Deformation of Solids*, Springer-Verlag, New York, 1977, pp. 1–15.

24. G. Regazzoni and F. Montheillet, "Influence of Strain Rate on the Flow Stress and Ductility of Copper and Tantalum at Room Temperature," *Conf. Ser. No. 70—Inst. Phys.*, J. Harding, Ed., 63–70 (1984).

25. N. Perrone and P. Bhadra, "A Simplified Method to Account for Plastic Rate Sensitivity with Large Deformations," *J. Appl. Mech.*, **46**, 811–816 (1979).

26. N. Perrone and P. Bhadra, "Simplified Large Deflection Mode Solutions for Impulsively Loaded Viscoplastic Circular Membranes," *J. Appl. Mech.*, **51**, 505–509 (1984).

CHAPTER 6

Dynamic Rupture of Shells

T. A. Duffey

APTEK, Inc.

Albuquerque, New Mexico

ABSTRACT

The relative lack of and need for suitable failure criteria for dynamically loaded shell structures composed of ductile elastic–plastic materials are discussed. Simplified methods of estimating ductile failure of dynamically loaded shells are proposed, using an in-plane strain-based criterion, as well as a transverse-shear failure criterion at supports (hard points). The inadequacy of using equivalent plastic strain as a failure criterion is demonstrated. The hard-point shear failure criterion is suggested to be reasonably insensitive to geometry: hard-point shear failure in cylindrical shells is found to be adequately predicted by a theory for shear failure in beams developed earlier by Jones. By using this theory as an approximation for cylindrical shells, the role of pressure-pulse loading details as well as the influence of support flexibility are examined.

1. INTRODUCTION

A large number of structural computer codes are currently available to calculate the linear or nonlinear, static or dynamic, response of shells or solids subjected to a variety of loading types [1–3]. Many of these codes have been compared with experimental data or exact solutions in the dynamic elastic–plastic range; and it has been demonstrated that, for complex geometries, materials, and loadings, the dynamic elastic–plastic response can be rather precisely calculated [4, 5].

In fact, the computer program user must often decide between a number of available nonlinear material behavior options (e.g., kinematic or isotropic hardening, etc.) in utilizing these codes. Calculation of plastic structural-response details, except in the most complicated cases, is at this point probably limited only by the availability of biaxial yield, strain hardening, and strain-rate-sensitivity data; and there are now available a wealth of papers providing such data for many engineering metals [6]. Prediction of the point of structural failure by material separation, however, can be a more formidable task.

The situation is similar with analytical and approximate solutions for dynamic plastic behavior of structures. Whereas an impressive number of solutions are available for predicting the dynamic plastic response [7], prediction of actual failure by material separation is less straightforward [8].

Intensely loaded structures are often constructed of ductile materials and are under stress states that do not lead to brittle failure. Thus, particularly for thin shells, considerable plasticity can occur before failure by material separation, so that the concepts of classical brittle fracture mechanics are not applicable for failure prediction. Rather, failure can occur in a tensile-stress field by the formation of a local plastic instability, for example, followed by material separation due to some mechanism such as the growth and coalescence of voids in the material [9]. Or shear failure can occur along boundaries of the shell, such as at hard points.

Surprisingly, given the large number of studies reported on biaxial plasticity (by illustration, a survey of the subject in Reference 6 contains 277 references), relatively little attention has been focused on ductile-material failure under multiaxial loading states. This relative lack of multiaxial-material-failure information compared to multiaxial-plasticity information is readily apparent in material-constitutive models currently available in elastic–plastic computer programs [2, 3]. Further, in the few dynamic plastic structural computer programs in which a failure criterion is included [10], the criterion is typically based upon equating the equivalent plastic strain to the failure strain in simple tension.

A convenient categorization scheme has been previously developed for failure modes of beams (see Figure 1) and provides some insight into general shell failure. As first observed experimentally by Menkes and Opat [11], the failure response of the beam due to the sudden intense loading falls into at least one of the following three modes:

Mode I: Large inelastic deformation
Mode II: Tearing (tensile failure) of outer fibers at or over the support
Mode III: Transverse-shear failure at the supports

While all three modes are interpreted in Reference 11 as failure, only modes II and III result in actual material separation. Mode I is considered a failure mode in Reference 11 in the sense that plastic deformations are so large that the structure has been rendered incapable of performing its intended function.

Presence of a given failure mode is dependent upon the impulse or initial velocity magnitude. As described in Reference 11, mode I occurs at lower impulse levels, when residual central deflection of the beam reaches or exceeds some prescribed value. The mode-II threshold occurs next, and is the impulse (or velocity) that first causes tearing.

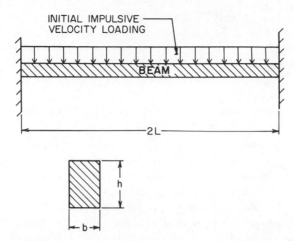

Figure 1. Impulse-loaded fully clamped beam.

Finally, at a higher impulse level, a pure well-defined shear failure occurs (mode III), with no significant deformation in the severed central section of the beam. Note that between threshold impulses for modes II and III, a combination of modes II and III characterize failure (shearing and tearing) in beams.

Since the emphasis in this chapter is failure by material separation, only modes II and III are discussed. Regarding mode II, Jones [12] successfully bounded the experimental beam data [11] for tearing of outer fibers at the supports by analytically determining the maximum strain occurring at the supports and equating this strain to the uniaxial failure strain of the ductile beams tested. As discussed in Appendix I, this is effectively a plastic tensile instability criterion of failure.

Now for loadings other than spatially uniform, or for other structural configurations, it may well happen that maximum strains occur at locations away from the boundaries. Thus, it appears important to monitor the strain field throughout the structure, using a multiaxial in-plane strain-based criterion of failure. The onset of failure is most probably associated with strain localization.

Regarding mode III, transverse-shear failure at supports, Jones [12] found that the threshold velocity necessary for the onset of complete through-shearing in beams with rectangular cross sections is constant for a given material, independent of length and beam thickness. It was also found analytically that the plastic deformation is localized near the supports, with the major central portion of the beam remaining straight. Indeed, these predictions [12] were generally observed in beam experiments [11]. Because deformations are so localized at supports and threshold velocity is independent of the beam geometry suggest that these mode-III beam results can apply, at least approximately, to hard-point shear failure in other shell structures.

In the cylindrical-shell test series reported in Reference 13, mode II was found not to exist, failure proceeding from large inelastic deformation directly to pure shear failure at the supports. The authors did acknowledge, however, that this observation may depend on the range in cylinders tested. Note that mode-III damage thresholds were found experimentally to be somewhat lower for cylinders than for beams [13], and it is emphasized in Reference 13 that conservative estimates for transverse-shear thresholds can be made for cylinders by experimenting with beams.

It is the purpose of this chapter to propose an approximate approach for predicting the onset of ductile failure in thin-shell structures. The approach consists of two phases:

1. Examining the calculated biaxial strain field in the plane of the shell as a function of time to determine if a limiting strain surface is reached. Such a surface can indicate, for instance, the onset of a tensile instability-type failure. This type of biaxial tensile failure can occur either in the vicinity of boundaries or hard points of the shell or in regions distant from such boundaries. It can be interpreted as a generalization of mode-II failure proposed initially in Reference 11.

2. Examining for the presence of shear failure at hard points of the structure. This is the mode-III failure also proposed in Reference 11.

First, the in-plane strain-based-failure criterion is discussed. Then the shear-failure mode at hard points is presented. Finally, the roles played by pressure-pulse details and support flexibility on the shear-failure mode are evaluated.

2.   BIAXIAL DUCTILE STRAIN LOCALIZATION AND FAILURE: MODE II

Strain localization and failure of uniaxially loaded specimens are discussed in Appendix I. While the concept of uniaxial ductile tensile instability has been understood for many years [14], work on instability of geometrical shapes loaded under a state of biaxial stress is of much more recent origin [15, 16]. Swift [15] examined the conditions for instability of plastic strain under plane stress for various biaxial stress ratios and conditions of applied loading, bringing out the important role played by strain hardening. Hill [16] examined the plastic bulging of a metal diaphragm loaded by lateral pressure (the bulge-test geometry). He pointed out differences in instability strains of different geometrical and loading situations.

In the above papers, loss of stability is assumed to take place when an increment in strain occurs with no simultaneous increase in the pressure or load (analogous to the load-prescribed situation in Appendix I), i.e., loss of stability of the shell as a whole. More recently, Marciniak and Kuczynski [17] analyzed the loss of stability (localization of strains) for sheet metal subjected to a range in biaxial tension ratios. For sheet-metal processing, they conclude that fracture depends only on a local discontinuity or concentration of strains (more analogous to the displacement-prescribed situation in Appendix I) rather than on loss of stability of the shell as a whole. Again, the importance of the strain-hardening exponent (as well as other parameters, particularly homogeneity) on loss of stability and limiting strain are demonstrated.

McClintock [9] developed a criterion for ductile fracture by the growth and coalescence of cylindrical holes in ductile metals, which provides an upper limit on ductility. Localization of this deformation just preceding fracture into shear bands was subsequently analyzed by Rudnicki and Rice [18], Anand and Spitzig [19], and others; and Theocaris [20] and Wilson and Acselrad [21] summarize efforts to date in the development of mathematical models and experimental observations of strain localization and subsequent ductile failure due to void-coalescence mechanisms.

Particular motivation for understanding and obtaining biaxial plastic tensile instability information was provided by the sheet-metal forming industry. A number of experimental studies have been performed on thin-sheet metals to determine their limiting strains at the onset of localized necking (beyond which material failure by other mechanisms soon occur).

The forming-limit diagram, as illustrated in Figure 2, is an experimentally determined curve in principal strain space for the locus of points representing the onset of localized necking, obtained, for example, by stretching an initially flat sheet specimen with a hemispherical punch [22]. Strains are monitored by measuring stretching of major and minor diameters of small circles that have been previously photoetched onto the surface of the sheet. Different strain ratios are achieved by using different lubricants between punch and plate; and by using less than a full-width plate.

Other techniques used for obtaining such forming-limit information include in-plane sheet stretching and hydraulic bulging; and a survey and comparison of data by different biaxial loading techniques is presented in Reference 23.

Limited testing to failure using thin-walled tubular specimens subjected to a combination of tension, torsion, and internal pressure has also been reported [24]. These data are also presented in the form of a failure curve in principal strain space, as shown, for example, for 6A1-6V-2Sn in Figure 3. As might be anticipated based on the uniaxial loading case in Appendix I, both forming-limit data and the thin-walled tubular data are highly dependent on the degree of strain-hardening.

In view of the similarity in shapes of Figures 2 and 3, it is interesting to determine if in fact the strain-based failure curves correspond to a general failure curve in stress space. Two such possibilities are examined, drawing in part upon a methodology outlined in Reference 24.

2.1. Von Mises Loading Curve

Assuming the final loading curve prior to failure in stress space is given by the von Mises ellipse:

$$\sigma_u^2 = \sigma_1^2 + \sigma_2^2 - \sigma_1\sigma_2 \tag{1}$$

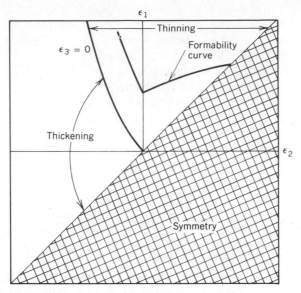

Figure 2. Typical forming-limit diagram.

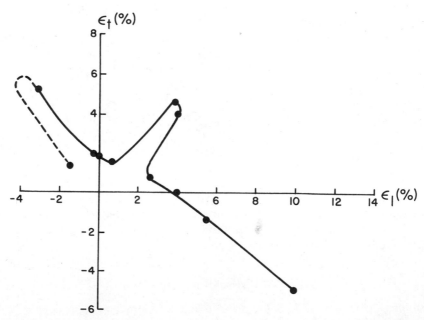

Figure 3. Titanium biaxial strain failure data from tension–torsion machine (data from Reference 24).

where σ_u is the effective stress at failure, a set of coordinate pairs along the ellipse $[\sigma_1(1), \sigma_2(1)], \ldots, [\sigma_1(n), \sigma_2(n)]$ are selected. Effective stress is next calculated using

$$\sigma_{\text{eff}} = (\sigma_1^2 - \sigma_1\sigma_2 + \sigma_2^2)^{1/2} \tag{2}$$

An equivalent stress–strain curve is assumed of the form

$$\sigma_{\text{eff}} = A\epsilon_{\text{eff}}^n \tag{3}$$

where the effective strain is

$$\epsilon_{\text{eff}} = (\epsilon_1^2 + \epsilon_1\epsilon_2 + \epsilon_2^2)^{1/2} \tag{4}$$

Effective strain is calculated using Equation (3). Finally, the strain components, ϵ_1 and ϵ_2, are determined using the Levy–Mises equations in integrated form for the condition of plane stress [25]:

$$\frac{\epsilon_1}{2\sigma_1 - \sigma_2} = \frac{\epsilon_2}{2\sigma_2 - \sigma_1} = \frac{\epsilon_1 + \epsilon_2}{\sigma_1 + \sigma_2} \tag{5}$$

Here a proportional stress state is assumed during loading up to the subsequent loading surface at which failure occurs.

The von Mises elliptical failure surface, when so mapped to strain space, is shown in Figure 4 for 6061-T6 aluminum.

2.2. Tresca Loading Curve

Now consider the final loading curve in stress space prior to failure as given by the Tresca Law. In the case of a biaxial stress state, $\sigma_3 = 0$, then in the positive–positive quadrant of the σ_1–σ_2 plane, the final loading curve prior to failure in stress space is given by

$$\sigma_1 = \sigma_u \qquad \sigma_1 > \sigma_2 > 0$$
$$\sigma_2 = \sigma_u \qquad \sigma_2 > \sigma_1 > 0 \tag{6}$$

(This is identical to the maximum principal-stress criterion in the positive–positive quadrant of stress space.)

The procedure for mapping this surface to strain space is similar to that just presented:

1. A set of coordinate pairs $[\sigma_1(1), \sigma_2(1)], \ldots, [\sigma_1(n), \sigma_2(n)]$ is selected that lie on the assumed failure surface given by Equation (6).
2. Effective stress for each coordinate pair is calculated using Equation (2).

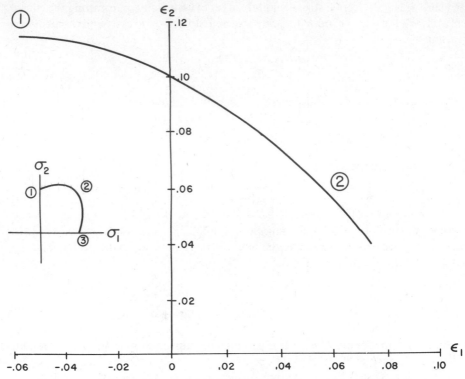

Figure 4. Von Mises failure surface for 6061-T6 aluminum mapped to strain space.

3. Effective strain is calculated using Equation (3).
4. Equation (4) then represents one equation in two unknowns, ϵ_1 and ϵ_2.
5. Equation (5), the Levy–Mises plastic stress-strain relations, provide the second equation in terms of ϵ_1 and ϵ_2 for each set.

Note that steps 2–5 are approximate since the plastic stress-strain relations and effective stress–effective strain definitions technically are associated with the von Mises criterion.

The results for one numerical example are plotted in Figure 5. Again, 6061-T6 aluminum is used, for which $n = 0.1$ and $A = 50,357$ psi.

2.3. Comparisons with Experimental Data

It is interesting to compare the general shapes of Figures 4 and 5 with experimentally measured biaxial strain-failure data. For the tension–torsion failure curve of Figure 3, failure occurs at rather low strain levels. The general shape of the failure curve is in excellent qualitative agreement with that of Figure 5 (Tresca or maximum principal-stress theory of failure), but bears little

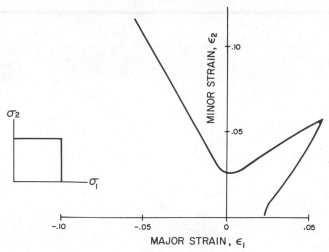

Figure 5. Tresca failure surface for 6061-T6 aluminum mapped to strain space using nonassociated flow rule.

resemblance to the von Mises shape in Figure 4. Additional biaxial strain-failure data is shown in Figure 6, from Reference 26. This data is at considerably higher failure strain levels and is based upon formability-type experiments, where a spherical punch is driven transversely into an initially flat clamped blank. Again, qualitative agreement of the general shape with the Tresca-type failure strain surface (Figure 5) is good, with the possible exception of the 304 stainless-steel data.

Mechanisms of failure in Figures 3 and 6 can be somewhat different in view of the large differences in failure strain ranges and the different test techniques employed. This lends some credence that ductile failure can be generally associated with a maximum principal-stress-failure theory or a Tresca-shaped theory (the two appear the same in the positive–positive quadrant of stress space).

Note that a similar mapping from stress space to strain space is performed in Reference 24 with similar results: the Tresca shape is in much better qualitative agreement than the von Mises shape.

Implications for a failure curve associated with the von Mises shape are that failure occurs when the distortion energy equals that corresponding to failure in simple tension. Effective stress, Equation (2), and effective strain, Equation (4), are constant at failure, independent of the biaxial stress or strain ratio during proportional loading. While these concepts may have strong intuitive appeal for describing ductile failure under multiaxial loading states, in fact, this von Mises elliptical shape simply does not adequately represent actual failure strain behavior.

Finally, it should be mentioned that the just-discussed formability and tension–torsion data are limited to quasi-static or low strain rates. Based on

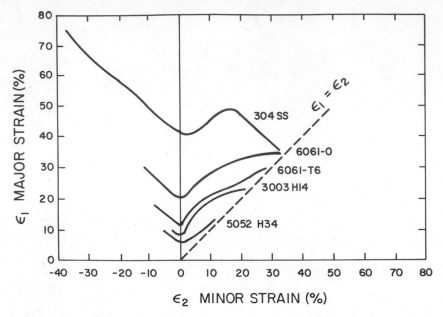

Figure 6. Formability curves for four aluminum alloys and one stainless-steel alloy (from Reference 26).

dome-bulging data [27] obtained from high-energy-rate forming using explosives, ductility does depend somewhat upon forming velocity, or strain rate. These variations are likely due to variations in the stability of the elongation. This stability (see Appendix I) is influenced by dependence of both flow stress and strain-hardening slope on strain rate. Further, the influence of strain-rate sensitivity on the forming-limit diagram is analyzed in Reference 28. It is found that an increase in exponent m, expressing strain-rate sensitivity of the material, affects the limit strains in much the same way as that of an increase in exponent n, representing strain-hardening.

3.   SHEAR FAILURE AT SUPPORTS: MODE III

Based on data from References 11–13, an explicit comparison of critical shear-failure thresholds for beams and cylindrical shells is presented for four combinations of cylinder diameter and hard-point separation distance in Figures 7–10. In the figures, critical specific impulse to cause shear failure is plotted as a function of structure thickness.

It can be seen that the beam experimental data reasonably approximates and generally bounds the cylindrical shell data from above. Thus, use of Reference 12 for shear-failure predictions of both beams and cylindrical shells seems appropriate, particularly over the wall thickness range of 0.187–0.375 in. Differences in

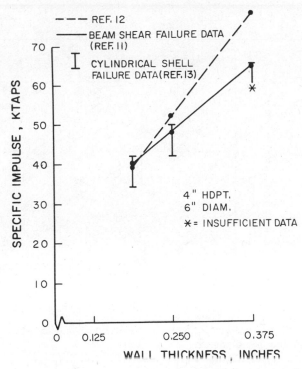

Figure 7. Critical impulse for shear failure in beams and cylindrical shells for 4 in. hardpoint separation and 6 in. diameter.

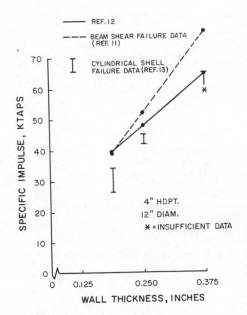

Figure 8. Critical impulse for shear failure in beams and cylindrical shells for 4 in. hardpoint separation and 12 in. diameter.

171

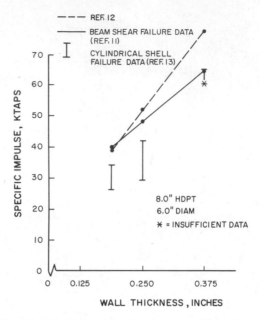

Figure 9. Critical impulse for shear failure in beams and cylindrical shells for 8 in. hard-point separation and 6 in. diameter.

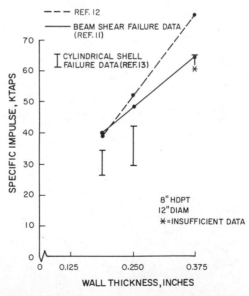

Figure 10. Critical impulse for shear failure in beams and cylindrical shells for 8 in. hard-point separation and 12 in. diameter.

slope between beam shear data and Equation (23) [12] may be due to decreased ductility or other material-property differences in the thicker sections.

It is particularly important to note that the definitions of shear thresholds for beams and for cylindrical shells in Refs. 11 and 13 differed somewhat, the shear threshold in beams being considerably more subjective than for cylinders.

Based on the reasonable agreement between cylindrical-shell experimentally observed failure levels and both theoretically predicted and experimentally observed failure levels for beams, it is here proposed that structural details are of lesser importance for the hard-point shear-failure mechanism; and that cylindrical-shell hard-point failure can be adequately approximated by a beam model for failure predictions. Energy arguments are presented in Appendix II, which further support use of a beam model at hard points in cylindrical shells. If this were not the case, a more complicated cylindrical-shell analysis would be necessary [29, 30].

4. INFLUENCE OF PULSE DURATION ON HARD-POINT SHEAR FAILURE

Based on comparisons in the previous section, it appears that critical impulse levels for cylindrical shells (and possibly other structures) can be conservatively bounded using beam-shear-failure analysis or experiments [11-13]. However, References 11–13 are restricted to the case in which loading can be idealized as purely impulsive, i.e., as a delta function in time. The purpose of this section is to extend the analysis of Reference 12 to evaluate the influence of pressure-pulse details on the hard-point shear failure of beams, keeping in mind that beam-failure analysis and experiments were found to bound cylindrical-shell failure from above.

4.1. Rectangular Pressure Pulse

Consider a strip of material along the crown of a cylindrical shell as modeled by the fully clamped beam subjected to a spatially uniform pressure pulse depicted in Figure 11. The pressure pulse is taken for simplicity as rectangular in time. The beam is composed of a rigid perfectly plastic material.

Jones [12] implies from the elementary bending-only solution to the rigid-plastic-beam problem of Figure 11 for the closely related case of ideal impulse loading that the transverse shear force in the beam reaches a maximum value at the supports. Therefore, shear sliding (plastic shear hinge) would be expected at the supports of beams of finite shear strength Q_0. The velocity field, taken from Reference 12, is shown in Figure 12, where plastic bending hinges form at A, B, C, and D, in addition to plastic shear hinges at A and D. As proposed in

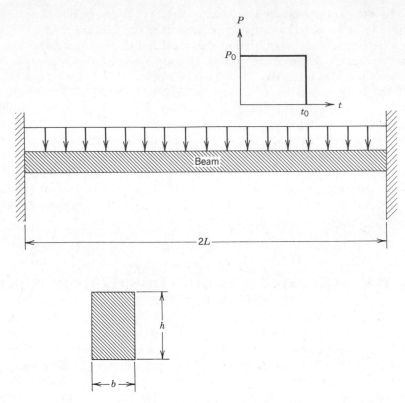

Figure 11. Fully clamped beam subjected to spatially uniform rectangular pressure pulse.

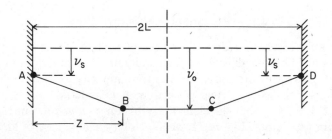

Figure 12. Velocity field for shear sliding at the supports (after Reference 12).

174

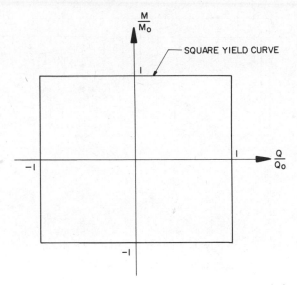

Figure 13. Square interaction curve for moment and shear.

Reference 12, a square interaction curve for moment and shear is assumed, as shown in Figure 13.†

Phase I: Shearing During Application of Pressure Pulse.   The free-body diagram of rigid link AB (of unknown length Z) is shown in Figure 14, which is identical to that of Jones [12] except that uniform pressure-pulse loading is applied. Note that Figure 14 represents phase I, during which the external pressure pulse is still acting.

Applying Newton's second law for center of mass (cm) motion of link AB,

$$-Q_0 + P_0 bZ = \rho h b Z \ddot{y}_{cm} \tag{7}$$

where ρ is the mass density, b is the beam width, h is the beam depth, and y_{cm} is the displacement of the center of mass.

Integrating Equation (7) with respect to time over the duration of the pressure pulse, and noting that the initial velocity of the beam is zero, the following results:

$$-Q_0 t_0 + P_0 bZt_0 = \rho h b Z (V_0)_{cm} \tag{8}$$

---

†As discussed in Reference 12, this square interaction curve circumscribes a more complicated interaction curve proposed by Drucker [31]. Results for the corresponding inscribing interaction curve utilized in Reference 12 are not presented here, but can be obtained by replacing the flow stress σ_0 by $0.7245\sigma_0$.

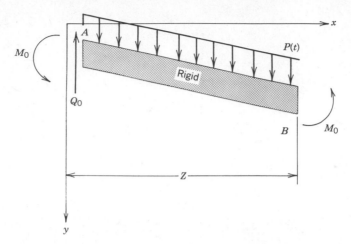

Figure 14. Free-body diagram of rigid link AB during phase I.

where $(V_0)_{cm}$ is the center-of-mass velocity of link AB. It is also assumed that link length Z is independent of time.

From kinematics,

$$(V_0)_{cm} = \frac{V_A + V_B}{2} \tag{9}$$

where V_A and V_B denote the velocity of end A and end B, respectively. Since the velocity of end B takes on the velocity of the center section of the beam (with length $2L^* = 2L - 2Z$), V_B is determined from the center-section velocity. By applying Newton's second law for the center section, the velocity of end B at time t_0, when the pulse ceases is

$$V_B = \frac{P_0}{\rho h} t_0 \tag{10}$$

Combining Equations (8)–(10),

$$(P_0 bZ - Q_0)t_0 = \rho hbZ \left(\frac{V_A(t_0)}{2} + \frac{P_0 t_0}{2\rho h} \right) \tag{11}$$

Next, applying Euler's equation about the center of mass of link AB and integrating with time:

$$(-2M_0 + Q_0 Z/2)t_0 = (1/12)\rho bhZ^3 w \tag{12}$$

where $w = (V_B - V_A(t_0))/Z$.

Combining Equations (10) and (12) results in

$$\left(-2M_0 + Q_0 \frac{Z}{2}\right) t_0 = \frac{1}{12} \rho bhZ^2 \left(\frac{P_0}{\rho h} t_0 - V_A(t_0)\right) \tag{13}$$

Equations (11) and (13) represent two equations in two unknowns, Z and $V_A(t_0)$. Their solution is

$$Z = 6M_0/Q_0 \tag{14}$$

$$V_A(t_0) = \left(\frac{P_0}{\rho h} - \frac{2Q_0}{\rho hbZ}\right) t_0 \tag{15}$$

First, considering Equation (14), this length of link AB agrees with that in Reference 12. As shown in Reference 12, using $Q_0 = \sigma_0 bh/3^{1/2}$ and $M_0 = \sigma_0 bh^2/4$, where σ_0 is the uniaxial flow stress, then

$$Z \simeq 2.6h \tag{16}$$

Thus, Z is very short, and is time-independent, which provides some justification for the assumption following Equation (8).

Equation (15) represents the velocity of end A, at the shear hinge, at the instant t_0 at which the pressure pulse ends. Inspection of Equation (15) reveals that the velocity at the shear hinge is linearly proportional to time (as long as the pressure pulse is acting). Note also that there is the following requirement on the pressure magnitude:

$$P_0 > 2Q_0/bZ \tag{17}$$

Using Equation (16) and $Q_0 = \sigma_0 bh/3^{1/2}$, this requirement becomes:

$$P_0 > \frac{4}{9}\sigma_0 \tag{18}$$

No yielding at the shear hinge occurs for a pressure below $(4/9)\sigma_0$.

At this point, a determination can be made if shear failure occurs while the pressure pulse is still acting. Failure is defined [12] as occurring when the shear hinge has displaced an amount equal to the thickness of the beam.† Assuming

---

†An alternate definition would be that failure occurs when the shear hinge has displaced an amount $kh, 0 < k \leqslant 1$, the argument being that failure occurs before complete through-shearing is reached due to the weakened condition of the partially sheared beam. This approach is used in Reference 32, where it is found that for the somewhat related but more complicated analysis of impact-loaded beams, transverse-shear failure predictions agree with experimental results for considerably smaller values of k. However, in view of the good agreement with experiments on impulse-loaded beams in Reference 12 using $k = 1$, it would seem unnecessary to introduce such a parameter here.

failure occurs at some time, $t_f \leqslant t_0$, the failure condition is

$$\int_0^{t_f} V_A \, dt = h \tag{19}$$

Integrating Equation (19) using Equation (15) (with t replacing t_0), the failure time can be shown to be

$$t_f = h \left(\frac{2\rho}{P_0 - \frac{4}{9}\sigma_0} \right)^{1/2} \tag{20}$$

Given that the requirement of Equation (18) is met, if t_f is found to be less than or equal to the pulse duration t_0, then failure in shear occurs before the pulse ends. Note that Equation (20) does not apply if $t_f > t_0$.

Phase II: Shearing Following Application of Pressure Pulse. If $t_f > t_0$, as determined by Equation (20), then failure does not occur in shear during Phase I. Failure then may or may not occur in Phase II.

In Phase II, $t > t_0$, and there is no further external pressure-pulse loading. The new free-body diagram of rigid link AB is given by Figure 14, with the pressure loading $p(t)$ omitted. Again, applying Newton's second law for rigid body AB,

$$-Q_0 = \rho h b Z \ddot{y}_{cm} \tag{21}$$

Integrating with respect to time,

$$\int_{t_0}^{t_s} (-Q_0) \, dt = \rho h b Z \left[(V_s)_{cm} - (V_0)_{cm} \right] \tag{22}$$

where

$$(V_0)_{cm} = \frac{V_A(t_0) + V_B(t_0)}{2} \tag{23}$$

Combining Equations (10), (15), (16), and (23), and the definition for Q_0 following Equation (15),

$$(V_0)_{cm} = \frac{t_0}{\rho h} \left(P_0 - \frac{2\sigma_0}{9} \right) \tag{24}$$

Referring again to Equation (22),

$$(V_s)_{cm} = \frac{1}{2} \left(V_A(t_s) + \frac{P_0 t_0}{\rho h} \right) \tag{25}$$

Combining Equations (22), (24), and (25) results in

$$V_A(t_s) = \frac{P_0 t_0}{\rho h} - \frac{4\sigma_0 t_f}{9\rho h} \qquad (26)$$

The time t_f^* at which the velocity of the shear hinge goes to zero is determined from Equation (26) to be

$$t_f^* = \frac{9}{4}\frac{P_0}{\sigma_0} t_0 \qquad (27)$$

Again, a determination can be made if shear failure occurs in this second response phase. Failure again is said to occur when the shear hinge has displaced an amount equal to the thickness of the beam. This failure condition is

$$\int_0^{t_0} V_A\, dt + \int_{t_0}^{t_f^*} V_A\, dt = h \qquad (28)$$

Using Equations (15) (with t replacing t_0) and (26), the failure condition stated in Equation (28) can be expressed in the following normalized form:

$$\bar{I}' = \frac{\bar{P}}{9\bar{P} - 4} \qquad (29)$$

where $\bar{I} = I_0/2h(2\rho\sigma_0)^{1/2}$, $\bar{P} = P_0/\sigma_0$, and $I_0 = P_0 t_0$.

Equation (29), which expresses the relationship between normalized pressure $\bar{P}$ and normalized impulse $\bar{I}$ for the case of marginal shear failure, is plotted in Figure 15. This curve is an isodamage curve for a rectangular pressure-pulse loading. For large values of pressure, failure is seen to be insensitive to the normalized specific impulse; whereas for large values of impulse, failure is pressure-insensitive.

Also shown in Figure 15 is the requirement of Equation (18). Yielding at the rigid-plastic shear hinge does not occur for applied pressures less than $(4/9)\sigma_0$, no matter how high the total impulse.

4.2. Exponential Pressure Pulse

Using similar assumptions, the shear-failure response to the exponential pressure pulse $P(t) = P_0 e^{-t/T_0}$ is analyzed here.

Applying Newton's second law for center-of-mass motion of link AB, Figure 14, with the exponential pressure pulse,

$$-Q_0 + P_0 e^{-t/T_0} bZ = \rho h b Z \ddot{y}_{cm} \qquad (30)$$

Integrating Equation (30) with respect to time and noting that the initial

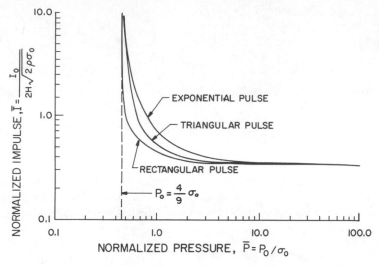

Figure 15. Comparison of isodamage curves for three pulse shapes.

velocity of the center of mass is zero, the following equation results:

$$-Q_0 t + P_0 b Z T_0 (1 - e^{-t/T_0}) = \rho h b Z V_{cm} \tag{31}$$

where $V_{cm} = (V_A + V_B)/2$ is the velocity of the center of mass at time t.

Since the velocity of end B is the same as the velocity of the center section of the beam (with length $2L^* = 2L - 2Z$), Newton's second law is applied to the center section to determine V_B:

$$\rho h b L^* \ddot{y}_B = P_0 e^{-t/T_0} b L^* \tag{32}$$

Integrating, V_B becomes

$$V_B(t) = \frac{P_0 T_0}{\rho h} (1 - e^{-t/T_0}) \tag{33}$$

Combining Equations (31) and (33),

$$V_A(t) = \frac{2}{\rho h b Z} \left[-Q_0 t + \frac{P_0 b Z T_0}{2} (1 - e^{-t/T_0}) \right] \tag{34}$$

Euler's equation remains unchanged from the previous section and is given by Equation (12). Solving Equations (34) and (12) simultaneously using Equation (33) results in

$$Z = 6 M_0/Q_0 \tag{35}$$

which agrees with Equations (14) and (16). Also, using the definitions

$$Q_0 = \sigma_0 bh/3^{1/2} \quad \text{and} \quad M_0 = \sigma_0 bh^2/4 \tag{36}$$

$$V_A = \frac{P_0 T_0}{\rho h} (1 - e^{-t/T_0}) - \frac{4\sigma_0}{9\rho h} t \tag{37}$$

Next, the time t^* at which the velocity at the shear hinge V_A approaches zero is determined using Equation (37). It is given by the following transcendental equation:

$$t^* = \frac{9\bar{P}T_0}{4} (1 - e^{-t^*/T_0}) \tag{38}$$

where $\bar{P} = P_0/\sigma_0$. Therefore, time t^* depends only on $\bar{P}$ and T_0.

The limiting condition for shear failure at the plastic shear hinge is given by

$$\int_0^{t^*} V_A(t)\,dt = h \tag{39}$$

Integrating using Equation (37), the condition for shear failure to just occur is given by

$$\frac{\rho h^2}{\sigma_0 T_0} = \left(\bar{P} - \frac{4}{9}\right) t^* - \frac{2}{9} \frac{t^{*2}}{T_0} \tag{40}$$

where t^* is given by Equation (38). Equations (38) and (40) define a pressure-impulse failure condition analogous to that of Equation (29) for a rectangular pulse. Noting that the total specific impulse for the exponential pressure pulse shown in Figure 15 is $I_0 = P_0 T_0$, Equations (38) and (40) have been evaluated numerically to determine an isodamage curve corresponding to an exponential pressure pulse. Results are presented in Figure 15. Results for an initially peaked triangular pressure pulse are also included in the figure.

Note that for short-duration loadings (large $\bar{P}$ in Figure 15), all curves in the figure approach the ideal impulse solution of Jones [12]. Further, inspection of the figure reveals that the specific impulse necessary to cause shear failure increases monotonically with pulse duration.

5. A COMPARISON WITH RELATED EXPERIMENTS

Experiments performed on cylindrical shells subjected to finite-duration blast loading are reported in Reference 33. While the emphasis of the work was buckling response, a few of the shells underwent the type of catastrophic failure of interest herein: actual material separation. Several simple single-layer cylindrical shells are reported that *appeared* to have failed in a mode associated with shear at the ends (hard points). These experimental data are compared in Figure 16 to the mode-III beam-shear-failure theory, revealing the following:

1. In all cases, the normalized pressures for the cylindrical-shell data lie well below the lower limit of $P_0/\sigma_0 = 4/9$ for pure shear failure to occur (in beams). Thus, pure shear failure would *not* be expected to occur.

2. The cylindrical-shell data shows a dependence on radius-to-thickness ratio a/h as well as material type. In particular, failure impulse-pressure combinations move upward and to the right in Figure 16 for decreasing radius-to-thickness ratios. Similar trends can be seen for the 6061-T6 aluminum and the AZ31B magnesium. If cylindrical-shell failure were predicted by mode-III beam-shear theory, all the data presented should collapse to a single curve for all radius-to-thickness ratios and materials in view of the normalization of the impulse and pressure values.

Keeping in mind the limited quantity of data available as well as the fact that the data points represent a level of failure beyond the point of imminent failure, it is clear that mode-III beam-shear-failure theory does not adequately predict the failures reported in Reference 33. This is consistent with the fact that experimental peak pressures, when normalized with the yield stress, lie below the range of applicability of the pure shear (mode-III) theory. It is believed that the failures reported in Reference 33 fall rather into a combined mode II–mode III category of failure (combined tearing or tensile failure of the outer fibers at the support along with shear failure at the support). Some supporting evidence is

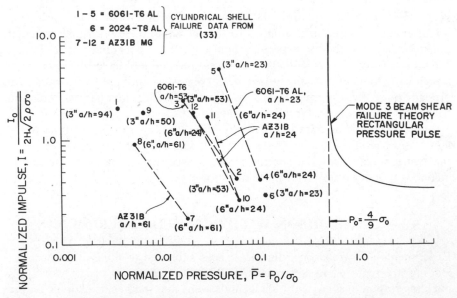

Figure 16. Comparison to failure data from Reference 33.

provided by photographs of the failed shells in Reference 33. These photographs imply more stretching before failure than the shells in Reference 13. Also, the fact that failure occurs at one end only on some shells in Reference 33 strongly implies that the rigid-link shear mechanism of failure modeled using a beam does not apply at these lower pressures below $(4/9)\sigma_0$. Failure appears to occur later in the response of the shell when normal forces have built up sufficiently to play a major role in failure. Unfortunately, no data appears to be available for pressures greater than $(4/9)\sigma_0$, for which the mode-III predictions would apply.

6. ROLE OF FLEXIBLE END STIFFENERS ON HARD-POINT SHEAR FAILURE

In practice, end constraints and stiffeners for cylindrical shells may not be perfectly rigid. It is the purpose of this section to approximately evaluate the role of hard-point flexibility on the shear-failure mechanism, again drawing upon similarities of shell- and beam-shear-failure levels.

As shown in Figure 17, a half-cosine impulse-loaded cylindrical shell with partial end constraint is modeled for shear failure as a beam subjected to uniform loading. The effect of the partial constraint is modeled by a stiff spring of spring constant K at each end of the beam. It is further assumed that only vertical deflections occur at the ends. Both ends of the beam are fixed against rotations. Thus, the ends of the beam deflect an amount Q_0/K, since the frequency response of the support is assumed to be much higher than the beam response to the impulse loading.

Utilizing the free-body diagram of Figure 14, but deleting the pressure

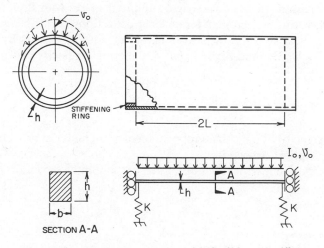

Figure 17. Model for cylinder with flexible end stiffeners.

loading, the integrated equations of motion are

$$V_0 - V_s = \frac{2Q_0 t_s}{\rho h b Z} \tag{41}$$

$$V_0 - V_s = \frac{6Q_0 Z - 24M_0}{\rho b h Z^2} t_s \tag{42}$$

The location of the shear hinge Z is given by Equation (14); and the time at which V_s reaches zero is determined from Equation (19) for $P_0 = 0$ as

$$t_s = \frac{V_0 \rho h b Z}{2Q_0} \tag{43}$$

Equations (41)–(43) are equivalent to those derived in Reference 12.

With flexible supports, shear failure occurs at the shear hinges A and D when point A (or D) travels a distance at least equal to the beam thickness plus the amount of support deflection δ. The limiting case for which severance just occurs with flexible supports is

$$\int_0^{t_s} V_s \, dt = h + \delta = h(1 + \beta) \tag{44}$$

where $\delta = Q_0/K$, K is the equivalent support stiffness, and $\beta = \delta/h$. Integrating Equation (44) using Equations (41)–(43) results in

$$V_0 = \left[\frac{2(1 + \beta)Q_0^2}{3\rho b M_0} \right]^{1/2} \tag{45}$$

Note that Equation (45) for $\beta = 0$ (rigid supports) agrees with Equation (23) of Reference 12. It can be seen that, for stiffener deflections of one beam thickness ($\beta = 1$), the critical velocity for hard-point shear failure increases by 41%.

For $Q_0 = \sigma_0 b h / 3^{1/2}$ and $M_0 = \sigma_0 b h^2/4$, the critical velocity can be written

$$(V_0)_{\text{critical}} = \frac{2(2)^{1/2}}{3} \left[\frac{(1 + \beta)\sigma_0}{\rho} \right]^{1/2} \tag{46}$$

The critical velocity with flexible supports depends on material properties (ρ, σ_0) and relative support deflection.

The corresponding critical impulse for shear failure at flexible supports is

$$(I_0)_{\text{critical}} = \frac{2(2)^{1/2}}{3} [\rho(1 + \beta)\sigma_0]^{1/2} h \tag{47}$$

Critical impulse curves for hard-point shear failure for cases of support

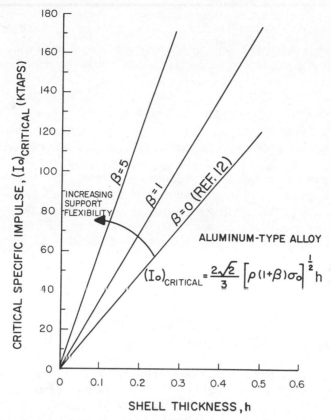

Figure 18. Influence of support flexibility on critical impulse.

deflections of $\beta = 0$, 1, and 5 are shown in Figure 18 for a hypothetical aluminum material. As one might predict, the more flexible the support, the higher the critical velocity and impulse necessary to cause failure at the support.

Again, it is assumed here that the support system responds statically to the applied shear load Q_0, i.e., that the response time of the support is much shorter than the response time of the beam.

7. SUMMARY OF RESULTS AND CONCLUSIONS

1. Availability of ductile-failure criteria in structural/shell computer programs is extremely limited. When available, it usually consists of defining failure as occurring when the effective plastic strain reaches the uniaxial failure strain value. This is in contrast to the universal and refined treatment of plasticity in these computer programs. A similar lack of failure criteria exists in

analytical and approximate approaches to ductile-plastic analyses of structures/shells.

2. It appears that ductile failure of dynamically loaded shells can be initiated as the result of a tensile-type failure (e.g., plastic tensile instability) either away from or in the vicinity of boundaries or hard points; or by shear failure at hard points. These types of failure are generalizations of the mode-II and mode-III failure mechanisms originally proposed for beams.

3. Forming-limit curves in principal strain space developed by the sheet-metal industry are available to predict the onset of localized necking instability for a variety of metals. A limited amount of other biaxial ductile failure curves are available as well, obtained by tension–torsion tests on thin-walled cylinders. Such curves appear useful in predicting the onset of mode-II ductile failure in structures. Using this approach, transient strain fields calculated by a structural/shell computer program would be monitored to determine when this in-plane biaxial strain limit is reached, which would indicate the onset of ductile failure in the shell.

4. It is observed, consistent with Reference 24, that the shape of the failure surface in strain space is strongly dependent on the shape in stress space. Further, the Tresca-shaped surface in stress space results in a strain-space biaxial in-plane strain-failure curve that is qualitatively very similar to experimental data, both for small (tension–torsion experiments) and large (formability experiments) strains.

Implications of the comparisons are that ductility, as measured by equivalent strain to failure, is a strong function of the biaxial stress or strain ratio. Use of a ductile failure strain value determined under conditions of uniaxial stress for determining failure under multiaxial loading states, such as might occur in shells, is found to be invalid for several metals examined.

5. It is shown in Appendix I that, beyond the limit load, uniaxially loaded elements or test specimens can exhibit one of two distinct instabilities, depending upon whether the test is one of load control or of displacement control. The instability associated with displacement-controlled loading depends on the size of a weakened region of the specimen. If larger than a certain size, no instability occurs under displacement-controlled loading. The instability associated with the load-controlled loading case (plastic tensile instability) occurs when increased capacity due to strain-hardening is exactly balanced by a decrease in cross-sectional area due to necking. Both instabilities are associated here with strain localization.

6. A detailed comparison of experimental shear-failure data for beams [11] and cylindrical shells [13] is made with beam-shear-failure predictions from Reference 12. It is found that, over the thickness range of the beam and shell experiments, the critical velocity (or impulse) developed in Reference 12 provides reasonable estimates for mode-III shell failure at supports (as well as for beams), generally bounding the shell and beam critical shear-failure impulses from above. Agreement of beam- and shell-shear-failure levels is not surprising in

view of the localized nature of this type of failure. Therefore, there appears to be justification for using the approach of Reference 12 for analyzing shear failure at boundaries and supports of cylindrical (and probably other) shells as well.

7. Since reasonable bounding estimates of shear failure in cylindrical shells are given by beam analysis, the beam-shear-failure analysis presented by Jones [12] was extended to evaluate the influence of pressure-pulse details on hard-point shear failure in structures. Rectangular, exponential, and triangular pressure-pulse shapes were considered. Isodamage curves for the three pulse shapes were then constructed. These curves provide the relationship between normalized impulse and normalized pressure for the case of imminent failure by pure shear at the supports. Finally, it was noted that for pressures less than $(4/9)\sigma_0$, no hard-point shear failure in this mode (mode III—pure shear) occurs. It was found that, for short pulse durations, pulse shape is unimportant; and failure values of impulse approached those for ideal impulse loading in Reference 12. Also, the specific impulse necessary to cause shear failure was found to increase monotonically with pulse duration.

Data from an earlier experimental program on cylindrical shells subjected to finite-duration pressure-pulse loading were located. All of these "shear failure" tests were performed with pressure magnitudes *below* the critical value of $(4/9)\sigma_0$, so that a mode-III hard-point shear failure would not be expected. Examination of photographs of the failed shells suggested that shell failure in that experimental program was not of the pure shearing type (mode III), but rather a combination of tearing and shearing. Additional confirmation was provided by the fact that, when plotted on the nondimensional pressure-impulse failure curve developed herein, the experimental data did not collapse onto a single curve, but rather showed a dependency on radius-to-thickness ratio and material type.

8. Again, the close parallel of shear failure at supports between beams and cylindrical shells is utilized to investigate the role of support flexibility. Based on a simple modification of the shear-failure analysis by Jones [12], it is found that as stiffness of supports (i.e., hard points) is decreased, critical velocity and specific impulse for hard-point shear failure increase. In particular, for a support deflection of one shell thickness, critical velocity and critical impulse both increase 41% over the values for rigid supports.

APPENDIX I
DUCTILE FAILURE IN ONE DIMENSION

LOAD-PRESCRIBED TENSILE TEST

Consider a test specimen subjected to quasi-static uniaxial loading. For a ductile material, the behavior is qualitatively as depicted in Figure 19. Proceeding from left to right in the figure, and noting that load and elongation are the quantities plotted, the initially unloaded sample (A) is loaded elastically to B, the tensile

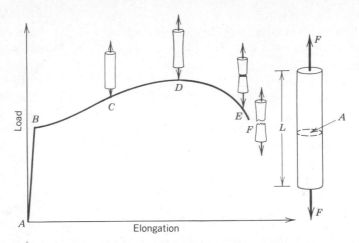

Figure 19.   Tensile loading (load-prescribed) of uniaxial specimen to failure.

yield point of the material. From B to D, the material continues to flow plastically and to strain-harden, carrying increased load. This increased load capacity is counteracted in part by reduction in area caused by thinning of the material in transverse and through-thickness directions. At D, the maximum load-carrying capability due to strain-hardening exactly equals the loss in capacity due to area reduction. Following this point of neutral stability, local necking occurs (E) followed at a later point by complete separation of material (perhaps due to growth and coalescence of voids in the material).

The point at which the instability forms is directly related to the degree of strain-hardening in the material. Strain-hardening effectively spreads out the deformation, preventing strain localization (up to a point). (Strain-rate sensitivity and material inertia have also been shown to influence the instability and failure points [34].)

Formation of this uniaxial instability is well known. It can be shown that, in the absence of inertia and strain-rate sensitivity, the onset of instability is given by [14]

$$d\sigma/d\epsilon = \sigma \qquad (48)$$

At this point of load maximum, the relative decrease in cross-sectional area exactly balances the relative increase in true stress caused by strain-hardening. Assuming the stress–strain curve can be written in the form $\sigma = K\epsilon^n$, then substituting into Equation (48) results in [35]

$$\epsilon^* = n \qquad (49)$$

where ϵ^* is the instability strain. Therefore, for materials with greater strain-hardening (higher n), the instability strain is higher.

DISPLACEMENT-PRESCRIBED TENSILE TEST

A decade ago, Bazant [36] analyzed a displacement-prescribed compressive test performed on a reinforced concrete specimen and illustrated a strain-localization mode of instability of the specimen in the strain-softening range using an energy approach. More recently, Schreyer and Chen [37] utilized equilibrium arguments to illustrate the presence of a similar instability for loading in tension.

Schreyer and Chen [37] consider a specimen of length $L = a + b$, consisting of a full-strength region A of length a, and a weakened region B of length b in series. Both materials harden up to their limit stresses, beyond which they strain-soften with slope β. The limit stress for region B is slightly less than that for region A.

If the stress on the element is such that the strain in region B exceeds the limit state, then strain-softening occurs (while region A unloads). This softening is assumed to occur uniformly over weakened region B.

It is found [37] that if weakened region B, which exhibits strain-softening, is smaller in length than

$$b^* = \beta L/(1 + \beta) \tag{50}$$

then instability occurs. However, if the length of the weakened region is greater than b^*, no instability occurs in a displacement-prescribed test.

The instabilities can both be envisioned on the same engineering stress–engineering strain curve shown in Figure 20. In region I, neither type of instability occurs; in region II, an instability occurs for the load-prescribed test; and in region III, both types of tests are unstable.

The displacement-type and load-type instabilities are intrinsically different: a

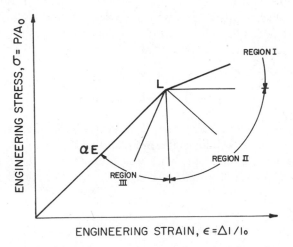

Figure 20.   Engineering stress–strain curve for uniaxial specimen in tension.

size effect is present for the displacement-type instability. For example, for a given cross-section tensile specimen, the longer the specimen, the more prone it is to a displacement-type instability.

In an actual dynamically loaded shell, consisting of an assemblage of such elements, an individual element may not be load- or displacement-controlled, but rather some combination of the two. Thus, it might be anticipated that, for example, formability limits discussed earlier depend to an extent on testing details. In fact, significant differences have been observed between forming-limit diagrams obtained for three metals by the in-plane and the punch-stretching techniques [23]. Further, the role of initial nonhomogeneity of the material on the forming-limit diagram is discussed by Marciniak and Kuczyński [17].

APPENDIX II
JUSTIFICATION OF THE BEAM MODEL

From Reference 38, the rate of energy dissipation D at a given location in the shell can be expressed as

$$D = M^{\alpha\beta}\dot{\kappa}_{\alpha\beta} + N^{\alpha\beta}\dot{\epsilon}_{\alpha\beta} + Q^{\alpha}\dot{\gamma}_{\alpha} \qquad (51)$$

where $M^{\alpha\beta}$, $N^{\alpha\beta}$, and Q^{α} are, respectively, bending, membrane, and transverse-shear-stress resultant components, $\dot{\kappa}_{\alpha\beta}$ are curvature rates, and $\dot{\epsilon}_{\alpha\beta}$ and $\dot{\gamma}_{\alpha}$ are strain rates. At the clamped boundary, Equation (51) [38] reduces to the following equation describing a plate strip or beam:

$$D = M^{xx}\dot{\kappa}_{xx} + N^{xx}\dot{\epsilon}_{xx} + Q^{x}\dot{\gamma}_{xx} \qquad (52)$$

By hypothesizing that those components of stress and strain rate that do not contribute to plastic dissipation will not affect failure [38], then it is not surprising that a beam model appears satisfactory for failure predictions of cylindrical shells near hard points.

ACKNOWLEDGMENTS

The author expresses his appreciation to Professor T. Wierzbicki and Professor N. Jones for numerous helpful suggestions, and to Professor H. L. Schreyer for discussions on instabilities. This work was supported by the Defense Nuclear Agency under Contract DNA 001-85-C-0264. Contract Technical Monitor was Dr. Michael Frankel.

REFERENCES

1. B. Fredriksson and J. Mackerle, *Structural Mechanics Finite Element Computer Programs*, 4th Ed., Advanced Engineering Corporation, Linkoping, Sweden, 1983.
2. A. K. Noor, "Survey of Computer Programs for Solution of Nonlinear Structural and Solid Mechanics Problems," *Comput. Struct.*, **13**, 425–465 (1981).
3. H. H. Fong, "An Evaluation of Eight General Purpose Finite-Element Computer Programs," *Proc. AIAA/ASME/AHS Struct., Struct. Dyn., Mater. Conf., 23rd*, Pap. No. 82-0699-CP (1982).
4. R. E. Ball, W. F. Hubka, N. J. Huffington, P. Underwood, and W. A. von Riesemann, "A comparison of Computer Results for the Dynamic Response of the LMSC Truncated Cone," *Comput. Struct.*, **4**, 485–498 (1974).
5. G. Yagawa, H. Ohtsubo, H. Takeda, Y. Toi, T. Aizawa, and T. Ikushima, "A Round Robin on Numerical Analyses for Impact Problems," *Nucl. Eng. Des.*, **78**, 377–387 (1984).
6. S. S. Hecker, "Experimental Studies of Yield Phenomena in Biaxially Loaded Metals," in *Constitutive Equations in Viscoplasticity: Computational and Engineering Aspects*, J. A. Stricklin and K. J. Saczalski, Eds., pp. 1–33, Am. Soc. Mech. Eng., New York, 1976.
7. N. Jones, "Recent Progress in the Dynamic Plastic Behavior of Structures. Part I," *Shock Vibr. Dig.*, **10**(9), 21–33 (1978).
8. J. F. Proctor, "Containment of Explosions in Water-Filled Right-Circular Cylinders," *Exp. Mech.*, **10**, 458 (1970).
9. F. A. McClintock, "A Criterion for Ductile Fracture by the Growth of Holes," *J. Appl. Mech.*, **35**, 363–371 (1968).
10. J. O. Hallquist, *User's Manuals for DYNA3D and DYNAP*, Rep. No. UCID-19156, Lawrence Livermore Lab., Livermore, CA, 1981.
11. S. B. Menkes and H. J. Opat, "Broken Beams," *Exp. Mech.*, **13**, 480–486 (1973).
12. N. Jones, "Plastic Failure of Ductile Beams Loaded Dynamically," *J. Eng. Ind.*, **98**, 131–136 (1976).
13. H. J. Opat and S. B. Menkes, *Hard Point Failure in Relationship to Lethality*, Tech. Rep. 4738, Picatinny Arsenal, Dover, NJ, 1974.
14. A. Considere, "L'Emploi du Fer et de l'Acier Dans les Constructions," *Ann. Ponts Chaussees*, **9**, 574–775 (1885).
15. H. W. Swift, "Plastic Instability Under Plane Stress", *J. Mech. Phys. Solids*, **1**, 1–18 (1952).
16. R. Hill, "A Theory of Plastic Bulging of a Metal Diaphragm by Lateral Pressure," *Philos. Mag.*, **41**(7), 1133–1142 (1950).
17. Z. Marciniak and K. Kuczynski, "Limit Strains in the Processes of Stretchforming Sheet Metal," *Int. J. Mech. Sci.*, **9**, 609–620 (1967).
18. J. W. Rudnicki and J. R. Rice, "Conditions for the Localization of Deformation in Pressure-Sensitive Dilatant Materials," *J. Mech. Phys. Solids*, **23**, 371–394 (1975).
19. L. Anand and W. A. Spitzig, "Initiation of Localized Shear Bands in Plane Strain," *J. Mech. Phys. Solids*, **28**, 113–128 (1980).
20. P. S. Theocaris, "Yield Criteria Based on Void Coalescence Mechanisms," *Int. J. Solids Struct.*, **22**, 445–466 (1986).

21. D. V. Wilson and O. Acselrad, "Strain Localization in Biaxially Stretched Sheets Containing Compact Defects. I." *Int. J. Mech. Sci.*, **26**, 573–585 (1984).

22. S. S. Hecker and A. K. Ghosh, "The Forming of Sheet Metals," *Sci. Am.*, **235**(5), 100–108 (1976).

23. A. K. Ghosh and S. S. Hecker, "Stretching Limits in Sheet Metals: In-Plane Versus Out-Of-Plane Deformation," *Metall. Trans.*, **5**, 2161–2164 (1974).

24. U. S. Lindholm, L. M. Yeakley, and D. L. Davidson, *Biaxial Strength Tests on Beryllium and Titanium Alloys.*, AFML-TR-74-172, Air Force Systems Command, Wright-Patterson Air Force Base, OH, 1974.

25. A. Mendelson, *Plasticity: Theory and Application*, Macmillan, New York, 1968.

26. R. Salzbrenner and P. Bortniak, *Formability Facility*, Rep. No. SAND82-0690, Sandia Lab, Albuquerque, NM, 1982.

27. A. A. Ezra, *Principles and Practice of Explosive Metalworking*, Garden City Press, Letchworth, Great Britain, 1973.

28. Z. Marciniak, K. Kuczyński, and T. Pokora, "Influence of the Plastic Properties of a Material on the Forming Limit Diagram for Sheet Metal in Tension," *Int. J. Mech. Sci.*, **15**, 789–805 (1973).

29. N. Jones, "The Influence of Large Deflections on the Behavior of Rigid-Plastic Cylindrical Shells Loaded Impulsively," *J. Appl. Mech.*, **37**, 416–425 (1970).

30. N. Jones and J. G. de Oliveira, "Impulsive Loading of a Cylindrical Shell with Transverse Shear and Rotatory Inertia," *Int. J. Solids Struct.*, **19**, 263–279 (1983).

31. D. C. Drucker, "The Effect of Shear on the Plastic Bending of Beams," *J. Appl. Mech.*, **23**, 509–514 (1956).

32. N. Jones, this volume, Chapter 5.

33. H. E. Lindberg, D. L. Anderson, R. D. Firth, and L. V. Parker, *Response of Reentry Vehicle-Type Shells to Blast Loads*, SRI Proj. FGD-5228, Stanford Res. Inst., Menlo Park, CA, 1965.

34. G. Regazzoni and F. Montheillet, "Influence of Strain Rate on the Flow Stress and Ductility of Copper and Tantalum at Room Temperature," *Conf. Ser.—Inst. Phys.*, **70**, 63–70 (1984).

35. W. Johnson, *Impact Strength of Materials*, Crane, Russak, New York, 1972, p. 141.

36. Z. P. Bazant, "Instability, ductility, and Size Effect in Strain-Softening Concrete," *Proc. Am. Soc. Civ. Eng.*, **102**(EM2), 331–344 (1976).

37. H. L. Schreyer and Z. Chen, "One-Dimensional Softening with Localization," *J. Appl. Mech.*, **53**, 791–797 (1986).

38. Personal communication with Professor T. Wierzbicki, Massachusetts Institute of Technology, September 15, 1987.

CHAPTER 7

Failure of Brittle and Composite Materials by Numerical Methods

A. de Rouvray and E. Haug
Engineering Systems International S.A.
94578 Rungis—Cedex, France

ABSTRACT

This chapter deals with the numerical prediction of the static failure of engineering brittle materials such as ceramics and composites, with an emphasis on composite plastic materials, such as fiber reinforced carbon/epoxy (CE) composite laminates in static tension and woven composite CE tissues in compression and bending.

Composite laminates, when loaded in static tension, exhibit complex heterogeneous damage and failure modes, such as matrix cracking, delamination, fiber breakage, and fiber-matrix interface debonding. The severity of damage depends on laminar strength and stiffness, ply orientation, stacking sequence, fiber-volume fraction, type of loading, cutout geometry, specimen size, and environmental conditions. Failure occurs when critical levels of loads or deformations are reached, related to damage-dependent failure criteria. When loaded in static compression, composites can exhibit the damage mode of local fiber buckling. This is especially true in composite fabrics, whose weave causes the fibers to bend near the intersections of the warp and weft threads. The associated redirection forces put a high stress on the matrix material, which can undergo compressive failure via slipping (plasticity) and/or crazing (fracturing). This in turn deconfines the fibers and leads to increases in curvature and to a relaxation with a decrease in fiber forces and ultimate failure. Analytical computations establish laminate-damage evaluations, i.e., matrix micro- and macrocracking, fiber buckling and rupture, and layer delamination. The initial, tangent and residual material properties are computed as well as damage initiation in critical zones of arbitrary composite laminate structures to evaluate stable and unstable growth, and thresholds for low-strain criteria. Emphasis is placed on the identification and calibration of reasonably intrinsic material parameters for failure criteria.

A preliminary discussion reviews several concepts of damage growth and fracture initiation as applicable, first, to brittle homogeneous materials, such as engineering ceramics, and, second, to composite materials, such as short-fiber reinforced composite ceramics and long-fiber composite laminates. An attempt is made to identify regimes and transitions for probabilistic vs. deterministic failure criteria, and to characterize fracture initiation and propagation with intrinsic material parameters even in the absence of a clearly defined single macrocrack, as is typically the case for composite materials as opposed to metals.

Several numerical examples simulate damage and fracture in the form of calibrations and predictions, using the specialized fracture- and damage-mechanics finite-element code PAM-FISS/Bi-Phase.

1. INTRODUCTION

As compared to metals or common alloys, composite materials of the resin-fiber type have attractive rheological features [Figure 1(a), Reference 1]. Graphite–epoxy composite laminates, for example, typically have specific moduli and ultimate stresses, respectively, about 1.5 and 3 times higher than steel and aluminum, as shown in the table in Figure 1 [2].

Unfortunately, this advantage vanishes if the design specifications do not allow internal damage, such as matrix microcracking or incipient delamination between layers. Indeed, in notched multilayered specimens [3], internal (stable) matrix microcracking can initiate at about 20% of the ultimate tensile strength of the composite, and the first fiber rupture at about 40% of this strength. Note in passing that the net strength, i.e., the average failure stress over the ligament, is typically 50 to 75% of the intact specimen strength (unnotched test pieces), which indicates a pronounced notch-sensitivity effect, typical of composite laminates [4].

An efficient design using composites should, in general, include a certain level of internal damaging, whether true or hypothetical, which should be calculated at its levels of initiation and propagation. There is no simple theory, however, like the von Mises or Tresca theories of plasticity in metals, that represents, even in an approximate fashion, damage accumulation in composites. One reason is associated with the following fundamental difference: damage tends to separate the composite material phases (fibers, matrix), which leads to a heterogeneous medium, where traditional continuum mechanics, although valid for lesser loads, become highly inadequate. Also, fracture usually materializes in the form of multiple interacting cracks difficult to characterize as a well-defined equivalent macrocrack to which apply standard fracture-mechanics theories as for metals.

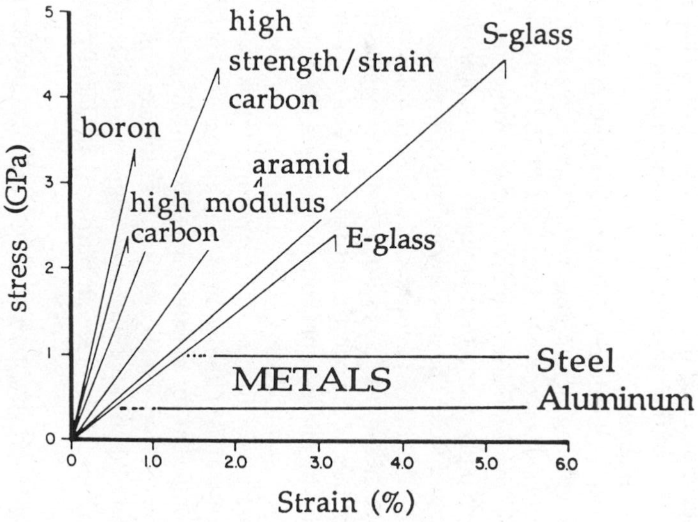

Table 1: The competitors

	Specific strength (miles)	Specific stiffness (miles)
2024-T3A1	11	1600
7075-T6A1	13	1600
7175-T73A1	11	1600
Ti6A1-4V	13	1600
300M steel	16	1600
G/E uni	65	5100
K/E uni	84	3500
Angle-plied properties		
	Specific strength (miles)	Specific stiffness (miles)
7175-T73A1	11	1600
G/E [0]	65	5100
G/E [0/±30]s	37	3200
G/E [0/±45]s	31	2300
G/E [0/±60]s	32	2100
G/E [0/±90]s	29	2100

Figure 1. Stress-strain curves for some typical fibers used in advanced composite materials. After Dorey [1].

195

Fiber-reinforced laminates are heterogeneous (matrix, fibers), anisotropic (stiffness and strength mainly in the fiber direction), nonlinear (matrix and fiber damage, fiber-matrix interface damage), and layered materials (stackup of unidirectional layers). The numerical prediction of the failure of fiber-reinforced composite laminates, therefor°, appears formidable.

Nevertheless, for certain purposes, long-fiber reinforced composite laminates can be viewed globally as a homogeneous material, e.g., when incorporated into a structural component or when functioning as the skin of a sandwich panel, and the concern of the structural engineer is to know or to evaluate its mechanical properties, such as stiffness, strength, thermal constants, fracture resistance, fatigue behavior, etc. For a given laminate, these properties can be evaluated by appropriate physical experiments. Although feasible, this avenue seems often impractical in view of the great number of tests that would have to be performed in order to account for the ever-varying stackups, geometries, and loading conditions under operational and accidental conditions.

According to the laws of mixture on the ply level and through assembly of the ply properties into laminate properties, certain mechanical properties of composite laminates, such as the overall linear stiffnesses, can be calculated from the properties of its constituent plies. Other laminate properties, such as fracture resistance or toughness, cannot readily be calculated at the ply level, however. The evaluation of, e.g., fracture resistance and associated toughness parameters of a single long-fiber-reinforced composite ply is neither straightforward, nor is it possibly very helpful, since the fracture behavior of a single isolated ply and the fracture behavior of the same ply when embedded in a stackup are likely to be fundamentally different.

Should the analyst then descend to the microscale of each single ply, i.e., the scale of material heterogeneity? This would imply evaluation of the fracture-resistance properties of layered stackups first on the constituent materials of a single ply, namely the fibers and the matrix, from which the fracture-resistance properties of a given ply due to the loadings it experiences when it is part of the stackup must be deduced. The answer can be both, yes and no.

If answered yes, the analyst adopts a *local microscopic view*, which means the mechanical properties must be evaluated on the single-ply level from the mechanical properties of the constituent-material phases. For this purpose, one must first identify the microscopic constituents of a single long-fiber-reinforced composite ply.

For the evaluation of linear stiffness parameters, knowledge of the linear stiffness coefficients and of the geometrical arrangement of the fiber and the matrix phase is sufficient for calculating equivalent single-ply coefficients with enough practical precision, e.g., according to simple laws of mixture. In general, however, even the evaluation of linear ply-stiffness coefficients is quite complex, and only approximate. For practical purposes, sufficiently accurate analytical solutions exist.

For the evaluation of single-ply-fracture toughness properties, knowledge of isolated fracture-resistance properties of the ply constituent-material phases (fibers, matrix) is generally not a broad enough basis, since it neglects, e.g., the influence of fiber–matrix interface failure on single-ply-fracture toughness. Knowledge of the fracture properties of the fiber–matrix interface must, therefore, be added to the knowledge of the fracture properties of the fibers and of the matrix, including the effect of severe confinement, as it may be introduced by sound multilayered stackups.

The affirmative answer, therefore, involves considerable analytical efforts even in simplified analytical approaches [5] and the resulting single-ply-fracture resistance parameters may not be representative for ply fracture when the ply is embedded into a composite stackup. The reason is that single-ply fracture may be influenced by matrix and fiber damages afflicted to the outer surfaces of single plies by adjacent plies, such as is the case when, e.g., a 90°-layer matrix transverse crack is arrested near the interface with a 0°-layer, in an intact or in a notched cross-ply tensile test piece [6]. The relatively small stress concentrations of such intralaminar matrix cracks may well be important enough to initiate fiber rupture near the surface of adjacent plies, which, in turn, has an influence on the final fracture of these plies. Knowledge of single-ply-fracture behavior must, therefore, be applied discriminately throughout the thickness of a single ply and the influences resulting from the structural aspect of the stackups must be considered.

In summary, to use single-ply-fracture knowledge, e.g., to predict the overall fracture behavior of notched multilayered long-fiber reinforced tensile test pieces, the evolution of intra- and interlaminar matrix damage and of the accumulated fiber damage in all adjacent plies must be traced simultaneously, and fracture must be evaluated discriminately throughout the thickness of each individual ply.

Although there is little hope to achieve this goal analytically, the use of numerical techniques and of modern high-speed computers or engineering work stations opens the avenue for benchmark calculations to be carried out that shed light on and even predict in some cases the complex fracture behavior of composite laminates. Modern numerical methods, such as the finite-element method, specially adapted and optimized fracture- and damage-mechanics computer codes, such as PAM-FISS [7], special material laws, such as the Bi-Phase model in the PAM-FISS code, and fracture-mechanics criteria, such as G (strain-energy release rate), J (Rice's integral), K (stress-intensity factor), Poe's (strain-intensity) criterion [8], the point-stress criterion [9], and the more recent family of the critical-damage-over-characteristic-distance $[D_c, r_c]$ damage-mechanics criteria [10], now permit the effective numerical treatment of complex composite fracture events.

Whereas a detailed numerical fracture prediction of composite laminates, based on single-ply material-fracture criteria seems feasible with advanced numerical models, rapid answers ask for simplified procedures, such as to evaluate the knockdown factor, which relates nominal ligament strength to the unnotched strength in notched specimens and which is a measure of notch sensitivity.

If the answer is no to the previous question whether or not the analyst should work up from single-ply fracture to complex laminate fracture, then the analyst chooses to adopt *a global view*. This implies that simplified analytical or numerical procedures and/or a considerably more extensive data base must be available, which fulfill the purpose of adequately describing the fracture and ultimate failure properties of the considered class or subclass of composite laminates.

In the simplest case, the aim is to evaluate the fracture behavior of a composite laminate on the homogenized equivalent laminate (HEL) level using simplified or even linear analysis. In order to overcome the limitations of such a simplified approach it is therefore necessary to introduce *validity limits*, such as for stackup type, specimen geometry and size, loading conditions, etc. These limits can either be established by tests or by detailed simulations.

Once the type of approach to the evaluation of fracture of fiber reinforced or woven materials has been chosen, the analyst must select the proper criterion to characterize fracture. For characterizing the evolution of a single large macro-crack in a fairly brittle material, the classical fracture-mechanics approach has been very successful. For brittle materials, linear elastic fracture mechanics (LEFM), which uses the critical strain-energy release rate or the stress-intensity factor based fracture criteria, is well suited to describe crack growth. For situations where a sizeable plastic zone exists in a confined region near the crack tip, elastic–plastic fracture mechanics (EPFM) can be used to describe the growth of the macrocrack.

Both approaches become inappropriate in cases where specimen failure is not linked to the growth of one single (or a few) isolated macrocrack(s), but rather to the initiation of fracture in a previously undamaged material. In many practical cases, fracture is a consequence of the nucleation, growth, and coalescence of a great number of microflaws, such as spherical material microvoids or micro-cracks, that extend over a critical region. In the classical example of an intact soft-metal tension rod, specimen failure occurs via fracture in the necking area, e.g., through the growth and coalescence of spherical microvoids, to form an immediately unstable macrocrack.

Such cases have recently been approached with great promise with the *microstatistical fracture-mechanics* (MSFM) method [11], which includes key measures of statistically averaged microscopic flaw behavior, such as flaw concentration, orientation, and size distribution, into the constitutive material laws. This approach becomes possible when the microscopic fracture processes governing the evolution of the microflaws can be statistically averaged such that their combined effect can be treated within the methods of classical continuum mechanics. The resulting increased nonlinearity of the local material behavior can now be handled with advanced computer models in the framework of numerical solutions. The exciting potential of MSFM resides in the fact that it links materials science with fracture or damage mechanics because it in-corporates microstructural variables into the constitutive laws of continuum mechanics. This theory also applies to cases inaccessible by classical fracture mechanics (FM), as well as to cases where classical FM does apply. The reason is that in the process zone near the tip of a macrocrack, fracture is initiated and propagated on the microstructural level by the very same mechanisms of microvoid nucleation, growth and coalescence that lead to failure, e.g., in the initially crack-free necking area of an intact tension rod.

It has been observed generally that nucleation of microflaws occurs near material heterogeneities of the size of the material grains. For single crystals, this scale is atomic; for polycrystalline materials, the grains have the size of the crystals; and for composite materials, the graininess can be on the scale of the long-fiber cross sections or of the short-fiber lengths, and flaw nucleation can occur near the interface between the fiber and the matrix materials or as local fracture of individual weak fibers, or as matrix cracks traveling around short fibers.

Contrary to classical continuum mechanics, MSFM or *damage mechanics* contains the notion of a smallest *material scale*, closely linked to the average grain size of the material, below which neither theory applies. MSFM works on volumes of solids greater than a given characteristic volume that is large enough to contain sufficiently many microflaws for the assumption of statistical flaw distribution to hold. The characteristic volume is also linked to the stress and strain gradients: stresses and strains calculated from continuum mechanics must vary slowly over the dimension of typical microflaws. If this is not the case, the flaws must be treated individually as single cracks.

In its original forms, MSFM describes flaw nucleation, growth, and coalescence as highly nonlinear rate processes in which internal damage kinetics interact in a rate- and history-dependent nonlinear fashion with the imposed load kinetics.

In a related but simpler form of the theory [10], the description of nucleation, growth, and coalescence of microflaws has been lumped into a damage function D that depends entirely on the classical fields of continuum mechanics, such as the stress, strain, or plastic strain fields. This damage function must be defined and calibrated from material to material, and can be evaluated pointwise in a solid, depending, e.g., on plastic strain or other irreversible damage measures. Fracture, i.e., coalescence of microflaws, occurs whenever a calibrated critical threshold value D_c of the damage function has saturated a characteristic material volume V_c found also by calibration. Both D_c and V_c are material constants within the framework of the model. This theory was applied successfully in metals to the quasi-static fracture prediction of cracked, notched, and unnotched tension and torsion specimens and was able to reproduce specimen size effects and ductile–brittle transitions. One appeal of this approach lies in the fact that no time-dependent rate laws have to be integrated for the description of fracture.

This chapter reports on the experiences gained at ESI in the field of numerical failure prediction of *brittle homogeneous and composite* material test pieces under *static* loads with the objective of identifying universal fracture criteria, based on *intrinsic material parameters*, which could be used safely in large numerical models under arbitrary geometries, scales, and loading conditions. Among the conventional fracture-mechanics criteria studied, none has been found universally applicable to predict alone composite fracture and failure in situations of varying stress concentrations and specimen or component sizes. Typically, for areas of high-stress concentrations, a *deterministic* damage-mechanics-based $[D_c, r_c]$-type *fracture* criterion has proven successful, whereas for areas of low-stress concentrations, a *probabilistic* Weibull-type *strength* criterion appeared well suited. In both cases, a detailed description of the local stress–strain field must be used, incorporating the effect of damage.

Weibull-type *strength criteria* link the often observed statistical scatter of strength data, especially in brittle materials, to randomly distributed material defects. If the stress–strain distribution is smooth over areas large with respect to the size of random flaws, strength depends on the most critical random flaw. In such areas, strength is likely to be inferior to the theoretical maximum and it scatters about an average value. In areas small with respect to the existing randomly distributed material defects, such as in the vicinity of a crack tip or a notch root of a cracked or notched tensile specimen, the local strength can be higher than, for example, the observed unnotched tensile strength of an intact specimen of the same size.

If applied to specimens to which in the strict sense LEFM does not or at best apply approximately, the $[D_c, r_c]$ criterion has been found superior in predictive power to the classical fracture-mechanics criteria, including for composite laminates to the Whitney and Nuismer's point-stress criterion. The criterion has been applied successfully to any fracture event, be it due to an existing sharp crack (typical fracture-mechanics situation) or due to the failure of an intact tensile rod. The criterion as such, therefore, encompasses classical strength analysis and fracture mechanics and it opens the road for a unified treatment of fracture and failure caused by material separation, when complemented by a probabilistic local strength description in the areas of low-stress gradients.

A combination of *deterministic* and *probabilistic* fracture criteria, therefore, appears reasonably universal to predict the pseudo-static failure of brittle homogeneous and composite materials.

The foregoing fracture-model concept is illustrated, first, on the static fracture-toughness characterization of two engineering zirconia ceramic materials, where microfracturing damage induces local phase changes, and, second, on a short-fiber reinforced silica composite ceramic, all materials with high thermal stability and typical of hot motor component applications. Then the same strength/toughness characterization methodology is applied to long-fiber reinforced plastic (LFRP) laminates, typical of aeronautical and space structural applications, where the above concept is rendered more complex due to the constitutive heterogeneity of the material both at the ply (i.e., fibers and matrix) and laminate (i.e., multilayered stackup) levels.

The practicality of the approach for complex composite laminates under arbitrary loading conditions is demonstrated in the form of a simplified version of the $[D_c, r_c]$ criterion, first, in an approximate fashion at the homogenized equivalent laminate (HEL) level of notched tensile composite test pieces using an analytical approach, and, second, in a more precise fashion using a detailed numerical model at the level of the load-bearing ply in cross-ply specimens. Compressive and bending test cases are also considered.

The following abbreviations and notations are used throughout this chapter and are summarized here for ease of reading.

Glossary

CE	Carbon-epoxy
CM	Continuum mechanics
CP	Cross-ply
CT	Compact tension
DENT	Double edge notched tensile
DM	Damage mechanics
EPFM	Elasto-plastic fracture mechanics
FE	Finite Element
FM	Fracture mechanics
HEL	Homogenized equivalent laminate
LEFM	Linear elastic fracture mechanics
LFRP	Long fiber reinforced plastic
ML	Multi-layered
MNT	Middle notched tensile
MSFM	Microstatistical fracture mechanics
PSZ	Partially stabilized zirconia
TTZ	Totally tetragonal zirconia
SENB	Single edge notch bending
UD	Unidirectional

Microflaws are natural or nucleated load-induced material microvoids or microcracks that occur at the scale of material heterogeneity or microstructural graininess.
Microflaws cause damage.

Damage is an abstract measure of microflaw nucleation and growth.
Damage precipitates fracture.

Fracture is local material separation due to microflaw growth and coalescence into a macroflaw or "crack" after a critical damage state has been reached. Fracture occurs at the scale of macrostructural material homogeneity, or at the scale of classical continuum mechanics. *Brittle* fracture is plane strain fracture predicted by linear elastic fracture mechanics. *Ductile* fracture is fracture that is not predicted by linear elastic fracture mechanics.
Fracture precipitates failure.

Failure is the global disability of a component or specimen to perform due to stable or unstable fracture growth; rupture means tensile failure.
Failure precipitates catastrophy.

Notations

a	half notch or crack length
c, α, β	material constants in Wilkins' damage function
d	material modulus (fracturing) damage
D	damage
E	elastic modulus
F	force
f, g	functions (fracturing)
G	strain energy release rate
J	Rice J-integral
K	Westergaard crack stress intensity factor; notch stress concentration factor
K^*	Craeger "Apparent" notch stress intensity factor
l	distance away from crack tip or notch root
m	Weibull modulus
p	hydrostatic pressure
P	probability
r	distance away from crack tip or notch root
S	deviatoric stress
t	thickness
T300/914C	fiber and matrix material of carbon-epoxy composites
V	volume
w	weight term; specimen width
Y	shape factor
ρ	notch root radius
σ	crack opening stress
ε	strain; plastic strain; equivalent strain
φ	coefficient of variation
Γ	Gamma function

Indices

b	bending
c	critical or characteristic
I,II,III	crack opening mode I, II, III
f	failure
i	initial
m	mean
n	net, nominal
o	far field; zero
p	plastic
s	symmetric
t	tensile
u	ultimate

2. GENERAL REMARKS

2.1. Composite Laminate Damage-Growth Mechanisms

In *static tension*, damage initiates and propagates in composite laminates in zones of high-stress concentrations, such as free edges around cutouts, joints, delamination edges, where micro- and macrocracks first develop in the matrix phase. In fiber-controlled laminates, this matrix damage does not precipitate catastrophic failure. On the contrary, fracture occurs only when the fiber phase of the load-bearing (e.g., 0°-) plies is sufficiently overstressed to reach fiber strength. However, stress concentrations in the fiber phase, although higher due to anisotropy, are possible only by shear transfer through the matrix phase, and, consequently, tend to be relieved by matrix cracking parallel to the fibers, which in turn is stabilized by the confinements introduced by the transverse plies.

This is a major difference with homogeneous materials such as metals, where a crack acts as a sharp stress riser and grows steadily, becoming more critical with increasing size, until catastrophic failure occurs when the crack becomes unstable. This opposite effect of microcracking damage on the maximum fiber-stress concentration for composites as compared to metals is qualitatively illustrated on Figure 2 for a notch root under tension, for a composite laminate, and for a homogeneous material.

A typical example of the evolution of matrix and fiber damage in LFRP composite laminates is reported [12] on $[0,90]_{2s}$ middle-notched tensile (MNT) cross-ply (CP), carbon–epoxy (CE) specimens, where the matrix damage, in the form of splitting, transverse and delamination cracks, plays a major role in the initiation and propagation of the 0°-ply fiber cracks, which ultimately lead to specimen failure. Once fiber breakage begins in some 0°-ply, it can be arrested due to a second matrix splitting, which can lead to stable propagation. This stable propagation continues up to the eventually "brittle" failure of the laminate, when the touchening effect due to the progressing matrix damage has peaked. Reference 12 contains X-ray radiographs of the $[0,90]_{2s}$ MNT specimen at 71 and 92% of the average failing load. Figure 3 shows the through-the-thickness damage in their $[0,90]_{2s}$ MNT specimen at 71% of average failure load. Self-similar asymmetric fiber breakage and major axial splits form in the outside 0°-plies and delamination occurs in the outer 0,90- and 90,0-interfaces. Symmetric transverse macrocracking due to fiber breakage occurs in the interior 0°-plies, where matrix damage appears to be less prominent (in the lowest 90°-layer, a nonself-similar through-the-fibers crack seems to develop, the existence of which might be due to an initial specimen flaw).

Under *fatigue loading*, the matrix cracks tend to grow even more, leading to accelerated failure for homogeneous materials, but increased fiber overstress relief for composite laminates and potentially improved residual strength (Figure 4) [13].

In *static compression*, composites can exhibit the damage mode of local fiber buckling. This is especially true in composite fabrics, the weave of which causes

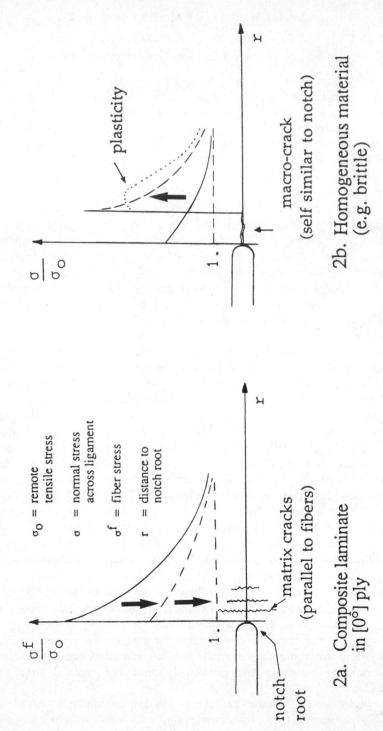

σ₀ = remote
     tensile stress

σ = normal stress
     across ligament

σᶠ = fiber stress

r = distance to
     notch root

$\frac{\sigma^f}{\sigma_0}$

matrix cracks
(parallel to fibers)

notch
root

2a. Composite laminate
     in [0°] ply

plasticity

$\frac{\sigma}{\sigma_0}$

macro-crack
(self similar to notch)

2b. Homogeneous material
     (e.g. brittle)

Figure 2. Axial stress concentrations at the notch root before and after crack initiation.

204

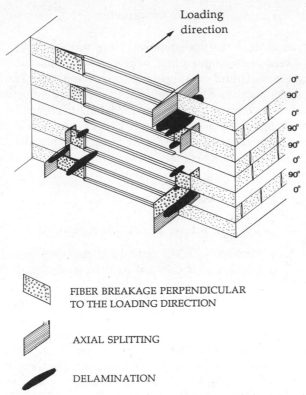

Loading
direction

0°
90°
0°
90°
90°
0°
90°
0°

FIBER BREAKAGE PERPENDICULAR
TO THE LOADING DIRECTION

AXIAL SPLITTING

DELAMINATION

Figure 3. Through-the-thickness damage in the $[0/90]_{2s}$ MNT (CE) specimens at 71% of average failing load [12].

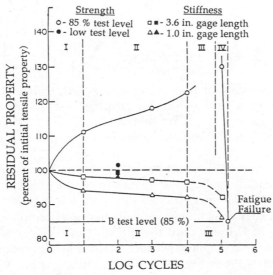

Figure 4. Stiffness and residual strength vs. life T300-5208 graphite/epoxy. (After G. Kress, DFVLR Braunschweig, 1/84.)

the fibers to bend near the intersections of the warp and weft threads. The associated redirection forces put a high stress on the matrix material, which can undergo compressive failure via slipping (plasticity) and/or crazing (fracturing). This in turn deconfines the fibers and leads to increases in curvature and to a relaxation with a decrease in fiber forces and ultimate failure.

Consequently, *damage-growth analyses* in LFRP composite laminates require matrix micro- and macrocracking models for intra- (parallel to fibers) and interlaminar (delamination) cracking as well as fiber-rupture models, which are qualified under high-stress gradient loadings, characteristic of typical defect areas.

2.2. Fracture-Mechanics Approach

Classical fracture mechanics (FM) assumes preexisting well-defined sharp cracks, e.g., due to fatigue, and it does not treat the problem of crack initiation inside initially intact solids. Crack *initiation* and *propagation* is treated by a discipline called *damage mechanics*, rather than by fracture mechanics, which only deals with the problem of the propagation of a single sharp macrocrack.

Linear elastic-stress concentrations found near notch roots or near crack tips (if a crack has already formed) can degenerate upon crack initiation into an infinite stress singularity, and crack propagation, therefore, has to be based on stress-intensity factors or other energetic parameters rather than on the magnitude of stress itself, which is not bounded.

In classical fracture mechanics, the following criteria are applied to situations of crack propagation:

$$K \geqslant K_c$$

$$G \geqslant G_c$$

$$J \geqslant J_c$$

where K = linear stress intensity factor (modes I, II, or III), G = strain-energy release rate per unit area of crack formation, and J = Rice J-integral for contours encompassing linear or confined-plastic crack-process zones. Subscript c indicates critical threshold values for the respective quantity for crack advance.

In LFRP composites, fracture mechanics criteria have been used with success to simulate intra- and interlaminar *matrix* cracks, such as splitting, transverse cracks, and delamination cracks in tension test pieces [14, 15]. These isolated large matrix cracks, although only indirectly responsible for specimen failure, seem to be well described by LEFM. It has been found, however, that *fiber* breaks leading to ultimate specimen failure, where a crack tip as such is ill-defined, are not well-covered by LEFM criteria. Further, the treatment of *ubiquitous systems* of matrix cracks near notch roots of multilayered composite laminates becomes forbidding when each single matrix crack is to be traced individually. As shown later, ubiquitous matrix cracking in composite laminates that accompanies fiber breaks can be treated with the help of a matrix constitutive fracturing law, which models matrix modulus damage (i.e., stiffness loss and strain-softening) as it would result from the discrete matrix crack system.

2.3. Damage-Mechanics Approach

Basic Description of the [D_c, r_c]-Damage-Mechanics Fracture Criterion Fracture of materials can be viewed as a consequence of the detailed history of micromechanical damage accumulation inside the material near critical areas of stress. If sufficiently damaged, the material separates locally through nucleation, growth, and coalescence of material microflaws and a stable or unstable macrocrack forms. The postulated $[D_c, r_c]$-damage-mechanics fracture criterion states that *"fracture, i.e., material macroscopic separation initiates when a cumulative measure of material damage D exceeds a critical value D_c over a characteristic material distance r_c."* The critical damage D_c and the characteristic distance r_c are phenomenological intrinsic *material constants* related to the locally accumulated micromechanical material damage at the onset of fracture. The critical distance r_c is equivalent to a critical volume V_c when the critical volume is assumed to be roughly spherical.

Damage Models Damage models of fracture relate identified fracture-inducing damage quantities to macroscopic material separation, that is, fracture. In order to calculate the detailed and local history of micromechanical damage of arbitrary components well up to the point of incipient failure, and beyond, complete *nonlinear constitutive modeling* and complete *nonlinear analysis* techniques with *high-speed computers* are a prerequisite.

Constitutive Relationships Constitutive relationships are models to describe physical material behavior. For a meaningful analysis of fracture, the constitutive relationship used in an analysis must be reaching far enough to describe the essential physics of the studied phenomenon up to the very moment of fracture. Therefore, the solution at or close to fracture is likely to contain already most or all key elements needed for the identification of a criterion of fracture initiation and propagation. The relevant damage parameters that satisfy the postulated fracture criterion can then be identified from a series of rupture tests that should be chosen to form an independent basis for most practical damage situations, and from the accompanying numerical simulations of these tests.

The idea that a critical damage must spread over a characteristic distance to induce local fracture is due to the basic phenomenological observations described next.

Stress-Gradient Effect   We see that the maximum fracture stresses calculated at the very notch root of notched specimens for stress profiles with lower stress gradients tend to be systematically lower (dashed line in Figure 5) than the maximum failure stresses calculated from specimens with higher stress gradients (solid line), namely,

$$\sigma_2^{max} \text{ lower than } \sigma_1^{max}$$

for

$$d\sigma_2/dr \text{ lower than } d\sigma_1/dr$$

This implies that brittle failure-stress profiles tend to intersect at some point near the notch root. It is this observation that leads to a first phenomenological fracture model, where macrocrack growth is said to occur when a *critical stress* σ_c extends over some *characteristic distance* r_c, where $[\sigma_c, r_c]$ are the coordinates of the point of intersection of the failure-stress profiles near the notch root. If this point is unique, the stress profile at rupture of a specimen with a crack also passes through the point (dotted line in Figure 5), and so does that of a theoretically intact smooth specimen, where it is assumed that the intact strength is not reduced by local defects, which for brittle materials is almost always the case.

The premise that such a unique point that characterizes fracture can exist stems from the basic concept that fracture results from the coalescence of previously grown microdefects. Such coalescence is assumed to occur if a

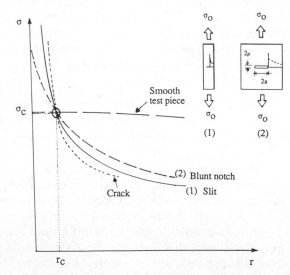

Figure 5. Brittle-fracture transverse-stress profiles over the ligament at fracture of middle-notched tensile specimens.

sufficiently large and approximately constant volume of material, of extent r_c, is critically damaged. The damage extends over a region that contains many grains or initial flaws, typically an order of magnitude larger than the basic grain size or microscale dimension.

The *damage function D*, in general, represents a weighted combination of damage factors responsible for defect growth and progressive coalescence over the given distance r_c. Coalescence of the defects to form a macrocrack is quantitatively achieved when the damage D reaches the critical level D_c.

Simplified Criterion   If in LEFM-type situations, the point (σ_c, r_c) is *unique* for specimens with different stress gradients, then the general $[D_c, r_c]$ criterion degenerates into the simple $[\sigma_c, r_c]$ criterion, where critical stress σ_c and characteristic distance r_c are material constants.

Damage Mechanics vs. Stress-Intensity Criteria   Note that the phenomenological implications of the two-parameter damage-mechanics $[\sigma_c, r_c]$-class failure criteria to describe the gradient or scale effects differ fundamentally from those of the one-parameter stress–strain intensity or energy-release (e.g., K or G) class criteria. In order to demonstrate this, it is instructive to extrapolate the notion of, say, Westergaard's critical stress-intensity factor K, which is well established in LEFM as a *constant* (i.e., intrinsic) material property for cracks, to notches with a notch root radius ρ, which is sufficiently large, with respect to the material microscale, as will be quantified later.

For *ideal cracks* $(\rho = 0)$ and for an infinite specimen width and isotropic material, the stress profile *at rupture* is given by Westergaard's formula as

$$\sigma_c(r) = \sigma_{ou}a^{1/2}/(2r)^{1/2} = K_{Ic}/(2\pi r)^{1/2} \qquad \text{crack-stress profile}$$

where

$$K_{Ic} = \sigma_{ou}(\pi a)^{1/2}$$

and where r is the distance away from the crack tip, a is the half-crack length, σ_{ou} is the ultimate applied far field stress, and $K_{Ic} = $ constant is the classical critical Westergaard stress-intensity factor for cracks.

For *notches* $(\rho \neq 0)$, the corresponding stress profile at rupture can be expressed using Craeger's formula as

$$\sigma_c(r) = [\sigma_{ou}a^{1/2}/(2r)^{1/2}]f(\rho/r) = [K_{Ic}^*/(2\pi r)^{1/2}]f(\rho/r) \qquad \text{notch-stress profile}$$

where

$$K_{Ic}^* = \sigma_{ou}(\pi a)^{1/2}$$

and where the function $f(\rho/r)$ depends mainly on the notch geometry, and

Craeger's "apparent" critical stress-intensity factor K_{Ic}^* has been introduced for notches by analogy to the critical stress-intensity factor for cracks. Note that for cracks, $K_{Ic}^* = K_{Ic}$ and $f(\rho/r) = 1$.

The stress-intensity-type K-criterion and the damage-mechanics-based $[\sigma_c, r_c]$ criterion can be connected by evaluating both stress profiles at the point (σ_c, r_c), which, according to the $[\sigma_c, r_c]$ criterion, is common to all stress profiles at fracture:

$$\sigma_c = K_{Ic}/(2\pi r_c)^{1/2} = [K_{Ic}^*/(2\pi r_c)^{1/2}]f(\rho/r_c)$$

This equality permits expression of the notch equivalent or "apparent" critical stress-intensity factor K_{Ic}^* in terms of the classical K_{Ic} for cracks, namely,

$$K_{app}(\sigma_c, r_c; \rho) = K_{Ic}^* = \mathbf{K}_{Ic}/f(\rho/r_c) = [\sigma_c(2\pi r_c)^{1/2}]/f(\rho/r_c)$$

Note, that if the $[\sigma_c, r_c]$ criterion is held valid, it follows, for a given material, that the apparent notch stress-intensity factor K_{Ic}^* is a function of ρ/r_c, the ratio of the notch root radius ρ relative to a *material size* constant r_c, i.e., K_{Ic}^* is no longer a constant, as postulated by the K_{Ic} criterion for cracks. It follows also that the $[\sigma_c, r_c]$ criterion is more general, since it applies to any notch geometry or stress profile. This criterion implies that only one unique point is in common between all fracture-stress profiles, whereas the K-type criteria imply matching fracture-stress profiles over the full ligament length.

Note also that the validity of the $[\sigma_c, r_c]$ criterion implies that the failure-stress profiles $\sigma_c(r)$ only depend on the notch radius ρ and are independent of the half-notch width a, i.e.,

$$\sigma_c(r) = [K_{Ic}/(2\pi r)^{1/2}]f(\rho/r)/f(\rho/r_c) = [\sigma_c(r_c/r)^{1/2}]f(\rho/r)/f(\rho/r_c)$$

This means that for a series of notched test pieces with constant ρ and varying a values, the previously derived equivalent constant apparent K toughness

$$K_{app}(\rho) = K_{Ic}^* = K_{Ic}/f(\rho/r_c)$$

appears like a material property, although it is not intrinsic and has a validity domain limited to constant-ρ values for all a values.

[σ_c, r_c] vs. Point-Stress Criterion   In the context of the homogenized equivalent laminate theory (HEL), the Whitney and Nuismer [9] point-stress criterion states for composite materials that fracture initiates as soon as the crack-opening stress exceeds the unnotched specimen strength σ_{ou} over a critical laminate distance l_c. If compared with the $[\sigma_c, r_c]$ criterion, this amounts to replacing the free parameter σ_c by a fixed material constant, namely, σ_{ou}, i.e., to state the $[\sigma_c, r_c]$ criterion in the less general form $[\sigma_{ou}, l_c]$. The $[\sigma_c, r_c]$ criterion, therefore, has one more freedom, available for calibration, namely, the mag-

nitude of the critical stress σ_c. The fact that σ_c may turn out different from, as pointed out earlier, and in fact greater than, σ_{ou} is not in direct violation with common sense, since the characteristic distance r_c is very small as compared to the specimen size. The high analytical stresses in excess of σ_{ou} occur, therefore, only very locally near the tip of a crack in a region of very high stress gradients, whereas σ_{ou} is a strength value that is reduced by Weibull effects over larger material volumes. In practice, l_c is an order of magnitude larger than r_c for usual LFRP laminates and other brittle materials, whereas σ_c can be one and one-half to twice as large as σ_{ou} for smooth and notched test pieces with large round holes.

The Scale Effect   The scale effect relates fracture stresses of self-similar specimens which differ only by their linear scale. The K_{Ic} criterion links the fracture stress of two precracked specimens with scales 1, 2 via the constance of K_{Ic} as follows:

$$\sigma_{ou}^2/\sigma_{ou}^1 = (a_1/a_2)^{1/2}$$

The $[\sigma_c, r_c]$ criterion leaves free the magnitude of the scale effect in notched specimens and lets it be a result of calibration, and correlated with ρ/r_c.

The Notch Root Radius Effect   The *damage-mechanics failure criteria* are based on the *actual* stress profiles (which need not be linear elastic), and, consequently, are applicable to *any* notch geometry, from round holes to blunt (or sharp) notches (i.e., slits) or cracks. That is to say, this class of damage-mechanics criteria is *applicable to any stress profile*, or stress riser, such as created by a notch, joint, free edge, local defect or delamination, etc. This is a major advantage over the stress-intensity class of failure criteria, which require either an ideal crack or a given stress profile with a given gradient.

Wilkins' Ductile Fracture Criterion   In brittle and ductile fracture of metals, the micromechanism leading to fracture is assumed to be material separation on the macroscale due to microvoid nucleation, growth, and coalescence over a characteristic material distance, as originally postulated by McClintock. As soon as this coalescence happens, a macrocrack opens and tends to propagate.

For ductile metal fracture, Wilkins [10] identified this damage quantity to be equal to the following *weighted average* of the plastic strain:

$$D = \int w_1 w_2 \, d\varepsilon_p$$

$$w_1 = [1/(1 + cp)]^\alpha$$

$$w_2 = (2 - A)^\beta$$

$$A = \max(S_2/S_3, \, S_2/S_1) \qquad S_1 > S_2 > S_3$$

where ε_p is the equivalent plastic strain, e.g., due to the von Mises plastic law, w_1 is a hydrostatic-pressure weighting term, w_2 is a shear (asymmetric strain) weighting term, p is the hydrostatic pressure, S_1, S_2, and S_3 are the principal stress deviators, and c, α, and β are material constants. By calculating profiles of this damage quantity near regions of high energy, for compact and notched tension tests, Wilkins was able to identify this particular damage quantity and to calibrate the law $[D_c, r_c]$. The resulting fracture model, calibrated once and for all for a given material (e.g., aluminum or steel) was successfully applied to the static strength predictions of smooth, notched, and cracked test pieces of arbitrary shapes and sizes, encompassing the brittle–ductile transition, crack initiation, and propagation, as well as the pseudo-dynamic loads of the Charpy test.

Effect of the $[D_c, r_c]$ Fracture Criterion in Composite Laminates   Figure 6 symbolizes the effect of the $[D_c, r_c]$ criterion in a low- and a high-gradient fracture situation of multidirectional composite laminates. If compared with metals, two basic differences must be noted, namely,

- the presence of laminate *subcritical matrix damage* prior to incipient transverse fracture across the critical ply
- the nature of the micromechanical *fiber transverse crack* tip damage in the critical ply

The subcritical matrix damage is due to the structural complexity of multidirectional composite laminates and it need not enter directly the $[D_c, r_c]$ fracture criterion. The damage $D(r)$ drawn over the ligament of the examples, Figures 6(*a*) and 6(*b*), rather symbolizes the micromechanical damage of the critical ply that is truly held responsible for the critical ply fracture. To what extent this damage can be simplified, e.g., be represented simply by the fiber stress or strain profile over the ligament of the critical ply, or be approximated by the linear elastic-stress profile $\sigma(r)$ of the homogenized equivalent laminate (HEL), remains to be clarified for each subclass of practical problems. If simplifications are made, the range of validity of the simplified fracture criterion must be identified.

Applications of the Simplified $[\sigma_c, r_c]$ Criterion to Quasi-Brittle Homogeneous Materials   The simplified $[\sigma_c, r_c]$-fracture criterion has been applied in gradual complexity to the ultimate static-failure prediction of a series of notched test pieces made of various homogeneous, heterogeneous, and composite globally brittle materials: first, to homogeneous zirconia ceramic materials; second, to short-fiber composite silica ceramics; third, to quasi-isotropic carbon–epoxy (CE) long-fiber composite laminates treated initially as elastic homogenized equivalent materials (HEL) at the laminate level, then as heterogeneous composite materials at the load-bearing ply level introducing nonlinear subcritical matrix damage prior to fiber failure.

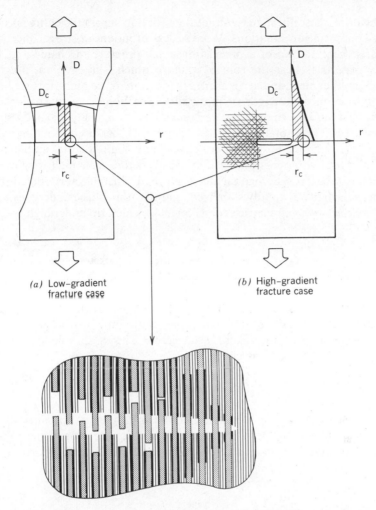

(a) Low–gradient
fracture case

(b) High–gradient
fracture case

(c) Micromechanical damage in the critical ply (after P.W. Beaumont)

Figure 6.   Effect of the $[D_c, r_c]$ criterion in (a) a low-gradient and (b) a high-gradient fracture situation in a multidirectional composite laminate.

3.   TOUGHNESS-TO-STRENGTH TRANSITIONS IN ELASTIC–BRITTLE MATERIALS

Fracture vs. Damage vs. Strength Concepts   Before tackling the more complex problem of the failure of heterogeneous composite materials, it is useful to clarify and demonstrate in some detail the concepts of damage and fracture mechanics on simpler homogeneous materials, and to introduce a promising theoretical framework to describe material-fracture initiation across the entire spectrum of stress-concentration situations. More or less pronounced stress concentrations can arise near the tip of a preexisting crack (fracture mechanics; LEFM for brittle materials), near the root of a sharp notch (slit) or blunt notch (hole) (damage mechanics), and at an arbitrary location in the high-stress regions of a smooth specimen (strength of materials). The homogeneous materials considered first are two engineering ceramic materials, namely, partially stabilized zirconia (PSZ) and totally tetragonal zirconia (TTZ), used in industrial applications that require both high toughness and high thermal stability, as for internal motor components. These materials derive their high toughness from the fact that the ZrO_2 tetragonal phase, normally unstable at room temperature, has been partially or totally stabilized, and that this phase undergoes a change accompanied with volumetric expansion at the crack tip under high mechanical strains.

Classical ASTM K-Toughness Determination   Crack-toughness characterization for such materials is rendered difficult by the fact that initial fatigue cracks cannot be propagated stably. Consequently, the ASTM K-toughness parameter, or apparent toughness K_{app}, which approximates the true K_{Ic} of cracks, is obtained from fracture-initiation measurements on *sharp notched or slit* (as opposed to fatigue precracked) test pieces. The K_{app} value, applying the standard ASTM formula, is given by

$$K_{app}(\text{ASTM}) = \sigma_u Y(a; \text{geometry})$$

We observe that K_{app} is not a material constant (i.e., not intrinsic to the material), but tends to be proportional to the square root of the notch root radius ρ for practical small ρ values (0.15 to 0.5 mm), i.e., proportional to the maximum stress σ_{max} at the notch root:

$$K_{app} \simeq \sigma_{max}$$

or

$$K_{app} \simeq A\rho^{1/2}$$

This implies that ideal cracks with $\rho = 0$ would have a vanishing or very low toughness, namely,

$$K_{app}(\text{ASTM}) \to 0 \qquad \text{for an ideal crack}$$

which is evidently erroneous, and which corresponds to the fact that the σ_{max} criterion is not applicable to slits or cracks.

Classical Unnotched Strength Determination after Weibull   At the other extreme, smooth test pieces without notches are also difficult to characterize at failure. In fact, their static tensile strength σ_{tu} is much influenced by the presence of initial micro or otherwise small defects, the instability of the most critical of which triggers catastrophic failure. This gives σ_{tu} a stochastic distribution, with a mean value inversely proportional to the volume of material tested (size effect).

For the brittle ceramics at hand, and many brittle materials, the Weibull theory provides a satisfactory framework, assuming statistical independence between the initial defects. According to this theory, the failure probability P_f of a unit volume of material V under a uniform tensile stress σ_t is given by

$$P_f = 1 - \exp[\Gamma^m(1 + 1/m)(\sigma_t/\sigma_{\text{ref}})^m V/V_{\text{ref}}]$$

where m is the Weibull modulus, V_{ref} is the reference volume, σ_{ref} is the mean failure strength of the reference volume, and Γ is the gamma function. The mean failure stress for the unit volume of material V is given by the expression

$$\sigma_{\text{tum}} = \int_0^\infty (1 - P_f)\,d\sigma = \sigma_{\text{ref}}(V_{\text{ref}}/V)^{1/m}$$

with a standard deviation

$$S = \sigma_{\text{tum}}[\Gamma(1 + 2/m)/\Gamma^2(1 + 1/m) - 1] \approx 1.2\sigma_{\text{tum}}/m$$

The Weibull modulus m characterizes the critical defect-size dispersion, that is, the material reliability as opposed to its toughness. A large Weibull modulus implies a small standard deviation, and a small size effect, i.e., a "good" material, but not necessarily a strong or tough one.

For a complex component or test piece geometry, the stress state is nonuniform, and the component reliability is the combined reliability of each of its constitutive elementary volumes V_i, namely,

$$1 - P_f = \prod_i (1 - P_f^i)$$

For an arbitrary material volume V under an arbitrary triaxial stress state σ, we obtain

$$1 - P_f = \exp[\Gamma^m(1 + 1/m)V_{\text{eff}}/V_{\text{ref}}]$$

In this expression, the normalized effective volume V_{eff} is given by the integral

$$V_{\text{eff}} = \int_V [(\langle\sigma_1\rangle + \langle\sigma_2\rangle + \langle\sigma_3\rangle)/\sigma_{\text{ref}}]^m\,dV$$

and $\langle\sigma_i\rangle$ is the positive part of principal stress σ_i for a normalized load Q_{norm}. Test-piece failure occurs at the load $Q = Q_{\text{norm}}(V_{\text{ref}}/V_{\text{eff}})^{1/m}$ under the standard deviation $S = 1.2Q/m$. The practical implementation of such a model is straightforward and was performed in the finite-element code PAM-FISS. Note that the fracture-triggering mechanism is the local stress σ, which implies a σ_{max}-based fracture criterion, and consequently tends to predict (as before with the ASTM K-toughness approach) a vanishing apparent toughness for specimens with sharp notches, namely,

$$K_{app}(\text{Weibull}) \sim \sigma_{tum} Y(a; \text{geometry})$$

$$K_{app}(\text{Weibull}) \rightarrow 0 \qquad \text{for an ideal crack}$$

Damage-Mechanics $[D_c, r_c]$-Toughness Determination Neither of the previous approaches, ASTM K toughness or Weibull strength, is therefore satisfactory for sharp notches in tough brittle materials. As an alternative, the simple form $[\sigma_c, r_c]$ of the $[D_c, r_c]$-fracture criterion can be applied, based on the following *three postulates*:

(i) Fracture initiates at the notch root, due to growth and coalescence of local subcritical microdefects, which practically sheds the influence of other remote near-critical initial defects.

(ii) Nonlinear effects in the process zone are negligible to the extent that the critical stress σ_c at distance r_c from the notch root is realistically predicted by the elastic failure-stress profile.

(iii) Fracture initiates at the root of a sharp notch when the elastic stress at distance r_c, namely, $\sigma(r_c)$, reaches the same value of σ_c as that induced by an ideal crack of toughness K_{Ic}, namely,

$$K_{Ic} = \sigma_c (2\pi r_c)^{1/2}$$

As mentioned earlier, for an ideal crack ($\rho = 0$), the stress profile normal to the crack plane at a small distance r from the crack tip is given by Westergaard's formula as

$$\sigma(r) = K_I/(2\pi r)^{1/2} \qquad \text{(Westergaard's stress profile)}$$

where K_I is Westergaard's stress-intensity factor in crack-opening mode I. Further, for a notch of root radius ρ, the normal stress profile for homogeneous materials is given by Craeger's formula [21] as

$$\sigma(r) = [K_I^*/(2\pi r)^{1/2}] f(\rho/r) \qquad \text{(Craeger's stress profile)}$$

or

$$\sigma(r) = [K_I^*/(2\pi \rho)^{1/2}] g(r/\rho)$$

with the functions $f(\rho/r) = (1 + \rho/r)/(1 + \rho/2r)^{3/2}$ and $g(r/\rho) = (r/\rho + 1)/(r/\rho + 1/2)^{3/2}$, and where K_I^* is Craeger's stress-intensity factor in notch-opening mode I. Evaluating Westergaard's and Craeger's stress profiles each at the point (σ_c, r_c), which is common to both stress profiles at the moment of specimen fracture, gives the apparent toughness of the notch assuming the material satisfies the previous three postulates:

$$K_{app}(\sigma_c, r_c) = K_{Ic}^* = K_{Ic}/f(\rho/r_c) = \sigma_c(2\pi r_c)^{1/2}(1 + \rho/2r_c)^{3/2}/(1 + \rho/r_c)$$

Note that this formulation has been independently proposed and validated by Kinloch [16] for polymers.

Discussion The $[\sigma_c, r_c]$ criterion departs from the previous two ASTM K-toughness and Weibull-strength models as it differs from a stress-intensity-based criterion, since

$$K_{app}(\sigma_c, r_c) = K_{Ic}/f(\rho/r_c)$$

and from a σ_{max}-based criterion, since the maximum stress at the notch root $(r = 0; g = 2^{3/2})$ is given by

$$\sigma_{max} = 2K_I^*/(\pi\rho)^{1/2}$$

that is,

$$K_{app}(\sigma_c, r_c) = 1/2(\pi\rho)^{1/2}\sigma_{max}^u$$

where σ_{max}^u is the ultimate maximum stress or the maximum stress at failure initiation. Note that $K_{app}(\sigma_c, r_c)$ no longer vanishes as the notch root radius ρ tends to zero, as for an ideal crack, for which σ_{max} becomes infinite and $K_{app}(\sigma_c, r_c)$ is formally indeterminate. In the limiting case of a vanishing notch root radius, we obtain

$$K_{app}(\sigma_c, r_c) = \sigma_c(2\pi r_c)^{1/2} \neq 0 \qquad \text{for an ideal crack}$$

Lower Limit for Small ρ Values When ρ becomes smaller than ρ^* such that

$$\rho^*/r_c = 1 + 5^{1/2} \simeq 3.2$$

then, mathematically, $K_{app}(\sigma_c/r_c)$ becomes smaller than $K_{Ic} = \sigma_c(2\pi r_c)^{1/2}$, implying that a sharp slit $(\rho < \rho^*)$ may turn out more critical than an ideal crack $(\rho = 0)$. By reasoning that such a sharp slit would immediately fracture and turn into a crack, K_{Ic} is taken as the lower practical limit for $K_{app}(\sigma_c, r_c)$ and the corresponding curves as a function of ρ are truncated for $\rho < \rho^*$ and $K_{app}(\sigma_c, r_c) \leqslant K_{Ic}$.

4. EXAMPLE OF HOMOGENEOUS AND COMPOSITE CERAMICS

Zirconia Ceramics   The previous three fracture criteria (ASTM K, Weibull strength, and $[\sigma_c, r_c]$) have been applied to PSZ and TTZ zirconia ceramics. Note that the PSZ has a larger grain size (ca. 10 to 30 μm) and a smaller mean ultimate bending strength ($\sigma_{bum} = 370$ MPa) than the TTZ material (respectively, ca. 1 μm and 870 MPa), as shown in Table 1.

Two types of test pieces have been tested, namely, SENB (single-edge notch bending) and CT (compact tension) test pieces, each with several notch sizes, including self-similar specimens in scale ratios of 1, 2, and 3, as shown in Figure 7. Four specimens were tested per result point. Calibration of the considered fracture criteria is conducted on a subset of the SENB test pieces, corresponding to the small scale 1, labeled S_{1i}, with several ρ values i. This calibration is then used to predict the ultimate strengths of the larger SENB (scales 2 and 3, labeled S_{2i} and S_{3i}) and CT (scale 1, labeled C_{1i}) test pieces.

As mentioned earlier, the ASTM procedure applied to notched test pieces yields stress-dependent apparent toughnesses, K_{app} (ASTM), which are plotted as the experimental data points with their scatter bar in Figure 8(a) and Figure 8(b), for PSZ and TTZ, respectively. The tendency for K_{app} (ASTM) to follow the σ_{max} criterion for medium sharp notches is shown by the dashed line $K_{app} = \sigma_{bum}/2(\pi\rho)^{1/2}$. Note the increasing discrepancies toward smaller and larger ρ values.

The calibration of the $[\sigma_c, r_c]$ criterion is given on Table 1, and plotted as the solid lines in Figure 8. Note the excellent calibration (S_{1i}) as well as prediction

TABLE 1   Fracture-Criteria Calibration for PSZ and TTZ Ceramics

	PSZ	TTZ
Grain size (Desmarquet) (μm)	10–30	<1
σ_{bum} (MPa) mean bending strength	370	870
Calibration scatter	narrow	broad
$[\sigma_c, r_c]$-Criterion Calibration		
σ_c(MPa)	368	877
$r_c(\mu$m)	50	~18
$K_{Ic} = \sigma_c(2\pi r_c)^{1/2}$(MPa-m$^{1/2}$)	6.5	9
ρ/r_c	2.5 → 9	8 → 25
ρ^* (μm) (crack upper limit $\rho^* = 3.2r_c$)	160	60
Weibull Criterion Calibration		
m modulus	19.3	11.6
σ_{ref} (MPa)	371	874
V_{ref} (mm^3)	0.19	1

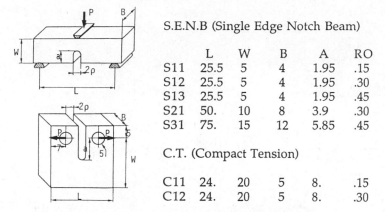

S.E.N.B (Single Edge Notch Beam)

	L	W	B	A	RO
S11	25.5	5	4	1.95	.15
S12	25.5	5	4	1.95	.30
S13	25.5	5	4	1.95	.45
S21	50.	10	8	3.9	.30
S31	75.	15	12	5.85	.45

C.T. (Compact Tension)

C11	24.	20	5	8.	.15
C12	24.	20	5	8.	.30

Figure 7. Zirconia ceramics test pieces (SENB and CT).

$(S_{2i}, S_{3i}; C_{1i})$ at all ρ values for the narrow-scatter PSZ material data points, and the comparatively mediocre ones for the broader scatter TTZ material data points. Incidentally, σ_c is very close to σ_{bum} for both materials.

The Weibull criterion calibration has been performed on smooth test pieces and the values are given in Table 1. As observed, the PSZ material has a higher Weibull modulus m and, accordingly, a lower σ_{bum} scatter than the TTZ material. Using the normalized effective volume as calculated by the PAM-FISS code for the SENB and CT test pieces allows us to draw the predicted mean (solid curve) and standard-deviation (dashed lines) values of K_{app} (Weibull) for the Weibull criterion, incorporating the effects of the high-stress concentrations at the notch for PSZ and TTZ materials, as shown in Figure 9.

Note the systematic underprediction of the Weibull criterion for the PSZ material, which behaves tougher than predicted, and the good correlation for the TTZ material, which behaves as predicted by the Weibull model.

Discussion   The previous results can be interpreted as follows:

(a) For the "good" PSZ material ($m = 19$), the measured K_{app} (ASTM) values for notched test pieces show little scatter, in fact, less than implied by the scatter on the smooth pieces ($1.2/m \simeq 6\%$). However, the effect of the notch-stress concentrations and the scale effect is uniquely (and closely) predicted by the $[\sigma_c, r_c]$ criterion, suggesting the validity of the defect growth and coalescence damage model, as opposed to the preexisting critical-defect instability model of the Weibull criterion. Note that the maximum-stress criterion does not apply, except fortuitously for a limited range of sharp notches.

(b) For the relatively "poorer" TTZ material ($m = 11$), the measured K_{app} (ASTM) values show a larger scatter, coherent with that implied by the Weibull model ($1.2/m \simeq 10\%$), and the notched mean values are well

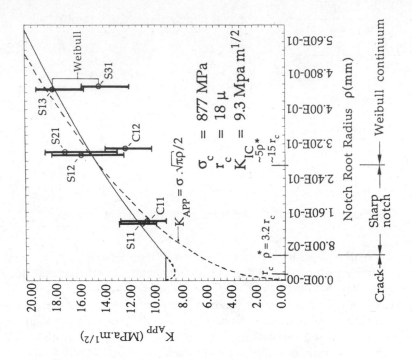

Fig. 8b : Zirconia TTZ

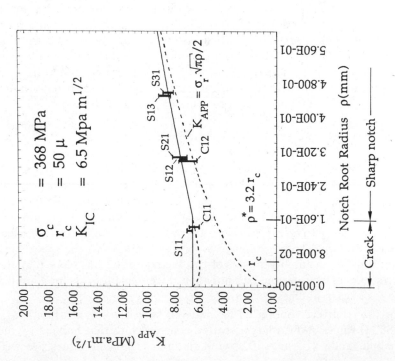

Fig. 8a : Zirconia PSZ

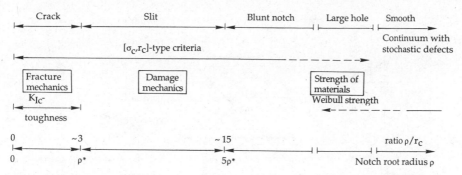

Figure 8. (a) and (b) Application of fracture criteria to zirconia ceramics. (c) Crack-to-notch continuum transition in terms of the characteristic material distance and validity ranges of criteria.

predicted by the Weibull model using the stress-normalized effective volume. However, the $[\sigma_c, r_c]$ criterion yields equally well or better the mean values of the smaller SEND test pieces S_{1i} and of the small C_{1i} test piece, but systematically overpredicts the toughness of the test pieces with large notches (e.g., $\rho > 5\rho^*$).

(c) These observations suggest the following fundamental interpretation of material toughness for fracture initiation in notched brittle test pieces, based on the quantification of the value of the notch root radius ρ with respect to the intrinsic characteristic material distance r_c:

(i) If $\rho < \rho^* \sim 3.2r_c$, then the notch behaves like a *crack* of deterministic toughness $K_{Ic} = \sigma_c(2\pi r_c)^{1/2}$, which is the lower-bound intrinsic toughness value of the material.

(ii) If $\rho^* \leqslant \rho \leqslant \sim 5\rho^*(\sim 15r_c)$, then the *sharp notch* behaves like a slit, of pseudo-deterministic apparent toughness $K_{app}(\sigma_c, r_c) = K_{Ic}/f(\rho/r_c)$, and the material may appear tougher than predicted based on its smooth static strength combined with the local maximum stress.

(iii) If $\rho \gg \rho^*$, the *blunt notch* behaves like a smooth continuum with a hole with a local stochastic strength predicted by its Weibull continuum parameters.

These three ρ regions, for cracks, slits, and blunt notches, are delimited by ρ^* and $5\rho^*$, as indicated in Figure 8(c).

(d) The characteristic material distance r_c seems typically proportional to the material grain size and of the order of 10 grain sizes.

Composite SiO$_2$/SiC Ceramics   The $[\sigma_c, r_c]$ criterion has been applied in a similar fashion to silica composite ceramics reinforced with silicon carbide short fibers. The ligament failure-stress profiles used for calibration on a series of notched CT test pieces with notch root radii of $\rho = 0.47, 0.30$, and 0.15 mm are

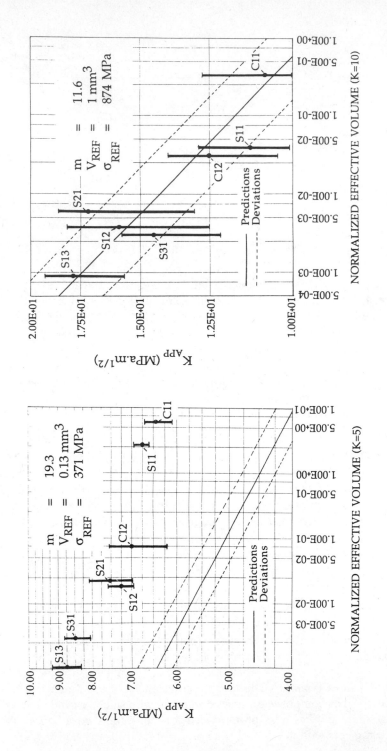

Figure 9. Application of Weibull's strength model to zirconia ceramics.

a. Zircone PSZ (large grains, $2.5 < \rho / r_c < 9$)

b. Zircone TTZ (fine grains, $8 < \rho / r_c < 25$)

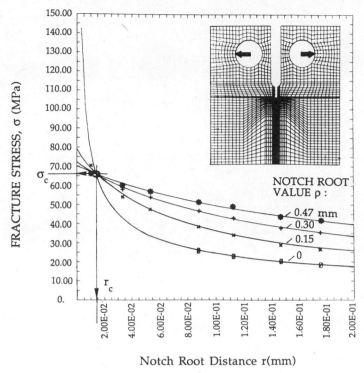

Figure 10. Calibration of the $[\sigma_c, r_c]$ criterion on short-fiber composite SiO_2/SiC ceramics.

shown in Figure 10, together with the extrapolation for the true crack-stress profile, $\rho = 0$, through the calibrated (σ_c, r_c) point. The excellent correlation supports the validity of the three damage-mechanics postulates made previously, as potentially applicable to short-fiber reinforced composite ceramics.

5. THE PAM-FISS/BI-PHASE COMPUTER MODEL

Bi-Phase Model The *Bi-Phase* model is a numerical material model for finite-element (FE) analyses adapted to unidirectional long-fiber reinforced composites or composite fabrics, as shown in Figure 11. The stiffness and the resistance of its elements are calculated by superimposing the effects of an orthotropic material phase (matrix minus fibers) and of a unidirectional material phase (fibers), with (or without) deformation compatibility. Each phase (fibers, matrix) is assigned a different rheological law: elastic–plastic/brittle orthotropic for the matrix phase and unidirectional elastic–brittle for the fibers. Upon incremental loading, the stresses are calculated separately in each phase and damage (matrix cracking, matrix slipping; fiber rupture) can propagate

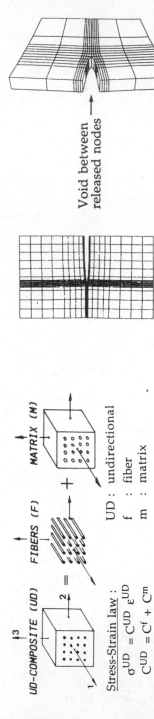

UD-COMPOSITE (UD) = FIBERS (F) + MATRIX (M)

Stress-Strain law :
$$\sigma^{UD} = C^{UD} \varepsilon^{UD}$$
$$C^{UD} = C^f + C^m$$

UD : undirectional
f : fiber
m : matrix

Known material properties :
$E_{11}^{UD}, E_{22}^{UD}, G_{12}^{UD}, \nu_{12}^{UD}$ = in-plane UD material constants

E_{true}^{f} = true fiber modulus

α = fiber volume fraction

Calculated quantities :
$$\nu_{21}^{UD} = \nu_{12}^{UD}\, E_{22}^{UD} / E_{11}^{UD}$$
$$N^{UD} = 1 - \nu_{12}^{UD}\, \nu_{21}^{UD}$$
$$E_{11}^{f} = \alpha E_{true}^{f}$$

Derived orthotropic matrix material constants :
$$E_{11}^{m} = E_{11}^{UD} - \varepsilon_{11}^{f}$$
$$E_{22}^{m} = E_{22}^{UD} / (1 + \nu_{12}^{2}\,(E_{22}^{UD} / E_{11}^{UD})\,(E_{11}^{f} / (E_{11}^{UD} - E_{11}^{f})))$$
$$\nu_{12}^{m} = \nu_{12}^{UD}$$
$$\nu_{21}^{m} = \nu_{21}^{UD} / (1 - E_{11}^{f}\, N^{UD} / E_{11}^{UD}) \neq \nu_{21}^{UD}$$
$$G_{12}^{m} = G_{12}^{UD}$$

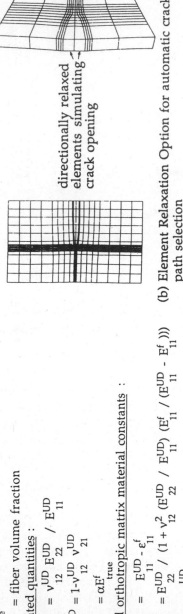

Void between released nodes

(a) Node Release Option along prespecified mesh lines

directionally relaxed elements simulating crack opening

(b) Element Relaxation Option for automatic crack path selection

Program PAM-FISS : Automatic/Arbitrary Crack Advance Option

Figure 11. Program PAM-FISS/Bi-Phase rheological model.

independently, based on the criteria chosen for each phase. A multidirectional laminate is modeled by stacking through the thickness several such elements with the fibers oriented in different directions with respect to a global reference frame.

Subsequently, the material model of the matrix phase has been augmented by an elastoplastic (von Mises) material law and by a modulus damage-fracturing material law. This law is also available for the fiber phase.

PAM-FISS Program   The PAM-FISS computer code contains the Bi-Phase model and is a specialized linear and nonlinear fracture-mechanics and stress FE analysis code [7]. It allows for arbitrary three-dimensional geometries, static and dynamic loading (including impact, contact, and interface sliding), and thermal loading (steady-state or transient temperature distributions, e.g., calculated by the thermal FE analysis code PAM-T3D).

The specialized *fracture-mechanics* options of the PAM-FISS program allow us

- to zoom on the areas of high-stress concentrations (notch roots, delamination crack tips) via mesh reduction and locally fine meshes
- to propagate crack or delamination fronts, in directions independent of the finite-element mesh orientation, by evaluation of the criteria for crack propagation in each finite element at each load increment [Figure 11(*b*)]
- to automatically evaluate and select from several toughness criteria: K (stress-intensity factor), J (Rice integral), G (total or partial strain-energy release rate), $[D_c, r_c]$ (critical damage over a characteristic distance)

The program is linked to interactive graphic pre- and postprocessors (PRE-3D; DAISY), which permit us to generate data and to exploit the results in a coherent fashion.

Modulus Damage-Fracturing Law   The modulus damage-fracturing law in PAM-FISS introduces a linearly varying fracturing damage (microcracks, crazing damage) $d(\epsilon)$ into the material modulus matrix, activated between a user-specified equivalent initial threshold strain ϵ_i and a specified final rupture strain ϵ_u such that

$$E(d) = (1 - d)E_0$$

where E is the current elastic modulus matrix of the resin phase or the axial modulus of the fibers, E_0 is the initial state of these quantities, and d is the fracturing damage, varying linearly between $d = 0$ for $0 < \epsilon < \epsilon_i$ and $d = d_{max}$ at $\epsilon = \epsilon_u$, as shown in Figure 12. The equivalent strain ϵ can represent volumetric strain (first invariant of the strain tensor) or represent shear strain e.g., second invariant of the deviatoric strain tensor) or it can be a chosen mixture of volumetric and shear strains.

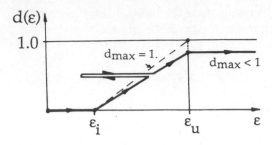

ε_i = initial (threshold) strain where fracturing begins

ε_u = ultimate strain for maximum damage
$d_{max} \leq 1.0$

(a) Linear fracturing damage law

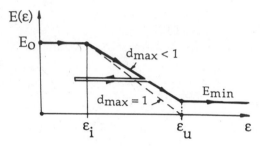

$E(\varepsilon) = (1-d(\varepsilon))\, E_o$

$E_{min} = (1-d_{max})\, E_o$

for $d_{max} = 1.0 : E_{min} = $ zero

(b) Evolution of the axial modulus with damage

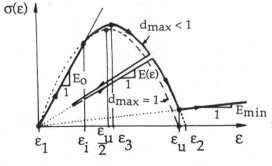

$\sigma(\varepsilon) = E(\varepsilon)\, \varepsilon$

$= (1-d(\varepsilon))\, E_o\, \varepsilon$

$= E_o\, \varepsilon\, (1 - \dfrac{d_{max}}{\varepsilon_u - \varepsilon_i} <\varepsilon - \varepsilon_i>)$

which is for $d_{max} = 1.0$:

$= E_o\, \varepsilon\, \dfrac{<\varepsilon - \varepsilon_i>}{\varepsilon_u - \varepsilon_i}$

(c) Fracturing law stress-strain diagram

Figure 12.   Fracturing material law in PAM-FISS Code.

A typical stress–strain diagram resulting from this model is also shown in Figure 12. The diagram illustrates the capacity of the model to introduce strain-softening at loading and modulus reduction at unloading, as it is, e.g., the case for microfracturing materials. The model is therefore well suited to describe subcritical matrix damage that occurs, e.g., in notched CE test pieces prior to and simultaneously with fiber cracks.

6. NUMERICAL SIMULATION OF MACROSCOPIC MATRIX CRACKS IN TENSILE TEST PIECES

The first applications shown deal with the propagation of subcritical macroscopic matrix cracks in UD, CP, and ML double-edge notched tensile CE (T300/914C) test pieces. The analyses demonstrate the general ability of the PAM-FISS code to deal with such problems, and they demonstrate the applicability of the critical strain-energy release rate (G_c) criterion to the simulation of the stable or unstable propagation of macroscopic matrix cracks. Such cracks occur in the form of intralaminar transverse matrix cracks (e.g., in the 90°-plies), splitting axial or off-axis matrix cracks (e.g., in the 0°-off-axis-plies), and as interlaminar matrix cracks (delamination cracks between individual plies).

Note, however, that macroscopic subcritical matrix cracks only create the correct boundary conditions for the critical load-bearing ply transverse fiber cracks, which are responsible for specimen rupture. Their detailed calculation can therefore be replaced ultimately by incorporating their equivalent action into an appropriate material law, such as the PAM-FISS fracturing material law.

Prediction of Matrix Splitting Composite materials, when they have free edges or defaults, experience damage very early. A basic problem is the choice of the strength or toughness function for such materials and the determination of its critical level in particular for matrix cracks that occur well before fiber rupture [3], [14].

The example of Figure 13 [14] deals with the identification of such a toughness criterion for macroscopic matrix crack advance, which predicts the initiation and propagation of intralaminar matrix cracks (matrix splitting) in *unidirectional* double-edge-notched tensile test pieces (UD-DENT). Depending on the orientation of the fibers with respect to the direction of pulling (0°, 10°, 45°, 90°), the crack can have four branches and it can be totally stable (0°) (i.e., the cracks propagate only upon load increase), or it can be initially stable and then unstable showing two branches (10°), or it is totally unstable having only one branch (45° and 90°).

For the same material and loading, it is essential that the prediction of these crack patterns be possible using one and the same intrinsic criterion, the nature of which remains to be determined. In the present case, the G_{Ic} criterion applies (fracture mode I strain-energy release rate). The work leading to this result is described in Reference 14. The G_{Ic} criterion uniquely and completely characterizes the modes of crack propagation and rupture of unidirectional tensile specimens. The critical value for the epoxy resin was found equal to 0.15 N/mm.

Prediction of Delamination Multilayered (ML) smooth (i.e., unnotched) $[0_2, 45, 0_2, -45, 0_2, 90]_s$ test specimens can be built such as to avoid premature delamination under static tensile loading. After a certain number of fatigue

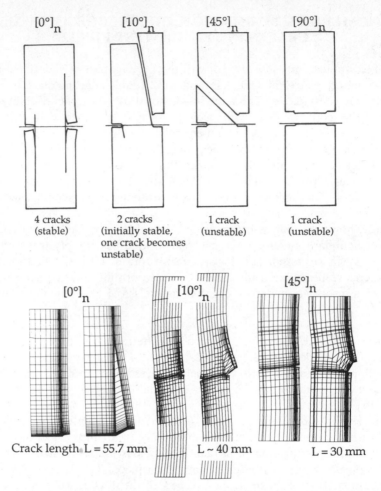

Figure 13. Automatic numerical prediction of stable/unstable crack advance in the matrix resin, parallel to the fiber direction, for various fiber orientations in unidirectional double-edge notched test pieces (UD-DENT).

cycles at about 50 to 70% of the ultimate loading, however, delamination sets in and it propagates in a stable fashion under subsequent static tensile reloading, toward and up to specimen rupture, as shown in Figure 14 [15, 17].

The G_{Ic}-toughness criterion identified before permits to assess the way the delamination crack initiates and propagates under static tension. The true risk of instability can be evaluated by calculating the function $G_I(L)$, Figure 14(b), for a hypothetical crack advance at some level of constant imposed axial displacements, where L is the distance of propagation of the delamination crack front normal to the free edge.

The first branch of the curve near the free edge is ascending steeply as a

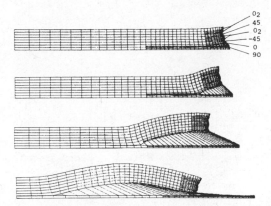

(a) ML Delamination Simulation ; Deformed Shapes after Delamination (Displacement Scale : 50) - PAM-FISS/BI-PHASE Model

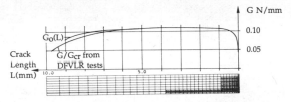

(b) Curves of the Strain Energy Release Rate vs delamination length $G(L)$ at longitudinal level ε_{xx} = 0.87 % (transverse slice model)

Figure 14. Delamination of the unnotched CE multidirectional DFVLR standard test piece $[0_2, +45, 0_2, -45, 0, 90]_s$.

function of L. This means that the static delamination crack tends to be unstable initially. Once propagated over a short distance of the order of magnitude of the laminate thickness, the $G(L)$ curve becomes shallow and it decreases farther on. This indicates that the strain energy released during (hypothetical) crack opening approaches a peak and then decreases (at constant loading) and that delamination is eventually stable under constant loading.

It is remarkable that the G_I criterion and the experimentally calibrated critical value $G_{Ic} = 0.15$ N/mm are identical to those found before for matrix-splitting on DENT test pieces. The work leading to these results is described in detail in Reference 15.

Prediction of Generalized Damage (Matrix Ubiquitous Cracking) The same multilayered $[0_2, 45, 0_2, -45, 0, 90]_s$ composite, tested in the form of a double-edge-notched tensile test piece (ML-DENT), develops a region of generalized subcritical matrix damage near the notch root, which spreads progressively as the axial tensile load increases. Other than in unidirectional UD-DENT specimens that clearly exhibit intralaminar matrix macrocracks (matrix splitting), the cracking pattern of the ML-DENT test piece is very diffuse (i.e.,

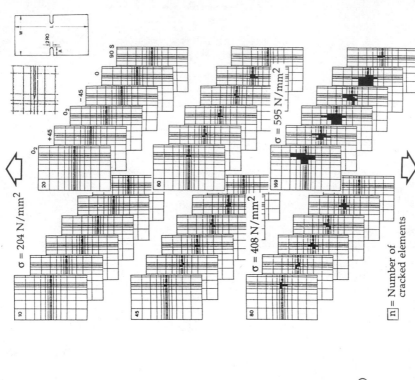

$\sigma = 204 \ N/mm^2$

$\sigma = 408 \ N/mm^2$

$\sigma = 595 \ N/mm^2$

90 S

0

−45

0_2

+45

0_2

20

60

169

10

45

80

$\boxed{n}$ = Number of cracked elements

(a) Matrix Crack Propagation (UD) & Damage Spread (ML) in Notched CE Tension Specimens (X-Ray pictures)

7.28 KN

8.04

11.86

10.2 KN

13.10

(b) Numerical simulation of the damage spread near the notch root by cracking of the first 169 elements for an axial stress growing from 204 N/mm^2 to 595 N/mm^2 ($\sigma_R \sim 820$ N/mm^2).

Figure 15. Matrix crack-damage spread in the multidirectional notched standard test piece. Experimental vs. numerical results (Pam-Fiss/Bi-Phase model).

230

ubiquitous), and it consists of a multitude of stable, short, intralaminar matrix macrocracks, as shown in Figure 15(*a*). These macrocracks initiate early at about 20% of the smooth laminate strength, and they propagate intermittently parallel to the fibers of each layer. These cracks can be viewed as intralaminar matrix macrocracks that are continually intercepted and arrested by the transverse fibers of the confining adjacent layers.

The first *fiber fractures* occur at a sensibly later stage at about 50% of the rupture strength of the considered composite, when the region of ubiquitous matrix damage has typically spread over a distance comparable to the notch length.

The same PAM-FISS/Bi-Phase crack advance model has again been used to simulate this ML-DENT test. The matrix of each element is fractured automatically as soon as G_I is greater than a critical value G_{Ic}, which is the same as the one reported before. About 100 elements experienced matrix cracks while the axial load increases from about 25 to 50% of the specimen rupture strength. At the tensile load observed experimentally for the first fiber ruptures near the notch root [Figure 15(*b*)], the region of matrix damage has been reproduced realistically in shape and extent.

Since the simulated subcritical matrix damage is not directly responsible for ultimate specimen failure, and since matrix macrocracks do not really open, the equivalent effect of the network of ubiquitous matrix cracks can be calculated more efficiently by using microfracturing material constitutive relations for the matrix phase. This level of ubiquitous and stable precracking, preceding fiber fracture, is in fact only a salubrious blunting of the stress concentration in the load-bearing ply fibers near the notch root.

Discussion: Admissible Stress State   Up to this point, the overall behavior of the specimen remains almost completely linear; the structure has effectively adapted via localized matrix ubiquitous macrocracking, demonstrating its *tolerance to damage*. Noting that the fibers remained mostly intact, we can admit that the state of stress is *admissible* although the material has undergone damage or rather has adapted. There exist analogies in other composite materials (e.g., reinforced concrete). An analogy can also be made with metals, where crack blunting occurs through dislocation and plastic flow, with hardening around crack tips. Matrix ubiquitous macrocracking plays a similar beneficial role in notched or damaged LFRP composites as plasticity in metals with respect to crack propagation.

7. NUMERICAL SIMULATION OF THE FAILURE OF WOVEN COMPOSITE COMPRESSIVE TEST PIECES

Simulation of Local Buckling   In regions of compression in the fiber direction, LFRP composites can exhibit the damage mode of local fiber buckling. This is especially true in composite fabrics, the weave of which causes the fibers to bend near the intersections of the warp and weft threads [18]. The associated redirection forces inflict a high stress on the matrix material, which can undergo compressive failure via slipping (shear plasticity) and/or crazing (tensile fracturing). This in turn deconfines the fibers and leads to increases in curvature and to a relaxation with a decrease in fiber forces.

This decrease in fiber forces weakens the intact specimen critical cross section in a typical compression or bending test by reducing the net resisting force in a compression test or by reducing the contribution of the compressive fibers to the resisting moment in a bending test. The result is a stress redistribution to other parts of the critical cross section. In bending tests, the tensile fibers tend to rupture subsequently, due to the overstress they receive after the compressive fibers have unloaded. This in turn can give rise to higher compression in the fibers that have not yet undergone buckling, and the repeated "domino" failure process of matrix failure, fiber buckling, and fiber fracture ultimately leads to total specimen failure.

Compressive Test   Figure 16(*a*) shows details of the three-dimensional finite-element mesh of the equilibrated three-layered carbon-epoxy Satin 5 fabric "G803 M10 Brochier" (50% warp, 50% weft) used for the simulation of the compression test, as well as a sequence of deformed shapes of the specimen, obtained with the PAM-FISS code and Bi-Phase material model.

The elastic properties of the fiber and matrix phase of this fabric are essentially the same as for the specimens studied before. The fiber volume fraction is 0.47. The overall axial elastic modulus of the three-layered fabric is equal to 46,026 N/mm^2, which is about 18% below the axial stiffness of a corresponding cross-ply with straight fibers. The matrix material was allowed to yield (von Mises plasticity) at a calibrated yield stress in order to simulate local matrix failure under the influence of the high stresses due to the fiber actions.

Deformed shapes are shown in Figure 16(*b*) and experimental and analytical axial force vs. axial overall strain curves are compared, and the local fiber buckling near the warp–weft intersections is clearly visible in Figure 16(*c*). The results have been achieved by calibrating the matrix elastic limit to 16 N/mm^2 and with a true fiber-rupture stress of 3000 N/mm^2.

Bending Test   Figure 17 gives a closeup on the finite-element mesh of the unidirectional three-layered carbon-epoxy Satin 1 fabric "G827 M10 Brochier" (98.5% warp, 1.5% weft) used for the simple-beam transverse bending test (roller supports). The individual phases of this tissue have rheological properties identical to those used for the compression test.

The calculated von Mises plastic strain contours of the matrix phase, plotted on the upper surface of each of the three layers, show a sudden tenfold increase when one of the compressive threads buckles.

The experimental and the calculated transverse load vs. transverse (midspan) displacement curves are compared. The failure in bending is interpreted as a succession of compressive fiber buckling due to matrix-yielding followed by tensile fiber ruptures due to subsequent stress redistributions.

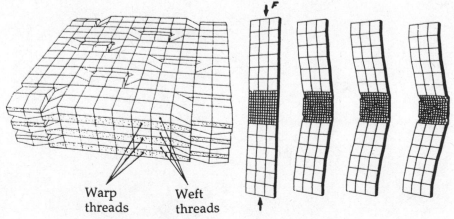

(a) FE Mesh Closeup
    (thickness scale : 5)

(b) Deformed Shape Plots
    (scale 10 : 1)

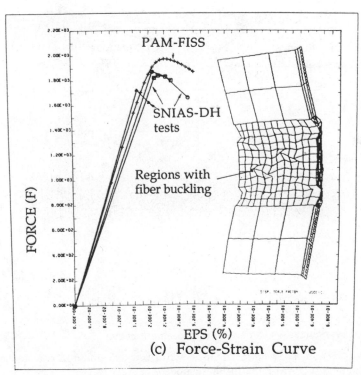

(c) Force-Strain Curve

Figure 16. Carbon-epoxy Satin 5 "G803 Brochier" equilibrated tissue (three layers) compression test (PAM-FISS/Bi-Phase model).

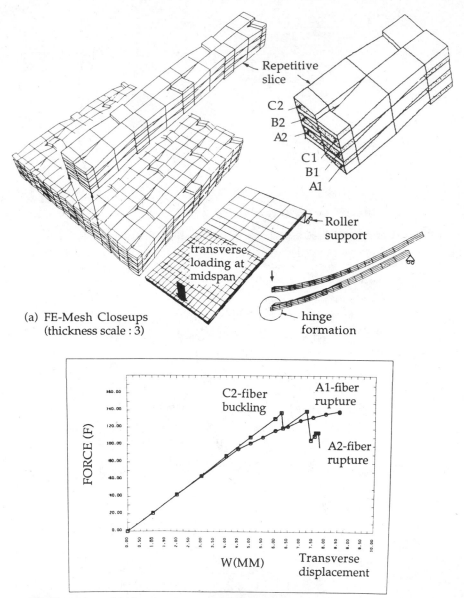

(a) FE-Mesh Closeups
(thickness scale : 3)

(b) Transverse load vs transverse displacement curve

Figure 17. Carbon-epoxy Satin 1 "G827 M10 Brochier" unidirectional tissue (three layers) bending test (PAM-FISS/Bi-Phase model).

8. NUMERICAL SIMULATION OF NOTCHED-LAMINATE FIBER FRACTURE AND ULTIMATE TENSILE-FAILURE PREDICTION

While in the previous examples of composite-laminate failure predictions, fracture simulation was based on stress-intensity-type fracture criteria for matrix macrocrack advance, and on local matrix shear and compressive failure with an ensuing deconfinement and local buckling of compressed fibers in the case of intact compression and bending test pieces for woven CE fabric, the examples of this section deal with the ultimate failure of middle-notched tensile (MNT) CE cross-ply (CP) and multilayered (ML) laminates, including fiber fracture in the load bearing $0°$ and off-axis plies.

As described in the previous section, the main challenge now is to identify and calibrate for composite materials an intrinsic local fracture criterion for the fiber phases, independently of the local loading and stress–strain conditions. This is best achieved by studying the fracture initiation and ultimate strength of notched tensile test pieces, where variations of notch geometry from round holes to slits, and of notch size from minute to large, allow us to produce arbitrary local *stress concentrators* and *stress gradients* in a controlled fashion at the notch root.

First, a series of MNT test pieces under quasi-static tensile loading with different stackups, scales, thicknesses, hole shapes (slits, notches, round holes), and sizes are examined by comparing test results at failure of the specimens with predictions gained from various fracture criteria, including the stress- or strain-intensity type K, G, and Poe criteria and the simplified damage-mechanics type $[\sigma_c, r_c]$- fracture criterion, applied on the *homogenized equivalent laminate* (HEL) level. In the latter, criterion damage is associated simply with the linear elastic ligament tensile stresses $\sigma(r)$ calculated at the level of the HEL, and the criterion postulates that laminate fracture begins when a critical threshold value σ_c is exceeded by the HEL stresses in the ligament stress profile, across a characteristic distance r_c. The critical stress and the characteristic distance are material constants and must be found by calibration. Since the HEL theory treats each laminate as a homogeneous equivalent material, we must expect to find different values of critical stress σ_c and characteristic distance r_c for each different stackup, which is a limitation associated with the HEL and simplified criterion. But the simplified $[\sigma_c, r_c]$ criterion, indeed, appears to be intrinsic with respect to the stress-gradient effect introduced by variations of the notch radius when all other laminate characteristics are kept constant, while the intensity-type criteria are not.

Next, the $[D_c, r_c]$ criterion is applied to the fracture prediction of the *load-bearing ply* in the form of an $[\epsilon_c^f, r_c^f]$-fracture criterion, which postulates that fiber fracture across the load-bearing ply occurs as soon as the linear ligament tensile fiber strain $\epsilon^f(r)$ exceeds a critical threshold value ϵ_c^f across a characteristic distance r_c^f. The critical strain and the characteristic distance are material constants that must be found by calibration. Since this approach directly recognizes the tensile fracture of the load-bearing plies, the calibrated values of ϵ_c^f and r_c^f must be universally applicable to all composite stackups when the individual plies are made from the same fiber and matrix materials to the same volume fraction.

The $[\epsilon_c^f, r_c^f]$ criterion requires, however, that the load-bearing-ply fiber tensile strains be calculated accurately inside the laminate by taking into account stress redistributions that result from subcritical matrix cracks. This generally involves a complete nonlinear calculation, where the laminates are modeled individually. The PAM-FISS/Bi-Phase material model permits such calculations to be carried out economically with modern computers, when matrix damage such as subcritical matrix cracks and interlaminar delamination cracks is simulated via the equivalent constitutive matrix modulus damage-fracturing law described earlier.

The work leading to these results is documented in detail in de Rouvray et al. [19, 20] and is briefly summarized now.

8.1. Experimental Data Base

Figure 18 contains experimental results obtained by P. W. Beaumont at the University of Cambridge Engineering Department (UC/ED), on the tensile failure of middle-notched tensile (MNT) cross-ply (CP) and multilayered (ML) carbon-epoxy (CE) test pieces of various scales and geometries.

The test pieces include the stackups CP $[90_m, 0_m]_{ns}$, where $n = 1, 2$ for the $m = 1$ specimens, made of material A, fabricated by DFVLR Braunschweig, West Germany, and where $n = 1, 2, 4$ for the $m = 2$ specimens, made of material B, fabricated by Fothergill-Rotorway, as well as a few multilayered test pieces, ML $[0, \pm 45, 90_{1/2}]_s$, fabricated by DFVLR.

Smooth Specimens   Of each material, stackup, and thickness, smooth (i.e., unnotched) specimens have been tested up to tensile failure, which resulted in a set of ultimate smooth-specimen tensile strengths σ_{ou}. The strength values of Table 2 are defined as specimen ultimate tensile force F_{tu} at catastrophic failure divided by specimen width w times thickness t, or *material strength*, namely,

$$\sigma_{tu} = F_{tu}/wt$$

The given values are the mean of three or more result points.

Notched Specimens   All ratios of specimen width to notch length were kept constant, namely, $w/2a = 3$, where $2a$ is the notch length, and the specimen sizes cover six different geometric scales s, from $s = \frac{1}{2}$ to $s = 4$, covering a scale span of 8. The notch geometries include five different stress risers: round holes ($a/\rho = 1$), blunt notches ($a/\rho = 3$ and 6), and slits ($a/\rho = 30$ and 60), covering an a/ρ span of 60. Table 3 summarizes the basic set of 26 notched specimens tested in the

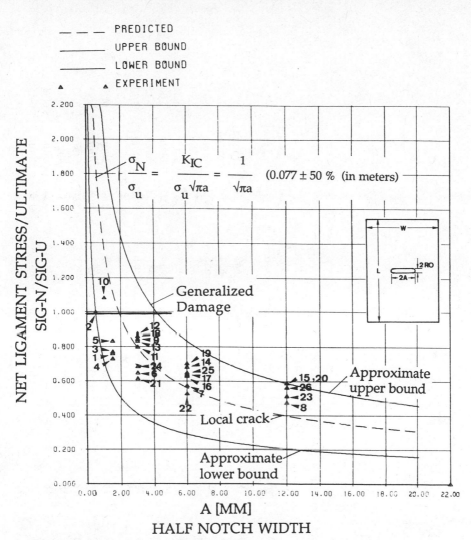

Figure 18. Experimental results on CP and ML middle-notched tension test pieces.

TABLE 2 Smooth-Specimen Tensile Strength

Laminate	$\sigma_{ou}(\text{N/mm}^2)^a$	$\epsilon_{tu}(\%_{oo})^b$	Material
CP $[90, 0]_s$	854	11.70	DFVLR
CP $[90_2, 0_2]_s$	945	12.00	Fothergill-Rotorway
CP $[90_2, 0_2]_{2s}$	884	12.25	Fothergill-Rotorway
CP $[90_2, 0_2]_{4s}$	921	12.19	Fothergill-Rotorway
ML $[0, \pm 45, 90]_s$	448	11.95	DFVLR

$^a\sigma_{ou}$ = smooth tensile strength (ultimate failure force/section area)
$^b\epsilon_{tu}$ = smooth specimen and 0°-ply fiber ultimate tensile strain

TABLE 3 Typical Experimental Notched-Specimen Failure Test Results (Cambridge University)

Laminate	No.	t (mm)	s	w (mm)	a (mm)	ρ (mm)	a/ρ	σ_{tnu} (N/mm^2)	$\dfrac{\sigma_{tnu}}{\sigma_{ou}}$	Notch Type	ϵ_{tnu} (‰)
CP $[90, 0]_s$	1	0.5	1	9	1.5	1.5	1	647	0.76	round	5.94
	2		1/3	3	0.5	0.5	1	852	1.00	round	7.82
	3		1	9	1.5	0.5	3	655	0.77	blunt	6.01
	4		1	9	1.5	0.25	6	621	0.73	blunt	5.70
	5		1	9	1.5	0.05	30	709	0.83	slit	6.51
	6		2	18	3.0	0.5	6	551	0.65	blunt	5.06
	7		4	36	6.0	1.0	6	487	0.57	blunt	4.47
	8		8	72	12.0	2.0	6	404	0.47	blunt	3.70
CP $[90_2, 0_2]_s$	9	1.0	2	18	3.0	3.0	1	796	0.84	round	6.70
	10		2/3	6	1.0	1.0	1	1022	1.08	round	8.64
	11		2	18	3.0	1.0	3	752	0.80	blunt	6.35
	12		2	18	3.0	0.05	60	827	0.88	slit	7.00
	13		2	18	3.0	0.5	6	784	0.83	blunt	6.63
	14		4	36	6.0	1.0	6	642	0.68	blunt	5.43
	15		8	72	12.0	2.0	6	556	0.59	blunt	4.71

Layup	No.		t	w	a	ρ	s	σ_{tmu}		notch	ϵ_{tmu}
CP [90₂, 0₂]₂ₛ	16		4	36	6.0	6.0	1	554	0.63	round	5.12
	17		4	36	6.0	2.0	3	561	0.63	blunt	5.19
	18	2.0	2	18	3.0	0.5	6	758	0.86	blunt	7.00
	19		4	36	6.0	1.0	6	618	0.70	blunt	5.71
	20		8	72	12.0	2.0	6	520	0.59	blunt	4.80
CP [90₂, 0₂]₄ₛ	21		2	18	3.0	0.5	6	563	0.61	blunt	4.96
	22	4.0	4	36	6.0	1.0	6	483	0.52	blunt	4.26
	23		8	72	12.0	2.0	6	471	0.51	blunt	4.14
ML [0, ±45, 90₁/₂]ₛ	24		2	18	3.0	0.5	6	306	0.68	blunt	5.41
	25	0.875	4	36	6.0	1.0	6	289	0.65	blunt	5.12
	26		8	72	12.0	2.0	6	252	0.56	blunt	4.48

The column headings are defined as follows:

t = thickness

s = scale with respect to specimen No. 1

w = specimen width

a = half-notch length

ρ = notch root radius

σ_{tmu} = net (ligament) tensile strength (force/ligament section)

σ_{ou} = unnotched material tensile strength (force/width × thickness)

ϵ_{tmu} = net ultimate tensile strain (‰) (σ_{tmu}/E)

$w/2a = 3$ = constant for all specimens

UC/ED laboratory. The numbers of each test in Table 3 correspond to the numbers indicated in Figure 18.

The notched-specimen tensile failure *net strength*, or nominal *ligament strength*, is characterized by the tensile net ultimate strength σ_{tnu} defined by dividing the notched-specimen ultimate tensile force F_{tu} by the ligament cross section $t(w - 2a)$ instead of specimen cross section tw, which gives:

$$\sigma_{tnu} = F_{tu}/t(w - 2a)$$

Values of the *ligament* over *material* strength ratios σ_{tnu}/σ_{ou}, or inverse *notch-sensitivity* factors, are plotted in Figure 18 over the half-notch length a. As pointed out by Beaumont, the values fall within the region bordered by the curves

$$\sigma_{tnu}/\sigma_{ou} = K_{Ic}/[(\sigma_{ou}(\pi a)^{1/2}] = (0.077 \pm 50\%)(\pi a)^{1/2}$$

where the half-notch length a is measured in meters. Both curves (solid lines) and the average curve (dashed line) are also shown on Figure 18. The apparently large $\pm 50\%$ *scatter* measured by the vertical distance between the upper and lower solid curves illustrates the so-called *high notch sensitivity* of LFRP laminates.

Note that the empirical stress-intensity-type failure model does not recognize the influence of the notch root radius ρ, and it predicts a vanishing ligament nominal strength as the notch length increases, introducing an apparent bias (segregation) in the strength/toughness prediction.

8.2. Simplified Fracture Criterion

The simplified $[\sigma_c, r_c]$ damage-mechanics criterion has been calibrated on a representative subset of test pieces of each stackup. Such a subset must contain test pieces for which the transverse-stress profiles over the ligament have largely different gradients to enhance stress-profile intersections, which is typically the case for self-similar test pieces at different scales, as shown in Figure 5. In such cases, the stress profiles at rupture clearly intersect, and the critical parameters at the point of intersection, σ_c, and r_c, can be evaluated.

Stress Profiles   The homogenized equivalent stress profiles over the ligament of a given stackup, in the simplified HEL approach, can either be found from a linear elastic finite-element calculation that uses orthotropic HEL properties, or from analytical formulas. One such formula for the stress profile over the ligament of a quasi-isotropic homogeneous equivalent material near the root of a notch with radius ρ is given by Craeger [21]:

$$\sigma(r) = [K_I^*/(2\pi r)^{1/2}]f(\rho/r)$$

where

$$f(\rho/r) = (1 + \rho/r)/(1 + \rho/2r)^{3/2}$$

and where K_I^* is Craeger's stress-intensity factor, and r is the distance away from the notch root. Note that other descriptions of the notch root stress profile can be used, particularly ones that take into account orthotropy [22].

Calibration The calibrated critical stresses σ_c and characteristic distances r_c are collected in Table 4 for all laminate stackups.

Prediction Figure 19 shows the prediction of the net ultimate tensile strength σ_{tnu}, normalized by the simplified-criterion critical stresses σ_c, and the nominal notch stress-concentration factors K_{tn}, drawn vs. the logarithm of the notch root radius ρ, normalized by the critical distances r_c of the simplified criterion, in the form of the curve

$$(\sigma_{tnu}/\sigma_c)K_{tn} = 2^{3/2}g(\rho/r_c)$$

where

$$g(\rho/r_c) = [(\rho/r_c)^{1/2}f(\rho/r_c)]^{-1}$$

The previous expression follows with the definition of the nominal notch-concentration factor K_{tn}, which is a function of a/ρ and which relates the maximum stress σ_{max} in the stress profiles near the notch root ($\rho \neq 0$) to the average (nominal) ligament stress σ_{tn} of the notched specimens in the form of the quotient

$$K_{tn} = \sigma_{max}/\sigma_{tn}$$

The maximum stress near the notch root follows from Craeger's formula

TABLE 4 **Critical Simplified [σ_c, r_c] HEL Fracture-Criterion Parameters**

Type	σ_c (N/mm²)	r_c (mm)
CP [90, 0]$_s$	1360	0.15
CP [90$_2$, 0$_2$]$_s$	1600	0.22
CP [90$_2$, 0$_2$]$_{2s}$	1500	0.25
CP [90$_2$, 0$_2$]$_{4s}$	1500	0.085
ML [0, ±45, 90]$_{1/2}$]$_s$	838	0.085

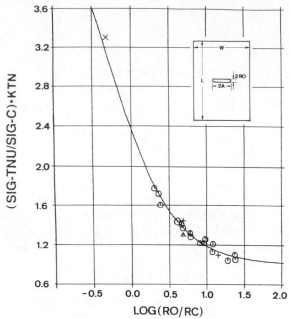

SIG-TNU = net ligament stress
SIG-C   = critical stress
KTN   = stress concentration factor
RO   = notch root radius
RC   = characteristic distance

Figure 19. Calibration and prediction with the $[\sigma_c, r_c]$ criterion and LFRP (HEL-theory).

evaluated at the distance $r = 0$ from the notch root, which gives

$$\sigma_{\max} = 2K_I^* / (\pi\rho)^{1/2} = K_{tn}\sigma_{tn}$$

By using the stress profile at rupture, K_I^* becomes critical, σ_{tn} becomes ultimate, and we obtain

$$K_{Ic}^* = \tfrac{1}{2}(\pi\rho)^{1/2}K_{tn}\sigma_{tnu}$$

Substitution of the point (σ_c, r_c) into Craeger's ultimate stress profile gives

$$\sigma_c = [K_{Ic}^* / (2\pi r_c)^{1/2}]f(\rho/r_c)$$

from which σ_{tnu} can be predicted in terms of K_{tn} and ρ by substituting the previous expression of K_{Ic}^*, and normalizing with respect to σ_c and r_c.

For better identification of the individual test pieces numbered 1 through 26,

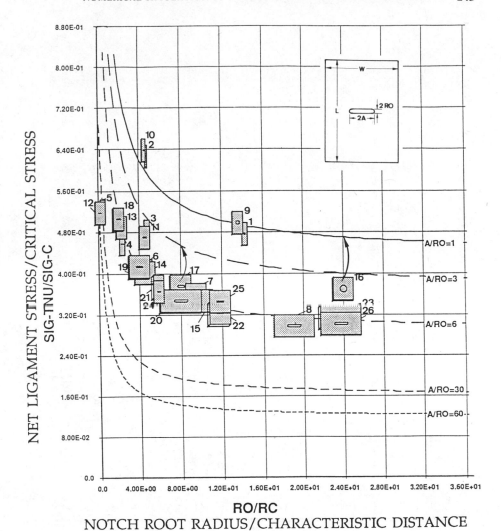

Figure 20. Calibration and prediction with the $[\sigma_c, r_c]$ criterion separated by different a/ρ-values (HEL-theory).

Figure 20 shows the predictions σ_{tnu}/σ_c separated for different notch concentrators a/ρ. In this case, the predicted curves are given by

$$\sigma_{tnu}/\sigma_c = (1/K_{tn})2^{3/2}g(\rho/r_c)$$

and curves for values of $a/\rho = 1.0$ (circular hole), 3.0, 6.0 (blunt notches), and 30 and 60 (slit) are indicated.

The measured experimental values are seen to coincide closely with the

TABLE 5 Calibrated and Predicted Ultimate Net Tensile Stresses and Coefficients of Variation for the Simplified $|\sigma_c, r_c|$-Fracture Criterion

Laminate	No.	r_c (mm)	a/ρ	ρ (mm)	$\log(\rho/r_c)$	$P = 2^{3/2}g_c$ calibrated (c) and predicted	σ_c (kN/mm²)	$E = \dfrac{\sigma_{tnu}}{\sigma_c}K_{tn}$ experiment	ϕ (%)
CP	1		1	1.5	1.155	1.130		1.094	−3.2
[90, 0]$_s$	2		1	0.5	0.678	1.370		1.441	+5.2
	3	0.105	3	0.5	0.678	1.370		1.300	−5.1
	4		6	0.25	0.377	1.687(c)	1.360	1.598	−5.3
	5		30	0.05	−0.322	3.161		3.305	+4.6
	6		6	0.5	0.678	1.370(c)		1.418	+3.5
	7		6	1.0	0.979	0.192(c)		1.253	+5.1
	8		6	2.0	1.280	1.098(c)		1.040	−5.3
CP	9		1	3.0	1.135	1.136		1.144	+0.7
[90$_2$, 0$_2$]$_s$	10		1	1.0	0.658	1.386		1.469	+6.0
	11		3	1.0	0.658	1.386		1.269	−8.4
	12	0.22	60	0.05	−0.643	3.945	1.600	3.747	−5.0
	13		6	0.5	0.357	1.715(c)		1.715	±0
	14		6	1.0	0.658	0.386(c)		1.404	+1.0
	15		6	2.0	0.959	1.201(c)		1.216	+1.3
CP	16		1	6.0	1.380	1.078		0.850	−21.1
[90$_2$, 0$_2$]$_{2s}$	17		3	2.0	0.903	1.227		1.010	−17.7
	18	0.25	6	0.5	0.301	1.796(c)	1.500	1.769	−1.5
	19		6	1.0	0.602	1.434(c)		1.442	+0.6
	20		6	2.0	0.903	1.227(c)		1.213	−1.1

CP [90$_2$, 0$_2$]$_{4s}$	21	0.085	6	0.5	0.770	1.304(σ	1.500	1.313	+0.7
	22		6	1.0	1.071	1.157(σ		1.127	−2.6
	23		6	2.0	1.372	1.080(σ		1.099	+1.8
ML [0, ±45, 90$_{1/2}$]$_s$	24	0.085	6	0.5	0.770	1.304(σ	0.838	1.278	−2.0
	25		6	1.0	1.071	1.157(σ		1.207	+4.3
	26		6	2.0	1.372	1.080(σ		1.053	−2.5

The column headings are defined as follows:

r_c = characteristic distance (mm)

ρ = notch root radius (mm)

$P = 2^{3/2}g_c$ = predicted value of $(\sigma_{tnu}/\sigma_c)K_{tn}$

g_c = Craeger's function of ρ/r_c

σ_c = critical stress (kN/mm^2)

E = experimental value of $(\sigma_{tnu}/\sigma_c)K_{tn}$

σ_{tnu} = ultimate net tensile stress (ligament tensile strength)

K_{tn} = notch nominal stress-concentration factor (kN/mm$^{3/2}$)

$\phi = (E - P)/P$ coefficient of variation (%)

245

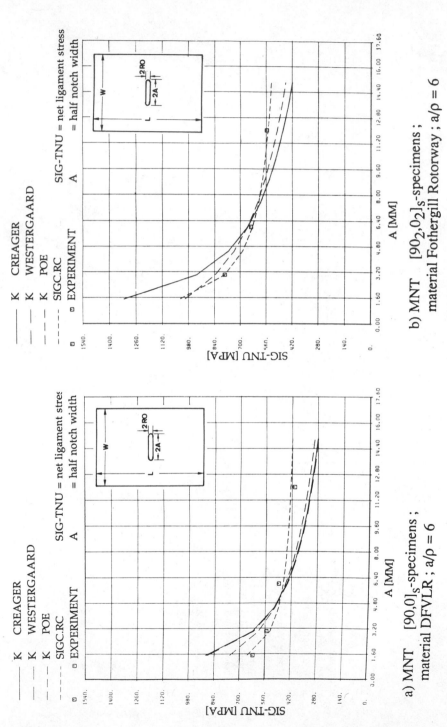

Figure 21. Comparison of different fracture-criteria calibrations.

a) MNT $[90,0]_S$-specimens ; material DFVLR ; a/ρ = 6

b) MNT $[90_2,0_2]_S$-specimens ; material Fothergill Rotorway ; a/ρ = 6

prediction curves. The calibration of the material parameters (σ_c and r_c) is conducted on the test pieces with blunt notches ($a/\rho = 6$) for each type of laminate. These same (σ_c, r_c) values are used to predict the ultimate strengths σ_{tnu} for specimens with other notch shapes (holes, notches, or slits).

Table 5 contains a list of calibrated and predicted ultimate nominal stresses and the coefficient of variation with respect to the experimental values. This coefficient varies *in average* by 4.4% or, if the "bad" prediction of test pieces 16 and 17 are omitted, by only 3.2%. The relatively poor performance of test pieces 16 and 17 is unexplained at present. It is possible that the proposed simplified [σ_c, r_c] criterion, which assumes a purely elastic and homogeneous equivalent laminate behavior, applies less well to large round holes, e.g., to large test pieces having low notch concentrators $a/\rho = 1.0$ or 3.0.

Comparison of Different Criteria   Figure 21 plots the curves of ultimate net stresses σ_{tnu} for the ligament for the half-notch length a of some specimens with $a/\rho = 6$, as calibrated with the [σ_c, r_c] criterion, with the K-stress criteria (Westergaards, Craeger, and G) and with Poe's [8] strain-intensity criterion. In the latter criteria, calibration was based on an average critical stress- (or strain-) intensity factor found by matching the predictions given by these criteria with the experimental values. The experimental values are also shown in the diagram. Clearly, the damage-mechanics [σ_c, r_c] criterion behaves best among all considered criteria, and the stress-intensity and also the energy-release-rate criteria behave less well. Among the latter criteria, however, Poe's fiber-strain-intensity criterion seems to lead to a somewhat better prediction of the ultimate tensile nominal stress σ_{tnu}, as it is seen in Figure 21.

Discussion   The proposed *simplified* [σ_c, r_c] criterion recognizes the physical effects of different stress gradients (stress-gradient effect) because it is built on the premise that fracture occurs when a critical damage, here σ_c, extends over a characteristic volume, here r_c, where the critical damage and the characteristic volume are constants for a given material. As a consequence, the criterion predicts that the stress profiles with concentrations of different severity at fracture pivot around the common point [σ_c, r_c]. Of course, the simplified criterion neglects all nonlinear effects such as matrix cracking, delamination, etc., when applied to equivalent homogenized composite laminates.

On the other hand, the stress/strain-intensity K-type criteria, based on a crack-growth instability physical model, predict that the stress profiles at failure are also identical for a given value of the notch root radius ρ, but imply that both their shape and critical intensity, $K_{Ic} = \sigma_{ou}(\pi a)^{1/2}$, are independent of ρ, assuming the effect of ρ negligible, and predicting quite a different gradient (or scale) effect in $a^{1/2}$. The previous calibrations and predictions show that the growth coalescence model is more likely to apply to LFRP notched composite laminates.

9. LOAD-BEARING-PLY FRACTURE ANALYSIS

In order to assess and to extend the validity range of the promising simplified $[\sigma_c, r_c]$-fracture criterion it is necessary to account for local nonlinear effects such as subcritical matrix damage, material heterogeneity, etc. The work carried out for the analysis of the load-bearing ply is described in detail in Reference 20 and the results are only summarized.

The detailed three-dimensional failure FE analyses, using the microfracturing-damage model for the matrix and the local $[\epsilon_c^f, r_c^f]$ criterion for fiber fracture, have been carried out on the two [90, 0] MNT specimens of the UC/ED experimental data base (Table 3, specimens 4 and 8). The specimens are labeled C0.5, 1, 0.5/6 (scale 0.5) and C414/6 specimen (scale 4) in Figure 22. The

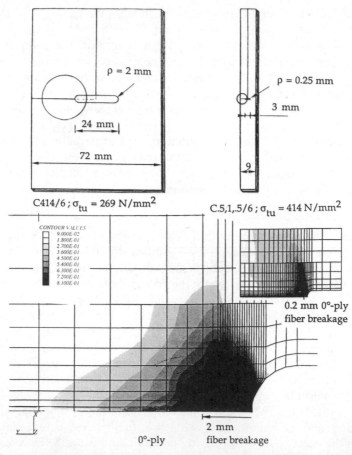

Figure 22. Load-bearing-ply analysis: damage extent or spread in the 0°-layers for the C0.5, 1, 0.5/6 and C414/6 specimen at the incipient catastrophic failure (PAM-FISS/Bi-Phase model).

analyses plausibly predict the principal physical phenomena during the fracture of the two specimens such as an initially unstable 0°-fiber transverse crack for the small specimen, inducing immediate specimen catastrophic failure, and a stable 0°-fiber transverse crack of about 2 mm before catastrophic failure, for the large specimen, as shown in Figure 22. Figure 23(*a*) and Figure 23(*b*) show the numerical and experimental stress–strain diagrams of the two specimens. The predicted ultimate stresses are very close to the experimental values (respectively, $\sigma_{tu} = 414$ and $269\,\text{N/mm}^2$), which indicates the capability of the damage-mechanics fracture model to predict pronounced notch sensitivity and scale effects. The fiber breakages are shown by the sudden drops in the stress and global modulus of the specimens, whereas the matrix damage does not create any evident nonlinearity on the curves.

The mechanism of the 0°-fiber transverse-crack propagation based on the fiber-stress profiles on the ligament and the matrix-damage maps during the fracture process of the specimens has been identified and can be summarized as follows.

Once the 0°-fiber transverse crack initiates, the existing matrix damage propagates along with it. If the splitting matrix damage increases during the 0°-fiber transverse-crack advance, blunting of this crack tends to become more severe as it propagates and more loading is required to advance this crack, i.e., the crack tends to be initially stable. As soon as the extent of the matrix-damage zone becomes constant with further 0°-fiber transverse-crack advance (i.e., steady-state "resistance" is reached), this crack ceases to be stabilized via the growth of crack-tip blunting, and the crack tends to become unstable for this test piece geometry, with the ensuing catastrophic overall specimen failure. The size of this transition zone defines the stable crack length at catastrophic failure.

The influence of specimen scale on the propagation mode of the 0°-fiber transverse crack is reflected by the size of the transition zone, approximately proportional to the notch root size, and can be summarized as follows: the larger the specimen, the larger is the size of the transition zone and, consequently, the longer is the stable crack at catastrophic failure (Figure 22).

It follows from these discussions that the three-dimensional detailed FE analysis of notched tensile test pieces give a powerful insight into the complex interplay of subcritical matrix damage with stable—and ultimately unstable—fiber breaking in the load-bearing plies. If calibrated on the fiber strain of the load-bearing ply, the proposed $[D_c, r_c]$-fracture criterion constitutes a far-reaching LFRP fracture-prediction tool, which can be applied to virtually any laminate and specimen geometry or scale.

Remark As shown earlier on the homogeneous ceramic specimens, this class of $[D_c, r_c]$ damage-mechanics fracture criterion applies to high stress-gradient areas, and it needs to be complemented by a Weibull-type model for low-gradient areas (i.e., smooth test pieces or large blunt notches and round holes).

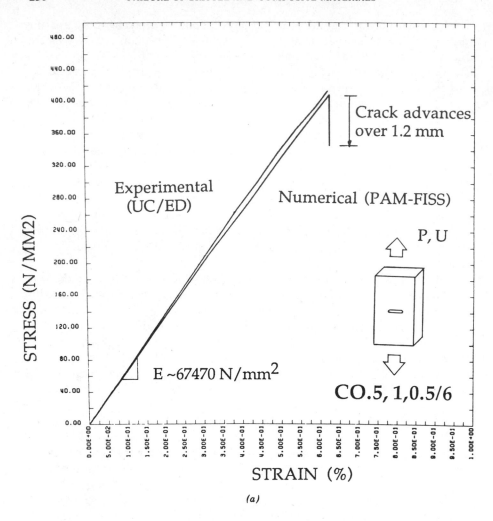

(a)

10. CONCLUSION

The thoughts and the work presented in this chapter lead to the general conclusion that the engineering capacity of the prediction of static fracture and ultimate failure of brittle homogeneous and composite materials in general and long-fiber-reinforced composite laminates in particular has progressed substantially. This progress is ascribed to two key elements, which are given by the introduction and the use of new damage-mechanics fracture criteria, reaching beyond the realm of classical fracture mechanics, and by the existence and the use of sophisticated numerical tools together with powerful modern computers, which permit the systematic and punctual detailed study of complex fracture events in an accurate and economic fashion.

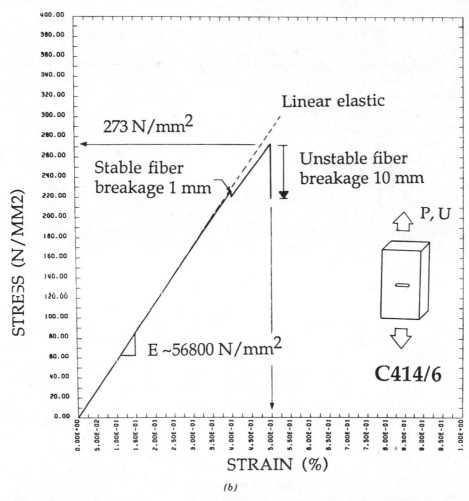

Figure 23. (*a*) Load-bearing-ply analysis: C0.5, 1, 0.5/6 specimen; stress ($\sigma = P/wt$) as a function of global strain (U/L); experiment/analysis (PAM-FISS/Bi-Phase model). (*b*) Load-bearing-ply analysis: C414/6 specimen; stress ($\sigma = P/wt$) as a function of global strain (U/L); experiment/analysis (PAM-FISS/Bi-Phase model).

Both elements together permit the rational construction of simplified fracture criteria, e.g., by producing catalogs of simplified criteria parameters for different laminate stackups, based on experiments or on detailed calculations using more universal fracture parameters for the load-bearing ply.

Classical fracture mechanics has been shown to successfully predict crack advance in homogeneous materials and matrix macrocrack advance in composite laminates, but it has failed to predict accurately events involving composite fiber fracture where clear crack fronts cannot easily be identified.

Since matrix macrocracks are not generally responsible for composite specimen failure, the role of classical fracture mechanics for composite failure prediction is limited.

The encouraging results obtained using the combination of deterministic $[D_c, r_c]$ damage-mechanics criteria in high-stress-gradient areas and of probabilistic Weibull-type strength criteria in low-stress-gradient areas for the prediction of brittle ceramic and composite material failure justifies further investigations of this approach, toward a general predictive numerical fracture model.

ACKNOWLEDGMENTS

The reported work has been carried out in part under European Space Agency (ESA/ESTEC) contracts with Dr. C. Stavrinidis as ESA's technical representative. The continued encouragement and the stimulus given by Dr. Stavrinidis is specially acknowledged. The work benefited from experiments performed by the Deutsche Forschungs-und Versuchsanstalt für Luft-und Raumfahrt (DFVLR), Braunschweig, under Drs. Bergmann, Prinz, Eggers and Kirschke, and from experiments performed at the University of Cambridge Engineering Department (UC/ED) under Prof. P. Beaumont, Mark Kortschot, and Mark Spearing. The work on ceramics and composite fabrics was due to contracts by the French Direction des Recherches et Etudes Techniques (DRET/SDR/G8) under Mrs. Levy and Mr. Grellier, jointly with Ceramiques Desmarquets, Mr. Torre, and Ecole des Mines, Mr. Lamon, and, respectively, AEROSPATIALE Helicopter Division under Mr. Mens, where the testing on composite fabrics was performed. The authors gratefully acknowledge the support and the stimulus received by all named institutions and representatives.

At ESI, Mr. F. Perié on ceramics and Dr. P. Dowlatyari on composite laminate plastics performed most of the numerical studies, and Mr. F. Vogel did the analytical investigations of the homogenized laminate model. All these essential contributions are gratefully acknowledged.

REFERENCES

1. G. Dorey "Impact and Crashworthiness of Composites Structures," in G. A. O. Davis, Eds., *Structural Impact and Crashworthiness*, Vol. 1, Elsevier Applied Science Publishers, London, Chapter 6.

2. J. W. Mar, "Fracture Longevity and Damage Tolerance of Graphite/Epoxy Filamentary Composite Material," *J. Aircr.* **21**(1) (1984).

3. L. Kirschke, "Schädigungsmechanismus in gekerbten CFK-Laminaten," DFVLR paper presented as DGLR-Lecture nb 82-003, Symposium Entwicklung und Anwendung von CFK-Structuren, May 26–27, 1982.

4. A. de Rouvray, E. Haug, and J. Dubois, "Failure Mechanisms and Strength Reduction in Composite Laminates with Cut-Outs—A 3-D Finite Element Numerical Autopsy," in I. H. Marshall, ed., *Composite Structures*, Applied Science Publishers, London and New York, 1983.

5. P. W. R. Beaumont, *Fracture Mechanics in Fibrous Composites*, University of Cambridge, Department of Engineering, Cambridge, U.K., 1980.

6. R. D. Jamison, "On the Interactionship between Fiber Fracture and Ply Cracking in Graphite/Epoxy Laminates," *ASTM Spec. Tech. Publ.*, **STP 907**, 252-273 (1986).

7. *PAM-FISS*, A specialized Finite Element Code for Fracture Mechanics Analyses, ESI-SA, Rungis-Cedex, France.

8. C. C. Poe, "A Unifying Strain Criterion for Fracture of Fibrous Composite Laminates," *Eng. Fract. Mech.*, **17**(2), 153–171 (1983).

9. J. M. Whitney and R. J. Nuismer, "Stress Fracture Criteria for Laminated Composites Containing Stress Concentrations," *J. Compos. Mater.*, **8**, 253–265 (1974).

10. M. J. Wilkins, R. D. Streit, and J. E. Reaugh, *Cumulative Strain Damage Model of Ductile Fracture: Simulation and Prediction of Engineering Fracture Tests*, Lawrence Livermore Lab., University of California, Livermore, 1980.

11. D. R. Curran, L. Seaman, and D. A. Shockey, "Dynamic Failure of Solids," *Phys. Rep.*, **147**(5 and 6), 253–388 (1987).

12. C. E. Harris and D. H. Morris, "Fracture Behaviour of Thick, Laminated Graphite/Epoxy Composites," *NASA* [*Contract. Rep.*] *CR*, **NASA-CR-3784** (1984).

13. G. R. Kress and H. W. Bergmann, "Fitigue Responses of Notched Graphite-Epoxy Laminates", *Forschungsber.—Versuchsanst. Luft- Raumfahrt*, **DFVLR-IB-131-84/04** (1984).

14. E. Haug and A. de Rouvray, *Toughness Criterion Identification for Edge Notched Unidirectional Composite Laminates*, Joint Inst. Adv. Flight Sci. (JIAFS), NASA. Georges Washington University, Arlington, VA, 1984.

15. E. Haug, A. de Rouvray, P. Dowlatyari, "Delamination Criterion Identification for Multilayered Composite Laminates," in *Composite Structures*, Paisley College of Technology, Paisley, Scotland, 1985.

16. A. J. Kinloch and R. J. Young, *Fracture Behaviour of Polymers*, Applied Science Publishers, London and New York, 1983.

17. R. Prinz, "Growth of Delamination under Fatigue Loading," DFVLR paper presented at the 56th Structures and Materials Panel Meeting of AGARD London, U.K., April 12–14, 1983.

18. E. Haug, P. Dowlatyari, and A. de Rouvray, "Numerical Calculation of Damage Tolerance and Admissible Stress in Composite Materials using the PAM-FISS Bi-PHASE Material Model," in *Eur. Space Agency* [*Spec. Publ.*] *ESA Sp.*, **ESA SP-238** (1986).

19. A. de Rouvray, F. Vogel, and E. Haug, *Investigation of Micromechanics for Composites*, Phase 1 Rep. Vol. 2, ESI Rep. ED/84-477/RD/MS (under ESA/ESTEC Contract), Eng. Sys. Int., Rungis-Cedex, France, 1986.

20. A. de Rouvray, P. Dowlatyari, and E. Haug, *Investigation of Micromechanics for Composites*, Phase 2a Rep., WP2, ESI Rep. ED/85-521/RD/MS (under ESA/ESTEC contract), Eng. Syst. Int., Rungis-Cedex, France, 1987.

21. M. Craeger, "The Elastic Stress Field Near the Tip of a Blunt Crack," Thesis, Lehigh University, Bethlehem, PA, 1986.

22. S. M. Bishop, *Stresses Near an Elliptical Hole in an Orthotropic Sheet*, RAE Rep. 72026, Royal Aircraft Establishment, U.K., 1972.

CHAPTER 8

Energy Absorption of Polymer Matrix Composite Structures: Frictional Effects

A. H. Fairfull

Matsel Systems Ltd.,
Liverpool, England

and

D. Hull

Department of Materials Science and Metallurgy
University of Cambridge
Cambridge, England

ABSTRACT

A brief review is given of the material, structural, and testing parameters that affect the response of tubes and cones to axial compression. Models of the axial crushing of fiber-reinforced plastic tubes have been concerned primarily with micromechanisms involving fracture. In this chapter, the frictional processes that occur in the crush zone and the interactions between fracture and friction that lead to the overall level of energy absorption are considered. The role of friction between specimen and platen has been investigated by crushing glass cloth/epoxy tubes against four hardened steel platens of different surface roughness. Coefficients of sliding friction against the surfaces were measured using a combined torsion and compression machine. The load fractions borne by the different regions of the crush zone were also measured and the results incorporated in a conceptual model. It is concluded that friction can account for more than half of the overall energy absorption of composite tubes. Frictional mechanisms within the crush zone can also account for the serrations on the load-displacement trace and can influence the degree of continuity of the crush debris. Crushing of conical shells includes an additional frictional process that contributes to the superior energy absorption of cones over similarly made tubes.

1. INTRODUCTION

Interest in the energy absorption of structures made from fiber reinforced polymer matrix composite materials has been concentrated primarily in the automobile and helicopter industries. Here the attractive features of composite materials such as high specific strength and stiffness offer potential savings in fuel costs and improved performance. Much is dependent on the development of suitable manufacturing technology and there have been many significant developments in recent years. For high-performance structural applications, the main candidate materials are carbon, aramid, and glass fibers in thermosetting polymers, principally epoxy, phenolic and polyester resins. One of the most demanding requirements on composite structures is their ability to absorb energy in a controlled way in crash conditions. The design aspects of this problem for cars and helicopters are described in References 1–7.

Unlike metals and thermoplastics used in structural applications, which undergo extensive plastic deformation after the onset of yielding, thermosetting polymers are brittle. Carbon and glass fibers are also brittle, so the ability of polymer matrix composites to deform in the conventional way is very limited. Structural energy absorption must be achieved in other ways. The primary processes involve microfragmentation in which large amounts of energy are absorbed by brittle fracture on a very fine scale.

The failure modes of an axisymmetric tube subject to axial compression are dependent on many geometrical, material, and test parameters. Providing buckling modes are avoided by suitable choice of the dimensions of the tube, i.e., length L, diameter D, and wall thickness t, there are three main types of failure in polymer matrix composites, namely, (i) brittle fracture of the tube at the compressive strength of the material, (ii) folding and hinging that spreads along the tube in a progressive manner from one end, and (iii) progressive crushing and microfragmentation. Each of these produces a characteristic load-displacement response. The first mode occurs in square-ended tubes and tubes with end loading supports [8] and results in catastrophic failure with little energy absorption. The folding mode is similar to the behavior of ductile metal and plastic tubes [9–11] and occurs in tubes made from aramid fibers and some thin-walled tubes made from glass and carbon fibers [12–14]. The progressive crushing mode requires a mechanism for triggering microfracture at one end of the tube before the compressive strength of the tube is reached. Various triggering methods have been used, including chamfers [8, 13, 15–18] and cones [19]. In general, the energy absorbed in the folding mode is less than in the progressive crushing mode.

Because mass considerations are so important, the mechanical response of tubes are usually quoted as specific crush stress σ_s (crush stress/specific gravity) and specific energy absorption S_s (energy absorbed per unit mass of material crushed). Both of these parameters require careful definition in relation to uniformity of loading, stroke efficiency, and geometry of the crush zone. Thus, for example, the speed of movement of the crush front in the folding mechanism is greater than the cross-head speed, whereas in the crushing mechanisms, the two speeds are the same [18]. However, the fragmented debris formed during crushing can interfere with crushing in the later stages because of packing inside the tube. This affects the stroke efficiency. Thus, σ_s and S_s are not simply related and reference must be made to the folding or crush geometry.

The material parameters affecting σ_s and S_s have been investigated extensively. The complexity and range of possible composite systems has meant that much of this work has been exploratory with little fundamental understanding of the mechanisms involved or of the factors that control σ_s and S_s. The main material variables studied are fiber properties, modulus and strain to failure, resin properties, modulus toughness and strain to failure, and fiber matrix interface properties [14, 20, 21]. Some of these properties are temperature- and strain-rate-dependent.

There are numerous possible variables in fiber arrangement and many of these have been explored, such as winding angle Φ in helically wound glass-fiber tubes [21] and prepreg layup carbon-fiber tubes [13], layup sequence in angle-wound carbon-fiber tubes [13] and in $0°/90°$ tubes made from glass fibers, carbon fibers, aramid, and hybrids of different fibers [21–22], fiber-volume fraction and degree of fiber-bundle dispersion in glass/polyester sheet molding compounds [23]. The highest values of S_s have been observed in carbon-fiber reinforced epoxy tubes using high-modulus fibers and resins and a balanced proportion of $0°$- and $90°$-oriented fibers on the inside and outside surfaces of the walls of the tube. Foam filling can also have an effect on the mode of failure and on the values of σ_s and S_s [20, 24, 25].

Most of the experimental evidence suggests that in epoxy-based materials, S_s increases with test speed [7, 26]. The opposite effect has been observed in some glass/polyester systems, but the effects are usually small and can be associated with either the strain-rate sensitivity of the resin or the glass fibers [21].

Recent systematic studies in glass cloth/epoxy tubes [27, 28] have confirmed previous work [16, 29] that showed that σ_s and S_s are not independent material parameters, but depend on tube dimensions. Values of S_s between 55 and 90 kJ/kg were observed in a series of tubes with different values of D, t, and t/D made from the same material. This raises a number of issues relating to scaling effects. These geometry effects become more pronounced when tube sections other than axisymmetric are used [3, 7, 30]. σ_s and S_s decrease when noncircular elements are introduced and some systematic variations have been reported. Similarly, part sections of circular tubes [17, 31] are less efficient energy absorbers. These geometrical factors are relevant to the design of energy-absorbing beam sections and some important work has been done on various beam geometries including sine wave, and rectangular and circular tubes with web stiffeners [32]. Attempts have been made to produce simple expressions relating the various contributions to energy absorption by different shaped sections.

The design of the trigger becomes very important when structures incorporating the progressive crushing mode have to be designed. A number of novel methods have been used. These include the development of triggers to give controlled load-bearing ability before the onset of progressive failure.

Off-axis crash conditions are of particular concern in car design. One approach has been to evaluate the response of composite cones to both axial and off-axis loads [32]. Fabrication of cones is an order of magnitude more difficult than tubes, particularly when controlled fiber orientations are required. So far, work has been limited to handlaid chopped-strand glass/polyester composites that provide a random in-plane fiber arrangement. Progressive crushing has been demonstrated with S_s comparable to tubes made from the same material. Off-axis progressive crushing has been produced and this is dependent on cone angle. A particular advantage of cones is that they are self-triggering.

The mechanisms of energy absorption in folding and progressive crushing are closely related to the basic properties of the fibers, matrices, fiber/matrix interfaces, and fiber architecture. They invariably involve intensive microfracture. A wide range of micromechanisms have been identified. Detailed studies have been made of the microfracture processes at the crush front and these have been correlated with crush-zone geometry and S_s and σ_s. Some attempts have been made to estimate the energy absorbed by integrating all the contributions to the fragmentation processes [21]. Reasonable agreement has been obtained. However, the understanding of micromechanisms and the measurement of fracture energies in composites is exceptionally complex and extensive research is still required.

The most unifying concept for correlating the effect of structural form, scaling factors, material variables, and fiber architecture with σ_s and S_s for the progressive crush mode is the geometry of the crush zone. This, in turn, is dependent on the microfracture processes. The approach offers some possibility for developing a more ordered assessment of the phenomena involved and their application to various aspects of design. A paper on this theme is in preparation. Some insights into their effects are considered in the present chapter. Thus, it is shown that changing the friction behavior at the crush front changes the crush-zone morphology through the changes brought about by the redistribution of stress and the effect of this on the active micromechanisms.

Perhaps the most important message from the work carried out to date is that with proper understanding of the progressive crush processes, structures can be designed and fabricated using composite materials that have S_s considerably greater than corresponding metal systems.

2. FRICTIONAL EFFECTS

As mentioned in the previous section, progressive crushing initiates with the formation of a well-defined crush zone, which then travels along the tube at an essentially constant level of axial load. The configuration of this steady-state crush zone is dependent on a number of material variables, including fiber and matrix types, layup sequence, and fiber orientations. One of the most common forms of crush zone is illustrated in Figure 1. This has been observed in tubes fabricated from SMC [23], glass cloth/epoxy [27], chopped-strand mat/polyester [23], filament-wound glass/polyester [21], filament-wound glass/epoxy [21], and pultruded glass/polyester [18], and is exhibited particularly by tubes incorporating axially orientated fibers. The main feature is an annular wedge of tightly compacted and highly fragmented debris that is forced axially through the tube wall. The wedge is formed during crushing of the trigger and subsequently deflects delaminated strips from the wall radially inward and outward in the form of continuous fronds.

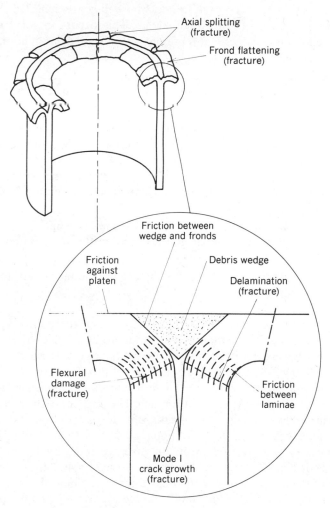

Figure 1. Schematic representation of the crush zone illustrating sources of energy absorption.

As part of a detailed analysis of the material and geometric factors affecting progressive crushing, a conceptual model of this crush-zone configuration, based on work with glass cloth/epoxy tubes has been developed [28]. Eight factors that contribute to energy dissipation were identified and are illustrated in Figure 1.

1. Propagation of a mode-I opening crack at the apex of the debris wedge. The crack extends ahead of the wedge by varying amounts, depending on the material.
2. Frictional resistance to penetration of the debris wedge between the internal and external fronds.
3. Extensive delamination in the fronds, where they are deflected through a small radius of curvature by the wedge. This occurs in some materials along well-defined planes that represent the limits of delamination damage within the tube wall.
4. Flexural damage at the delamination limits in the form of multiple transverse cracking through the individual plies of the fronds.
5. Frictional resistance to sliding between adjacent plies as they pass through the deflection arc of the crush zone.
6. Frictional resistance to internal and external fronds sliding across the crush platen. With cylindrical tubes, there is no relative movement between the debris wedge and the crush platen, since after its initial formation, the wedge has been shown to remain essentially unchanged [28].
7. Propagation of axial splits between fronds. The spacing of the splits, and hence the number of fronds, is governed primarily by the initial external curvature of the tube.
8. Multiple longitudinal cracking through the individual plies of the fronds, facilitating their transverse flattening.

In summary, for this mode of progressive crushing, there are five fracture mechanisms, involving the creation of new surfaces, and three friction processes that result in energy dissipation. Previous models have concentrated on fracture processes, and Berry [33] and Keal [21] have accounted for approximately one-third of the overall energy absorption of glass cloth and filament-wound glass reinforced tubes, respectively, by this approach. The purpose of the present work is to establish independently the contributions due to friction, using experimental techniques where possible. Frictional resistance to crush debris sliding across the platen is the only mechanism not governed solely by the properties of the tube, since it depends also on the nature of the surface of the platen. The greater part of the experimental work was designed to determine the extent to which the surface condition of the platen affected the overall energy-absorption performance.

3. EXPERIMENTAL DETAILS

3.1. Crush Platen Preparation

Four platens of different surface roughness were produced from nonshrinking oil-hardening (NSOH) steel:

1. Ground surface produced by precision toolroom grinding after full hardening; this surface is typical of standard testing-machine platens.
2. Polished surface, produced by hardening and grinding, followed by lapping and polishing with 6-micron diamond paste.
3. Sandblasted surface; produced by grinding and sandblasting before hardening.
4. Cross-milled surface; produced by milling perpendicular rows of grooves before hardening, as illustrated in Figure 2.

Before each test, the sandblasted and crossmilled platens were cleaned by further sandblasting, and all four surfaces were degreased using a proprietary solvent.

3.2. Composite Tube Preparation

Test specimens were prepared from plainwoven glass cloth/epoxy tubestock of 50 mm internal diameter. The warp and weft directions were parallel to the hoop and axial directions, respectively. For the majority of experiments, tubestock manufactured by Tufnol Ltd. using the wet layup process was used. These tubes had a mean glass volume fraction of 0.48 and were machined to give a wall thickness of 2.4 mm, representing approximately 15 plies. For the experiments involving load-fraction determination (see Section 3.5), tubes fabricated from prepreg were used. These had very similar glass cloth dimensions and crush-zone configurations to the wet layup variety, but were used here because of their

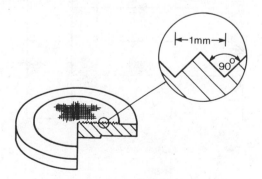

Figure 2. Detail of cross-milled platen.

more reproducible debris wedge dimensions and position. Two batches, having mean glass-volume fractions of 0.36 and 0.40, respectively, were produced and machined to the same wall thickness.

3.3.   Energy-Absorption Determination

The first part of the experimental work involved axially crushing the glass cloth/epoxy tubes against each of the four platen surfaces. This was carried out on a servohydraulic testing machine manufactured by ESH Testing Ltd., having both rotary and axial actuators, but operated here to provide simple axial compression. From the tubestock manufactured by wet layup, three lengths were each cut into four test specimens, one to be crushed against each of the platen surfaces. This approach eliminated any variation between tube batches. The specimens were machined to a length of 135 mm, chamfered at one end with a 45° outside chamfer and crushed at 20 mm/min. Each test was stopped before the internal debris reached the mounting spigot.

3.4.   Determination of Coefficients of Friction

From the same type of tube and the ESH torsion/compression machine, the coefficients of sliding friction μ_3 between the composite tubes and each of the platen surfaces were measured. The tubes were first chamfered at one end and crushed for 1 cm to establish the crush zone. They were then mounted in the testing machine with the crushed end bearing against the selected platen surface, as illustrated in Figure 3.

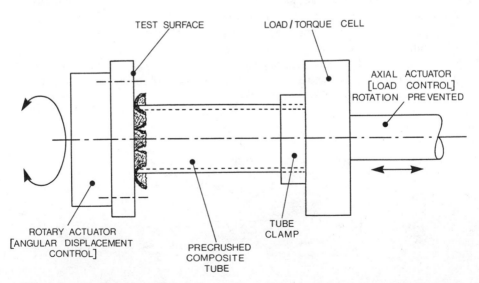

Figure 3.   Configuration of a tube in a test machine for measurements of friction at the crush zone.

Values of μ_3 were first measured at constant levels of axial load, maintained in 5 kN intervals by the axial actuator. Rotation of the crush platen at a fixed rate thus set up a torque in the specimen due to frictional resistance to sliding at the platen surface, enabling μ_3 to be calculated from load and torque values. The rate of rotation was selected such that the sliding speed at the tube wall centerline was equal to the axial speed of 20 mm/min used in quasi-static crush testing. In addition to constant-load testing, tests were also carried out with the axial load increasing (and decreasing) linearly with time, giving a continuous plot of μ_3 vs. axial load. The maximum load was governed by the onset of crushing due to the combined torsion/compression loading. This occurred at values of axial load less than the conventional crush load, and at lower loads with the rougher surfaces.

The crush platen was rotated through 50° for each test. Between tests, the axial load was relaxed before resetting the rotary actuator position, thus preventing anomalous torque readings or accumulated damage due to reversing of the sliding direction against the specimen. Since there is no relative motion between the annular debris wedge and the platen during steady-state crushing, some friction tests were carried out with the wedge removed. However, no significant differences in torque level were observed.

3.5. Determination of Crush-Load Distribution

Since only the fronds are involved in frictional energy dissipation between specimen and platen, any analysis of the role of friction in the crushing of cylindrical tubes requires an estimate of the portion of the total axial load that is supported by the fronds. This was determined for the present work using the prepreg glass cloth/epoxy tubes described previously, and two custom-built test rigs that operated as follows.

The geometrical configuration of the first rig is shown in Figure 4(a). A solid cylindrical platen is attached directly onto the load cell by a threaded bolt. The diameter of this platen corresponds to the region of interest in the crush zone so that the load cell measures the total axial force transmitted by this region. The outer casing is bolted to the machine cross head, and forms a close sliding fit around the inner platen. By suitable shimming to ensure that the casing and inner platen surfaces are flush to one another when supporting the full crush load, it is possible to crush the specimen normally as if against a continuous surface. Two sets of these split platens were produced, with the split positions corresponding to the inner and outer edges of the debris wedge, respectively, as illustrated in Figure 4(b).

The main disadvantage of this type of rig is that the load fraction borne by that material that spreads radially outward from the central cylinder could only be determined by algebraic summation. A second rig, shown in Figure 5, was built that measured the load borne by this external area directly. The principle of presenting a "continuous" surface to the specimen is retained, but in this case, the outer annular platen is screwed onto the cell. The inner solid cylinder is

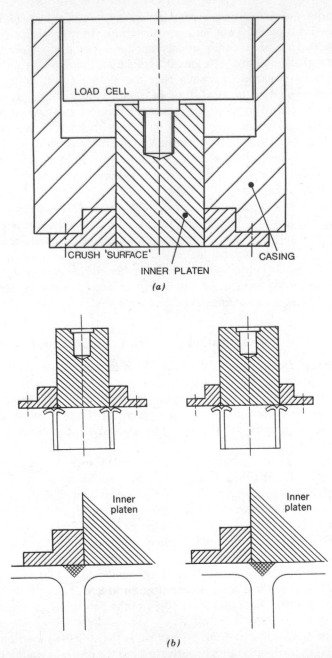

Figure 4. (*a*) General arrangement of concentric platen fixture to measure the load on internal fronds. (*b*) Dimensions of inner platen to omit or include annular debris wedge.

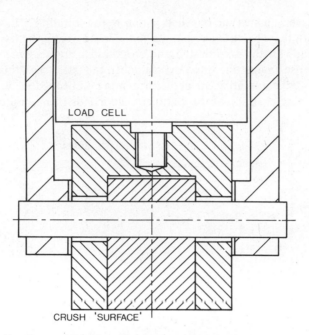

LOAD CELL

CRUSH 'SURFACE'

Figure 5. Concentric platen fixture to measure loads on external fronds.

attached to the casing by a round bar passing through the wall of the measuring platen. Shimming was used to ensure a ridge-free crush zone.

These rigs were installed in a Schenck-Trebel RM50 screw-driven testing machine. The specimens were mounted on a spigot bolted to the base plate, and crushed for 20 mm at 20 mm/min against each platen combination, including a conventional platen to give the overall crush load.

4.   RESULTS

4.1.   Energy-Absorption Determination

The force-displacement traces obtained from crushing against the three smoother platens shared similar features, as shown in Figures 6(a)–(c). In particular, during initial crushing of the chamfered region, there was no overshoot of load above the steady-state level. This was not the case with the cross-milled platen, as shown in Figure 6(d). About 10 mm of crush was required before the load settled to its mean level. This occurred with all three specimens tested against this surface.

The mean crush loads from the various specimen and platen combinations are shown in Table 1. Only the polished platen produced mean crush loads that were significantly different than those against the "standard" ground surface.

Crush-load levels against the polished platen were typically 7% lower, which is consistent with there being less resistance to the fronds sliding across the surface. The crush loads for the sandblasted and cross-milled surfaces were marginally lower than those against the standard platen. In the latter case, this was due to the annular wedge and adjacent debris becoming embedded in the grooves of the platen during crushing of the chamfer. The fronds then abraded a smooth

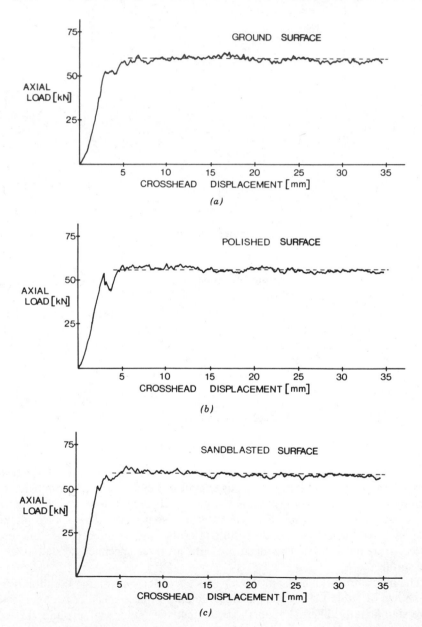

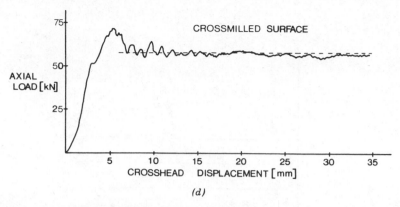

(d)

Figure 6. Typical load-displacement traces of tubes crushed against different surfaces: (a) ground, (b) polished, (c) sandblasted, and (d) cross-milled.

TABLE 1   Effect of Platen Surface on Mean Crush Loads

Specimen Reference Number	Platen Surface	Crush Load (kN)	Normalised Crush Load
1/1	Ground	58.5	1.00
1/2	Polished	54.3	0.93
1/3	Sandblasted	57.5	0.98
1/4	Cross-milled	56.0	0.96
2/1	Ground	53.4	1.00
2/2	Polished	50.0	0.94
2/3	Sandblasted	54.0	1.01
2/4	Cross-milled	52.5	0.98
3/1	Ground	53.5	1.00
3/2	Polished	49.3	0.92
3/3	Sandblasted	52.5	0.98
3/4	Cross-milled	53.0	0.99

path across the packed debris during the next few millimeters crush distance, the resulting steady-state coefficient of friction being coincidentally similar to that against the ground platen.

4.2. Determination of Coefficients of Friction

Figure 7 shows the torque vs. rotation curves obtained from the constant-load torsion tests. There was very little variation of torque with increasing angular displacement for all surfaces, apart from slight torque peaks at the beginning of some tests. Repeated tests at the same value of axial load gave some scatter in

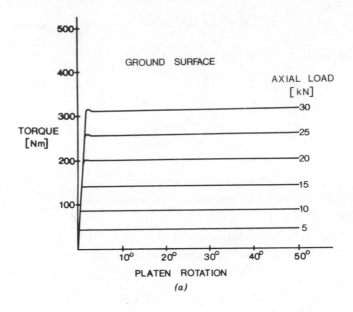

(a)

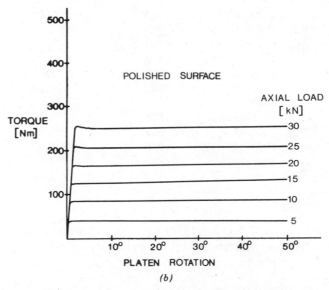

(b)

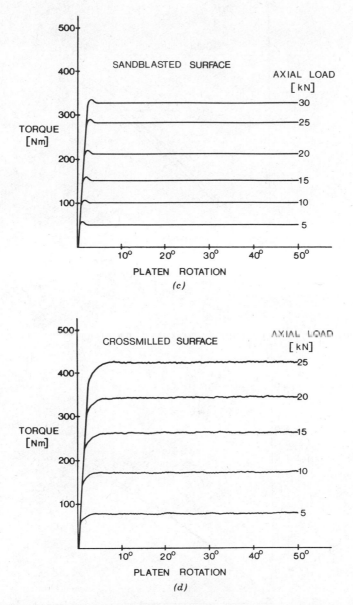

Figure 7. Torque-rotation curves for tubes pressed against different surfaces at constant axial compressive loads: (a) ground, (b) polished, (c) sandblasted, and (d) cross-milled.

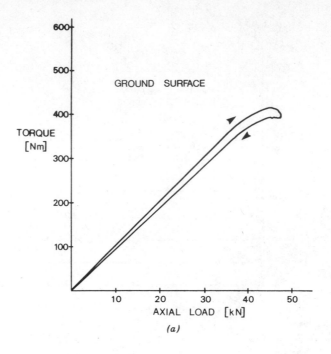

(a)

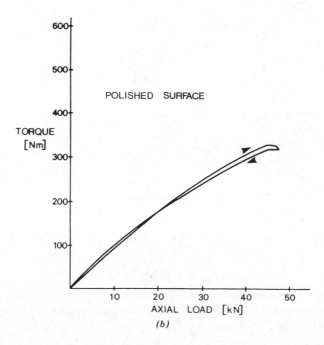

(b)

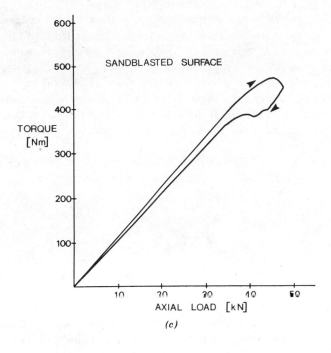

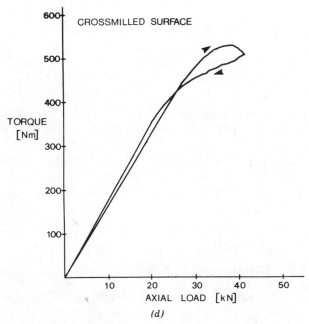

Figure 8. Torque-load curves for tubes pressed against different surfaces: (a) ground, (b) polished, (c) sandblasted, and (d) cross-milled.

torque levels, however, and each step increase in load did not necessarily result in an exactly proportional increase in torque. In combination with the torsional loading, constant axial loads above 35 kN (30 kN for the cross-milled platen) produced progressive crushing of the tubes.

Figure 8 shows the variation of torque with axial load in continuously varying axial-load tests. Relaxation of the axial load was initiated just after the onset of progressive crushing. If the load was relaxed before the tube began to crush, the loading and unloading curves were parallel and very reproducible in cyclic loading tests.

In both of these tests, the coefficient of sliding friction μ_3 is obtained from the instantaneous torque τ and the axial load P from the relation

$$\mu_3 = \frac{\tau}{P\bar{r}} \tag{1}$$

where $\bar{r}$ is the mean radius of the crush zone. The characteristics of the variations of μ_3 for each platen surface can be summarized as follows:

1. *Ground Surface.* The continuous loading curves showed μ_3 to be independent of axial load up to the point of progressive crushing. Constant-load tests gave similar values, but with greater scatter. Overall, mean values of μ_3 were between 0.35 and 0.39, with less variation than this for each individual specimen.

2. *Polished Surface.* The continuous loading curves showed μ_3 to decrease gradually with increasing axial load, although this trend was not present in the constant-load results. The mean coefficient of friction by either method was significantly lower than for the ground surface, being in the range of 0.26 to 0.30.

3. *Sandblasted Surface.* The continuous loading curves showed μ_3 to be independent of axial load, although each constant-load curve showed a slight torque peak at the beginning of the test. Mean values of μ_3 were between 0.38 and 0.41.

4. *Cross-Milled Surface.* This surface was significantly rougher than the other three surfaces and gave the least stable results. The constant-load tests gave level torque/rotation curves and the continuous loading curves showed a transition to crushing and abrasion at about 25 kN axial load. In the linear regions of the curves, the mean values of μ_3 were between 0.65 and 0.70.

4.3.   Determination of Crush-Load Distribution

Table 2 summarizes the load fractions supported by the inner fronds, debris wedge, and external fronds for the two batches of prepreg glass cloth/epoxy tubes, as measured by the split-platen test rigs. The load fraction on the internal fronds was measured directly using the first rig (Figure 4). The corresponding values for the debris wedge were determined from measurements of the load

TABLE 2   Load Distribution Across the Crush Zone

Specimen Reference Number	Mean Vf	Load Fraction on Internal Fronds	Load Fraction on Debris Wedge by Algebraic Summation	Load Fraction on External Fronds	
				by Algebraic Summation	by Direct Measurement
4/1		0.12	0.63	0.25	0.20
4/2		0.14	0.70	0.16	0.19
4/3	0.40	0.13	0.66	0.21	0.19
4/4		0.13	0.68	0.19	0.18
Mean		0.13	0.67	0.20	0.19
5/1		0.17	0.56	0.27	0.20
5/2		0.18	0.62	0.20	0.16
5/3	0.36	0.20	0.60	0.20	0.19
5/4		0.18	0.61	0.21	0.19
Mean		0.18	0.60	0.22	0.19

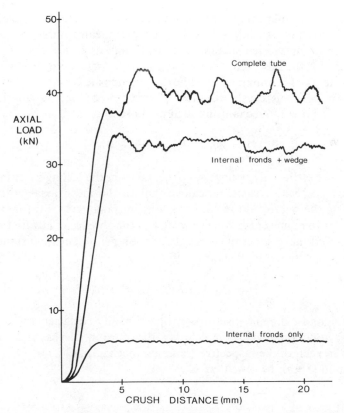

Figure 9. Load-displacement traces for different parts of the crush zone.

273

fraction in the internal fronds and debris wedge combined by subtracting the load fraction on the inner fronds measured separately. The load fraction on the external fronds was determined by direct measurement using the second rig (Figure 5) and also by subtraction. Considering the slight variation in debris-wedge position as crushing proceeds, there was good agreement between the load fractions obtained by the two methods. It is clear from the data in Table 2 that, in all cases, the annular debris wedge supported between 55 and 70% of the axial crush load.

A typical set of load-displacement traces using the first rig is shown in Figure 9. In tests with the measuring platen in contact with the internal fronds and the debris wedge, the curve shows a similar distribution of low-amplitude serrations to those observed on the complete tubes. In contrast, the load-displacement trace from the fronds alone were very smooth.

5. DISCUSSION

5.1. Frictional Components of Overall Crush Load

This chapter is primarily concerned with a general assessment of the effects of μ_3 on overall energy absorption. For this purpose, mean values from the experimental results are sufficient and a detailed analysis of the micromechanisms accounting for the observed values of coefficient of friction is not included. For steady-state axial crushing of a cylindrical composite tube, the overall energy absorption per unit crush distance W is given by the summation of the contributions due to mechanisms such as those illustrated in Figure 1. Thus,

$$W = W_{(1)} + W_{(2)} + \cdots + W_{(8)} \tag{2}$$

This is equal to the energy input to the tube from the testing machine per unit crush distance, which is numerically equal to the mean axial crush load P. Combining the microfracture contributions into one term, P_M (the axial force required for the same crush zone geometry in the absence of any friction), leaves the three frictional sources of energy dissipation per unit crush distance, each of which is proportional to P. Thus,

$$P = P_M + \mu_1 a P + \mu_2 b P + \mu_3 c P \tag{3}$$

where a, b, and c are dimensionless proportionality coefficients determined by the crush-zone geometry and defining the portion of the total axial load involved in each of the respective frictional mechanisms.

Equation (3) can be rewritten as

$$P = \frac{P_M}{1 - (\mu_1 a + \mu_2 b + \mu_3 c)} \tag{4}$$

An increase in any of the coefficients of friction thus increases the overall crush load P in accordance with equation (4), the limit being when the bulk compressive strength of the tube wall is reached.

In the case of friction between fronds and platen, the coefficient c is simply the combined load fraction supported by the internal and external fronds. With the necessary assumption that altering one parameter does not affect the overall crush-zone configuration, the effect of this mechanism can be examined in isolation by considering the other friction contributions to be constant, i.e.,

$$P = \frac{P_M}{1 - (k + \mu_3 c)} \tag{5}$$

By taking a typical value of c as 0.35 from Table 2, and mean values of μ_3 as 0.37 and 0.28 for ground and polished platens, respectively, then from Equation (5), against the ground platen

$$P_1 = \frac{P_M}{0.87 - k} \tag{6}$$

and against the polished platen,

$$P_2 = \frac{P_M}{0.90 - k} \tag{7}$$

From the results of the crush tests, however,

$$P_2/P_1 = 0.93 \tag{8}$$

Solving Equations (6)–(8) yields $k = 0.43$, or

$$P = \frac{P_M}{0.57 - \mu_3 c} \tag{9}$$

Inserting the assumed values for μ_3 and c,

$$P_1 = 2.27 P_M \tag{10}$$

and

$$P_2 = 2.13 P_M \tag{11}$$

Thus, according to this model, even for the smoothest platen used, the frictional contributions account for more than half of the overall crush load. This supports the results of Keal [21] and Berry [33], who arrived at similar conclusions from consideration of the fracture mechanisms. Consideration of equations such as Equation (4) or (9) also explains why crushing against platens with similar values of μ_3 (such as the ground and sandblasted surfaces tested in this work) yields barely detectable differences in mean crush load levels.

5.2. Implications for the Crushing of Conical Shells

All of the foregoing applies to cylindrical tubes, where only the fronds, carrying a fraction of the total crush load, are involved in frictional-energy dissipation against the crush platen. This is not the case in the crushing of conical shells, where, by definition, as crushing progresses from the smaller end, the annular debris wedge must increase in diameter and thus move radially across the crush platen. If, as is the case with cylindrical tubes, the wedge carries the majority of the crush load, then there will be a substantial additional contribution to energy absorption that is not available with tubes. This is supported by the results of Price and Hull [32], who have shown that conical shells have superior specific energy-absorption capabilities than cylindrical tubes of the same materials, in this case, chopped-strand glass/polyester.

5.3. Other Sources of Frictional Energy-Dissipation

As mentioned in the introduction, frictional energy is dissipated by other processes within the crush zone illustrated in Figure 1.

Frictional resistance to penetration of the debris wedge is difficult to measure in isolation. From the results of Table 2, however, this is expected to be an important mechanism, since the wedge carries the majority of the crush load. Fairfull [28] fabricated tubes with a double midwall interlayer of PET film, such that during progressive crushing, the debris wedge passed between two smooth surfaces of PET. The overall appearance of the crush zones of these tubes was almost identical to those of tubes of conventional layup, but the mean crush load was reduced by 23% and this is attributed almost entirely to the reduced load supported by the debris wedge.

The force-displacement traces from the tubes incorporating PET film were significantly less serrated than those of standard tubes. Taken in conjunction with the observations from the load-fraction measurements in the present work, this indicates that for tubes that produce continuous debris fronds, the apparently random serrations superimposed on the load-displacement trace are primarily due to stick-slip frictional resistance to debris-wedge penetration. This is not the case in tubes that crush by the production of discrete fragments; here the serrations are thought to correspond to dislodgement of fragments.

Frictional resistance to sliding between adjacent laminae formed by delamination and interply splitting as the fronds bend through the sharp radius at the crush front has an important effect on the degree of continuity of the resulting crush debris. The combination of incomplete delamination and interply friction can lead to several laminae acting together as a single beam, exhibiting tensile and compressive damage on the top and bottom faces, respectively. Once again, it is difficult to determine the effect of this on energy absorption, since the reduction in interply frictional dissipation is offset by increased tensile and compressive damage. In general, however, brittle material and high interply friction favor the formation of discrete fragments, whereas flexible material and low friction produce continuous fronds. Materials with high interlaminar strength also produce fragments, but this involves a different crush-zone geometry.

6. CONCLUSIONS

1. The energy-absorbing capacity of composite components that undergo progressive crushing involving sliding of crush debris across the platen depends on the coefficient of friction between composite and platen. Thus, data on specific energy absorption must include reference to the surface conditions of the platen.

2. With cylindrical tubes, which crush by the formation of an annular debris wedge with internal and external fronds, there is no significant movement of material within the wedge as crushing progresses. Only the fraction of the axial load borne by the fronds is therefore involved in frictional-energy dissipation between specimen and platen. This is not the case with conical shells. Here the annular debris wedge increases in diameter as crushing progresses, producing an additional frictional contribution to the overall energy absorption that is not available with tubes.

3. The mean coefficients of sliding friction between the crush zones of glass cloth/epoxy tubes and hardened steel platens are 0.37, 0.28, 0.40 and 0.64 for ground, polished, sandblasted, and cross-milled platen surfaces, respectively, when measured quasi-statistically by rotating the tube against each surface.

4. When crushing a composite component against a surface of substantial macroscopic roughness, the platen surface can become filled with crush debris during initial crush-zone formation, leading to a reduction in the coefficient of friction between specimen and platen.

5. In the crushing of glass cloth/epoxy tubes, the annular debris wedge supports the majority of the crush load. Frictional resistance to penetration of the wedge, therefore, represents a significant contribution to overall energy absorption. By reducing the coefficient of friction between wedge and fronds, reductions in crush load of over 20% have been observed.

6. For tubular specimens producing continuous fronds, the axial loads supported by the fronds are essentially constant. The majority of the serrations on the overall load-displacement trace are due to stick-slip frictional resistance to debris-wedge penetration.

7. Frictional resistance to sliding between delaminated plies in the crush zone is influential in determining the degree of continuity of the crush debris. For

example, a high coefficient of friction and brittle material favor the formation of discrete fragments.

8. The various frictional processes account for more than 50% of the total energy-absorption capacity of glass cloth/epoxy tubes, even when crushed against a very smooth platen.

ACKNOWLEDGMENTS

This work was supported by the Polymer Engineering Directorate of the Science and Engineering Research Council, British Petroleum plc, Ford Motor Company Limited, and Pilkington Brothers plc, and the authors are grateful for their help and encouragement.

REFERENCES

1. C. L. Magee and P. H. Thornton, "Design Considerations in Energy Absorption by Structural Collapse," *SAE Trans.*, **87**, 2041–2055 (1978).
2. P. H. Thornton, H. F. Mahmood, and C. L. Magee, "Energy Absorption by Structural Collapse," in M. Jones and T. Wierzbicki, Eds., *Structural Crashworthiness*, Butterworth, London, 1983, pp. 96–117.
3. P. Beardmore and C. F. Johnson, "The Potential of Composites in Structural Automotive Applications," *Compos. Sci. Technol.*, **27**, 251–282 (1986).
4. H. Vogt, P. Beardmore, and D. Hull, "Energieabsorption von faserverstarkten Kunststoffkomponenten im Karosseribau," *Kunstst. Probl. Automobil.*, *VDI Conf.*, Mannheim (1987).
5. J. D. Cronkhite, T. J. Hass, V. L. Berry, and R. Winter, *Investigation of the Crash Impact Characteristics of Advanced Airframe Structure*, USARTL-TR-79-11, September 1979.
6. J. D. Cronkhite and V. L. Berry, *Investigation of the Crash Impact Characteristics of Helicopter Composite Structure*, USA AVRADCOM-TR-82 D-14, February 1983.
7. G. L. Farley, "Energy Absorption of Composite Materials and Structures," *43rd Am. Helicopter Soc. Ann. Forum*, May 1987.
8. D. Hull, "Axial Crushing of Fibre Reinforced Composite Tubes," in N. Jones and T. Wierzbicki, Eds., *Structural Crashworthiness*, Butterworth, London, 1983, pp. 118–135.
9. J. M. Alexander, "An Approximate Analysis of the Collapse of Thin Cylindrical Shells under Axial Loading," *Q. J. Mech. Appl. Math.*, **12**, 10–15 (1960).
10. P. H. Thornton and C. L. Magee, "The Interplay of Geometric and Material Variables in Energy Absorption," *J. Eng. Mater. Technol.*, **99**, 114–120 (1977).
11. W. Abramowicz and N. Jones, "Dynamic Progressive Buckling of Circular and Square Tubes," *Int. J. Impact Eng.*, **4**, 243–270 (1986).
12. D. W. Schmeuser and L. E. Wickliffe, "Impact Energy Absorption of Continuous Fibre Composite Tubes," *Trans. ASME*, **109**, 72–77 (1987).

13. G. L. Farley, "Energy Absorption of Composite Materials," *J. Compos. Mater.*, **17**, 267–279 (1983).

14. G. L. Farley, "Effect of Fiber and Matrix Maximum Strain on the Energy Absorption of Composite Materials," *J. Compos. Mater.*, **20**, 322–334 (1986).

15. P. H. Thornton, "Energy Absorption in Composite Structures," *J. Compos. Mater.*, **13**, 247–262 (1979).

16. P. H. Thornton and P. J. Edwards, "Energy Absorption in Composite Tubes," *J. Compos. Mater.*, **16**, 521–545 (1982).

17. P. H. Thornton, "Effect of Trigger Geometry on Energy Absorption in Composite Tubes," *ICCM Proc. Int. Conf. Compos. Mater.*, *1985*, pp. 1183–1199 (1985).

18. D. Hull, "Energy Absorption of Composite Materials under Crush Conditions," in T. Hayashi, K. Kawata, and S. Umekawa, Eds., *Progress in Science and Engineering of Composites*, Vol. 1, ICCM-IV, Tokyo, 1982, pp. 861–870.

19. C. M. Kindervater, "Energy Absorbing Qualities of Fiber Reinforced Plastic Tubes," *Natl. Spec. Meet. Compos. Struct. Am. Helicopter Soc.*, March 1983.

20. D. Hull and co-workers, unpublished work.

21. R. Keal, "Post Failure Energy Absorbing Mechanisms of Filament Wound Composite Tubes," Ph.D. Thesis, University of Liverpool, Liverpool, 1983.

22. G. L. Farley, "Energy Absorption in Composite Materials for Crashworthy Structures," *ICCM ECCM*, **3**, 3.57–3.66 (1987).

23. P. Snowdon and D. Hull, "Energy Absorption of SMC under Crash Conditions," *Fibre Reinf. Compos. Conf.*, *Plast. Rubber Inst.*, *1984*, pp. 5.1–5.10 (1984).

24. P. A. Kirsch and H. A. Jahnle, "Energy Absorption of Glass-Polyester Structures," *SAE Pap.* **810233** (1981).

25. P. H. Thornton, "Energy Absorption by Foam Filled Structures, SAE Trans., **89**, 529–540 (1980).

26. J. Berry and D. Hull, "Effect of Speed on Progressive Crushing of Epoxy-Glass Cloth Tubes," *Conf. Ser. Inst. Phys.* **70**, 463–470 (1984).

27. A. H. Fairfull and D. Hull, "Effects of Specimen Dimensions on the Specific Energy Absorption of Fibre Composite Tubes," *Proc. Int. Conf. Compos. Mater.*, *6th*, *1987*, pp. 3.36–3.45 (1987).

28. A. H. Fairfull, "Scaling Effects in the Energy Absorption of Axially Crushed Composite Tubes," Ph.D. Thesis, University of Liverpool, Liverpool, 1986.

29. G. L. Farley, "Effect of Specimen Geometry on the Energy Absorption Capability of Composite Materials," **20**, 390–400 (1986).

30. J. N. Price and D. Hull, "Crush Behavior of Square Section Glass Fiber Polyester Tubes," *Advanced Composites Conference*, Detroit, 1988, ASM International.

31. J. N. Price and D. Hull, "The Crush Performance of Composite Structures," in I. H. Marshall, Ed., *Composite Structures*, Elsevier, Amsterdam, 1987, pp. 2.32–2.44.

32. J. N. Price and D. Hull, "Axial Crushing of Glass Fibre-Polyester Composite Cones," *Compos. Sci. Technol.*, **28**, 211–230 (1987).

33. J. P. Berry, "Energy Absorption and Failure Mechanisms of Axially Crushed GRP Tubes," Ph.D. Thesis, University of Liverpool, Liverpool, 1984.

CHAPTER 9

The Mechanics of Deep Plastic Collapse of Thin-Walled Structures

Tomasz Wierzbicki and Wlodek Abramowicz

Department of Ocean Engineering
Massachusetts Institute of Technology
Cambridge, Massachusetts

ABSTRACT

The objective of this chapter is to develop a realistic and mathematically tractable shell model capable of describing large shape distortion, large displacements and rotations, and large plastic strains in thin-walled structures subjected to compressive loads. An exemplary shell element containing a single fully developed buckle is isolated from the rest of the shell and the assumptions regarding the interior and exterior of this element are formulated and discussed.

The model incorporates two simple concepts that distinguish the present method from all other approximating techniques in structural mechanics. One concept is consideration of local deforming subregions with floating rather than fixed boundaries. The other is consideration of a condition of kinematic continuity that relates the velocity and displacement fields and thus provides a natural and convenient means for continuously updating the deformed configuration of the shell. Both concepts dramatically reduce the number of degrees of freedom, thus rendering some closed-form solutions possible.

The model presented gives a qualitative explanation of several important phenomena observed in thin shells, such as softening, stiffening, progressive folding, touching, and locking. It also yields simple and reliable predictions regarding force levels and deformed shapes in a number of important engineering problems.

A computerized version of the present shell model, called a superfolding element (SE), holds the potential of forming a basis for constructing a hybrid (SE/FE) shell model for efficient crash calculations of thin-walled structural components.

281

1. INTRODUCTION

A complete description of the failure process of thin-walled structures under predominantly compressive loads involves a number of stages and raises far more questions and problems than merely determination of the bifurcation or ultimate load of a shell. The lateral deflections considered in the classical elastic–plastic buckling and postbuckling theories usually do not exceed several thicknesses of the shell. This limitation reflects restrictions on the existing theory rather than an inability of the structure to deform any further. In fact, there are numbers of situations where displacements are allowed to exceed two orders of magnitude beyond the wall thickness and even approach the linear dimension of the structure before touching of the walls occurs. Examples of such situations are given later in this chapter. Some of the most general questions regarding the sequence of events in structural collapse that must be posed and answered are:

(1) If a structure softens after having buckled, i.e., exhibits a generally decreasing load-deflection characteristic, will it regain its strength at a certain point? Possible scenarios are illustrated in Figure 1 and labeled A through E.

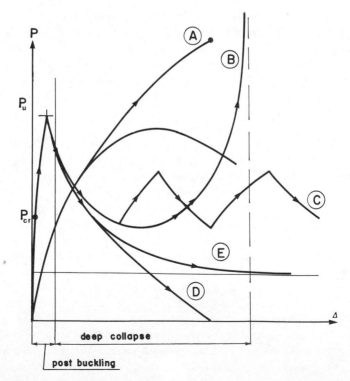

Figure 1.   Possible equilibrium paths in the process of deep collapse.

(2) If so, will the stiffening phase lead to "locking," *B*, or will the structure get into a "limit-cycle" response, *C*? An example of the limit-cycle response is the progressive collapse of prismatic columns.

(3) If the structure does not stiffen, will it lose its strength entirely, *D*, or will it approach a steady-state configuration, *E*?

(4) Finally, at what point of the deformation process will rupture of the wall occur due to the transition from overall compressive to localized tensile strains?

To gain an understanding of the qualitative features of the equilibrium paths is invaluable; however, of equal importance, is the ability to predict the magnitudes of forces and the deformed shapes of various members associated with deep structural collapse.

The answer to all these questions is not trivial and must be determined in order to assure the safe and economical design of certain types of structures. New methods and solution techniques have emerged in the literature to tackle the multitude of problems associated with large strains, curvatures, rotations, and displacements of severely deforming shells. These can generally be classified into experimental, numerical, and analytical techniques.

The objective of this chapter is to present a step-by-step development of a mathematically tractable shell model for efficient calculation of forces and displacements of thin-walled structural components subjected to compressive loads. This shell model provides a general understanding of failure mechanisms in the problem of deep plastic collapse of plate and shell structures. It also yields highly desirable closed-form solutions to a wide range of practical problems.

Historically, a great number of solutions to particular boundary-value problems were derived in the literature on the basis of some ad hoc assumptions. Furthermore, some of the early treatments of this class of problems were semiempirical. This has supported the opinion that the analytical methods of tackling crash problems lack generality and rigorousness. We prove in this chapter that this opinion is unjustified and that various previous solution techniques can now be generalized into a well-defined and coherent discipline-crushing mechanics.

A great deal of information on the crush performance of sheet-metal structures has been provided by tests on scale models and prototypes. These studies have been conducted by and/or for the automotive, shipbuilding, and aircraft industries, and are related to the question of collision protection and safety. Because of space limitations, the present chapter does not survey the multitude of experimental data presently available in the literature. However, reference to tests are often made throughout to justify various approximating assumptions and to help formulate the problem. New experiments on laterally loaded tubes are described in great details in the next chapter.

Several research groups around the world reported recently on spectacular applications of numerical techniques for crash predictions [1, 2]]. The success of the finite-element (FE) codes has been partially attributed to the development of an efficient finite shell element and partially to the enormous growth of the power of supercomputers. PAM-CRASH, DYNA 3D, and ABAQUS are examples of the most successful commercially available finite-element codes specifically designed for crash and impact analysis of land, sea, and air vehicles. Notwithstanding their great potentials and generality, the FE codes have introduced their own problems, which are associated with modeling complexity, an unavoidable conflict between accuracy and computing costs, and the inability to provide a qualitative understanding of the mechanisms of the collapse process. The advances made in FE codes are not discussed in this chapter. This is a vast area requiring an extensive separate treatment. An interesting survey of computational methods as applied to crashworthiness of vehicles has been presented at the symposium on the Application of Supercomputers in the Automotive Industry [3], sponsored by Gray Research, Inc. A thorough validation of either of the previously mentioned codes as applied to the problem of component crashing has not yet found its way to the literature. Some encouraging results have already been obtained in the laboratories of the automotive industry, but are still held proprietary at this time. We have found that the concept of a special folding element developed in the course of the present research constitutes an important connection of the present method with the general-purpose FE codes. This aspect of the problem is briefly covered at the end of this chapter.

The material within this chapter is organized as follows. We start with a general formulation of the problem, including the statement on equilibrium, constitutive behavior, and strain-displacement relations. This is followed by a description of localization of plastic flow and formulation of boundary and loading conditions in a representative shell element or superfolding element (SE). Then the choice of displacement field is thoroughly discussed. The key element in constructing a proper set of approximating functions is the concept of propagating plastic hinges or hinge lines. This concept is first explained by example of a ring, and then is extended to two-dimensional problems of general shell geometry. Further, a number of new issues are raised in connection with the development of the present theory, such as stiffening, touching, and locking, and a connection between shear and axisymmetric deformation is discussed. In the final and most substantial part of this chapter, some prominent qualitative and quantitative features of symmetric and nonsymmetric problems of deep structural collapse are described. This includes, among others, the progressive folding of multicornered columns, local lateral denting of circular tubes, and propagating buckle in pipelines. We conclude the chapter with a discussion of the effect of strain-hardening and rupture initiation in the shell.

2. GENERAL FORMULATION

Most of the existing solutions obtained in the area of crushing mechanics are based on some arbitrary assumptions, and, unlike buckling and postbuckling theory, lack generality and rigor. In a series of recent publications [4–6], an attempt was made to put the present solution technique on a firm theoretical foundation. In this section, we outline the basic methodology and make a clear distinction between well-established principles of mechanics and approximating assumptions made in deriving the theory.

2.1. Constitutive Behavior

We consider a rigid/perfectly plastic isotropic time-independent material, defined by the flow stress σ_0. Elastic strains are neglected altogether. The justification of this assumption stems from the fact that in the considered class of problems, the average strains in plastically deforming regions are two orders of magnitude larger than maximum elastic strains (see Sections 9 and 10). The important effect of strain-hardening is accounted for in an iterative way by suitably adjusting the magnitude of the flow stress, depending on the average value of the plastic strain. This procedure is explained in Section 9.

In some simple applications, such as the crushing of rings [7] and the bending of a plastic strip [8], a rigid linear strain-hardening model of the material was directly used in writing the governing equations. The equivalence of both methods was discussed by Bhat and Wierzbicki [7].

Assuming the usual normality–convexity properties of the yield function, the plastic dissipation is uniquely defined by the strain-rate tensor $\dot{\varepsilon}$. The present class of problems is characterized by very large strains, displacements, and rotations, and, therefore, proper measures of stresses and strain should be taken. In our approach, the Cauchy stress tensor σ and the conjugate velocity strain-rate tensor $\dot{\varepsilon}$ are used. Furthermore, the corotational yield condition is assumed to ensure that the constitutive equations are not affected by the rotation of the shell element.

2.2. Velocity-Strain Rate Relation

The starting point of the present method is the selection of a suitable kinematically admissible velocity field

$$\dot{\mathbf{u}} = \dot{\mathbf{u}}(\mathbf{x}, \xi, \beta, \alpha) \tag{1}$$

where $\mathbf{x}$ is the position vector, β is a vector of free parameters (degrees of freedom), ξ is a vector of input parameters describing initial configuration, and α is a nondimensional time-like parameter. The function $\dot{\mathbf{u}}$ should be piecewise continuous, and the resulting lines of slope discontinuities, which are admitted in the theory of rigid-plastic solids, are called the plastic-hinge lines. At hinge lines, the generalized kinematic variable is the relative rate of rotation $[\Omega]$. A distinguished feature of the present method is that hinge lines can travel with respect to material points as the deformation process progresses.

In the current deformed configuration, the relation between the strain rates

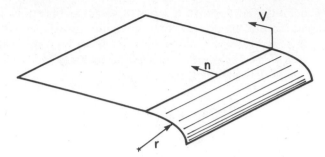

Figure 2. Illustration of the concept of a moving hinge that leaves behind a curved surface.

and velocities is linear. In the continuously deforming regions, the strain rate is defined as a symmetric gradient of the velocity field (the so-called velocity strain). Examples of calculations are given in the subsequent section. At the hinge line, the relative rate of rotation is found from the kinematic continuity condition [4, 9, 10].

It is important for the further exposition of the theory to explain this concept by a simple example. Consider a straight hinge line sweeping down a flat thin sheet with a constant velocity V. By definition, there must be a jump in the rate of rotation $[\Omega]$ on both sides of the hinge. The rotatory motion is superimposed on translatory motion, and, as a result, the hinge leaves behind a curved single-curvature surface of the local radius r, as shown in Figure 2. The condition of kinematic continuity relates all the kinematic variables involved, and has the form [11]

$$[\Omega] + V[\kappa] = 0 \qquad (2)$$

where $[\kappa]$ is the jump in curvature on both sides of the hinge, $[\kappa] = \kappa^+ - \kappa^-$. In our case, $\kappa^+ = 0$ and $\kappa^- = 1/r$. If, in addition, the jump in the rate of rotation at the hinge is defined as a relative rotation, then $[\Omega] = \Omega$ and Equation (2) simply transforms to

$$\Omega = V/r \qquad (3)$$

The propagation of circumferential hinges in axisymmetric shells is described by the same equation. Despite its simplicity, Equation (3) is of great consequence in developing the present theory. It provides a means for integrating the velocity field that calculates the displacement field $u(x, \xi, \beta, \alpha)$ and, therefore, updates the deformed configuration in a continuous manner.

2.3. Equilibrium

The equilibrium of a single structural element is expressed via the global balance of rates of energies:

$$\dot{W}_{ext}(\dot{\mathbf{u}}) = \dot{W}_{int}(\dot{\varepsilon}) \qquad (4)$$

where $\dot{W}_{ext}$ is the rate of external work, and $\dot{W}_{int} = \int_V \sigma\dot{\varepsilon}\, dV$ is the instantaneous rate of energy dissipated at the plastically deforming region. The general form for the rate of work of external forces is given in the next section and the expression for the dissipated energy in shells is discussed in Section 4.

If the class of considered velocity fields is sufficiently broad so as to include the actual velocity, Equation (4) is equivalent to local equilibrium. In practice, however, the kinematically admissible velocity field can contain only a finite number of functions, so that local equilibrium is satisfied only to within a certain acceptable error. For example, the accuracy of the FE method is related to the total number of degrees of freedom of the nodal points. In the Ritz or Galerkin method, accuracy is controlled by the number of terms in the Fourier expansion of the solution.

We are suggesting here a different approach to the construction of a set of approximating functions, which appears to be particularly suited for the present class of problems. The structure is subdivided into a small number of elements, whose boundaries are not fixed but rather are floating. In such a formulation, the velocity field, with a limited number of degrees of freedom can result in an accuracy only attainable by the FE method with a very dense mesh and hundreds of thousands of regular elements.

The following elementary example explains the main point of our method. Consider an inextensible beam or string of length L in which we would like to propagate a single symmetric "plastic" flexural wave with the shape described by three arcs of the same radius. This problem can be easily described in terms of our model with four propagating plastic hinges (see Figure 3). Hinge A imposes a positive curvature $[\kappa]_A = 1/\infty - 1/r = -1/r$, and hinge B reverses the curvature from positive to negative $[\kappa]_B = 1/r - (-1/r) = 2/r$. The remaining two hinges act just in the opposite direction. The relative rotation rates at the subsequent hinges are $\Omega_A = \Omega,\ \Omega_B = -2\Omega,\ \Omega_C = 2\Omega,\ \Omega_D = -\Omega$, respectively. In order to preserve the same shape of the wave, the hinge velocity with respect to the material point (the tangential velocity) must be equal at all hinges, $V_A = V_B = V_C = V_D = V$. Writing the condition of kinematic continuity, Equation (2), we obtain

$$\Omega - V/r = 0 \quad \text{at } A \text{ and } D \qquad -2\Omega + 2V/r = 0 \quad \text{at } B \text{ and } C \qquad (5)$$

Clearly, the above conditions are satisfied identically. Consequently, there is only one degree of freedom, the radius r.

The same problem can be solved in an approximate way by means of the FE method. In this method, the relative rotation between elements occurs at nodal points, which are stationary with respect to material points. By denoting the length of an element by Δl, the total number of elements is $N\Delta l = L$, Figure 3(c). The corresponding number of degrees of freedom is $3N$ (two displacements and a rotation at each node). The choice of N depends on the required accuracy,

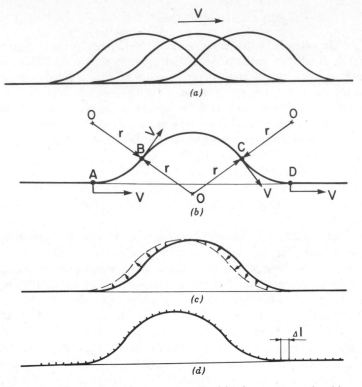

Figure 3. Flexural plastic wave described by four propagating hinges.

but one would probably need hundreds of elements to reproduce adequately the displaced shape if $r \ll L$.

We can conclude that the number of degrees of freedom in the previous example is reduced by two orders of magnitude if the concept of elements with floating boundaries is used. This feature explains the tremendous success in using the present method to describe shape functions of severely distorted shells. In all existing applications, we were able to describe the current deformed shape of structures by means of two to three free parameters. With such a degree of simplification, the derivation of a highly desirable closed-form solution has become a real possibility.

3. BOUNDARY CONDITIONS AND EXTERNAL WORK

Our modeling concept of deep plastic collapse and large shape distortion of shells involves the assumption of a localized zone of plastic deformation. This zone is contained between two conceptual cuts, illustrated in Figure 4, where a slice of prismatic tube/column of the total length 2η is shown. Between the two

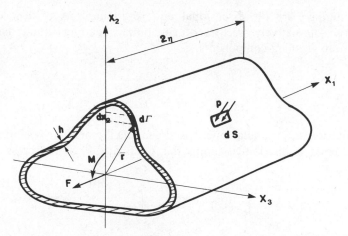

Figure 4. A conceptual cut separating plastically deforming and rigid regions in the shell.

plane sections, the shell can undergo arbitrarily large displacements. Outside the deforming zone, the shell is assumed to be rigid. The rigid parts can, however, be subjected to rigid-body translation described by the vector $\dot{\mathbf{u}}^0$ and rigid-body rotation described by the vector $\dot{\boldsymbol{\psi}}^0$. In rigid-body dynamics, the external loads are the global cross-sectional forces $\mathbf{F}$ and moments $\mathbf{T}$. Therefore, the work of external forces should, in general, include the six terms.

$$\mathbf{F}\dot{\mathbf{u}}^0 + \mathbf{T}\dot{\boldsymbol{\psi}}^0 \tag{6}$$

Let us fix the coordinate system $\mathbf{x}$ with two axes (x_2 and x_3) laying on the conceptual cut and one (x_1) perpendicular to it.

In the examples considered in this chapter, the global shear and twist are taken to be zero:

$$F_2\dot{u}_2^0 = F_3\dot{u}_3^0 = T_1\dot{\psi}_1^0 = 0 \tag{7}$$

These components of the external work lead directly to the development of shear stress and strains in the plastically deforming part of the shell that can not be described by the present computational model. The orientation of axes x_2 and x_3 within the cross section can be chosen such as to induce bending on one plane only, so that $T_3\dot{\psi}_3^0 = 0$. In order to satisfy Equation (7), either static or kinematic boundary conditions should be specified. Consequently, the components of the velocity vector are

$$\dot{\mathbf{u}}^0 = \{\dot{u}^0, 0, 0\}$$
$$\dot{\boldsymbol{\psi}}^0 = \{0, \dot{\psi}^0, 0\} \tag{8}$$

where $\dot{u}^0$ and $\dot{\psi}^0$ are the extensional and rotational rate of displacement, respectively. Alternatively, vectors of generalized loading should have the following nonvanishing components:

$$\mathbf{F} = \{F, 0, 0\}$$
$$\mathbf{T} = \{0, M, 0\} \tag{9}$$

from the interior side, the axial force F and bending moment are equilibrated by internal stresses σ distributed along the boundary Γ of the cross section according to

$$F = \int_\Gamma h\sigma\, d\Gamma \qquad M = \int_\Gamma h\sigma x_2\, dx_2 \tag{10}$$

where the correspondence of the increments $d\Gamma$ and dx_2 is explained in Figure 4.

In addition to the rate of work exerted on the edges, the external work must contain the term describing the contribution of lateral pressure p acting on the lateral velocity $\dot{w}$. From the previous considerations, the expression for the rate of external work follows directly:

$$\dot{W}_{\text{ext}} = F\dot{u}^0 + M\dot{\psi}^0 + \int_S p\dot{w}\, dS \tag{11}$$

The last term on the right-hand side of Equation (11) includes both distributed and concentrated transverse loads. For example, the ring load applied at $x = x_0$ is obtained by introducing the Dirac delta function $p = P_1\delta(x - x_0)$, so that the last integral transforms into

$$\int_\Gamma P_1\dot{w}(x = x_0)d\Gamma \tag{12}$$

Assuming that the rate of displacement is constant under the knife load $\dot{w} = \dot{\delta}$, Equation (12) further reduces to

$$\int_\Gamma P_1\dot{w}\, d\Gamma = P\dot{\delta} \tag{13}$$

where $P = \int_\Gamma P_1\, d\Gamma$ is the total lateral load.

With these assumptions, the final formula for the rate of external work is

$$\dot{W}_{\text{ext}} = F\dot{u}^0 + M\dot{\psi}^0 + P\dot{\delta} \tag{14}$$

The above function should be specified depending on the type of structure, loading, and end conditions. A number of different particular cases can be recovered from the general form of Equations (14) or (11), and these are discussed in the following sections.

Should a free edge exist along the x axis of the prismatic structure, the boundary conditions of Equation (14) must be formulated on this edge. Two situations need special consideration. At the free edge (in open-section members), all components of generalized forces vanish. Alternatively, at the plane of symmetry, the in-plane displacement vanishes, whereas the out-of-plane displacements might be arbitrary. In addition, the slope is zero as well. In either case, the boundary conditions are satisfied identically, and consequently there is no contribution of longitudinal edge to the rate of external work.

4. FOLDING MECHANISMS AND INTERNAL DISSIPATION

In rigid perfectly plastic shells, the rate of internal energy dissipation results, in general, from the continuous and discontinuous velocity fields

$$\dot{W}_{\text{int}} = \int_S (M_{\alpha\beta}\dot{\kappa}_{\alpha\beta} + N_{\alpha\beta}\dot{\epsilon}_{\alpha\beta})\,dS + \sum_{i=1}^{n} \int_{\mathscr{L}^i} M_0^{(i)}[\Omega]^{(i)}\,dl \tag{15}$$

where $\mathscr{L}^i$ is the length of the ith hinge line, n is the total number of stationary or moving hinge lines, and $M_0 = h^2\sigma_0/4$ is the fully plastic bending moment (per unit length).

In the continuously deforming zones, bending moments $M_{\alpha\beta}$ and membrane forces $N_{\alpha\beta}$ are defined in reference to the current variable thickness, according to

$$M_{\alpha\beta} = \int_{-h/2}^{h/2} \sigma_{\alpha\beta}z\,dz \qquad N_{\alpha\beta} = \int_{-h/2}^{h/2} \sigma_{\alpha\beta}\,dz \tag{16}$$

where the components of generalized stresses are related by a corotational yield condition.

The corresponding generalized conjugate strain-rate tensors are curvature rate $\dot{\kappa}_{\alpha\beta}$ and extension rate $\dot{\epsilon}_{\alpha\beta}$. These are linear operators of the velocity field $\{\dot{w}, \dot{u}_\alpha\}$.

Considerable simplification is now achieved by converting the present complex problem [Equation 15] into a set of simple one-dimensional problems. In most of the previous analyses of thin shells subjected to crush loading, these simplifications were based on some ad hoc assumptions. We now rationalize these assumptions by (i) identifying the directions of principal stresses within the shell, and (ii) assuming inextensibility of the material in either or both principal directions.

An intrinsic property of all severely deformed (crushed) thin shells is the appearance of folds or crests. These are zones of localized curvatures and extensions, as can be seen in Figure 5. A single fully developed fold forms, in

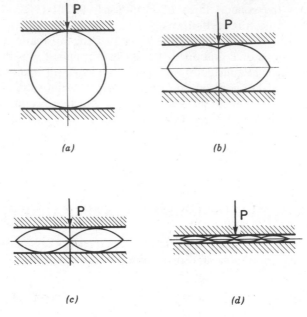

(a)   (b)

(c)   (d)

Figure 5. Subsequent collapse stages of a ring with stationary plastic hinges crushed between rigid plates.

general, a double curvature shell that is characterized by a large radius R and small radius r. Radii R and r can be held constant or can change as the deformation process progresses. Furthermore, folds once formed always stay in one plane and thus can be approximated by a section of a cylinder or a toroidal surface [10]. The axes of the so-formed surface of revolution are subjected to large translations and rotations that are tracked down during the folding process. In the convective coordinate system, associated with a given shell segment, the deformation is assumed to consist of radial plastic flow. Thus, not only the shell geometry is assumed to be axisymmetric, but so is the instantaneous velocity field.

The integrand in the first integral of Equation (15) has, in general, six components. In view of the rotational symmetry and the assumed radial plastic flow, all nondiagonal components of the strain-rate tensor vanish identically, $\dot{\kappa}_{\theta\phi} = \dot{\varepsilon}_{\theta\phi} = 0$, where θ and ϕ denote, respectively, the meridional and circumferential coordinates of the shell of revolution. Note that θ and ϕ are now principal directions. It is conceivable that the shell optimizes itself by developing such a flow pattern that eliminates shear. This is accomplished by suitable rotations of shell elements as to follow the direction of principal stresses. A justification of these two assumptions comes from an extensive experimental evidence regarding the observed failure patterns. Also, the solutions based on the previously mentioned assumptions generally give very good agreement with test results as far as the force level and the geometry of the deformed section are concerned. A rather convincing evidence of our hypothesis is the example of shear buckling of plates or cylindrical shells. The structure escapes from the high energy-consuming uniform shear state into the pattern of diagonal grooves. The grooves consist of suitably rotated sections of toroidal shells. A further study is needed to provide additional justification of such an interpretation.

A third simplifying assumption is the inextensibility of plastic flow in either of the principal directions. Three cases should be considered, depending on whether the extensibility is assumed in the direction normal or tangential to the propagating hinge line or even in both directions. This distinction gives rise to the following three basic deformation mechanisms:

1. *Inextensional Mechanism.* Inextensibility of the shell (or plate) element is assumed in both principal directions.
2. *Quasi-Inextensional Mechanism.* Inextensibility is assumed in the direction of radial plastic flow, i.e., in the direction perpendicular to the hinge lines.
3. *Extensional Mechanism.* Inextensibility is assumed only in the tangential direction to the direction of the hinge line.

These mechanisms are now considered separately, and, in each case, the expression for the internal rate of energy dissipation, Equation (15), is specified.

Inextensional mode applies to flat or single-curvature shells undergoing cylindrical bending. An example of such a deformation mode is provided by Figure 2. By setting the curvature and extension rate in the circumferential direction equal to zero, $\dot{\kappa}_{\phi\phi} = \dot{\epsilon}_{\phi\phi} = 0$, and assuming inextensibility in the perpendicular direction, $\dot{\epsilon}_{\theta\theta}$, the internal energy dissipation reduces to

$$\dot{W}^I_{\text{int}} = \int_S M_0 \dot{\kappa}_{\theta\theta} \, dS + \int_{\mathscr{L}} M_0[\Omega] \, dl \qquad (17)$$

where M at the plastically deforming zone S should be at yield.

The first term in Equation (17) represents the dissipation due to a continuous change of curvature of the strip, whereas the second term describes discontinuous dissipation at the stationary or traveling hinge lines. This type of deformation mechanism is often referred to as *rolling deformation* [12]. Examples of structures undergoing the inextensional collapse mode are given in two subsequent sections.

Quasi-inextensible mechanism refers to the deformation in the toroidal shell. With $\dot{\epsilon}_{\theta\theta}$ vanishing due to assumed meridional inextensibility, there are still three components of the generalized strain contributing to the continuous dissipation. Further simplifications are obtained by treating the smaller radius of the toroidal shell constant, $r = $ constant, which leads to $\dot{\kappa}_{\theta\theta} = 0$. The examination of the folding pattern of various shells reveals that the latter assumption is not far from reality. The term in Equation (15) describing the continuous dissipation now takes the form

$$\int_S (M_{\phi\phi}\dot{\kappa}_{\phi\phi} + N_{\phi\phi}\dot{\epsilon}_{\phi\phi})\, dS \tag{18}$$

where the generalized stresses are related by means of the corotational yield condition of the shell, represented by two intersecting parabolas $|M_{\phi\phi}/M_0| + (N_{\phi\phi}/N_0)^2 = 1$. It follows from the flow rule that if $R/r > 2$, which is always the case, the stress profile on the toroid is always confined to one point on the yield surface, $N_{\phi\phi} = N_0$, $M_{\phi\phi} = 0$. Consequently, the rate of energy dissipation due to plastic flow over the toroidal shell is reduced to

$$\dot{W}_{int}^{II} = \int_S N_0 \dot{\epsilon}_{\phi\phi}\, dS + \int_L M_0 [\Omega]\, dl \tag{19}$$

The above neat result is due to Abramowicz and Sawczuk [13]. The term "quasi" has been proposed to emphasize the fact that extensions are present, not in the direction of the radial flow, but perpendicular to it. Examples of the application of the quasi-extensional mode are given in Section 8.

Extensional mechanisms are probably the most obvious deformation modes of single-curvature shells, where the only extensions exist in the direction of generator, θ or x.

$$\dot{W}_{int}^{III} = \int_S N_0 \dot{\epsilon}_{\phi\phi}\, dS \tag{20}$$

The extensional mechanism is explored further in connection with the denting of tubes (Section 7) and the collapse of a box column (Section 8).

In summary, it should be noted that in the final expression for the rate of internal energy dissipation in all three plastic failure mechanisms described, one-dimensional states prevail. In addition, the bending and extensional actions have been effectively decoupled, thus preparing a ground for developing closed-form solutions to a wide class of engineering problems.

Why are all the introduced simplifications leading to the decoupling permissible? In our opinion, the class of problems defined as "deep structural collapse" is primarily controlled by large global geometry changes and these have been properly and accurately taken into account. The loss of accuracy resulting from neglecting certain terms in the expression for $\dot{W}_{int}$ are often second-order effects.

5. STIFFENING, TOUCHING, AND LOCKING

A correct description of deep collapse and large shape distortion of thin shells must include proper consideration of the contact problem. This necessity has been independently recognized by the developers of FE methods. Some of the most advanced FE codes developed specifically for crash analysis of thin shells, such as DYNA-3D, PAM-CRASH, and ABAQUS, are all equipped with contact elements.

We show in this section that the contact problems and several associated phenomena can be handled by the present theory in a natural and elegant way. As a vehicle to demonstrate such a capability, we use three examples. In addition, the results of this section provide some answers to the questions posed in the Introduction.

5.1. Plastic Ring as a Locking Mechanism

Consider the simplest example of a rigid–plastic ring crushed between two flat plates, as shown in Figure 5(a). Let the ring respond by developing a system of four stationary plastic hinges. The solution of this problem was given by de Runtz and Hodge [14], and the relationship between the load P and deflection δ is

$$P = \frac{4M_0}{R}\left[1 - \left(\frac{\delta}{2R}\right)^2\right]^{-1/2} \tag{21}$$

According to the assumed model, first touching occurs at $\delta = R/2^{1/2}$ or $\delta = R\cos\pi/n$, where n is the number of activated hinges, as can be seen in Figure 5(c). Further deformation is possible by activating four additional hinges. The ring stiffens in a discontinuous manner until the next touching occurs at $\delta = R\cos\pi/8$ etc., as shown in Figure 5(d). This process of subdivision can be carried out for a long time, until an infinite force is developed and deflection reaches the ring radius. The resulting force-deflection curve is plotted later in Figure 7 by a solid line.

5.2. Progressively Stiffening Ring

An example of continuous stiffening leading to locking was worked out by Wierzbicki and Bhat [15] in conjunction with the study of buckle initiation in undersea pipelines. This problem is of particular interest as it provides a good illustration of the inextensional folding mechanism described in the previous section.

Consider a ring of a unit width subjected to the same loading as before. However, the failure mechanism is now different. It consists of four plastic hinges moving outwardly with velocity V with respect to the vertical symmetry axis (refer to Figure 6). The hinges separate flat regions of the length $2b$ each, from uniform but continuously shrinking arcs of current radii r. The inextensibility condition $4b + 2\pi r = 2\pi R$ provides the relation between the velocity of hinge propagation and the rate of change of radius:

$$V = \dot{b} = -\frac{\pi}{2}\dot{r} \tag{22}$$

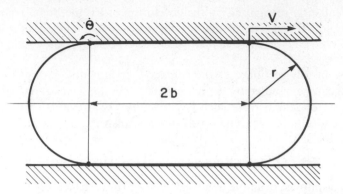

Figure 6. Failure mode of a ring with four traveling hinges and two arcs with continuously changing curvatures.

The rate of change of curvature is $\dot{\kappa}_{\theta\theta} = -\dot{r}/r^2$. Denoting by the crush distance $\delta = R - r$, the rate of external work is

$$\dot{W}_{\text{ext}} = 2P\dot{\delta} \qquad (23)$$

whereas the two terms in Equation (17) describing the internal energy dissipation take the form

$$\dot{W}_{\text{int}} = 2\pi r M_0 \frac{\dot{r}}{r^2} + 4M_0 \frac{\pi}{2} \frac{\dot{r}}{r} \qquad (24)$$

where the continuity condition of Equation (11) was used to calculate the rotation rate at the hinge. The force-deflection relationship can now be obtained from the global equilibrium statement, Equation (4). The final result is

$$P = \frac{2\pi M_0}{R} \left[1 - \frac{\delta}{R} \right]^{-1} \qquad (25)$$

The normalized crushing force is plotted in Figure 7 vs. dimensionless deflection (dashed line). This plot is compared with the previous solution described by Equation (21), showing a similar qualitative behavior. More realistic deformation patterns, including the so-called "dog bone" collapse mode, are discussed in the next section.

It should be noted that stiffening, with or without touching, is produced by increased resistance to bending due to large geometry changes. The work-hardening properties of the material can enhance the trend, but are not solely responsible for this effect. Because of the stiffening properties, rings and short cylinders are used as efficient energy-absorbing devices in laboratory and highway applications. This aspect of the problem has been extensively studied in

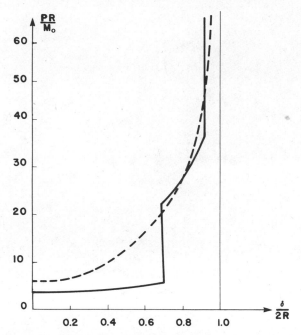

Figure 7. Stiffening characteristics of rings with stationary hinges (solid line) and moving hinges (dashed line).

a series of papers by Carney [16]. Also, a more detailed treatment of laterally loaded tubes can be found in the next chapter.

5.3. Softening Followed by Stiffening

The most instructive problem illustrating a transition from a softening to stiffening response of thin axisymmetric shells is provided by a progressively collapsing tube. Metal tubes with low radius-to-thickness ratios are known to collapse into the axisymmetric or "concertina" modes, with the force deflection fluctuating around a mean value [17]. The present example not only illustrates the quasi-inextensional deformation mode, but also provides much needed insight into the complex problem of local stiffening of shell structures subjected to crush loading.

The computational model consists of a system of traveling hinge circles and sections of conical surfaces undergoing circumferential extension. This is illustrated in a cross-sectional plane view in Figure 8. Initially, three hinge circles are formed at a distance $2H$ apart. The upper and lower hinges travel, respectively, downward and upward, whereas the central hinge splits immediately into two hinge circles traveling in the opposite direction. As the deformation proceeds, the length of a central portion of the cross section shrinks

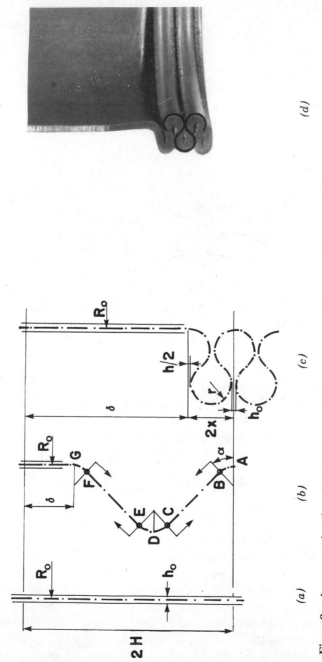

Figure 8. A computational model of the axisymmetric collapse of a tube with moving hinges and the actual pattern of folds in a copper tube.

and the lengths of circular arcs described by the angle α increases. The angle starts with $\alpha = 0$, passes through $\alpha = \pi/2$, and reaches a final value $\alpha_f = 5\pi/6$, at which point touching occurs. The final theoretical profile of the tube is compared with the experimental profile, showing very good agreement when the wall thickness is not taken into account, as shown in Figure 8(c) and Figure 8(d). The variation of the crushing force F with the angle α is given by the following formula:

$$\frac{P}{M_0} = \frac{4\pi + 3^{1/2}R/hH^2(1 - \alpha/\alpha_f)|\cos \alpha|}{H(1 - \alpha/\alpha_f)\sin \alpha} \tag{26}$$

where $H = 2.67(h/2R)^{1/2}$ is the optimum length of the folding wave normalized with respect to wall thickness h. Equation (26) and the relation between the angle α and crush distance u^0,

$$u^0 = 2H\left\{1 - \left[\frac{6}{5\pi}\sin \alpha + \left(1 - \frac{6}{5\pi}\alpha\right)\cos \alpha\right]\right\} \tag{27}$$

furnish the parametric representation of the force-deflection characteristics of the tube. This result is due to Wierzbicki and Bhat [18].

The force starts first at infinity due to the inextensibility (infinite compressive stiffness) of the shell in the direction of force application. Then it diminishes, reaches a minimum $\alpha = \pi/2$, stiffens up again, and locks when touching occurs at $\alpha = \alpha_f$. The plot of the normalized force vs. normalized displacement is shown in Figure 9 by a solid line. For comparison, the solution based on Alexander's concept of stationary plastic hinges [19] is shown by a dashed line. Clearly, this latter solution does not account for the stiffening phase and locks on touch.

The structure of Equation (26) with the term $|\cos \alpha|$ helps us understand the mechanism under which switching occurs from the softening into the stiffening behavior. The normal direction to the conical surface, described by the unit normal vector with horizontal component $\cos \alpha$, changes sign at $\alpha = \pi/2$. For smaller angles, the cone is pushed outward, whereas for $\pi/2 < \alpha < 5\pi/6$, the motion is reversed and the entire bulged section must be brought closer to the tube axis. This, in conjunction with the variable geometry of the shell, leads to the increasing instantaneous force. The axial force has been cut off in Figure 9 by the squash load for the tube, $F = 2\pi Rh\sigma_0$.

These examples suggest that theoretically (i.e., for a shell without thickness), stiffening and/or locking can occur actually prior to touching. In reality, the finite thickness of the shell induces a premature contact and together with the work-hardening of the material further reduces the peaks and smothers the force-deflection characteristics of the tube.

In the case of multicorner prismatic columns, the interaction of folds with smaller and larger radii complicates the problem still further. We return briefly to this topic in Section 8.

In summary, the contact phenomenon is controlled in the present approach

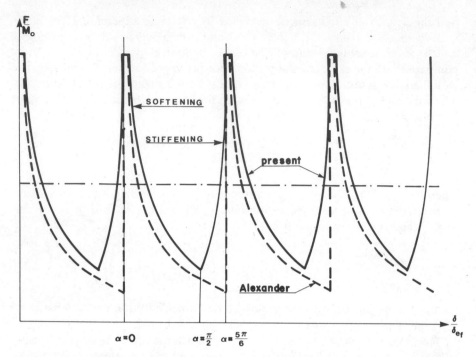

Figure 9. Load-deflection characteristics of the axially compressed tube with moving (solid line) and stationary (dashed line) plastic hinges.

by means of a single scalar parameter α_f. By contrast, the touching or locking conditions in numerical codes are determined through a complex iterative procedure that must be repeated at each time step.

6. PROPAGATING BUCKLES IN PIPELINES

In this and the subsequent two sections, we discuss some important industrial applications of the present method and illustrate the main concepts developed earlier in the chapter. Emphasis is placed on the formulation of a given problem and specification of the governing equations from the general form of the external rate of energy dissipation, Equations (11) and (14), and internal rate of dissipation, defined by Equations (17), (19), or (20). For details of the derivation and solution, the reader is referred to the original publications. We start with the analysis of the steady-state collapse of an infinite tube under external pressure, known as the problem of propagating "buckles" in pipelines.

Several authors presented various arguments as to the general formulation for the propagating pressure p_p. For example, Chater and Hutchinson [20] put forward a concept of the "Maxwell Line" of equilibrium in the pressure–volume consideration, and Palmer and Martin [21] and later Kyriakides et al. [22]

defined p_p as the average pressure to deform a ring from the initial to the final configuration. While all the above arguments are certainly correct, we demonstrate that the equation for the propagating pressure can be derived directly from the global statement of equilibrium, Equation (4), governing the affected tube section.

Consider a transition zone between the underformed and deformed parts of the tube, and denote the out-of-plane displacement of the generic point of the shell by $w(x, 0)$. The unconfined tubes buckle in the so-called "dog bone" shape, which has two axes of symmetry (see Figure 10). For that reason, the bending component of the rigid-body velocity field is identically zero, $\dot{\psi}^0 = 0$. In steady-state buckle propagation, the length of the deforming zone η is constant. Therefore, the relative velocity $\dot{u}^0$ between two sections vanishes as well. (The corresponding axial reaction force F may not be zero.) Equation (11) for the rate of external energy dissipation then reduces to

$$\dot{W}_{\text{ext}} = p \int_S \dot{w} \, dS = p \int_0^{2\pi} R \, d\theta \int_0^{\eta} \dot{w} \, dx \tag{28}$$

In the steady-state process, the vertical and horizontal components of the velocity field are related by the current slope of the deflection line, $w' = dw/dx$; thus,

$$\dot{w} = Vw'$$

where V is the propagation velocity and is constant for all points in the transition zone. Introducing the last expression into Equation (28) and performing the integration, we find that

$$\dot{W}_{\text{ext}} = pV \int_0^{2\pi} R[w|_{x=0} - w|_{x=\eta}] \, d\theta \tag{29}$$

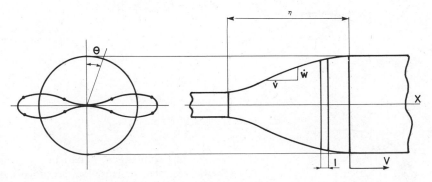

Figure 10. Definition of the main parameters in the process of buckle propagation in pipelines.

Clearly, the integral represents the difference in areas enclosed by the initial and final ring shapes.

$$\dot{W}_{\text{ext}} = pV[A_0 - A_f] \tag{30}$$

In a different approach, we can integrate the rate of external work done on the unit-width ring over the whole deformation process:

$$W_{\text{ext}} = \int_0^{t_f} \dot{W}_{\text{ext}}\, dt = p \int_0^{2\pi} R[w_0 - w_f]\, d\theta \tag{31}$$

which, apart from the constant-velocity term, is identical to the previously obtained expressions of Equations (29) and (30). The equivalence of space and time integration in steady-state processes is nothing new. We present it here to reach consistency within the chapter, and to show all the steps involved in reducing the general concept into specialized equations.

In view of the previous discussion, the rate of internal work for the entire plastically deforming region is also calculated by integrating a unit-width ring over the entire deformation history:

$$W_{\text{int}} = \int_0^{t_f} \dot{W}_{\text{int}}\, dt = \int_0^{\alpha_f} \dot{W}_{\text{int}}(\alpha)\, d\alpha \tag{32}$$

where the time-like parameter α is defined in Figure 11.

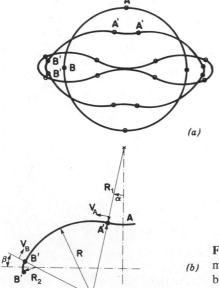

(a)

(b)

Figure 11. Subsequent stages of the symmetric sectional collapse in the process of buckle propagation and the current geometry of the section.

We mentioned earlier that all rings undergo inextensional deformation (mechanism (1)). Why do we believe that this mechanism applies in the case of buckle propagation? The main argument is that in steady-state processes, there must not be any axial deformations. If extensions (or compressions) were present, then corresponding strains are accumulated, changing the length of the tube. On the other hand, shear deformations do not vanish, but their effect should, in our opinion, be small for thin tubes. This can be demonstrated [15] by constructing a simplified model of a transition zone from construction paper that has a very high shear rigidity. This model is shown in Figure 12. The actual amount of shear in the deformed pipe can be easily measured by introducing an orthogonal grid on the lateral surface and then measuring the distortion of angles. The authors are not aware of any experimental results on this topic.

In the absence of any strong evidence about large shear deformations, we neglect the shearing strains in the analysis and reduce the computational model to a one-dimensional inextensional ring. The rate of energy dissipation is given by Equation (17). For steady-state processes, the length and time integration can be interchanged to yield

$$W_{int} = V M_0 \left[\int_0^{\alpha_t} \left(\int_0^{2\pi} R \dot{\kappa}_{\theta\theta} \, d\theta + \sum_i [\Omega]^{(i)} \right) d\alpha \right] \tag{33}$$

which can be interpreted as a total energy dissipated by a unit-length ring from the initial shape, $\alpha = 0$, to the final shape at touch, $\alpha = \alpha_f$.

Depending on the chosen collapse mechanism, various solutions can be obtained. Palmer and Martin [21] assumed four stationary plastic hinges

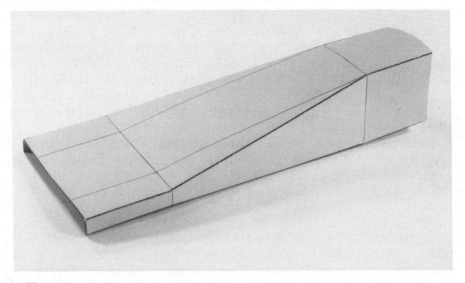

Figure 12. A "frozen" model of propagating buckle with no shear deformations.

[shown in Figure 5(b)], and through straightforward calculation, obtained a first estimate on the propagation pressure:

$$p_p = \pi\sigma_0(h/2R)^2 \tag{34}$$

A slightly lower magnitude of the numerical coefficient is obtained by considering a more realistic mechanism with eight moving plastic hinges [15]. The difference is that the coefficient π is now replaced by 3. A real improvement is obtained by introducing the work-hardening properties of the material. This has been done in two different ways by Croll [7], and Wierzbicki and Bhat [15]. The solution by the latter authors is expressed in terms of elementary functions, and has the form

$$p_p = 3\sigma_0\left[1 + 4\left(\frac{E_ph}{6\sigma_0R}\right)^{0.7}\right]\left(\frac{h}{2R}\right)^2 \tag{35}$$

where E_p is the plastic work-hardening modulus. For $E_p \to 0$, this result reduces to the solution for a rigid–perfectly plastic tube. The corresponding shapes of the rings at the undeformed, intermediate, and final deformed stages are shown in Figure 11. These curves resemble the experimentally observed shapes very well.

To conclude this section, we make the following three remarks. First, we would like to acknowledge that large deformations of elastic–plastic rings were studied numerically in the already mentioned paper by Charter and Hutchinson [20] and by Kyriakides et al. [22]. In addition, the second author performed a number of well-planned and well-executed experiments. However, the magnitude of the propagation pressure, calculated by means of the FE method, turned out to be not any better than the present analytical prediction of Equation (35) for both steel and aluminum pipes.

Second, we attempted without success to apply the present inextensional ring model to solve the problem of confined buckle propagation. One of the difficulties is that the deformed tubes assume a U shape, so that one axis of symmetry is lost and shear deformation becomes dominant. Calladine [23] looked at this problem from a different perspective and obtained some encouraging results.

Finally, it should be mentioned that the problem of initiation and arrest of buckles is governed by a combination of inextensional and extensional modes, and thus are much more difficult to treat. Some new light into this problem is shed by the analysis of tube denting, which is covered in the next section.

7. DENTING OF TUBES UNDER COMBINED LOADING

Our primary objective in this section is to show the effect of boundary conditions on the denting characteristics of a tube. The secondary objective is to illustrate the fully extensional deformation mechanism for shells. A reader interested in a literature survey on the problem of tube indentation and the residual strength of dented tubes is referred to Elinas and Valsgard [24] and Wierzbicki and Suh [5]. Also, some other aspects of the denting problem is discussed in Chapter 14 of this book. The experimental aspect of this problem is discussed by Reid in Chapter 10.

Consider a circular tube subjected to lateral concentrated loading causing a local plastic dent. At the same time, the tube is loaded by an axial force and bending moment applied to the tube axis far away from the dent. If no external torque acts on the tube, the rate of external work, defined by Equation (14), becomes

$$\dot{W}_{\text{ext}} = F\dot{u}^0 + M\dot{\psi}^0 + P\dot{\delta} \tag{36}$$

Because either the generalized force or the displacement rate can be prescribed at the boundary, there are eight possible combinations of boundary conditions altogether, and these are defined in Table 1.

We assume that the axial force and bending moment are fixed at a constant value. On such a prestressed tube, we superimpose an increasing lateral force P or lateral displacement δ. We believe that the previous formulation approximates the real-world situation, where tubular members carry normal operational loads and can be subjected to accidental loads in the lateral direction.

The resistance of the tube to an external load comes from the crushing of rings (inextensional bending) and the extension of longitudinal fibers or generators (extensional mechanism). It is assumed that these mechanisms do not interact with each other through the yield condition; also, the shearing deformations are neglected. Thus, the total expression for the rate of internal energy dissipation is

$$\dot{W}_{\text{int}} = \int_S M_0 \dot{\kappa}_{\theta\theta} \, dS + \int_i M_0 [\Omega]^{(i)} \, dl + \int_S N_0 \dot{\epsilon}_{xx} \, dS \tag{37}$$

where the first two terms correspond to the deformation of rings, and the last term describes the energy dissipated in the generators. The energy associated with ring deformations is described only briefly because this aspect of the problem was extensively covered in previous sections.

A realistic deformation mode of a ring is shown in Figure 13 at various stages of the crushing process. It consists of a flat portion under the "knife" load and three arcs with increasing or decreasing radii. The rate of energy dissipated per a unit width of the ring is

$$\dot{W} = \frac{M_0}{R} \dot{w}_0 f(w) \tag{38}$$

where w_0 and $\dot{w}_0$ denote, respectively, the deflection and velocity of the point in

TABLE 1

No.	Prescribed			
1	$\dot{\delta}$	$\dot{u}_0 = 0$	$\dot{\theta}_0 = 0$	
2	$\dot{\delta}$	N	$\dot{\theta}_0 = 0$	
2a	$\dot{\delta}$	$N = 0$	$\dot{\theta}_0 = 0$	
3	$\dot{\delta}$	$\dot{u}_0 = 0$	M	
4	$\dot{\delta}$	N	M	
4a	$\dot{\delta}$	$N = 0$	M	
4b	$\dot{\delta}$	N	$M = 0$	
4c	$\dot{\delta}$	$N = 0$	$M = 0$	

the ring intersecting the vertical axis of symmetry. The function $f(w_0)$ is monotonically increasing with the initial value $f(0) = 1$. Note that the present ring with only one axis of symmetry dissipates twice as much energy than a similar ring with two axes of symmetry, such as shown in Figure 6.

An interesting aspect of the present analysis is the interaction of inextensible and extensible mechanisms through the functions w_0 and $\dot{w}_0$. These functions represent, respectively, the transverse displacement and velocity of the so-called "leading" generator. In order to perform the integration of Equation (38) with

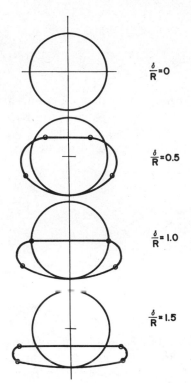

$\frac{\delta}{R}=0$

$\frac{\delta}{R}=0.5$

$\frac{\delta}{R}=1.0$

$\frac{\delta}{R}=1.5$

Figure 13. Subsequent stages of the asymmetric sectional collapse in the process of tube indentation.

respect to the axial coordinate x, the functions $w_0(x)$ and $\dot{w}_0(x)$ must now be specified.

Based on experimental observations, we assume that the deformation zone, defined by the width 2η is spreading as the depth of the dent δ increases. The boundary of the plastic region can be treated as a hinge circle propagating outwardly with the velocity $V = \dot{\eta}$. Each generator is subjected to rotation, and the resulting instantaneous velocity field is linear

$$\dot{w}_0 = \dot{\delta}(1 - x/\eta) \tag{39}$$

The displacement w_0 is obtained by integrating Equation (39) with respect to time in the domain with moving boundaries. It was shown in Section 2 [Equation (3)] that a hinge moving with a constant velocity V leaves behind a circular region. However, in our problem, there is a square-root dependence between δ and η [5]. The resulting displacement field is then parabolic:

$$w_0(x) = \delta(1 - x/\eta)^2 \tag{40}$$

Substituting Equations (39) and (40) into Equation (38) and integrating, the total rate of energy dissipated by the rings becomes

$$\dot{W}_{ring} = 2M_0 \frac{\eta}{R} \dot{\delta} Y(\delta) \tag{41}$$

where the function $Y(\delta)$ was found to be only slightly larger than unity for all values of δ.

The extensional deformations, defined by the last integral in Equation (37), can be easily evaluated by interchanging the order of integration. Consider first the energy dissipated by a single generator:

$$\dot{W}_{gen} = 2 \int_0^\eta N_{xx} \dot{\varepsilon}_{xx} \, dx \tag{42}$$

where $N_{xx} = \sigma h$.

According to the flow rule for the rigid–perfectly plastic material, $\sigma = \sigma_0 \text{sign}\, \dot{\varepsilon}$. The definition of the axial strain rate involves a nonlinear term

$$\dot{\varepsilon}_{xx} = d\dot{u}/dx + \dot{w}'w' \tag{43}$$

where the axial displacement u is due to the global translation u^0 and rotation $\zeta\psi_0$, defined in Section 3. The strain resulting from the axial displacement applied at the end of plastically deforming zone $x = \pm\eta$ is assumed to be uniform. Hence,

$$\dot{\varepsilon}_{xx} = \dot{w}'w' + \frac{1}{\eta}(\dot{u}^0 + \zeta\dot{\psi}^0) \tag{44}$$

Integrating Equation (43) in the limits $(0, \eta)$, the rate of energy dissipation of a unit-width generator becomes

$$\dot{W}_{gen} = 2N_0 \left| \frac{w\dot{w}}{\eta} + \dot{u}^0 + \zeta\dot{\psi}^0 \right|$$

It should be noted that the strain rate $\dot{\varepsilon}_{xx}$ does not change sign along a generator, but can change sign from positive to negative, etc., as we move from one generator to the other. Actually, there might be one, two, or even three crossing points. Consequently, the membrane force $N = \sigma h$ alternates several times along the tube circumference from compressive to tensile. Using the general definition, Equation (10), the total axial force F and bending moment M of the tube cross section are expressed by

$$F = 2\sigma_0 hR \int_0^\pi \text{sign}\, \dot{\varepsilon}_{xx} \, d\theta \qquad M = 2\sigma_0 hR \int_0^\pi \zeta \, \text{sign}\, \dot{\varepsilon}_{xx} \, d\zeta \tag{45}$$

These relations, together with Equation (38) furnish a system of nonlinear equations relating overall sectional rotation and extension rate with the corresponding generalized forces. A detailed explanation of the previous procedure for all eight boundary conditions defined in Table 1 was given by Wierzbicki and Suh [5]. The total energy dissipated in generators is obtained by integrating $\dot{W}_{\text{gen}}$ with respect to the tube circumference.

Figure 14 shows a set of load-indentation characteristics of the dented tube for various constant values of the pretension or precompression. The plotted results were obtained for a tube restrained from the global rotation. The axial load F has been normalized with respect to the so-called squash load, $F_{\text{sq}} = 2\pi Rh\sigma_0$. The crushing strength of the tube is seen to diminish with the magnitude of the compressive load. Furthermore, a precompression larger than $F/F_p = 0.6$ introduces the question of stability. For example, the tube brought in compression up to 70% of its squash load can tolerate in a stable way dents equal to one-fourth of the radius. By increasing the lateral load or the indentation depth still further, a spontaneous collapse of the tube takes place.

A simple closed-form solution, valid for $F/F_p > 0.6$, has the form

$$\frac{P}{M_0} = 16 \left\{ \frac{\pi}{3} \frac{D}{h} \frac{\delta}{R} \left[1 - \frac{1}{4}\left(1 - \frac{F}{F_p}\right)^{3}\right] \right\}^{1/2} \tag{46}$$

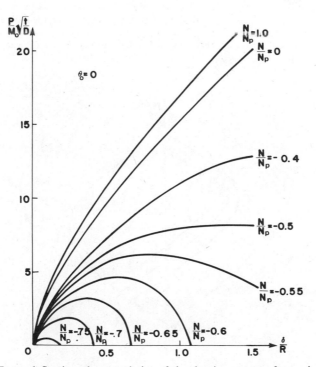

Figure 14. Force-deflection characteristics of the denting process for various values of the axial force.

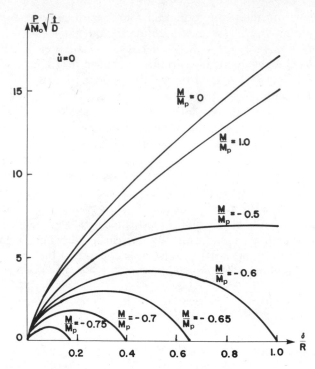

Figure 15. Force-deflection characteristics of the denting process for various values of the global bending moment.

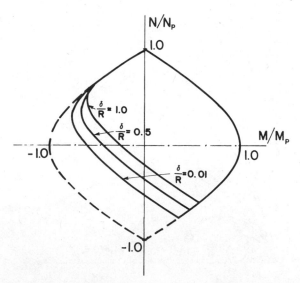

Figure 16. Softening cross-sectional behavior of the dented tube subjected to combined loading by bending moment and axial force.

A qualitatively similar picture is obtained for a tube subjected to preloading by a bending moment normalized with respect to the fully plastic bending moment of the circular tube $M = \sigma_0 D^2 h$, as depicted in Figure 15.

Perhaps the most interesting results regarding the denting-tube problem involve the three-dimensional interaction surface. The projection of this surface on the $M-N$ plane is shown in Figure 16. It is seen that as the depth of the dent increases, the interaction curve shrinks and shifts, displaying the complex character of a work-softening structure. The problem that remains to be answered is whether the normality property of the flow rule applies to the previous interaction curve. The normality rule was assumed, rather than proven, in the study by de Oliveira et al. [25] on the same subject.

8. COLLAPSE OF MULTICORNER PRISMATIC COLUMNS

Progressively collapsing multicorner prismatic columns are examples of structures in which all three basic folding mechanisms, described in Section 4, are present. Here we restrict our attention to columns with an even number of corners. More general structural shapes are discussed in Section 10.

In the case of sections with an even number of corners, it is possible to distinguish a representative angle element called the basic folding element (BFE), as shown in Figure 17(*a*).

In general, we can distinguish eighteen elements that can be grouped in the following four categories:

1. *Rigid Elements*. There are four trapezoidal elements, denoted by 5 in Figure 17(*b*), that translate and rotate. The boundaries of these elements change due to propagating hinge lines.

2. *Elements Undergoing Inextensional Deformations*. There are seven sections of horizontal cylindrical surfaces and two sections of inclined conical surfaces. The horizontal cylinders, denoted by 1 in Figure 17(*b*), have floating boundaries, but fixed centroidal axes. The inclined cylinders, denoted by 2 in Figure 17(*b*), have floating boundaries and floating centroidal axes.

3. *Quasi-Extensional Elements*. Quasi-extensional deformations are confined to the section of a single toroidal surface with moving boundaries and floating axis. This element is denoted by 3 in Figure 17(*b*).

4. *Extensional Elements*. Extensional mechanisms consist of two cones marked by 4 in Figure 17(*b*).

Due to displacement and slope continuity, the geometry of all these elements are fully described by only three free parameters, defined by the vector

$$\beta = \{H, r, \alpha^*\}$$

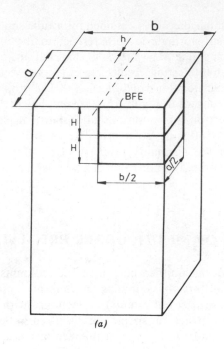

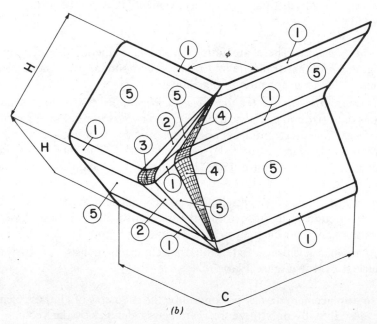

Figure 17. (*a*) Basic dimensions of a rectangular column and a representative basic folding element (BFE). (*b*) The BFE composed of 18 individual shell elements.

where H is the half-length of the BFE, r is the characteristic radius of curvature, and α^* is the parameter describing the contribution of extensional mechanisms to the total energy dissipation. The angle of rotation of rigid trapezoidal elements, α, was chosen to represent the time-like parameter.

Components of the vector of input parameters are the width C of the BFE, the angle ϕ of the corner element, and wall thickness h. Also, the stress–strain characteristics of the material enters the solution through the energy equivalent flow stress (see Section 9). We now specify the expression for the rate of external work. First, we observe that $\dot{\psi}_0 = 0$ due to the symmetry of the column geometry and loading. Furthermore, the transverse distributed pressure p is assumed to be zero. Therefore, Equation (14) for the rate of external work exerted on the plane normal to the tube axis (Γ_1 boundary) simplifies to

$$\dot{W}_{\text{ext}} = F\dot{u}_0 \tag{47}$$

As discussed in Section 3, no work is exerted on the edges of the BFE cut by the symmetry planes that are parallel to the axis of the tube [Γ_2 boundary in Figure 17(b)].

Turning now to the rate of internal work, we would expect that all eighteen elements identified in Figure 17(b) contribute to the dissipation. In other words, both quasi-extensional and extensional deformations would coexist within the deforming BFE. This suggests a "parallel-elements" computational model in which all mechanisms contribute simultaneously to the crushing resistance of the BFE. Such a model is interesting from the theoretical viewpoint, but leads to excessive computational complexity. Instead, we consider a model with "elements in a series," in which the quasi-inextensional mode of deformation, denoted 3 in Figure 17(b), persists during the first phase of deformation up to the certain intermediate configuration α^*, whereupon the extensional mode 4 takes control of the crushing process. Such a generalized folding mode contains as special cases all of the folding modes of columns considered earlier in the literature [10, 17, 26, 27].

For example, by setting the parameter α^* equal to zero, we obtain a combination of inextensional and extensional modes in which all extensional deformations are confined to two stationary conical surfaces. At the other extreme, when α^* equal to α_f, the deformation pattern is controlled by inextensional and quasi-inextensional modes.

In general, $0 < \alpha^* < \alpha_f$ and mechanisms 1 and 4 compete with one another. Actually, the word "compete" requires further clarification. One would expect that that mechanism is activated, which leads to the least value of the instantaneous crushing force. We have already used the minimum postulate for the instantaneous force in Section 7 to calculate the unknown current length of the dented zone η. In the problem of progressive folding of tubes, the vector of unknown parameters has three components. Furthermore, the unknown parameters (H, r, α^*) are all assumed to be constant with respect to the time-like parameter α. Therefore, they should be determined from the balance of total energies rather than the rate of energies. Integrating Equation (47) with respect to time in the limits from $t = 0$ to $t = t_f$, $\alpha = \alpha_f$, and using the mean-value theorem, we obtain

$$W_{\text{ext}} = \int_0^{t_f} \dot{W}_{\text{ext}} \, dt = F_m \int_0^{t_f} \dot{u}_0 \, dt = F_m u_{ef} \qquad (48)$$

In a similar way, we integrate the rate of internal energy. Since the total dissipation is obtained as a result of two different folding modes acting in "series," the integral in Equation (48) splits into two parts:

$$W_{\text{int}} = \int_0^{\alpha^*} \dot{W}_{\text{int}} \, d\alpha + \int_{\alpha^*}^{\alpha_f} \dot{W}_{\text{int}} \, d\alpha \qquad (49)$$

where the switching point parameter α^* denotes the configuration at which the extensional mode of deformation takes over the quasi-inextensional mechanism.

The mean crushing force F_m can now be calculated by equating Equations (48) and (49) to give

$$\frac{F_m}{M_0} = \left(A_1 \frac{r}{h} + A_2 \frac{C}{H} + A_3 \frac{H}{r} + A_4 \frac{H}{h} + A_5 \right) \frac{2H}{u_{ef}} \qquad (50)$$

where the constants $A(\phi, \alpha^*)$, $i = 1, \ldots, 5$ are known functions of ϕ and α^*.

A reader interested in calculations leading to Equation (50) is referred to a recent paper on this subject [6]. The expression of Equation (50) is a function of the vector of unknown parameters $\boldsymbol{\beta}$. These parameters can be determined from the minimum condition

$$\partial F_m / \partial \beta = 0 \qquad (51)$$

In general, the solution to Equation (51) cannot be found in closed form. There exist, however, at least two special cases for which this solution can be expressed in terms of elementary functions. The first example is the purely quasi-inextensional mechanism ($\alpha^* = \alpha_f$). In this case, coefficients A_4 and A_5 vanish and the remaining coefficients $A_1 - A_3$ become functions of the initial included angle ϕ only. The expression for F_m, Equation (50), now depends on two unknown parameters, H and r, and the simple closed-form solution to Equations (48) and (49) takes the following form [6]:

$$F_m = 3M_0(C/h)^{1/3}[A_1(\phi)A_2(\phi)A_3(\phi)]^{1/3}2H/u_{ef} \qquad (52)$$

Another special case of Equation (50) is described by $\alpha^* = 0$. In this case, the BFE is composed of inextensional and purely extensional mechanisms. The coefficients A_1 and A_3 in Equation (50) vanish, whereas the remaining coefficients, A_1, A_4, and A_5, become functions of ϕ only.

The mean crushing force is now a function of one unknown parameter H. After minimization, the final expression for the mean crushing force can be written as

$$F_m = M_0[8.07(\pi - \phi/2)^{1/2}(C/h)^{1/2}]2H/u_{ef} \tag{53}$$

In general, however, α^* is not known a priori, and together with the two remaining unknowns, H and r, must be found as a part of the solution. A numerical solution technique was used to solve the general case, and this aspect of the problem is extensively discussed elsewhere [6].

Figure 18 shows the dependence of an optimum value of the switching point parameter α^* on the value of included angle for few typical C/h aspect ratios. This figure reveals some interesting properties of the crushing process. First, we observe that the extensional deformation controls the crushing processes of thick obtuse angle elements ($\alpha^* < \pi/2$), whereas acute angle elements collapse predominantly in a quasi-inextensional mode ($\alpha^* = \pi/2$). Second, for acute angle elements, $\alpha^* = 90°$ regardless of the current aspect ratio. This means that the final stage of deformation ($\pi/2 < \alpha < \alpha_f$) is always controlled by the extensional mode of deformation. However, since the terminal stage of the crushing process ($\alpha > \pi/2$) gives only a small contribution to the total energy dissipation, we can admit that a quasi-inextensional mode should provide a reasonable estimate of the mean crushing force for all acute corner elements. This observation explains

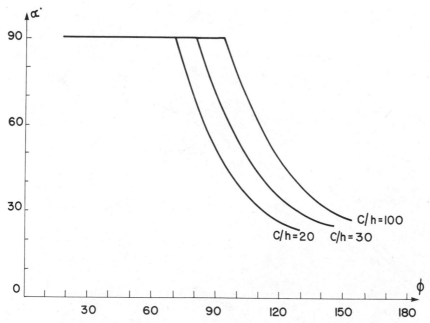

Figure 18. The dependence of the switching point parameter α^* on the central angle for several C/h aspect ratios.

the success of the quasi-inextensional modes in predicting the mean crushing force of rectangular and square tubes [10, 27].

In Figure 19, a comparison is made between the solutions based on the extensional mode (dashed line) and mixed modes (solid line) of deformation for a few typical C/h aspect ratios. The mean crushing force is seen to increase with ϕ until it reaches another branch. Then it falls. Both families of curves intersect at ϕ equal to approximately 120 degrees. This is an important result, suggesting that a hexagonal element ($\phi = 2\pi/3$) provides the highest energy-absorption capabilities among all other angled elements. Specifically, a hexagonal element absorbs up to 40% more energy than a right-angle element of the same weight. This conclusion has been confirmed by recent experimental results [28].

In the limits when the central angle ϕ approaches zero or 180 degrees, the energy-absorption capacity of an angle element drops rapidly to the limiting value corresponding to the inextensional bending along three stationary hinge lines.

In concluding the present section, we present accurate and practical formulas for the mean crushing strength of rectangular and hexagonal columns:

$$F_m = 12.16\sigma_0 h^2 (C/h)^{0.37} \qquad \text{(rectangle)} \tag{54}$$

$$F_m = 20.23\sigma_0 h^2 (C/h)^{0.4} \qquad \text{(hexagon)} \tag{55}$$

These formulas were derived on the basis of the presented generalized folding

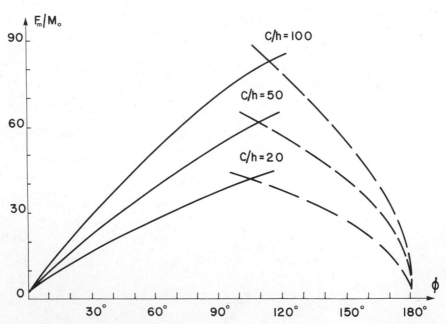

Figure 19. Normalized mean crushing force of the BFE vs. central angle. Intersection of solid and dashed lines indicates a change in the folding mode.

mechanisms. The fractional power in both equations, greater than $\frac{1}{3}$, reflects the larger contribution of extensional deformation modes, compared to Equation (53), which was based on the quasi-inextensional mode.

The generalization of the present results to the case of multicorner elements, with or without flanges, is straightforward and can be found in the authors' recent publications [6, 29]. Other topics such as strain-rate effects, mode transition, etc. were discussed by Abramowicz and Jones [7, 27]. An extensive discussion of the application of the present method to the calculations of real-life automotive structures can be found in a recent report [29].

9. ENERGY EQUIVALENT FLOW STRESS

The present theory has been developed for a perfectly plastic material characterized by a constant but otherwise arbitrary flow stress σ_0. For real materials, however, the flow stress is not constant, but depends on the current value of strain. Thus, the question arises as to what is the most representative value of the flow stress that should be used in the present calculations? The question is very relevant because for deep-drawn and other constructional steels, the difference between the initial yield stress σ_y and the ultimate strength σ_u can exceed 30%.

Generalization of the present analysis to strain-hardening materials is simplified by the fact that in each of the deformation mechanisms, defined earlier in this chapter, one-dimensional states prevail. There is a difficulty, however, due to the presence of many different deforming zones within a given structure. For example, there were eighteen elements in the BFE shown in Figure 17(b). In each of these elements, strains (and stresses) grow differently as the folding of the element proceeds. We suggest an iterative solution to this problem. A starting point is the solution where the yield stress is taken to be the same in all regions. This solution provides a complete description of the kinematics of the problem. One can then calculate the variation of strains and strain rates in the course of the deformation process. On that basis, maximum and average strains are calculated in each of the deforming regions. The stresses are then adjusted, according to the strain-hardening curve. In the case of the mean crushing-force calculations, the minimization procedure, defined in Equation (51), must be repeated to reach the improved solution and so on.

The general procedure is now illustrated by the example of a progressively collapsing rectangular column. For simplicity, the quasi-inextensional mode is considered, for which $\alpha^* = \pi/2$. Equation (50) for the mean crushing load still applies with a redefined value of the average flow stress that is now a function of the strain level in each of the deforming zones.

$$F_m = \frac{h^2}{4} \left(\sigma_0^{(1)} A_1 \frac{r}{h} + \sigma_0^{(2)} A_2 \frac{C}{H} + \sigma_0^{(3)} A_3 \frac{H}{r} \right) \frac{2H}{u_{\text{ef}}} \tag{56}$$

where the weighted average stress is defined by

$$\sigma_0^{(i)} = \frac{2}{(\varepsilon_0^{(i)})^2} \int_0^{\varepsilon_0^{(i)}} \sigma(\varepsilon) \, d\varepsilon \tag{57}$$

The weighting term in Equation (57) for the average stress arises from the change in the integration from the space of moments to the space of stresses. Thus, the second and third terms in Equation (56), describing, respectively, the energy dissipated in an inextensional bending at horizontal and inclined plastic hinges, are exact. In the zone of continuous deformation, described by the first term in Equation (56), the dependence of the average flow stress on the strain is more complicated from that given by Equation (57). This is due to the fact that not only strains but also the area of integration increases with deformation. Still, this important effect is accounted for by Equation (57) through the presence of the weighting function. Another advantage of this approximation is that we have only one expression for $\sigma_0^{(i)}$ in all regions of plastic deformation.

Minimizing, as before, the right-hand side of Equation (56) with respect to the vector of unknown parameters $\boldsymbol{\beta}$, a counterpart of Equation (52) is obtained in the form

$$F_m = \frac{3}{0.73} \frac{\sigma_u h}{4} (\alpha_1 \alpha_2 \alpha_3)^{1/3} (A_1 A_2 A_3)^{1/3} (C/h)^{1/3} \tag{58}$$

where the effective crush distance $u_{\text{ef}} = 0.732H$, and $\alpha_i = \sigma_0^{(i)}/\sigma_u$ is the average stress normalized with respect to the ultimate stress of the material, σ_u.

Defining the energy equivalent flow stress as follows

$$\sigma_0 = \sigma_u (\alpha_1 \alpha_2 \alpha_3)^{1/3} \tag{59}$$

we can represent the solution for the work-hardening tube by an expression formally analogous to the case of the perfectly plastic tube. This is given by Equation (52) with σ_0 defined by Equation (59). The energy equivalent flow stress reflects the joint effect of different stress levels attained in various regions of plastic deformation in the structure.

The problem has thus been reduced to the determination of the maximum plastic strain in each of the deforming regions. In the regions swept by the moving plastic hinges, the material is bent to the "rolling" radius, and the maximum strain is reached in the outer fibers, $z = \pm h/2$,

$$\varepsilon_{\max} = h/2\rho \tag{60}$$

In the horizontal hinge lines, the average radius $\rho = R$ is proportional to the effective crush distance and equals $R = 0.54H$. The half-length of the folding wave for the rectangular column is given by Abramowicz and Wierzbicki [6, 10]:

$$H/h \cong (C/h)^{2/3} \qquad (61)$$

Therefore, the maximum bending strain becomes

$$\varepsilon_0^{(2)} = 0.93(h/C)^{2/3} \qquad (62)$$

The inclined plastic hinges suffer larger strains because the material is bent to a smaller radius, $\rho = r$, where

$$r/h = 0.72(C/h)^{1/3} \qquad (63)$$

The resulting maximum bending strain is

$$\varepsilon_0^{(3)} = 0.69(h/C)^{1/3} \qquad (64)$$

In the zone of extensional deformation, the maximum strain depends on the geometry of the toroidal surface. A good estimate for $\varepsilon_0^{(3)}$ is

$$\varepsilon_0^{(3)} = r/R = 1.3(h/C)^{1/3}$$

By estimating $\varepsilon_0^{(i)}$, the average stresses $\sigma_0^{(i)}$ and normalized average stresses α_i can be found from the known stress–strain characteristics of the material, according to Equation (57). The final value of the energy equivalent flow stress follows then from Equation (59).

From the structures of the previous equations, we observe that σ_0 depends on three factors:

1. strain-hardening characteristic of the material
2. thickness-to-width ratio of the column C/h
3. central angle of BFE, ϕ, which was taken in the previous considerations to be equal to $\pi/2$

An example was run to illustrate the effect of the C/h ratio on the magnitude of the energy equivalent flow stress. By approximating the stress–strain relation for a typical mild steel by the power law $\sigma_0^{(i)}/\sigma_u = [\varepsilon_0^{(i)}/\varepsilon_u]^{0.1}$, and using the previous equations, the following formula was obtained:

$$\sigma_0/\sigma_u = 1.12(h/C)^{0.044} \qquad (65)$$

where the ultimate strain of material was $\varepsilon_u = 0.3$. As shown in Figure 20, the dependence of σ_0/σ_u on C/h is very weak and in most practical cases falls within the range

$$0.9\sigma_u < \sigma_0 < 0.95\sigma_u \qquad (66)$$

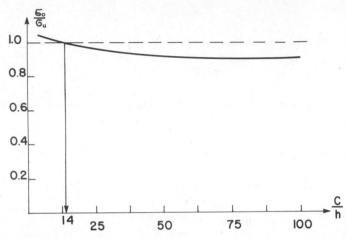

Figure 20. The dependence of the energy equivalent flow stress on the C/h aspect ratio.

Similar calculations were also performed for different deformation modes and central angles ϕ. It has been found that in all cases, the estimate provided by the inequality of Equation (66) is correct. The previous values of the equivalent flow stress are therefore recommended for practical use [29]. It should be cautioned, however, that the approximation provided by Equation (66) is valid for common mild steels only. For other materials, the difference between σ_0 and σ_u can be significant. In those cases, a reliable estimate of σ_0 must depend on the actual stress–strain diagram.

10. RUPTURE OF SHELLS UNDER COMPRESSIVE LOADS

Failure of sheet-metal structures in tension under predominantly compressive loads might seem to be a very unlikely possibility. However, it has been shown in the preceding sections that the structure "escapes" uniform compressive states and is, instead, into inextensional bending (mechanism 1) and/or localized tension/compression zones (mechanism 2). Bending of shell walls induces tensile strains. Plastic flow of the material over the toroidal surface induces tensile or compressive strains, depending on the direction of plastic flow. If the maximum tensile strain ε_{max} in either of the previously mentioned mechanisms, including purely extensional mechanism (3) also, exceeds the critical strain to rupture, $\varepsilon_{max} = \varepsilon_c$, the fracture of the shell wall could occur. This can lead to a dramatic reduction of the crushing strength and the premature collapse of the entire structure.

Prediction of the rupture strains is considerably simplified by the fact that we have to deal with one-dimensional states in either of the deforming zones. We illustrate this aspect of the problem by giving two examples.

The first example concerns the problem of propagating buckles in pipelines, previously discussed in Section 6. Excessive bending strain during the formation of a tube section can lead to the so-called "wet buckle." Wet buckle is the name given to any kind of severe damage of an offshore pipeline that has caused rupture leading to flooding of the pipeline [30]. Wet buckles are very undesirable because they require not only replacement of the damaged section of the tube, but also dewatering and cleaning the underwater pipeline filled with dirt. The maximum (circumferential bending) strain is caused by the curvature change from the initial value R to the final value R_2:

$$\varepsilon_{max} = (R/R_2 - 1)h/2R \tag{67}$$

The solution for R_2 was worked out by Wierzbicki and Bhat [15] for the linear work-hardening model of the material. This model is believed to give better predictions for the local curvatures than the rigid–perfectly plastic model. The final expression for ε_{max} is

$$\varepsilon_{max} = \frac{h}{2R} \left[2.31 \left(\frac{F_\mu h}{6R\sigma_0} \right)^{-0.265} - 1 \right] \tag{68}$$

The above equations can be used with an appropriate critical-strain ductile-fracture criterion to predict the initiation of a wet buckle during quasi-static buckle propagation. Simplified approaches such as this have been successful in predicting the plastic failure of plastic beams [31] and the rupture of ship plating due to hydrodynamic wave load [32]. Also the tensile failure in the purely extensional model (mechanism 3) was reported by Hayduk and Wierzbicki [26] in their studies on the compressive strength of cruciforms. The cruciforms were made by joining two right-angle elements, and in some specimens rupture occurred at the joint.

Finally, the combined effect of all three types of plastic mechanisms on the wall rupture of box columns can be estimated from the examination of Equation (65). Rupture occurs whenever the flow stress σ reaches the ultimate stress σ_u. Equating the right-hand side of Equation (65) to unity and solving it for C/h, we can estimate the critical magnitude of the aspect ratio to be $C/h = 14$. Tubes with the aspect ratio lower than 14 should exhibit fracture at one or several points in the cross section. Clearly, this is a very crude but useful estimate. In reality, rupture occurs whenever the maximum rather than the average strain reaches a critical value in any of the deforming zones.

11. SUPERFOLDING-ELEMENT ANALYSIS AND THE FINITE-ELEMENT METHOD

In Section 2, we illustrated some computational advantages of the present analytical methods as compared to classical finite-difference or finite-element approaches. We used the simple one-dimensional problem of a propagating flexural wave to demonstrate that approximating functions defined over regions with moving boundaries can dramatically reduce the number of degrees of freedom. The subsequent sections expanded this simple idea by showing applications to a few real-world problems. In the present exposition of the material, we have *purposely emphasized the analytical aspect* of the problem.

The question arises if there is any link of the present method with the numerical codes for large inelastic deformation of shells, and if these codes could actually benefit from our findings. The answer is definitely yes. What has so far been accomplished in this chapter is a step-by-step development of the concept of a special finite element with built-in knowledge of the folding process. Such an element, specifically designed for crash applications, is called a *superfolding element* (*SE*). It derives its name from the relatively large, or super, size that has been designed to capture the behavior of a typical buckle or fold in a compressed shell. A single superfolding element can then replace hundreds of regular finite elements, as illustrated in Figure 21.

The present version of the SE has been developed under the stringent symmetry requirements discussed in Section 3, and, therefore, it lacks the

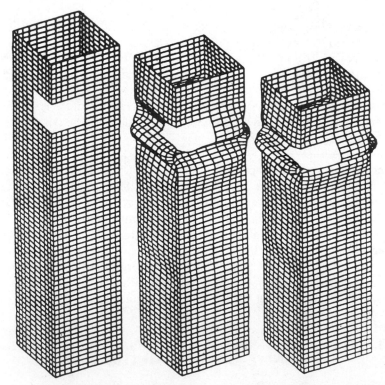

Figure 21. Superfolding element and a mesh of conventional finite elements in a square column subjected to axial compression (courtesy of Engineering Systems International).

generality needed in general-purpose FE codes. However, after suitable generalization, the SE holds the potential of being incorporated into standard FE codes to form a new generation of hybrid codes. Our idea is to introduce the SE in localized areas with sharp changes of curvature, while the remainder of the structure is represented by the mesh of regular finite elements.

The development of a computational model incorporating superfolding elements and conventional shell elements into a single shell model should consist of the following steps:

1. Identification of SE and determination of its location within the structure, size, orientation, and the point in time at which the SE should be switched on and off.
2. Determination of the global constitutive equation, i.e., nonlinear stiffness of SE, expressed as a relation between nodal forces and nodal displacements.
3. Evaluation of the finite-element solution of the crash problem, with the local structure removed and the nodal membrane forces and bending moments replaced by the nonlinear "stiffnesses" of the SE.

Successful completion of all these stages, while still retaining the appealing computational simplicity of the model, is an ambitious task requiring considerable intellectual and modeling efforts. However, early experience in this regard of some investigating teams has been very encouraging. A parallel SE/FE model was proposed by Haug et al. [33]. In this model, a group of regular finite elements was removed and replaced by independently determined crash properties. This approach has been utilized with success in an implicit module of the PAM-CRASH finite-element analysis package.

In our opinion, the concept of the superfolding element may have an impact in four different areas of computational mechanics. First, the new folding element captures the most essential features of the plastic deformation mechanisms in shells that collapse with large local curvatures and shape distortions. These features include the localization of plastic deformations and traveling zones of high curvature changes. Our observations and results may lead to the development of more efficient and reliable curved shell elements of arbitrary size. For example, in our SE, bending and membrane energies are properly handled, which prevents the element from developing excessive stiffening known as "membrane locking."

Furthermore, because the new element is large enough to accommodate one full buckling lobe in a given shell segment, one SE could replace hundreds of standard finite elements. For example, when calculating the crushing force of a rectangular box column, as discussed in Section 8, 1 SE replaced 60 regular elements. In this case, the introduction of SE leads to a *dramatic reduction of the total computational time*, at the expense of missing perhaps unimportant details that the regular FE method would have provided. Our experience shows that with the help of a prototype version of the SE, main parameters describing the crushing process of a square box column can be obtained in a fraction of a second on an IBM PC, with no appreciable loss of accuracy. Similar calculations would take hours on a Cray supercomputer using one of the most advanced FE codes [1].

In crashworthiness applications, the proposed method enjoys considerable *modeling simplicity* and allows the computational model to grow with the concept of a structure on the drawing board. Our methods can then be of great help in the preliminary design stage of prototypes when major dimensions of members have not yet been fixed and the concept of the structure undergoes frequent modifications. For example, in work accomplished for the automotive industry, the entire front end of a car was successfully modeled by means of 80 superfolding elements (the prototype version) [34]. This is two orders of magnitude less than usual modeling practice employing regular FE [35]. The saving in computational time is even larger than that of the simple prismatic column case.

Finally, an attractive feature of the SE is that it can be easily implemented on *personal computers*. Since the calculations per se take practically no time, the maximum available power of the present generation of personal computers could be used for pre- and postprocessing (three-dimensional graphics, solid surface modeling, etc.).

It can be concluded that the new concept offers considerable advantages in terms of computational efficiency, accuracy, and modeling simplicity, as compared to most advanced FE codes.

12. CONCLUSIONS AND DIRECTIONS OF FUTURE RESEARCH

In this chapter, we presented a systematic development of a simple and accurate shell model accounting for large shape distortion, large displacements, rotations, and large plastic strains of thin-walled structural components. We described the conceptual rather than historical evolution of analytical methods designed to tackle the problems of deep structural collapse. A detailed derivation of solutions to specific boundary-value problems and the comparison of our predictions with experimental data was tactically removed from the present text. This aspect of the analysis was extensively discussed in a number of authors' recent publications that are referenced in this chapter. Instead, emphasis was placed on the justification of the approximating assumptions made and on the discussion of fundamental properties of the solutions.

We believe that further progress in the description of catastrophic or progressive failure of shells can only be accomplished by gaining a deeper insight into the mechanics of the plastic folding process of sheet-metal structures.

A rather high degree of generality was achieved by identifying within the structure the basic folding element (BFE) and thoroughly discussing the properties of the exterior and interior of this element. A great deal of effort was also made to show that a number of seemingly unrelated problems, such as the lateral crushing of tubes and rings, propagating buckles in pipelines, local denting of tubes and axial crushing of columns, etc., all fall as special cases of a more general procedure.

The most important observations and results obtained in the course of this research can be summarized as follows:

1. The zones of local plastic deformation in shells can be assumed to be surrounded by a rigid structure undergoing rigid-body translations and/or rotations. This assumption restricts the class of displacement functions at the exterior of the element. The size of the local plastic zones can be taken as fixed or variable, depending on the problem.

2. The interior of the element can be subdivided into a number of simple plastic mechanisms acting "parallel" or in "series." In each of the simple elements, the state of stress is predominantly uniaxial, but the orientation of principal axes is subjected to rotation.

3. The boundaries of these simple elements are "floating." The concept of moving boundaries of the elements considerably simplifies the form of the approximating functions and dramatically reduces the number of degrees of freedom.

4. The condition of kinematic continuity relating displacement and velocity fields provides an interconnection between various elements within the shell and a convenient scheme for the time integration of the velocities to continuously update the configuration.

5. The effect of strain-hardening can be accounted for in an iterative way with the first iteration giving good results.

The in-depth discussion of all simplifying assumptions presented in the preceding sections has helped explain the limitations of existing approaches and define areas of future research.

The generalization of the present model can be sought at two different levels. In the first level, the existing solutions can be improved without making major changes in the set of approximating assumptions. For example, the combined loading (compression and bending) was studied in Section 7 only with respect to circular tubes. A similar problem should now be solved for any prismatic tube, including those with rectangular or square section. Furthermore, the analysis of the stiffening phase of the folding process will be improved if the effect of the deformation-induced imperfections are properly accounted for. This calls for relaxing the existing assumption that one fold is formed at a time and admitting the existence of two or even three folds in the progressively crushed tube. Also, the important problem of stability of progressive collapse should be properly addressed. The work on some of the above problems is currently under way at MIT, and the results will be communicated in a separate publication.

In the next level of generalization, the present method can be used to develop a superfolding element. In view of the fact that the SE will be incorporated in the specialized finite-element codes, it is no longer necessary for the solution to have an analytical form. Instead, the global constitutive equation for the super element, i.e., the relation between nodal forces and nodal displacements, can be presented in the form of a subroutine. Such a formulation opens the way for further extension of the present shell model. This will include: (i) relaxing symmetry requirements of the element, (ii) embedding the element in an elastic rather than rigid structure, (iii) adding inertia forces, and, finally, (iv) considering an arbitrary initial curvature of the element. The inclusion of all these effects into the present model to form a new family of hybrid codes is a difficult task. Research in all the above areas should be actively pursued in view of the great potential in future implementation of the superfolding element into existing FE codes to form a hybrid shell model for efficient crash calculations of arbitrarily shaped shells.

ACKNOWLEDGMENTS

The authors would like to express their appreciation to a number of colleagues who contributed to various stages of the present theory. In particular, thanks are due to Dr. S. Bhat of Amoco Production Company and Ms. J. Jones-Oliveira for suggesting a number of improvements to the present manuscript. Dr. Bhat was actively involved in our team work and was behind the developments of some new ideas and solutions to various members subjected to crash loads. Dr. M. Suh and Dr. J. de Oliveira assisted us in solving the problem of a tube under combined loading. Professor N. Jones directed the project on theoretical and experimental investigations of the crash behavior of circular and square tubes conducted at the University of Liverpool. Finally, Dr. E. Haug helped us in establishing a relation of the present theory with some of the most advanced finite-element codes develped for crash application.

The work reported herein was sponsored by the joint MIT–Industry Crashworthiness Consortium and by a special grant from the Ford Motor Company Scientific Research Laboratory in Dearborn, Michigan.

REFERENCES

1. L. Hallquist and D. Benson, "Dyna 3D, A Computer Code for Crashworthiness Engineering," *Proc. Int. FEM Cong., 1986*, pp. 169–188 (1986).
2. E. Haug and T. Scharnhorst, "FEM Crash, Berechnung eines Farzeugfrontaufpralls," *VDI-Ber.*, **613**, 479–505 (1986).
3. C. Marino (Ed.), *Numerical Techniques, Experimental Validations of Structural Impact and Crashworthiness Analysis with Supercomputers for the Automotive Industry*, A Computational Mechanics Publication, 1986.
4. S. Bhat, "Analysis of Large Plastic Shape Distortion of Shells," Ph.D. Thesis, Massachusetts Institute of Technology, Cambridge, MA, 1985.

5. T. Wierzbicki and S. Suh, "Denting of Tubes Under Combined Loading," *Int. J. Mech. Sci. 30* (314), 229–248 (1988).

6. W. Abramowicz and T. Wierzbicki (1988), "Axial Crushing of Multi-Corner Sheet Metal Columns," *J. Appl. Mech.* (in print).

7. J. G. A. Croll, "Buckle Propagation in Marine Pipelines," *Proc. Int. Offshore Mech. Arct. Eng. Symp. 4th*, pp. 433–507 (1985).

8. W. Abramowicz, "The Effective Crushing Distance in Axially Compressed Thin-Walled Metal Columns," *Int. J. Impact Eng.*, **1**, 309–317 (1983).

9. W. Abramowicz, "Mechanics of the Crushing Process of Plastic Shells," Ph.D. Thesis, Institute of Fundamental Technological Research Reports, 1981 (in Polish).

10. T. Wierzbicki and W. Abramowicz, "On the Crushing Mechanics of Thin-Walled Structures," *J. Appl. Mech.*, **50**, 727–739 (1983).

11. H. G. Hopkins, "On the Behavior of Infinitely Long Rigid-Plastic Beams under Transverse Concentrated Load," *J. Mech. Phys. Solids*, **4**, 38–52 (1955).

12. Y. Ohokubo, T. Akamatsu, and K. Shirasawa, "Mean Crushing Strength of Closed-Hat Section Members," *SEA Pap.*, **790060** (1979).

13. W. Abramowicz and A. Sawczuk, "On Plastic Inversion of Cylinders," *Res. Mech. Lett.*, **1**, 525–530 (1981).

14. J. A. de Runtz and P. G. Hodge, "Crushing of Tubes Between Rigid Plates," *J. Appl. Mech.*, **30**, 381–395 (1963).

15. T. Wierzbicki and S. Bhat, "Initiation and Propagation of Buckles in Pipelines," *Int. J. Solids Struct.*, **22**(9), 985–1005 (1986).

16. J. F. Carney III, S. R. Reid, and S. L. K. Drew, "Energy Absorbing Capacities of Braced Metal Tubes," *Int. J. Mech. Sci.*, **25**(9–10), 649–668 (1983).

17. W. Abramowicz and N. Jones, "Dynamic Axial Crushing of Square Tubes," *Int. J. Impact Eng.*, **2**(2), 179–208 (1984).

18. T. Wierzbicki and S. Bhat, "A Moving Hinge Solution for Axisymmetric Crushing of Tubes," *Int. J. Mech. Sci.*, **28**(3), 135–151 (1986).

19. J. M. Alexander, "An Approximate Analysis of the Collapse of Thin Cylindrical Shells Under Axial Loading," *Q. J. Mech. Appl. Math.*, **13**(1), 10–15 (1960).

20. E. Chater and W. Hutchinson, "On the Propagation of Bulges and Buckles," *Harv. Univ. Rep.* **MECH-44** (1983).

21. A. C. Palmer and J. H. Martin, "Buckle Propagation in Submarine Pipelines," *Nature (London)* **1**, 46–48 (1975).

22. S. Kyriakides, M. K. Yeh, and D. Roach, "On the Determination of Propagation Pressure of Long Circular Tubes," *J. Pressure Vessel Technol.*, **106**, 150–159 (1984).

23. C. R. Calladine, (1985), "Analysis of Large Plastic Deformations in Shell Structures," *Proc. IUTAM Symp. Inelastic Behav. Plates Shells, 1985* (1985).

24. C. P. Ellinas and S. Valsgard, "Collisions and Damage of Offshore Structures: A State-of-the-Art," *J. Energy Resour. Technol.*, **107**, 297–314 (1985).

25. J. de Oliveira, T. Wierzbicki, and W. Abramowicz, "Plastic Behavior of Tubular Members Under Lateral Concentrated Loading," *Det Norske Veritas Tech. Rep.*, **82-0708** (1982).

26. R. Y. Hayduk and T. Wierzbicki, "Extensional Collapse Modes of Structural Members," *Comput. Struct.*, **18**(3), 447–458 (1984).

27. W. Abramowicz and N. Jones, "Dynamic Progressive Buckling of Circular and Square Tubes," *Int. J. Impact Eng.*, **4**(4), 243–269 (1986).

28. W. Abramowicz, S. Imielowski, and A. O. Wasowski, *Quasi-Static Axial Crushing of Multicorner Metal Columns*, Inst. Fundam. Technol. Res. Rep., 1985 (in Polish).

29. T. Wierzbicki and W. Abramowicz, *Manual of Crashworthiness Engineering*, Vol. I, Tech. Rep., Center of Transportation Studies, Massachusetts Institute of Technology, 1987.

30. S. Kyriakides and C. D. Babcock, "Prediction of Wet Buckles in Offshore Pipeline," *Proc. Mar. Technol. Conf.*, *1980*, pp. 439–444 (1980).

31. N. Jones, "Plastic Failure of Ductile Beams Loaded Dynamically," *J. Eng. Ind.* Vol. 98, No. 1, 131–136 (1976).

32. T. Wierzbicki, C. Chryssostomidis, and C. Wiernicki, "Rupture Analysis of Ship Plating due to Hydrodynamic Wave Impact," *Proc. Ship Struct. Symp.*, *1984*, pp. 237–256 (1984).

33. E. Haug, F. Arnaugeau, L. Dubois, and A. de Rouvray, "Static and Dynamic Finite Element Analysis of Structural Crashworthiness in the Automotive and Aerospace Industries," in N. Jones and T. Wierzbicki, Eds., *Structural Crashworthiness*, Butterworth, London, 1983, pp. 175–217.

34. W. Abramowicz and A. O. Wasowski, *Computerized Analysis of Head-on Impact of Subcompact Car With Rear Engine*, Fiat Plant Internal Rep., 1986 (in Polish).

35. E. Haug et al., "Numerical Techniques, Experimental Validations of Structural Impact and Crashworthiness Analysis with Supercomputers for the Automotive Industry," *Proc. Int. Conf. Supercomput. Appl. Automot. Ind.*, *1986*, pp. 127–146 (1986) (organized by Cray Research, Inc., C. Marino, Ed. A Computational Mechanics Publiciation).

ADDITIONAL READING

B. Budiansky, "Theory of Buckling and Post-Buckling Behavior of Elastic Structures," *Adv. Appl. Mech.*, **14**, 1–65 (1974).

J. W. Hutchinson, "Plastic Buckling," *Adv. Appl. Mech.*, **14**, 67–144 (1974).

N. Jones and W. Abramowicz, "Static and Dynamic Axial Crushing of Circular and Square Tubes," in S. R. Reid, Ed., *Metal Forming and Impact Mechanics*, Pergamon, Oxford, 1985, pp. 225–247.

W. T. Koiter, "Elastic Stability and Postbuckling Behavior," *Proc. Symp. Nonlinear Probl.*, *1963*, pp. 257–275 (1963).

S. Kyriakides and C. D. Babcock, "Experimental Determination of the Propagation Pressure of Circular Pipes," *J. Pressure Vessel Technol.*, **103**, 328–336 (1981).

S. Kyriakides and C. D. Babcock, "Buckle Propagation Phenomena in Pipelines," in J. M. T. Thomson and G. W. Hunt, Eds., *Collapse: The Buckling of Structures in Theory and Practice*, Cambridge Univ. Press, London and New York, 1983, pp. 75–91.

S. R. Reid and T. Y. Reddy, "Axially Loaded Metal Tubes as Impact Energy Absorbers," in L. Bevilacqua, R. Feijór, and R. Valid, Eds., *Inelastic Behaviour of Plates and Shells*, Springer-Verlag, Berlin and New York, 1986, pp. 569–595.

T. C. T. Ting, "On the Solution of the Nonlinear Parabolic Equation with a Floating Boundary Arising in the Problem of an Impact in a Beam," *Q. Appl. Math.*, **21**(2), 133–150 (1963).

T. Wierzbicki, "On the Formation and Growth of Folding Modes," in J. M. T. Thompson and G. W. Hunt, Eds., *Collapse: The Buckling of Structures in Theory and Practice*, Cambridge Univ. Press, London and New York, 1983.

T. Wierzbicki, "Crushing Analysis of Metal Honeycombs," *Int. J. Impact Eng.*, **1**, 157–174 (1983).

CHAPTER 10

Denting and Bending of Tubular Beams Under Local Loads

S. R. Reid and K. Goudie
Department of Mechanical Engineering
University of Manchester Institute of Science and Technology
Manchester, England

ABSTRACT

A review of work on the local indentation of cylindrical shells is provided as a background to an account of an experimental study of the indentation and bending of tubular beams. The growth of the local dent and transition to a global bending mode of deformation is described. The experimental results are compared with a rigid–plastic analysis due to de Oliveira, Wierzbicki, and Abramowicz. A semiempirical model is provided that contains the effects of both elastic and plastic deformation in the shell.

1. INTRODUCTION

The local indentation of a cylindrical shell by an indenter is a generic problem that arises in many areas of structural mechanics and in safety considerations of process plants and power plants. It is a particularly challenging theoretical problem involving large plastic deformations that are governed by complicated interactions between bending, stretching, and shearing within the shell wall local to the dented region. This region grows during the indentation process. In addition to these local interactions, there are equally complex structural interactions between the local deformation in the dented region and global bending (and possibly membrane) modes of deformation of the shell that depend on the support conditions at its ends. Several attempts have been made to construct relatively simple models for the plastic deformation. However, few of these have been tested directly by comparison with experimental data. One aim of this chapter is to provide some such data.

331

The experimental work that is presented was conducted in order to gain an understanding of the nature and extent of the damage produced in the bracing members of a steel offshore structure as a result of a ship collision. This problem has had a lot of attention over recent years with regard to estimating the impact-energy-absorption capacity of these structural members and to assessing the residual strength of damaged components. Some of this work is reviewed in the following sections.

Although the main emphasis of this chapter is on the behavior of tubular beams loaded transversely by wedge-shaped rigid indenters, this being a simple experimental model for ship collision, it should be noted that there are other areas in which the same fundamental problem of shell indentation is to be found. Damage to piping systems of various sorts is a major interest in both the process plant industry and in the power plant industry. Gas pipes, oil pipelines, and high-pressure steam pipes in both nuclear and conventional power stations are all susceptible to damage and failure as the result of accidental loading due to digging, dropped objects, or missile impact, depending upon their environment. In each case, the ability of the component to resist failure is of interest to those who need to provide protective systems or to determine suitable dimensions for the components to ensure that the damage does not impair the serviceability of the system. Some of the work in these areas involves impact by projectiles of various shapes and sizes traveling over a wide range of velocities. Here the main emphasis is concerned with defining the ballistic limit, a set of conditions within which the target (pipe) is not penetrated. This essentially dynamic problem is excluded from detailed consideration here, the emphasis being placed on quasi-static loading of cylindrical shells.

In the next section, a literature survey is provided as background to the problem. There then follows a brief account of some experimental work conducted by the authors. The results of this are then discussed with reference to two analytical studies, one a preliminary closed-form rigid-plastic analysis, the other a semiempirical elastic–plastic analysis. Finally, some suggestions are made regarding the areas in which attention should be focused in order to produce a more comprehensive theoretical model of this phenomenon.

2. REVIEW OF THE LITERATURE

There are two main areas of interest with regard to the deformation of tubular structural components following impact loading, the energy-absorption capacity of the tube, and the residual strength of the member. In each case, the response is governed by the local and global deformation produced as the result of the impact.

References 1 and 2 contain descriptions of work that examined, in a preliminary way, various aspects of the behavior of simply supported tubular beams loaded laterally at their center by a rigid indenter. The sequence of deformation mechanisms shown in Figure 1 were identified. The influence of the

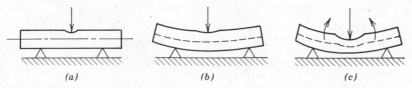

Figure 1. Three-phase collapse sequence for a simply supported tube [1]: (*a*) denting, (*b*) denting and bending, and (*c*) collapse.

diameter-to-thickness ratio D/t was examined experimentally, and it was observed that the larger the ratio, the more extensive (axially) is the dented portion of the tube. In order to gain an insight into the local deformation mode, a further study [3] examined the crushing of tubes of various lengths between opposed wedge-shaped indenters.

De Oliveira [4] first drew attention to the need to examine the collision protection of offshore structures, contrasting the lack of work in this field with the extensive research on various aspects of the collision protection of vehicles [5]. He cited References 1–3 as providing an insight into the modes of deformation of tubular members of steel offshore structures subjected to ship collisions. De Oliveira adapted the simple plastic work calculations contained in References 1–3 in order to estimate the energy absorbed in the indentation (denting) phase of the deformation. Reference was also made to the indentation mechanism described by Morris and Calladine [5] for shells loaded through a boss, which, at that stage, was considered to be too complex for the present problem. De Oliveira produced an analysis of the global bending component of the deformation by extending the plastic beam analysis by Hodge [6] to the case of tubular beams. The significant effect of the axial flexibility of the end supports of the beam was emphasized.

De Oliveira drew attention to the need for experimental data concerning various aspects of the deformation. These included an extension of the Watson et al. [2, 3] work to other end conditions than the simply supported conditions considered in References 1 and 2, an examination of the bending strength of partially deformed cross sections to supplement the work of Sherman [7] and Thomas et al. [1], and acquisition of general experimental data to test the validity of theoretical models.

Soreide and Amdahl [8] and Soreide and Kavlie [9] provided both experimental data and some numerical results for the behavior of horizontally free (but end rotations constrained) and fully built-in tubular beams under transverse central loading. Their results were discussed by Reid [10], and Figure 2 shows the significant effect of axial restraint on the large deformation behavior of such tubular beams. In Figure 2, the load is nondimensionalized with respect to P_0, the nominal plastic limit load for a fixed-end beam of the same section as the undeformed tube.

$$P_0 = \frac{8M_0}{L} = \frac{8D^2h\sigma_y}{L} \qquad (1)$$

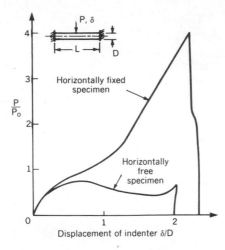

Figure 2. Effect of axial constraint on the load-deflection curves of centrally loaded tubular beams.

where M_0 is the fully plastic moment for the tube section, L is the length of the beam, D is its mean diameter, h is the wall thickness, and σ_y is the yield stress of the material. As Figure 2 shows, because of local denting, the axially free tube never attains the ideal collapse load, but goes unstable at a lower load. The load sustained by the axially fixed tube only exceeds P_0 as a result of the membrane tension generated at indenter displacements in excess of the diameter of the tube.

Soreide et al. [8, 9, 11] used finite-element methods and simple plastic beam theory similar to that described by de Oliveira [4] in order to model the behavior of the beams. Although the finite-element methods were quite successful, they were also time-consuming and costly, and alternative simpler methods are required for design purposes. Soares and Soreide [11] noted that their rigid–plastic method was valid for tubular beams with D/t less than 35 for which local indentation was of minor importance. For larger D/t values, the local indentation reduces the plastic section modulus of the beam. They commented that "if the local deformation can be predicted, the reduced plastic moment can be incorporated in the present formulation."

Because of the important role of local indentation, a number of experimental studies have been made [3, 12–14] and attention has been directed toward constructing a theoretical model using plastic shell analysis [15]. Some of the results of the work by Goudie [14] are presented in the next section for the case of tubular beams. The other experimental work has focused on the problem of a tube pinched between opposed wedge-shaped indenters, a problem that does not contain the complications of global bending response. This work is not considered here. Analytically, the indentation problem is a difficult one since it involves the growth of the dented region as the deflection is increased. Such a phenomenon requires the formulation of a model for traveling plastic regions within which both bending and membrane deformation occurs. A related problem is that of a propagating buckle in a submerged pipeline, which has attracted considerable attention over recent years [16–18], and this is discussed elsewhere [19]. It seems that there are still a number of open questions regarding the nature of the analytical model for this problem. The same comment applies to those attempts that have been made to model indentation in the problem under consideration. Kinematically admissible rigid–plastic mechanisms have been constructed [15] that utilize minimization techniques that cannot be considered rigorous in the context of problems involving large changes of geometry affected by traveling plastic zones. Attention has been drawn to this problem in a wider context by Calladine [20]. Nevertheless, these analytical models do provide some insight into the way in which the tube carries the load and they are discussed in relation to the experimental data presented.

For thin cylindrical shells under concentrated loads, Lukasiewicz [21] has described a number of interesting results produced by considering isometric transformations of the shell wall. These transformations involve no stretching within the major parts of the shell wall. For thin shells (typically, $D/t \geqslant 200$), there are only limited regions in which the shell has a double curvature that requires a nonisometric deformation mechanism. Such results can provide some insight into the solution of indentation problems involving thicker shells.

Although the residual strength of dented members is not of prime interest here, it is of great practical importance. Normally, the degree of denting is small compared with the gross damage with which we are concerned. Useful and interesting treatments of the calculation and assessment of residual strength are to be found in References 22–25.

3. EXPERIMENTAL STUDY OF TUBULAR BEAMS LOADED BY A WEDGE-SHAPED INDENTER

As a development of the work described by References 1–3, several series of experiments were performed in which the geometry of the deforming tube was correlated with the load and indenter deflection. Work on pinched tubes [12, 13] is reported elsewhere. Here we consider aspects of the behavior of tubular beams [14]. No attempt was made to measure strains within the tubes, rather the main emphasis was placed upon following the axial growth of the dent, the reduction in height of the central cross section, and the partitioning of the indenter displacement between this section distortion and transverse beam-type deflection due to global bending of the tubular beam. Tests were performed on simply supported tubes and tubes whose ends were both axially and rotationally constrained. The latter are described as fully fixed tubes.

All the tests were performed on mild-steel seamed tubes of 50.8 mm outside diameter and 1.6 mm wall thickness. In each case, the yield stress for the material was determined by performing a tube-flattening test between flat plates, as described by Reddy and Reid [26]. For the tube specimens referred to in the following sections, the uniaxial yield stress $\sigma_y = 375\,\text{MPa}$.

3.1.   Tests on Simply Supported Tubes

The rig used to test the simply supported tubes is shown in Figure 3. The tube was supported on two rollers to ensure that no significant end constraints could develop. The axes of these rollers were bolted to a rigid beam, which in turn was placed across the base plate on an Instron 1185 testing machine. The load was applied to the center of the beam through a wedge-shaped indenter that had a 4 mm radius nose to prevent penetration of the tube wall. Tests were performed on tubes of three lengths, 305, 457, and 1410 mm. The longer beams afford the opportunity to examine in particular the influence of elastic deformation on the behavior.

The indenter load-displacement $(P-\delta_T)$ curves for the three tubes are given in Figure 4(a). In all cases, the load reaches a maximum value after which the tube becomes unstable, the deformation continuing under conditions of reducing load. The tests were performed under displacement control and so this unstable behavior could be followed. Three further tests on identical specimens were performed in which the indenter displacement was incremented by 5 mm up to 20 mm displacement and thereafter by 10 mm increments. After each increment, the load was removed and the height of the loaded section D^* and the magnitude of the deformed tube diameter at 10 mm intervals along the top generator were measured using vernier calipers. In addition, the vertical displacement of the bottom generator, δ_B, was measured using a dial gauge and lever system, which is shown later in Figure 5(b).

Figure 3.   Test rig for loading simply supported tubes.

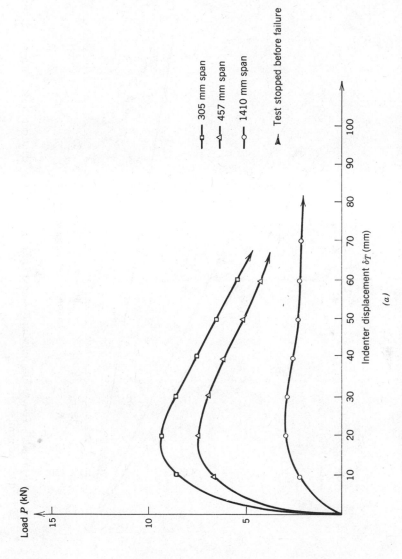

Figure 4a. Test results for simply supported tubes. (*a*) Load-deflection curves.

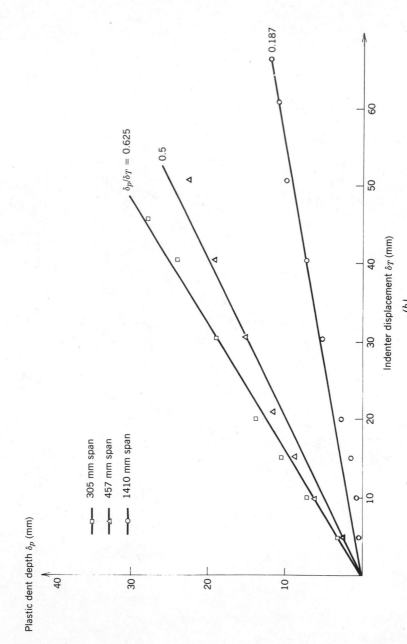

Figure 4b. Test results for simply supported tubes. (b) Plastic dent depth vs. indenter displacement.

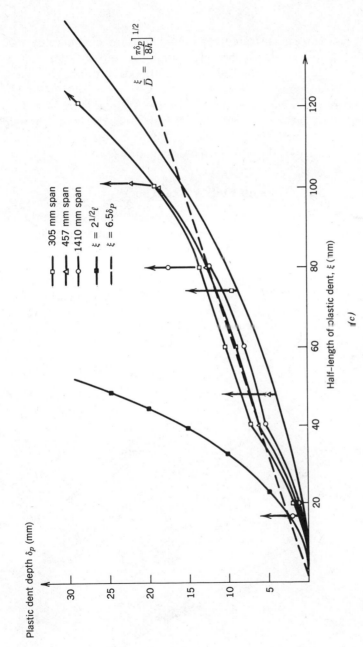

Figure 4c. Test results for simply supported tubes. (c) Dent depth vs. dent half-length. Experimental data and predictions of rigid–plastic analyses.

The plastic dent depth δ_p is defined as

$$\delta_p = D - D^* \tag{2}$$

The comparison between the measured values of δ_p and the difference $\delta_T - \delta_B$ showed that the shape of the deformed central section did not change appreciably when the load was removed at these large deflections. The plastic dent depth δ_p is plotted against δ_T for each of the three tubes in Figure 4(b). The half-length of the plastic dent, ξ, is plotted against δ_p in Figure 4(c). It is interesting to note that whereas for the two longer tubes the dent ceases to propagate axially at a certain stage of the deformation, the central section continues to reduce in height throughout the deformation.

3.2. Tests on Fully Fixed Tubes

A special rig was designed and built in which to provide axial and rotational constraint on the tubes during transverse loading at their centers. It is shown in Figure 5(a). The means by which lower generator deflection was measured is shown in Figure 5(b), and the end support blocks are shown in Figure 5(c). Each tube was plugged at its ends and then clamped between the support blocks. These blocks were then sandwiched between thick plates through which four threaded bars passed and were secured with nuts. The bars were 32 mm

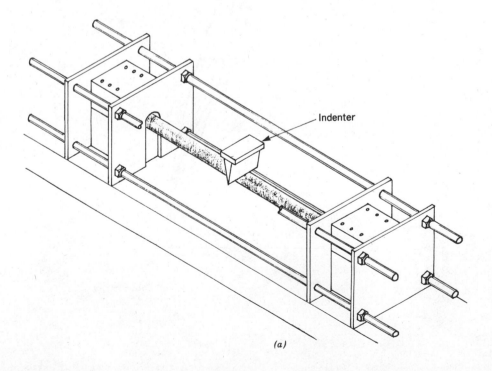

(a)

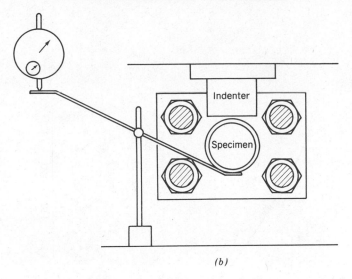

(b)

Figure 5. (*a*) A rig for fully fixed tube tests. (*b*) Layout for measuring lower-generator displacement using a dial gauge. (*c*) End-support blocks for fully fixed tubes.

diameter and the vertical separation between the bars was 86 mm. These provided the axial constraint and also the resistance to rotation of the support blocks. This rig also provided the means of imposing an initial axial tension or compression on the tube by suitably adjusting the fixing nuts shown. Here only the tests performed on tubes under zero preload are considered; the effects of initial tension and compression are discussed elsewhere. The results of the tests given in Figure 6 concern the response of 305, 457, and 1410 mm long specimens as in the simply supported tests. The loads vs. deflection, dent depth vs. indenter displacement, and the dent depth vs. dent half-length curves are given in Figures 6(*a*)–(*c*), respectively.

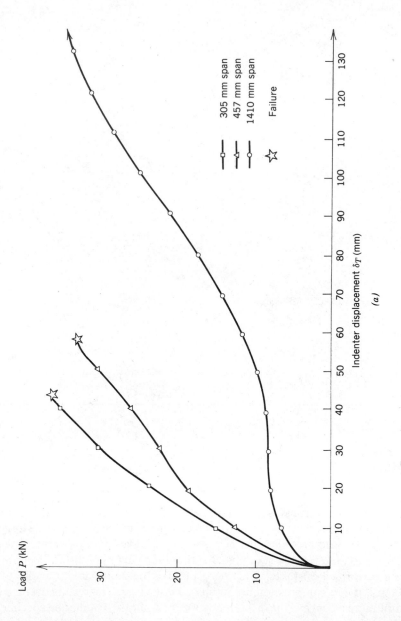

Load P (kN)

Indenter displacement δ_T (mm)

305 mm span
457 mm span
1410 mm span

☆ Failure

(a)

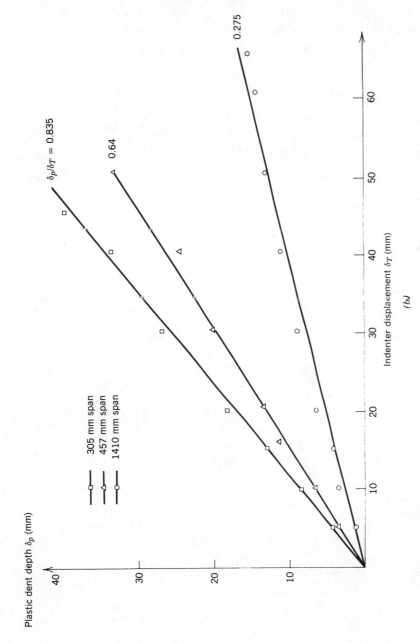

Figure 6a, b. Test results for fully fixed tubes. (*a*) Load-deflection curves and (*b*) Plastic dent depth vs. indenter displacement.

343

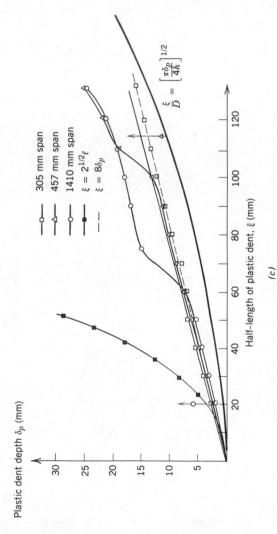

Figure 6c. Test results for fully fixed tubes. (c) Dent depth vs. dent half-length. Experimental data and predictions of rigid-plastic analyses.

344

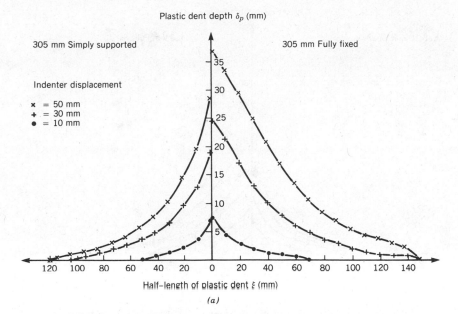

(a)

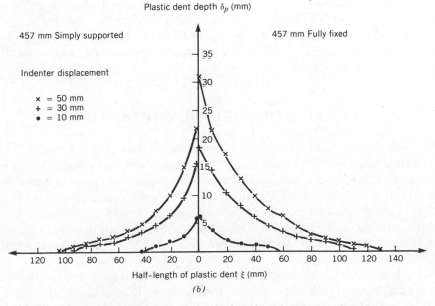

(b)

Figure 7a, b. Axial growth of plastic dent for (*a*) 305 mm, (*b*) 457 mm.

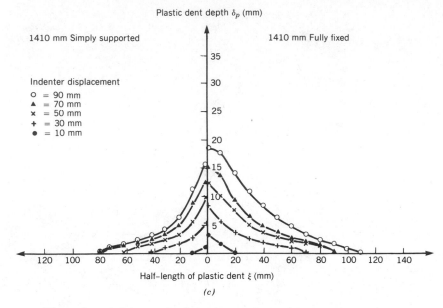

Figure 7c. Axial growth of plastic dent for (*c*) 1410 mm long tubes.

3.3. Dent Geometry

An indication of the variation in dent depth along the tube for each of the tubes tested is provided in Figure 7 for each tube length. For the purpose of comparison later with denting models, the width of the flat portion of the central section 2ℓ defined in Figure 8(*a*) was also measured at each stage, and the results are shown in Figure 8(*b*).

4. OUTLINE OF RIGID–PLASTIC ANALYSIS

The first comprehensive analytical treatment of the problem under consideration based on the assumption of rigid–plastic behavior is that provided by de Oliveira, Wierzbicki, and Abramowicz [15]. While it is recognized that this treatment is under revision [27], this original approach is examined in order to explore the degree of agreement available with the results of the experiments previously described. Only an outline account of the theory is provided here and the reader is referred to the very comprehensive report that contains all the details.

For simplicity, the mode of deformation is assumed to consist of two distinct phases, a denting phase and a global bending phase. During the denting phase, the top generators of the tubular beam are deformed as the load is applied. The

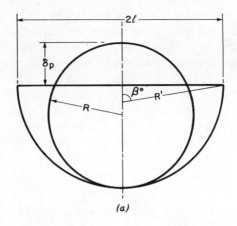

(a)

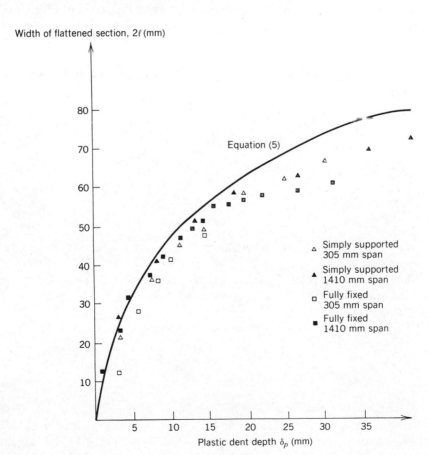

Figure 8. (a) Definition of dent width 2ℓ. (b) Comparison between experimental and theoretical variations of dent width.

dent grows axially as the central section is flattened, and it is in this facet of the problem that de Oliveira et al. [15] have made important contributions. The analysis first requires a model to be developed for the shape of this central section. Subsequent to the denting phase, this geometry change governs the change in the moment-carrying capacity there as well as the subsequent interaction between bending moment and axial force which significantly effects the postcollapse bending phase. This second phase is analyzed by adapting the plastic-beam analysis due to Hodge [6]. The three components, central-section geometry, denting force analysis, and global bending analysis, are summarized in the following sections.

4.1. Central-Section Geometry

In common with Thomas et al. [1], de Oliveira et al. [15] adopted the flattening section model, shown in Figure 8(a), due originally to Pippard and Chitty [28]. The deformation at this section and indeed throughout the shell is assumed to be inextensional in the circumferential direction. Thus, a flat of length 2ℓ is formed at a plastic dent depth δ_p, the radius of curvature of the curved portion of the section increasing from R to R'. Simple geometry leads to the following equations:

$$R' = \frac{\pi R}{\pi - \beta + \sin \beta} \tag{3}$$

$$\delta_p = \frac{R(2 \sin \beta + \pi - 2\beta - \pi \cos \beta)}{\pi - \beta + \sin \beta} \tag{4}$$

and

$$\ell = R' \sin \beta \tag{5}$$

As shown in Figure 8(b), Equation (5) is reasonably well-supported by the experimental data from the tests performed, although it begins to deviate beyond $\delta_p = 20$ mm.

The fully plastic bending moment for this section shape is given by

$$M_p = \frac{\pi^2(2 \sin \alpha - \sin \beta + \sin \beta \cos \beta)}{2(\pi - \beta + \sin \beta)^2} M_0 \tag{6}$$

where $M_0 = \sigma_y D^2 h$, and $\alpha = \frac{1}{2}(\pi + \beta - \sin \beta)$.

As Figure 9 shows, the reduction in M_p with δ_p is close to that of an initially square-section tube of side a, where $a = \pi D/4$. De Oliveira et al. [15] utilize this equivalent square-section tube extensively in their denting analysis, as it makes the computation of the stretching plastic work somewhat simpler. Furthermore,

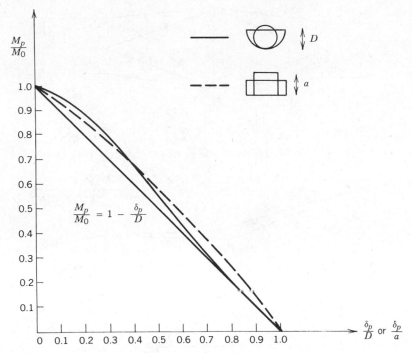

Figure 9. Reduction in fully plastic moment with dent depth (from de Oliveira et al. [15]).

as Figure 9 also shows, the simple formula

$$M_p = (1 - \delta_p/D)M_0 \qquad (7)$$

provides an excellent approximation for each of these section shapes.

4.2 Denting Analysis

Figure 10 shows a typical denting mechanism as assumed by de Oliveira et al. [4, 15] and by the present authors as described in the next section. In its simplest form, the length of tube either side of the loaded section is assumed to undergo a kinematically admissible mode of deformation in which the plastic hinges AB, AD, CB, and CD move through the tube, ξ increasing with the indenter displacement. Plastic bending work is done in these hinges and in the hinge BD of length 2ℓ. Since the generators in the triangular portion increase in length, membrane stretching occurs. This latter contribution was ignored by de Oliveira [4] and its neglect leads to inaccurate predictions for the dent length.

In Reference [15] the rate of plastic work dissipation is equated to the rate of working of the external force P. This results in an equation containing ξ, which

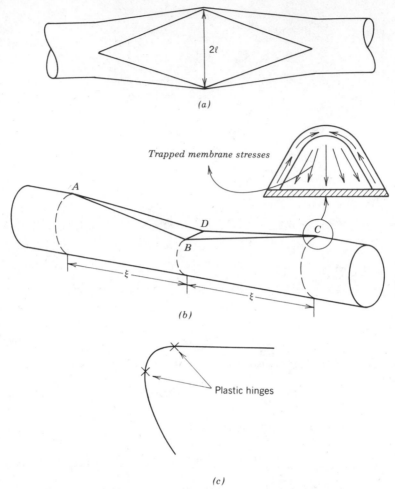

Figure 10. (a) Indentation model. (b) Hinge pattern in dented region. (c) Toroidal knuckle regions along AB, etc.

is then minimized according to the same procedures as used in applications of the upper-bound theorem of limit analysis. The calculations are somewhat simplified by using the equivalent square-section tube. The details are not presented here, but they result in the following two equations for the optimized dent length and for the force corresponding to a given deflection δ_p:

$$\xi = D \left[\left(\frac{\pi \delta_p}{4h} \right) \left[1 - \frac{1}{2} \left(\frac{N}{N_0} - 1 \right)^2 \right] \right]^{1/2} \qquad (8)$$

$$P = \frac{4M_0}{D} \left[\left(\frac{\pi h \delta_p}{D^2} \right) \left[1 - \frac{1}{2} \left(\frac{N}{N_0} - 1 \right)^2 \right] \right]^{1/2} \qquad (9)$$

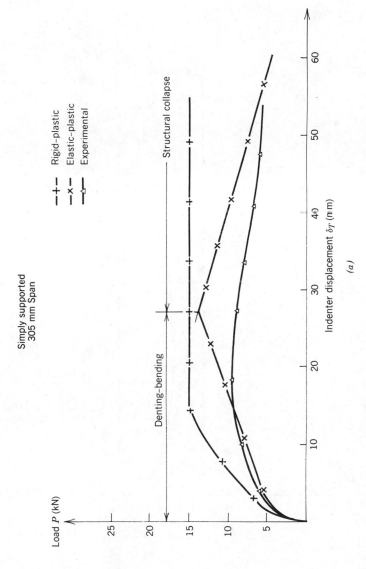

Figure 11a. Comparison between experimental load-deflection curves and theoretical predictions for simply supported tubes of length (*a*) 305 mm.

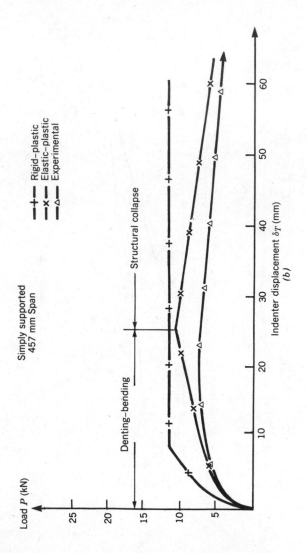

Simply supported
457 mm Span

Rigid–plastic
Elastic–plastic
Experimental

Structural collapse

Denting–bending

Load P (kN)

25
20
15
10
5

Indenter displacement δ_T (mm)
10   20   30   40   50   60

(b)

352

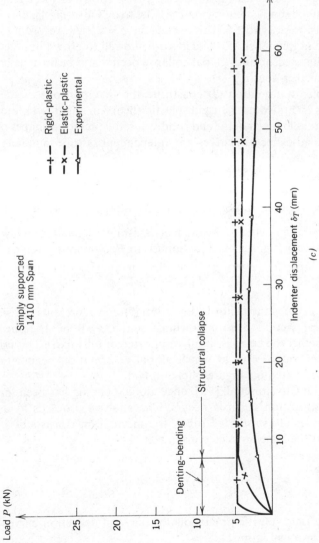

Figure 11b, c. Comparison between experimental load-deflection curves and theoretical predictions for simply supported tubes of lengths (*b*) 457 mm and (*c*) 1410 mm.

where N is the axial force generated in the tube, and $N_0 = \pi D h \sigma_y$ is the fully plastic value for this force. We note that for a free-end tube, $N/N_0 = 0$, and for a fully fixed tube, $N/N_0 = 1$.

During the denting phase, the dent length grows according to Equation (8). The results corresponding to the simply supported tests are shown in Figure 4(c) and those for the fixed-end tests in Figure 6(c). Also shown in these graphs are the predictions of the simple denting analysis due to de Oliveria [4], in which all membrane effects were ignored. This results in $\xi = 2^{1/2}\ell$. According to the model developed in Reference 15, the dents are allowed to grow until the length of the tube is exhausted or until global collapse occurs and beam-type bending ensues with no further denting.

The predictions of Equation (9) constitute the first portions of the load-deflection curves identified as the rigid–plastic theory in Figures 11 and 12. It should be noted that for each of the end conditions, the predicted shapes of these portions of the load-deflection curves are independent of tube length.

4.3. Global Collapse and Post-collapse Behavior

Global collapse occurs when a beam mechanism is generated. This is determined by equating P given by Equation (9) to P_c, which is given as

$$P_c = 4M_0(C_c + C_s)/L \tag{10}$$

where $C_c = (1 - \delta_p/D)$ is the factor in Equation (7) that represents the weakening of the central section due to crushing, and $C_s = 0$ or 1, respectively, depending on whether the beam is simply supported or fully fixed. This criterion determines δ_p^*, the value of δ_p at which global deformation begins, and the relevant points are noted on Figures 4(c) and 6(c).

According to de Oliveira et al. [15], once global bending has been initiated, no further central-section collapse occurs. The analysis proceeds as a beam analysis. For the simply supported tubes, the mechanism consists of a single central hinge and the load remains constant at

$$P = P_c = \frac{4M_0}{L}(1 - \delta_p^*/D) \tag{11}$$

which allows the rigid–plastic predictions of the load-deflection curves to be completed, as shown in Figure 11.

For the fully fixed tubes, what is essentially an adaptation of the theory due to Hodge [6] to account for the change in section under the load leads to

$$P = \frac{4M_0}{L}\left[\left(2 - \frac{\delta_p^*}{D}\right)\cos\left[\frac{\pi N}{2N_0}\right] + \frac{\pi N}{N_0}\frac{w}{D}\right] \tag{12}$$

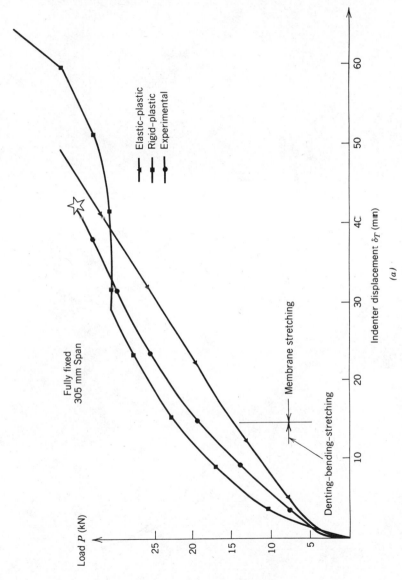

Figure 12a. Comparison between experimental load-deflection curves and theoretical predictions for fully fixed tubes of length (*a*) 305 mm.

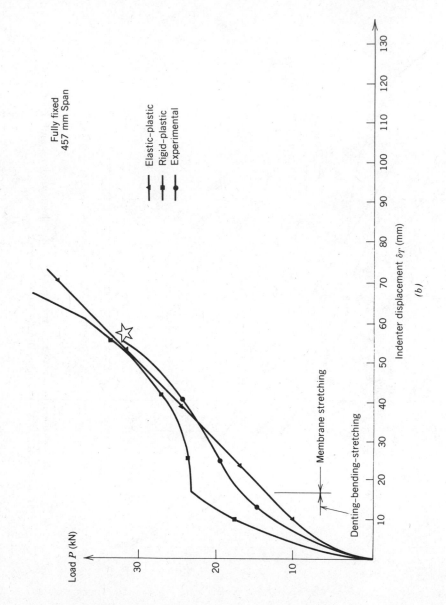

(b)

Load P (kN)

Indenter displacement δ_T (mm)

Fully fixed
457 mm Span

Elastic–plastic
Rigid-plastic
Experimental

Membrane stretching

Denting–bending–stretching

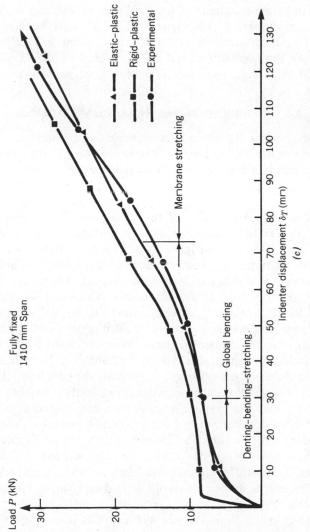

Figure 12b, c. Comparison between experimental load-deflection curves and theoretical predictions for fully fixed tubes of lengths (b) 457 mm and (c) 1410 mm.

where w is the indenter deflection in excess of δ_p^*. The axial force satisfies the following equation:

$$\frac{N}{N_0} = \frac{\pi}{4C_c}\frac{w}{D} \tag{13}$$

Thus, proceeding from equation (13) for a given extra deflection w, N/N_0 can be determined, and Equation (12) then provides P. This bending–stretching deformation continues until $N = N_0$, at which point a purely membrane response ensues, the slope of the load-deflection curve being $4\pi\sigma_y Dh/L$. These equations lead to the rigid–plastic load-deflection curves shown in Figure 12.

4.4. Comments on the Rigid–Plastic Analysis

The theory developed in Reference 15 is extensive and the previous brief account presents it only in its simplest form. For instance, some of the effects of strain-hardening are explored and it is shown how they tend to reduce the dent length. Notwithstanding the neglect of these effects, the rigid–plastic analysis provides a very useful first approximation to the large deformation of the tubular beams tested. More recent developments of the nonstrain-hardening version of the analysis [27] have included the more detailed modeling of the traveling hinges AB, AD, CB, and CD shown in Figure 10(b). This figure shows these lines to be those of slope discontinuity, which is not possible since the material must undergo bending and unbending as it flows through this region. The "edges" AB, etc., must in fact be more akin to toroidal or "knuckle" regions, as shown in Figure 10(c). Similar regions have been alluded to by Calladine [20] in the context of the running buckle in a pipeline. Such regions equilibriate the axial stresses generated within the dented region. Wierzbicki and Suh [27] have also considered toroidal regions of this type in the context of the present problem.

The rigid–plastic analysis tends to overestimate the dent length for both end conditions [see Figures 4(c) and 6(c)], although it clearly provides a much more realistic estimate than the simple ring-collapse mechanism due to de Oliveira [4]. One reason for this in the case of the fixed-end tubes is the finite stiffness of the tube-support rig. The curve in Figure 6(c) comes from Equation (8), with $N/N_0 = 1$. This implies full rigidity. One of the other variants in Reference 15 is to include the axial and rotational stiffnesses of the end supports. This is explored by the authors in the context of a related experimental program to that described herein [29].

Another observation from Figures 4(c) and 6(c) is that the rigid–plastic theory significantly underestimates the dent depth at which global collapse occurs. Figure 11 shows that the theory overestimates the load in the denting phase for simply supported tubes and, because of the assumption that denting ceases when global collapse occurs, it does not follow the postcollapse instability. Furthermore, the conclusion that the indenter displacement at which global collapse occurs reduces with tube length is not borne out in the experiments. Experimentally, the maximum load occurs at indenter displacements that increase from 17 mm for the shortest tube to 23 mm for the longest tube.

Figure 12 shows that the main features of the load-deflection curves of fixed ended tubes are reproduced by the rigid-plastic theory, the greatest error occurring in the initial denting phase.

5. SEMI-EMPIRICAL ELASTIC–PLASTIC THEORY

The theory to be outlined was constructed in response to the comparison previously made between the experimental data and the rigid–plastic theory. The results for both end conditions indicate that the predicted response is too stiff in the initial denting phase. This led to a consideration of elastic effects in the beam, starting from the beginning of the loading process. The theory is described in broad terms only, the analytical details being presented in References 14 and 29.

As noted earlier, when the displacement of the lowest generator is measured as well as the indenter displacement, it is clear that there is a degree of global deformation from the beginning of the loading process. The relationship between the local deformation represented by δ_p and the global deformation reflected in δ_B clearly depends on both the plastic collapse mechanism operating in the dented region and the elastic bending stiffness of the entire tube.

One simple way of representing the response of the tube is to think of it in terms of two springs in series, one of stiffness K_1, representing the crushing/denting stiffness, and the other K_2, representing the elastic stiffness of the entire tube. The latter could result from both elastic bending and elastic stretching of certain parts of the tube during deformation. Clearly, both K_1 and K_2 vary during the deformation. Equilibrium leads to

$$P = \delta_p K_1 = \delta_B K_2 \tag{14}$$

and the indenter displacement δ_T is given by

$$\delta_T = \delta_p + \delta_B \tag{15}$$

Once a denting mechanism has been defined, the load can be calculated in terms of the plastic dent depth. This is formulated in much the same way as for the rigid–plastic analysis and contains contributions from the hinges at the edges of the diamond-shaped dent shown in Figure 10(b), from axial stretching within the dent and from circumferential unbending of the region below the dent. The resulting force equation contains ξ, the dent half-length.

5.1. Simply Supported Tubes

For the simply supported tubes, the resulting P–δ_p relationship can then be converted into a P–δ_T relationship by using the experimental δ_p–δ_T relationship shown in Figure 4(b) together with the simple δ_p–ξ relationship

$$\xi = 6.5\delta_p \tag{16}$$

which is a reasonable approximation for the dent-length growth during the initial denting phase. These two substitutions remove the minimization stage of the calculation [which results in the optimum dent length indicated in Figure 4(c) in the rigid–plastic theory] and also automatically incorporates the elastic element of the indenter displacement through the δ_T–δ_p relationship. It does, however, make the theory semiempirical. Global collapse is identified as in the rigid–plastic analysis, and at this point, the dent length ceases to increase. However, the δ_T–δ_p relationship shows that beyond the onset of global collapse, the dent depth continues to increase and so the postcollapse load is calculated from

$$P = \frac{4M_0}{L}\,(1 - \delta_p/D) \tag{17}$$

The δ_T–δ_p relationship again allows this to be converted into an indenter load-displacement relationship.

The results of this process are shown in Figure 11, from which it can be seen that for the two shorter tubes, there is a much better agreement in the early denting phase than is produced by the rigid–plastic analysis because of the automatic inclusion of the elastic effects through the δ_T–δ_p relationship.

5.2. Fully Fixed Tubes

The initial denting phase is analyzed as previously described with two exceptions. The first is that the dent-length relationship used is now

$$\xi = 8\delta_p \tag{18}$$

in accordance with Figure 6(c). Second, an additional contribution to the applied force arises because of the axial elastic stretching of the generators in the region where the circumferential unbending occurs. Thus, together with the δ_T–δ_p relationships derived from Figure 6(b), the load-deflection curve for this denting–(elastic) bending–stretching phase can be deduced.

The global collapse and postcollapse behavior was analyzed by adapting the elastic–plastic analysis of a fully fixed beam due to Campbell and Charlton [30]. Instead of Equation (10), the following equations were used, together with the δ_T–δ_p relationships of Figure 6(b), to determine the point of global collapse.

$$\delta_c = \delta_p/2 + \delta_B \tag{19}$$

$$P_c^e = \frac{4}{L} \left[(C_s + C_c)M + N\delta_c \right] \tag{20}$$

$$N = \frac{AE\pi^2}{4} \left(\frac{\delta_c}{L} \right)^2 \tag{21}$$

and

$$\frac{M}{M_0} = \cos\left(\frac{\pi}{2} \frac{N}{N_0} \right) \tag{22}$$

Here δ_c is an approximation to the mean generator displacement, and E is the Young's modulus of the material. Equation (20) incorporates the effect of axial tension generated elastically during the early phase of the deformation. As before, $C_c = (1 - \delta_p/D)$, but $C_s = 0.78$ is used here, reflecting the reduced moment-carrying capacity of the clamped ends as measured in a separate set of tests [14].

The identification of the point of global collapse terminates the denting–bending–stretching phase. There then follows, in general, a plastic bending-membrane phase in a manner similar to the rigid–plastic analysis, except utilizing the Campbell and Charlton equations and allowing for the continued increase in dent depth indicated in Figure 6(b). Finally, a fully membrane phase is included, this having the form:

$$P = \frac{4\pi\sigma_y Dh}{L} \delta_c \tag{23}$$

with the use of δ_c tending to reduce the slope of these final sections of the load-deflection curves compared with the rigid–plastic solutions.

The results of applying this theory to the fully fixed tube tests are shown in Figure 12. For the two smaller tubes, the plastic bending–stretching phase is absent and the predicted curves take on more the shapes of the experimental curves with no collapse load identifiable. The longest tube does have this intermediate phase of deformation and matches the experimental data closely.

6. DISCUSSION AND CONCLUSIONS

The experimental data contained in Reference 14, of which that previously described is only a sample, should prove useful in providing a basis for comparison with theoretical models for this complex problem. The closed-form rigid–plastic analysis produced by de Oliveira et al. [15] has established a basis from which an accurate model of the mode of deformation of such tubes will emerge. It remains to be seen whether or not the denting mechanism they describe is sufficiently accurate or, indeed, whether the minimization process they use can be justified.

The purpose in constructing the semiempirical theory was to check whether all the major deformation mechanisms had been identified. The experimentally deduced relationship between dent depth and dent length has been used to remove any doubts surrounding minimization. Furthermore, the effects of elasticity have been included in a pragmatic way by utilizing the $\delta_T - \delta_p$ relationships. The theory is not and is not intended to be predictive. Rather, it has shown that if the geometry of the deforming tube is represented with reasonable accuracy, the major characteristic of the behavior, i.e., the load-deflection curve can be deduced with reasonable accuracy. That having been achieved, it now focuses the attention of structural analysts on to more accurate ways in which the two key relationships δ_p vs. ζ and δ_T vs. δ_p can be deduced theoretically.

ACKNOWLEDGMENTS

The authors are indebted to Prof. T. Wierzbicki for helpful discussions and correspondence regarding his theoretical work. The work was conducted under SERC research contract number GR/D/6127.7.

REFERENCES

1. S. G. Thomas, S. R. Reid, and W. Johnson, "Large Deformations of Thin-Walled Circular Tubes under Transverse Loading. Part I," *Int. J. Mech. Sci.*, **18**, 325–333 (1976).

2. A. R. Watson, S. R. Reid, W. Johnson, and S. G. Thomas, "Large Deformations of Thin-Walled Circular Tubes under Transverse Loading. Part III," *Int. J. Mech. Sci.*, **18**, 501–509 (1976).

3. A. R. Watson, S. R. Reid, and W. Johnson, "Large Deformations of Thin-Walled Circular Tubes under Transverse Loading. Part II," *Int. J. Mech. Sci.*, **18**, 387–397 (1976).

4. J. G. de Oliveira, "Simple Methods of Estimating the Energy Absorption Capability of Steel Tubular Members used in Offshore Structures," Rep. SK/R50, University of Trondheim Norwegian Institute of Technology, Div. Mar. Struct., 1979, also "The Behaviour of Steel Offshore Structures under Accidental Collisions," *Proc.—Annu. Offshore Technol. Conf.*, Pap. No. 4136 (1981).

5. A. J. Morris and C. R. Calladine, "Simple Upper-Bound Calculations for the Indentation of Cylindrical Shells," *Int. J. Mech. Sci.*, **13**, 331–343 (1971).

6. P. G. Hodge, Jr., "Post-Yield Behaviour of a Beam with Partial end Fixity," *Int. J. Mech. Sci.*, **16**, 385–388 (1974).

7. D. R. Sherman, "Tests on Circular Steel Tubes in Bending," *J. Struct. Div., Am. Soc. Civ. Eng.*, **102**, 2181–2195 (1976).

8. T. H. Soreide and J. Amdahl, "Deformation Characteristics of Tubular Members with Reference to Impact Loads from Collisions and Dropped Objects," *Norw. Marit. Res.*, **10**, 3–12 (1982).

9. T. H. Soreide and D. Kavlie, "Collision Damage and Residual Strength of Tubular Members in Steel Offshore Structures," in R. Narayanan, Ed., *Shell Structures; Stability and Strength*, Elsevier Applied Science Publishers, London, 1985, pp. 185–220.

10. S. R. Reid, "Metal Tubes as Impact Energy Absorbers," in S. R. Reid, Ed., *Metal Forming and Impact Mechanics*, Pergamon, Oxford, 1985, pp. 249–269.

11. C. G. Soares and T. H. Soreide, "Plastic Analysis of Laterally Loaded Circular Tubes," *J. Struct. Div., Am. Soc. Civ. Eng.*, **109**, 451–467 (1983).

12. G. A. King, "Plastic Deformations of Point Loaded Cylindrical Shells Having End Restraint," B.Sc.(Eng.) Hons. Thesis, Department of Engineering, University of Aberdeen, Aberdeen, Scotland, 1984.

13. G. C. Montgomery, "Deformation of Punch Loaded Tubes," M.Sc. Thesis, University of Aberdeen, Aberdeen, Scotland, 1985.

14. K. Goudie, "Experimental Study of the Gross Deformation of Tubular Beams," M.Sc. Thesis, University of Manchester Institute of Science and Technology, Manchester, England, 1986.

15. J. de Oliveira, T. Wierzbicki, and W. Abramowicz, "Plastic Behaviour of Tubular Members under Lateral Concentrated Loading," *Nor. Veritas Tech. Rep.* **82-0708** (1982).

16. A. C. Palmer and J. H. Martin, "Buckle Propagation in Submarine Pipelines," *Nature (London)*, **254**, 46–48 (1975).

17. S. Kyriakides and C. D. Babcock, "Buckle Propagation Phenomena in Pipelines," in J. M. T. Thompson and G. W. Hunt, Eds., *Collapse: The Buckling of Structures in Theory and Practice*, Cambridge Univ. Press., London and New York, 1983, pp. 75–91.

18. T. Wierzbicki and S. U. Bhat, "On the Initiation and Propagation of Buckles in Pipelines," *Int. J. Sol. Struct.*, **22**, 985–1005 (1986).

19. S. Kamalarasa and C. R. Calladine, "Buckle Propagation in Submarine Pipelines," *Int. J. Mech. Sci.*, **30**, 217–228 (1988).

20. C. R. Calladine, "Analysis of Large Plastic Deformations in Shell Structures, in L. Bevilacqua, R. Feijón, and R. Valid, Eds., *Inelastic Behaviour of Plates and Shells*, Springer-Verlag, Berlin and New York, 1986, pp. 69–101.

21. S. Lukasiewicz, "Inelastic Behaviour of Shells under Concentrated Loads," in L. Berilacqua, R. Feijón and R. Valid, Eds., *Inelastic Behaviour of Plates and Shells*, Springer-Verlag, Berlin and New York, 1986, pp. 537–567.

22. C. S. Smith, W. L. Somerville, and J. W. Swan, "Residual Strength and Stiffness of Damaged Steel Bracing Members," *Proc—Annu. Offshore Technol. Conf.*, Pap. No. 3981 (1981).

23. J. Taby and T. Moan, "Collapse and Residual Strength of Damaged Tubular Members," *BOSS '85*, Pap. B8 (1985).

24. S. Durkin, "An Analytical Method for Predicting the Ultimate Capacity of a Dented Tubular Member," *Int. J. Mech. Sci.*, **29**, 449–467 (1987).

25. L. A. Pachecho and S. Durkin, "Denting and Collapse of Tubular Members: A Numerical and Experimental Study," *Int. J. Mech. Sci.*, **30**, 317–331 (1988).

26. T. Y. Reddy and S. R. Reid, "On Obtaining Material Properties from the Ring Compression Test," *Nucl. Eng., Des.*, **52**, 257–263 (1979).

27. T. Wierzbicki and M. S. Suh, "Indentation of Tubes under Combined Loading," *Int. J. Mech. Sci.*, **30**, 229–248 (1988).

28. A. J. S. Pippard and L. Chitty, "Experiments on the Plastic Failure of Cylindrical Shells," *Civ. Eng. War*, **3**, 2–29 (1948).

29. S. R. Reid and K. Goudie, "Experimental Study of the Local Loading of Constrained Tubular Beams," *Int. J. Mech. Sci.* (to be published).

30. T. I. Campell and T. M. Charlton, "Finite Deformation of Fully Fixed Beams Comprised of a Non-Linear Material," *Int. J. Mech. Sci.*, **15**, 415–428 (1973).

CHAPTER 11

Dynamic Bending Collapse of Strain-Softening Cantilever Beams

J. B. Martin
Applied Mechanics Research Unit
University of Cape Town
Cape Town, Republic of South Africa

ABSTRACT

A simple analysis of rigid–plastic cantilever beams subjected to a transverse impulse on a tip mass is presented for the case where the hinge rotation is limited due to buckling or tearing of a thin-walled cross section. The problem presents some inherent difficulties in view of the softening characteristics of the cross section. The appropriate manner to deal with softening in simple-beam theory is reviewed, and a moment-hinge rotation relation is assumed for each hinge. This, in turn, implies the choice of discrete hinge positions along the beam; a discrete solution of the impulsively loaded cantilever problem is given and extrapolated to infer some general conclusions for the continuous case.

1. INTRODUCTION

Considerable insight into the response of ductile structures to very large dynamic loads has been gained by idealizing the material as rigid and perfectly plastic. One of the pioneer studies in this area was that carried out by Parkes [1], who considered cantilever beams subjected to a transverse impulsive load applied to a mass at the tip of the beam. In applying the idealization to a beam, the actual moment-curvature relation, shown in Figure 1(*a*), is replaced by a relation of the form shown in Figure 1(*b*). It is argued that for very large dynamic loads, the energy that can be stored elastically is small compared to the total energy input and the energy dissipated in plastic work. For this reason, the elastic response can be sensibly neglected, and the material is treated as one that is essentially viscous.

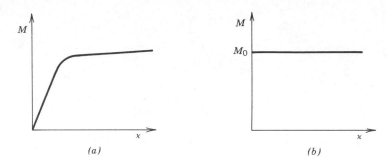

Figure 1. Plastic moment-curvature relationships.

The idealization is, of course, a very simple one; further, Parkes and other early workers formulated their problems as small-displacement problems. Two features that are important in structures subjected to very large dynamic loads are nonlinear geometric effects and the dependence of the yield stress on strain rate. Both these effects have subsequently been incorporated into studies of the dynamic response of ductile structures; nevertheless, the essential features of the response are captured by the rigid–plastic small-displacement analysis of the kind presented by Parkes.

A further concept that has been helpful in understanding the essential nature of the problem is that of modal behavior [2]. In its simplest form, the mode concept demonstrates that an impulsively loaded structure (i.e., a structure where initial velocities are prescribed at time $t = 0$, and loads are zero for $t > 0$) deforms in such a way that its velocity field changes to a mode form. A mode form is defined as a velocity field that can be written as the product of separate functions of space and time.

The mode concept has two important implications. First, it demonstrates a degree of stability of the response: problems that are identical except for small differences in the initial velocity field have solutions that converge on each other rather than diverge. This property is essential if we are to have confidence in the solution in the design sense, since initial conditions in practical situations are usually not precisely defined.

Second, the mode concept offers a means of determining a simple one-degree-of-freedom approximation to the solution. If the mode shape can be determined, the equation of motion reduces to an ordinary differential equation in one variable. This concept has been used with success in a variety of problems to predict the response, including geometric and visco-plastic effects, and has thus become a vehicle to extend the applicability of the simple-beam idealization.

The simple rigid–plastic theory for dynamically loaded structures is restricted, however, by the assumption of unlimited ductility. In beams that are made up of thin-walled sections, for example, ductility at a plastic hinge is limited due to geometric changes in the deformed section. The moment-rotation relation for the hinge is of the form shown in Figure 2(*a*); moment increases with rotation until a peak value is reached, and then decreases, indicating strain-softening. Recently, Jones and Wierzbicki [3] have considered a particular case of a free softening beam subjected to an impulsive load, representing a space vehicle in orbit. They suggested a rigid-softening idealization of the form shown in Figure 2(*b*), and were able to draw some interesting preliminary conclusions about the essential nature of the problem.

In this chapter, we set out to extend the exploratory study of Jones and Wierzbicki on dynamically loaded strain-softening beams. To do so, we follow the work of Parkes [1] and consider the case of a cantilever beam with a mass attached to its tip, subjected to a transverse impulse. Our object is to carry out simple calculations that reveal the essential nature of the problem, rather than to attempt to solve a realistic problem in detail.

Before doing so, however, we review briefly the essential concepts involved in the analysis of softening beams. These concepts have been the subject of some controversy. In one of the earliest references to the analysis of softening beams, Wood [4] pointed to what he saw as a paradox in the simple theory of bending, in that the analysis of a uniformly loaded fixed-end beam where the moment-curvature relation had a falling branch produced apparently puzzling results. The point is pertinent because we have shown in Figure 1 moment-curvature relationships, whereas moment-rotation relationships appear in Figure 2. We summarize the arguments, which indicate that it is appropriate to use the moment-rotation relationship when softening takes place.

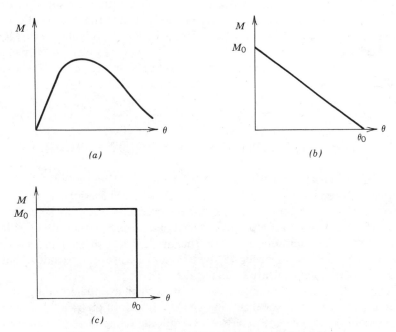

(a)

(b)

(c)

Figure 2. Softening moment-rotation relationships.

Proceeding to the dynamic analysis, we adopt a simpler idealization than Jones and Wierzbicki [3]; we assume that the moment-rotation characteristic has the form shown in Figure 2(c). This form retains the primary characteristic that the work that can be dissipated at a hinge is limited; in practical terms, it is appropriate when softening is due to tearing of the thin-walled section rather than local buckling. The idealization has the advantage that the calculations are simpler than for the case shown in Figure 2(b), without losing the essential nature of the effect of softening. In the spirit of Parkes [1], we also retain the small-displacement assumptions, even though these are clearly inappropriate when we anticipate deformation of the order of the dimensions of the structure. For the cantilever, in particular, the inclusion of geometric effects provides more accurate solutions, but does not add new features to the problem.

We consider then the case of a massless cantilever with an attached tip mass. This problem is of one degree of freedom, and is very simple. It shows, however, an important characteristic that can be readily anticipated; the response of the beam to energy inputs that exceed what can be dissipated at the hinge at the base is unlimited, or catastrophic.

The problem of real interest, however, is the case where the mass of the beam is taken into account. There is a basic problem in dealing with this case; Parkes [1] used the moment-curvature relation of Figure 1(b), whereas the relation shown in Figure 2(c) is a moment-rotation relation for a localized hinge. It is not immediately clear how to treat in a simple way deformation that is continuous along the beam under the assumptions that lead to the adoption of the softening characteristic of Figure 2(c). However, Parkes's continuous solution can be approximated by a discrete model of the beam in which the mass of the beam is lumped and deformation is restricted to localized hinges. It is shown that it is the nature of this approximation that a two-element model is of particular importance, and we consider the two- and three-degree-of-freedom cases in detail. We can use this case to draw some conclusions about the significance of modal behavior.

2. THE ANALYSIS OF STRAIN-SOFTENING BEAMS

It has been remarked that Wood [4] suggested that the analysis of beams with a softening moment-curvature relationship led to paradoxical results. His interest arose in the context of reinforced concrete beams, where softening results from the degradation of the material due to microstructural cracking. In subsequent work on reinforced concrete beams [5], the concept of a finite-length hinge was introduced in order to overcome the apparent paradox that was associated with a concentrated hinge.

The essential issue has become clearer as a further understanding has been gained in recent years of localization of deformations in physically or geometrically unstable materials: the behavior of strain-softening beams falls within this framework. It is now established that the use of local strain measures (curvature in the case of the beam) within some spatial discretization (such as an assumed hinge length) leads to numerical results that depend on the spatial discretization. Nonlocal strain measures [6] can be adopted to avoid sensitivity of the analysis to the spatial discretization.

Alternatively, we can look more carefully at the nature of the localization phenomenon, and adopt a suitable idealization consistent with simple-beam theory. In this approach, used by Maier [7, 8], Wierzbicki, Xirouchakis, and Choi [9], and Jones and Wierzbicki [3], the use of the local strain measure, the curvature, is restricted to stable hardening behavior. Softening is located at a plastic hinge and is represented by a moment-rotation relation.

In order to demonstrate the essential points in the behavior, consider the simply supported beam subject to a central point load, as shown in Figure 3. We consider the equilibrium relation between load P and central displacement δ, assuming that the central displacement is increased monotonically.

Suppose first that the moment-curvature relationship has the bilinear hardening form shown in Figure 4(a). The behavior is elastic until load P reaches the value $4M_0/\ell$. Thereafter, the load continues to increase with increasing δ; the curvature distribution is shown in Figure 4(b), and can be straightforwardly integrated to obtain the displacement. The load-displacement relation is of the form shown in Figure 4(c).

Now suppose that the section is elastic and perfectly plastic, with the moment-curvature relation shown in Figure 5(a). Again, the behavior is elastic until $P = 4M_0/\ell$. Thereafter, as is well known in plastic theory, a hinge forms at the center of the beam and further deformation occurs at constant load, as shown in Figure 5(b). It is instructive to ask what happens to the curvature at the center of the beam; the plastic hinge, of infinitesimal length, has a finite rotation, and, therefore, the curvature is infinite. This is, in effect, a localization of the deformation; we cannot conventionally deal with the curvatures in the zone of localized deformation, and we represent the deformation of the zone by a discontinuity in the generalized displacements. Given this idealization, we can carry out an equivalent analysis by assuming that the moment-curvature relation is elastic for the entire beam, and that there exists a central hinge with a rigid-plastic moment-rotation analysis of the form shown in Figure 5(c).

Third, let us consider the softening case. In view of our interpretation of the behavior of the perfectly plastic beam, we might expect that the softening moment-curvature relation must be defined for very large curvatures. We

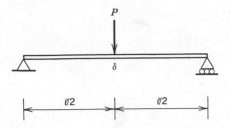

Figure 3. A simply supported beam.

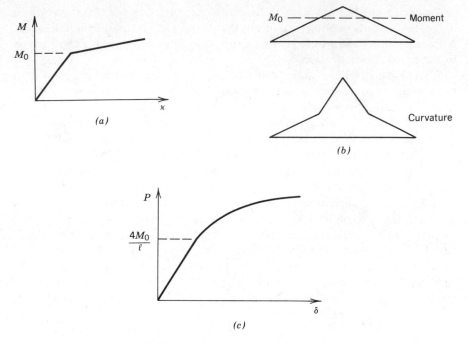

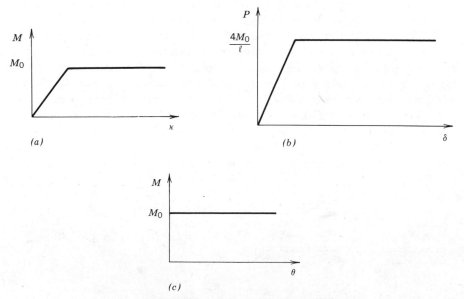

Figure 4. A hardening beam.

Figure 5. A perfectly plastic beam.

arbitrarily adopt the relation shown in Figure 6(a); the beam is elastic until a peak moment M_0 is achieved. Thereafter, the moment falls linearly with increasing curvature until a residual moment M_{00} is reached. Ductility is then unlimited for the residual moment.

As the beam is loaded, the behavior is elastic until the load reaches the value $4M_0/\ell$. If the central displacement is further increased, it is evident from the perfectly plastic case that a hinge must form at the center; the curvature must become infinite, and hence the moment falls to the residual value M_{00}. The equilibrium load-displacement diagram thus has the form shown in Figure 6(b). It exhibits a jump at the peak load; with respect to applied loads, the beam is unstable at this point, and a small dynamic disturbance would lead to unlimited deformation.

We note that we could carry out the analysis in an equivalent way by assuming that the entire beam was elastic, and that there was a hinge at the center with a moment-rotation characteristic of the form shown in Figure 6(c).

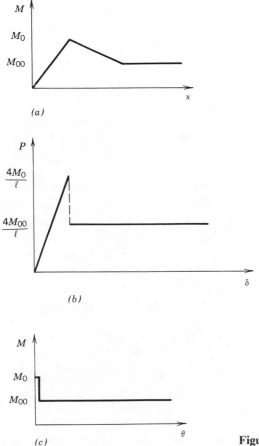

Figure 6.   A softening beam.

There are some important points that can be learned from this analysis. First, softening behavior must be associated with localization of deformation if simple-beam theory is to be employed; indeed, the perfectly plastic case, interpreted as the transition from hardening to softening, itself exhibits this localization.

Second, it is the nature of the moment-rotation relation at the hinge that defines the structural behavior. It is not suggested that the moment-rotation relationship of Figure 6(b) is appropriate: we would normally expect this relationship to have the form shown in Figure 2(a) or 2(b). What is important is that the "plausible" moment-curvature relationship of Figure 6(a) does not lead, in the presence of localization, to a plausible moment-rotation relationship. It is thus suggested that the physical behavior should be modeled at the level of the moment-rotation relation, not the moment-curvature relation.

If this is done, simple-beam theory leads to a consistent formulation of structural problems, as shown by Maier [7, 8]. The structural behavior exhibits softening and instability, but the analysis can be carried out within the conventional mechanical framework.

However, the introduction of hinges with a given moment-rotation relation implies that we must introduce into the problem, from the beginning of the analysis, some spatial discretization in which we locate the positions of the potential hinges. In static analysis of beam and frame structures, this does not present very great difficulties, since we can anticipate the positions of peak moment. In dynamic problems, however, particularly in cases where in perfectly plastic analysis we expect that the hinge can move continuously along the beam, the implications of the discretization are not as clear. It is this point that is one of the major motivations for the present study.

3. ONE-DEGREE-OF-FREEDOM CANTILEVER

Consider the beam shown in Figure 7(a), consisting of a massless cantilever of span ℓ with an attached tip mass G. The transverse displacement of the mass is $v(t)$; at time $t = 0$, we are given $v(0) = 0$; $dv/dt(0) = \dot{v}_0$. The displacement at time t is shown in Figure 7(b); deformation is localized at a hinge at the base where the rotation is $\theta(t)$.

The moment-rotation characteristic is given in Figure 2(c), and, for $\dot{\theta} > 0$, is described by

$$M = M_0 \qquad \text{for } \theta \leqslant \theta_0$$

$$M = 0 \qquad \text{for } \theta > \theta_0 \qquad (1)$$

The deformation field is consistent with monotonic deformation, $\dot{\theta} > 0$, and we proceed with the analysis on the assumption that $\theta < \theta_0$.

The free-body diagram is shown in Figure 7(c); taking moments about the

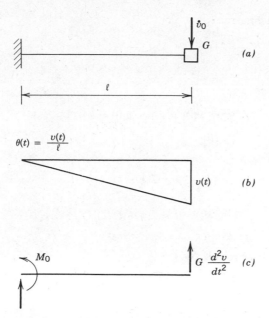

Figure 7. A one-degree-of-freedom cantilever.

base,

$$G\ell \frac{d^2v}{dt^2} = -M_0 \tag{2}$$

It is convenient to introduce dimensionless variables u and τ defined by

$$u = \frac{2M_0}{G\ell^2 \dot{v}_0^2} v \tag{3a}$$

$$\tau = \frac{M_0}{G\ell \dot{v}_0} t \tag{3b}$$

Then

$$\frac{du}{d\tau} = \frac{2}{\dot{v}_0} \frac{dv}{dt} \tag{3c}$$

$$\frac{d^2u}{d\tau^2} = \frac{2G\ell}{M_0} \frac{d^2v}{dt^2} \tag{3d}$$

By using (˙) to indicate differentiation with respect to τ, the equation of motion,

Equation (2), becomes

$$\ddot{u} = -2 \tag{4a}$$

with

$$\dot{u}(0) = 2 \qquad u(0) = 0 \tag{4b}$$

The solution is

$$\dot{u} = 2(1 - \tau) \tag{5a}$$

$$u = 2(\tau - \tau^2/2) \tag{5b}$$

The solution holds provides

$$v/\ell \leqslant \theta_0 \tag{6a}$$

or

$$u \leqslant 1/\beta \tag{6b}$$

where

$$\beta = \frac{G\dot{v}_0^2}{2M_0\theta_0} \tag{6c}$$

The parameter β is the ratio of the input kinetic energy to the maximum plastic work that can be dissipated in the hinge. It can also be shown that

$$\theta/\theta_0 = \beta u \tag{7}$$

Assuming that inequality of Equation (6a) holds, we see that the velocity $\dot{u}(\tau)$ decreases linearly, and becomes zero at time τ_f, where

$$\tau_f = 1 \tag{8}$$

Then

$$u_f = u(\tau_f) = 1 \tag{9}$$

We can now easily interpret the full solution. If

$$u_f = 1 \leqslant 1/\beta \qquad \text{or} \quad \beta \leqslant 1 \tag{10}$$

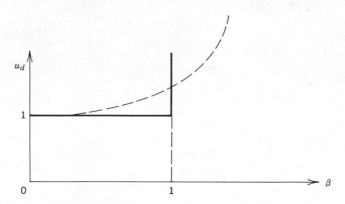

Figure 8. Relationship between u_d and β.

the complete solution is described by Equations (5a) and (5b) and the final displacement u_d is given by

$$u_d = u_f = 1 \tag{11}$$

If $\beta > 1$, the hinge rotation θ reaches the value θ_0 at time $\tau < \tau_f$, with $\dot{u} > 0$. The resisting moment drops to zero, and motion continues with $\dot{u}$ constant. The final displacement u_d thus becomes infinitely large. The relationship between u_d and β is shown by the solid line in Figure 8; the catastrophe occurs for $\beta > 1$.

Figure 8 is a highly idealized picture; for a more realistic moment-rotation curve, we might expect something in the nature of the dashed line. Nevertheless, the solid line captures the essential features.

4. DISCRETE APPROXIMATION TO PARKES'S PROBLEM

Parkes [1] considered the problem shown in Figure 9(a). The cantilever beam has mass m per unit length. The solution is quite straightforward for a rigid–plastic moment-curvature-rate relation of the form

$$M = M_0 \qquad \text{for } \dot{\kappa} > 0$$

$$-M_0 \leqslant M \leqslant M_0 \qquad \text{for } \dot{\kappa} = 0 \tag{12}$$

$$M = -M_0 \qquad \text{for } \dot{\kappa} < 0$$

The response is divided into two phases. In the first phase, the velocity field has the form shown in Figure 9(b), with a hinge located at distance λ from the tip. The moment in the stationary part of the beam is M_0, and as a result, the shear force at the hinge is zero. Linear momentum is preserved, and the angular-

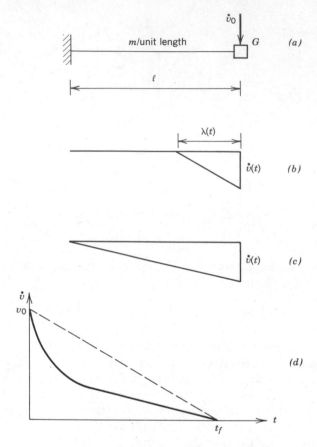

Figure 9. Solution to Parkes's problem.

momentum equation shows that $d^2v/dt^2 < 0$, $\dot{\lambda} > 0$. The hinge thus travels along the beam. The second phase begins when the hinge reaches the fixed end; the velocity field is then as shown in Figure 9(c). This is a modal phase, with the shape of the velocity field remaining constant. The acceleration of mass G is negative and constant, and the beam rotates with a hinge at its base until motion ceases.

The velocity history at the tip is shown in Figure 9(d). During the first phase, the velocity decreases rapidly, and becomes linear in the second phase. The deceleration magnitude in the second phase depends on the mass ratio, $\gamma = m\ell/G$, decreasing as the mass ratio increases, whereas the final time t_f is independent of the mass ratio. The solution for the massless case is recovered for $m = 0$, and is shown by the dashed line.

Because of the need to introduce a spatial discretization to deal with softening behavior, we seek a discrete approximation to Parkes's problem. We divide the beam into, say, n equal intervals, and lump the masses at the boundaries of these intervals, as shown in Figure 6(a). We limit the formation of hinges to the base end of each element, thus discretizing the deformation pattern. For the present, we assume that each hinge is rigid and perfectly plastic, with unlimited ductility.

We choose to lump the entire mass of the beam at the interior modes, so that, referring to Figure 10(a),

$$g = \frac{m\ell}{n-1} \qquad (13)$$

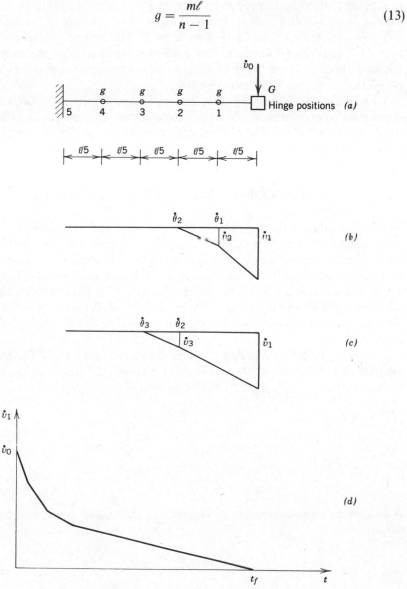

Figure 10.   Discrete approximation to Parkes's problem.

This is not the most accurate representation, but it has the advantage of computational simplicity in comparing solutions for different values of n. In the limit, as $n \to \infty$, we recover the case of a continuously distributed mass.

The traveling-hinge phase of the continuous solution is present in the discrete model, but is represented by the deformation of two hinges. The velocity field in the first phase is shown in Figure 10(b). In this phase, $d^2v_1/dt^2 < 0$ and $d^2v_2/dt^2 > 0$, with hinges at positions 1 and 2. The phase ends then $\dot{v}_2 = \dot{v}_1/2$, and hinge 1 unloads. Immediately, another hinge forms at position 3, and the second-phase velocity field is shown in Figure 10(c). During the second phase, $d^2v_3/dt^2 > 0$, with $d^2v_1/dt^2 < 0$, $d^2v_2/dt^2 < 0$. The second phase ends when $\dot{v}_3 = \dot{v}_1/3$, and hinge 2 unloads. This pattern is repeated until the cantilever is rotating about the base hinge as a rigid body, when the beam enters a modal phase that continues until motion ceases.

The form of the tip velocity–time history is shown in Figure 10(d); we obtain a piecewise approximation, since the accelerations are constant in each phase. As we increase the number of elements, we converge monotonically onto the continuous solution.

5. TWO-DEGREE-OF-FREEDOM MODEL

We now introduce softening [Figure 2(c)] into the discrete problem. It is evident from Figure 10 that two-degree-of-freedom motions describe the transient phases, and thus these are of particular significance. For this reason, we consider first the particular case of a two-mass approximation, as shown in Figure 11(a). The tip mass is denoted by G, and the central mass by g. The cantilever is divided into two equal lengths ℓ_1; we choose an arbitrary length because we can apply the first phase of this analysis for any value of n. The initial velocity of the tip mass is $\dot{v}_0$.

The velocity field for the first phase is shown in Figure 11(b). The free-body diagrams for the two elements are shown in Figures 11(c) and 11(d), respectively. Assuming that $\theta_1 \leqslant \theta_0$, $\theta_2 \leqslant \theta_0$, the equations of motion are

$$G\ell_1 = \frac{d^2v_1}{dt^2} = -M_0 \tag{14a}$$

$$g\ell_1 \frac{d^2v_2}{dt^2} = +M_0 \tag{14b}$$

We adopt the same dimensionless variables as before, and introduce further the parameters

$$g = bG \qquad \ell_1 = \alpha\ell \tag{15}$$

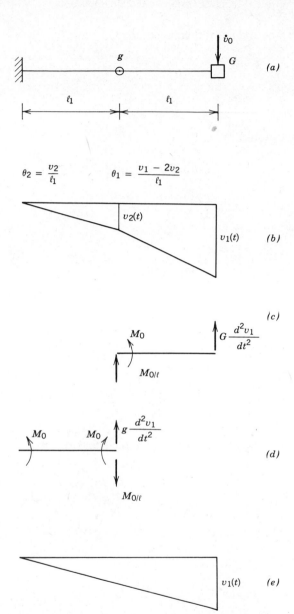

Figure 11. A two-degree-of-freedom model.

Equations (14a) and (14b) then become

$$\ddot{u}_1 = -2/\alpha \tag{16a}$$

$$\ddot{u}_2 = +2b\alpha \tag{16b}$$

The initial conditions are

$$\dot{u}_1(0) = 2 \qquad u_1(0) = 0$$
$$\dot{u}_2(0) = 0 \qquad u_2(0) = 0 \tag{16c}$$

We also note that

$$\frac{\theta_1}{\theta_0} = \frac{\beta}{\alpha}(u_1 - 2u_2) \qquad \frac{\theta_2}{\theta_0} = \frac{\beta}{\alpha}u_2 \tag{17}$$

The solution, provided $\theta_1 \leqslant \theta_0$ and $\theta_2 \leqslant \theta_0$, is given by

$$\dot{u}_1 = 2\left(1 - \frac{\tau}{\alpha}\right) \qquad \dot{u}_2 = \frac{2}{b\alpha}\tau$$

$$u_1 = 2\left(\tau - \frac{\tau^2}{2\alpha}\right) \qquad u_2 = \frac{1}{b\alpha}\tau^2 \tag{18}$$

The phase ends, in the absence of catastrophic behavior, when $\dot{\theta}_1 = 0$ or $\dot{u}_1 = 2\dot{u}_2$. From Equations (18), this occurs at time τ_1 given by

$$2\left(1 - \frac{\tau_1}{\alpha}\right) = \frac{4}{b\alpha}\tau_1$$

or

$$\tau_1 = \frac{\alpha b}{b + 2} \tag{19a}$$

Then

$$u_1(\tau_1) = \frac{\alpha b}{b + 2} \tag{19b}$$

$$u_2(\tau_1) = \frac{\alpha b}{(b + 2)^2} \tag{19c}$$

Hence,

$$\frac{\theta_1(\tau_1)}{\theta_0} = \beta \frac{b^2}{(b+2)^2} \qquad \frac{\theta_2(\tau_1)}{\theta_0} = \frac{\beta b}{(b+2)^2} \tag{20}$$

Finally,

$$\dot{u}_1(\tau_1) = \frac{4}{b+2} \qquad \dot{u}_2(\tau_1) = \frac{2}{b+2} \tag{21}$$

We continue the analysis for the case $\alpha = 0.5$, since the second phase applies only to this case. Motion continues with the beam rotating as a rigid body about the base, as shown in Figure 11(e). By taking moments about the base, the equation of motion is

$$(\tfrac{1}{4}g + G)\frac{d^2v}{dt^2} = -\frac{M_0}{\ell} \tag{22}$$

In dimensionless form, this becomes

$$\frac{b+4}{4}\ddot{u}_1 = -2 \tag{23}$$

With the initial conditions given by Equations (19a) and (21), the velocity is

$$\dot{u}_1 = \frac{8}{b+4}(1 - \tau) \tag{24}$$

It is clear that motion ceases for $\dot{u}_1(\tau_f) = 0$, when $\tau_f = 1$. We can also see very simply that

$$u_1(\tau_f) - u_1(\tau_1) = \frac{b+4}{(b+2)^2} \tag{25}$$

It follows then that

$$\frac{\theta_2(\tau_f)}{\theta_0} - \frac{\theta_2(\tau_1)}{\theta_0} = \frac{\beta(b+4)}{(b+2)^2} \tag{26}$$

and from Equations (20),

$$\frac{\theta_1(\tau_f)}{\theta_0} = \beta \frac{b^2}{(b+2)^2} \qquad \frac{\theta_2(\tau_f)}{\theta_0} = \frac{2\beta}{b+2} \tag{27}$$

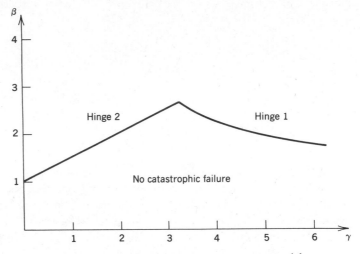

Figure 12. Safe domain for a two-mass model.

If catastrophe is to be avoided, we must have

$$\beta \geqslant \frac{(b+2)^2}{b^2} \quad \text{and} \quad \beta \geqslant \frac{b+2}{2} \quad (28)$$

The envelope of these conditions is plotted in Figure 12; note that for the case $n = 2, \gamma = b$. We see that for γ larger than about 3.25, failure of the hinge nearest the tip mass can occur; for smaller values of γ, the base hinge is the limiting case.

6. THREE-DEGREE-OF-FREEDOM MODEL

Consider now the three-element model shown in Figure 13(a). The general velocity field is shown in Figure 13(b). At time $t = 0$, the initial conditions are $\dot{v}_1(0) = \dot{v}_0$ and $\dot{v}_2(0) = \dot{v}_3(0) = 0$.

In the first phase, $\dot{v}_3 = 0$, and we can use the analysis for the first phase of the two-element model, with $\alpha = \frac{1}{3}$. Thus, at the end of the first phase, we have

$$\tau_1 = \frac{b}{3(b+2)} \qquad 2\dot{u}_2(\tau_1) = \dot{u}_1(\tau_1) = \frac{4}{b+2} \qquad \frac{\theta_1(\tau_1)}{\theta_0} = \beta \frac{b^2}{(b+2)^2} \quad (29)$$

In the second phase, hinge 1 is inactive; $\dot{\theta}_1 = 0$ and

$$\dot{v}_2 = \frac{\dot{v}_1 + \dot{v}_3}{2} \quad (30)$$

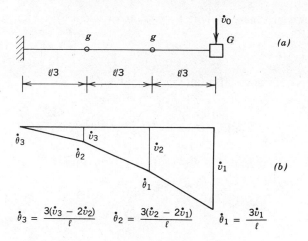

$$\dot{\theta}_3 = \frac{3(\dot{v}_3 - 2\dot{v}_2)}{\ell} \qquad \dot{\theta}_2 = \frac{3(\dot{v}_2 - 2\dot{v}_1)}{\ell} \qquad \dot{\theta}_1 = \frac{3\dot{v}_1}{\ell}$$

Figure 13. A three-mass model.

The equations of motion are set up in a form identical to that of the first phase, and are

$$(b + 4)\frac{d^2v_1}{dt^2} + b\frac{d^2v_3}{dt^2} = -\frac{6M_0}{G\ell}$$

$$(b + 4)\frac{d^2v_1}{dt^2} + 3b\frac{d^2v_3}{dt^2} = 0 \tag{31}$$

By uncoupling, and introducing dimensionless variables, these equations become

$$\ddot{u}_1 = -\frac{18}{b + 5}$$

$$\ddot{u}_3 = \frac{6(b + 2)}{b(b + 5)} \tag{32}$$

With initial conditions

$$\tau_1 = \frac{b}{3(b + 2)} \qquad \dot{u}_1(\tau_1) = \frac{4}{b + 2} \qquad \dot{u}_3(\tau_1) = 0 \tag{33}$$

the solution is

$$\dot{u}_1 = \frac{2}{b + 5}(5 - 9\tau)$$

$$\dot{u}_3 = \frac{6(b + 2)}{b(b + 5)}\left[\tau - \frac{b}{3(b + 2)}\right] \tag{34}$$

The phase ends at time τ_2, where $\dot{\theta}_2(\tau_2) = 0$, or when

$$\dot{u}_1 = 3\dot{u}_3 \tag{35}$$

Substituting Equations (34) into (35), we find

$$\tau_2 = \frac{4}{9}\frac{b}{b+1} \qquad \dot{u}_1(\tau_2) = 3\dot{u}_3(\tau_1) = \frac{2}{b+1} \tag{36}$$

We find also, on integrating from $\tau = 0$ to $\tau = \tau_2$, that

$$\frac{\theta_2(\tau_2)}{\theta_0} = \frac{4\beta b}{3(b+2)(b+1)} \tag{37}$$

In the third and final phase, the beam rotates about the base; we have $\dot{\theta}_1 = \dot{\theta}_2 = 0$, and

$$\dot{v}_3 = \dot{v}_1/3 \qquad \dot{v}_2 = 2\dot{v}_1/3 \tag{38}$$

The equation of motion is

$$\left(1 + \frac{5b}{9}\right)\frac{d^2 v_1}{dt^2} = -\frac{M_0}{G\ell} \tag{39}$$

In dimensionless parameters, this becomes

$$\ddot{u}_1 = -\frac{18}{9 + 5b} \tag{40}$$

and with the initial conditions of Equations (36), the solution is

$$\dot{u}_1 = -\frac{18}{9 + 5b}(1 - \tau) \tag{41}$$

We again see that motion ceases for $\dot{u}_1(\tau_f) = 0$, $\tau_f = 1$. By integrating from time $\tau = 0$ to $\tau = 1$, the rotation of the base hinge is

$$\frac{\theta_3(\tau_f)}{\theta_0} = \frac{2\beta}{3}\frac{b+3}{(b+1)(b+2)} \tag{42}$$

We can now set down conditions to avoid catastrophic failure at any one of the three hinges. Noting that for the case $n = 2$, $\gamma = 2b$, the conditions that

$$\frac{\theta_1(\tau_f)}{\theta_0} \leqslant 1 \qquad \frac{\theta_2(\tau_f)}{\theta_0} \leqslant 1 \qquad \frac{\theta_3(\tau_f)}{\theta_0} \leqslant 1 \tag{43}$$

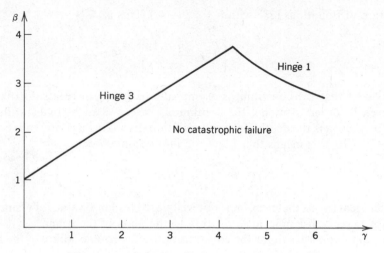

Figure 14.   Safe domain for a three-mass model.

are, from Equations (29), (37), and (42),

$$\beta \leqslant \frac{(\gamma + 4)^2}{\gamma^2} \quad \text{or } \beta \leqslant \frac{3(\gamma + 4)(\gamma + 2)}{8\gamma} \quad \text{or } \beta \leqslant \frac{3(\gamma + 2)(\gamma + 4)}{4(\gamma + 6)} \quad (44)$$

The envelope of these conditions is plotted in Figure 14. It can be seen that the constraint imposed by $\theta_2(\tau_f) \leqslant \theta_0$ is not active. On comparing Figures 12 and 14, it is noted that the constraint imposed by hinge 1 is less restrictive in the three-mass model, and the constraint imposed by the hinge at the base is of a very similar form.

7.   INFERENCES FOR THE CASE OF THE CONTINUOUSLY DISTRIBUTED MASS

It would appear that we can draw from the discrete-mass approximations, with a reasonable degree of confidence, inferences regarding the behavior of the cantilever with continuously distributed mass. A major point of concern is whether the modal behavior can give a reasonable prediction of the response of the beam, or whether softening results in localization and catastrophic failure occurs before the mode is established.

If catastrophic failure does occur before the mode is established, the lump-mass model indicates that it will occur at the hinge closest to the tip mass. This is consistent with Parkes's results [1], which show that the curvature due to the traveling hinge increases monotonically from the fixed end to the tip. In order to avoid catastrophic failure at the hinge nearest the tip, we must satisfy the first

386   COLLAPSE OF CANTILEVER BEAMS

condition of Equations (28), with $b = \gamma/(n-1)$. This is

$$\beta \leq \left[1 + \frac{2(n-1)}{\gamma}\right]^2 \tag{45}$$

It is evident that the constraint is asymptotic to 1 as γ increases, for fixed n. However, in order to model the continuous case, we might realistically put $n = \ell/u$, where u is the depth of the cross section. By taking a conservative value of, say, $n = 10$, this means that Equation (45) becomes

$$\beta \leq (1 + 18/\gamma)^2 \tag{46}$$

In physical terms, the problem is not well posed for large values of γ, since the attached mass is insignificant in relation to the mass of the beam.

We can obtain a result for the constraint on catastrophic failure of the hinge at the fixed end directly from Parkes's results. From Equation (11) of Parkes's paper [1], the final rotation of the base hinge is

$$\frac{\theta_f}{\theta_0} = \frac{4\beta}{3} \frac{\gamma + 3}{(\gamma + 2)^2} \tag{47}$$

Thus, catastrophic failure does not occur for

$$\beta \leq \frac{3(\gamma + 2)^2}{\gamma + 3} \tag{48}$$

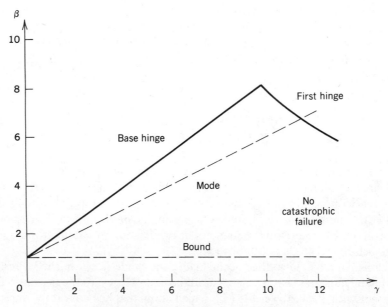

Figure 15.   Safe domain for large W.

This result is plotted in Figure 15, together with the constraint of Equation (46). It is seen that localized failure would be expected only for γ larger than about 10; this is at the limit of the physical reasonableness of the concentrated tip mass problem.

If we use the mode approximation of Martin and Symonds [2], in which it is assumed that the beam rotates about the base with the initial velocity modified to preserve the initial angular momentum, we find that the condition of Equation (48) is replaced by

$$\beta \leqslant 1 + \gamma/2 \tag{49}$$

This result is plotted as the upper dashed line in Figure 15, and is seen to be a reasonable approximation to the actual constraint.

We can thus infer that for the cantilever problem, the simple-mode approximation can be used as a basis to *design* a beam for which catastrophic failure will not occur.

8. BOUNDS ON HINGE ROTATIONS

We note briefly that we can use the result of Martin [10] to compute an upper bound on the rotation of any hinge in the discrete beam. The result states that if we choose any set of equilibrated forces on moments acting on the structure, such that the yield condition is satisfied, the product of the static forces and the actual final displacements of the impulse loaded rigid-plastic beam is bounded by the initial kinetic energy.

In this case, we choose a pair of self-equilibrating external moments of magnitude M_0 acting on either side of the position of the ith hinge, as shown in Figure 16. The upper bound theorem then gives

$$M_0\theta_i = \tfrac{1}{2}G\dot{v}_0^2 \tag{50a}$$

or

$$\theta_i/\theta_0 \leqslant \beta \tag{50b}$$

This result translates into a lower bound on the safe region in the design problem, shown by the dashed line in Figure 15. Clearly, the result is conservative for $\gamma > 0$; it is of interest to note, however, that the bound is the same for each hinge position, and that it is exact for $\gamma = 0$ and $\gamma \to \infty$.

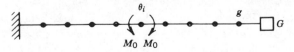

Figure 16. Static external loads for determining bounds.

9. CONCLUSIONS

We have attempted a very simple analysis of cantilever beams of a softening material for the case of a transverse impulse load applied to a tip mass. In this case, we have fairly strong evidence that localized nonmodal catastrophic behavior will not occur, and that we can use modal behavior as the basis for design. This is a reassuring result in that it permits us to use a variety of established techniques to carry out computations for more realistic models of the structure, and to include large displacements and rate sensitivity in the analysis.

We must note, however, that we cannot infer that localized nonmodal catastrophic behavior is always exluded. The case of a cantilever with two attached masses, where the mass of the beam is small compared to the attached masses, as shown in Figure 10, is clear evidence of localized failure for certain ratios of masses. Further study of the general behavior of dynamically loaded rigid–plastic–softening beams is thus clearly warranted.

REFERENCES

1. E. W. Parkes, "The Permanent Deformation of a Cantilever Struck Transversely at its Tip," *Proc. R. Soc. London, Ser. A*, **228**, 462–476 (1955).

2. J. B. Martin and P. S. Symonds, "Mode Approximations for Impulsively Loaded Rigid-Plastic Structures," *J. Eng. Mech. Div., Am. Soc. Civ. Eng.*, **92**(EM5), 43–66 (1966).

3. N. Jones and T. Wierzbicki, "Dynamic Failure of a Free-Free Beam," *Int. J. Impact Eng.* **6** (3), 225–240 (1987).

4. R. H. Wood, "Some Controversial and Curious Developments in the Plastic Theory of Structures," in J. Heyman and F. A. Leckie, Eds., *Engineering Plasticity*, Cambridge Univ. Press, London and New York, 1968, pp. 665–691.

5. P. LeP. Darvall, "Load-Deflection Curves for Elastic-Softening Beams," *J. Struct. Div., Am. Soc. Civ. Eng.*, **110**(ST10), 2536–2541 (1984).

6. Z. P. Bazant, F.-B. Lin, and G. Pijaudier-Cabot, "Yield Limit Degradation: Nonlocal Continuum Model with Local Strain," in D. R. J. Owen, E. Hinton, and E. Onate, Eds., *Computational Plasticity*, Pineridge Press, 1987, pp. 1757–1780.

7. G. Maier, "On Softening Flexural Behaviour on Elastic-Plastic Beams," (in Italian), *Rend.—Ist. Lomb. Accad. Sci. Lett., A*, **102**, 648 (1968); republished in English in Studi di Richerche, Vol. 8, Corso di Perfizionamento per le Construzioni in Cemento Armato, Politecnico di Milano, 1986.

8. G. Maier, "On Structural Instability due to Strain Softening," in H. Leipholz, Ed., *Instability of Continuous Systems*, Springer-Verlag, Berlin, 1971.

9. T. Wierzbicki, P. Xirouchakis, and S. K. Choi, "Flexural Failure of Softening Ice Sheets," *Int. Symp. Offshore Mech. Arct. Eng., 5th, 1986* (1986).

10. J. B. Martin, "Impulsive Loading Theorems for Rigid-Plastic Continua," *J. Eng. Mech. Div., Am. Soc. Civ. Eng.*, **89**(EM2), 107–132 (1963).

CHAPTER 12

Failure of Bar Structures Under Repeated Loading

Taijiro Nonaka and Satoshi Iwai†
Disaster Prevention Research Institute
Kyoto University
Uji, Kyoto, Japan

ABSTRACT

With the main attention focused on steel structures of bar members, structural failure due to the repeated action of loads is discussed after classification into three types: continued plastic deformation, instability due to changes in geometry, and structural fatigue. Various aspects of plastic failure due to variable repeated loads are first discussed on the basis of the classical theory of perfect plasticity as it applies to a simple-truss example. Next, consideration is given to an interactive phenomenon between plasticity and geometrical change, causing hysteretic instability on an axially and repeatedly loaded bar member. Finally, a literature survey is presented principally on experimental aspects of low-cycle fatigue from the point of view of overall structural-failure behavior.

1. INTRODUCTION

Structural failure due to the repeated action of loads is attributed to plastic or unstable behavior of the structure. Failure by fracture is also possible as caused by fatigue phenomena or brittle behavior. With failure modes classified into continued plastic deformation, instability due to geometrical changes, and structural fatigue, problems of repeated loading are discussed in this order, although these are not independent. In fact, most critical conditions are encountered when these interact. It is understood that cyclic plastic-strain

†The first author is responsible for the contents throughout. The second author contributed to Section 4.

accumulation can cause instability of metals. Accumulated tensile strain leads to necking and eventually to fracture. Notwithstanding instability failure, this and other types of material failure are not of direct concern here. Column or local instability due to accumulated compressive strain is briefly discussed in connection with structural fatigue. Main consideration is given to steel structures of bar members.

Various aspects of structural behavior under repeated loading are discussed with regard to a simple example of an elastic–perfectly plastic truss, based on an Rzhanitzin–Prager type of geometrical representation. This is preceded by introducing relevant terminology employed in the theory of perfect plasticity.

2. PLASTIC FAILURE

2.1. Definitions

The term *perfectly plastic* indicates the absence of the second-order effects in the plastic range, such as hardening and viscosity, and, at the same time, implies the presence of ample ductility. Since the stresses in a structure of perfectly plastic material cannot exceed the yield limit anywhere, the loads applied to the structure cannot be increased indefinitely. When the loads attain certain critical magnitudes, the perfectly plastic structure *collapses plastically*, and indefinitely increasing deformation makes the structure incapable of supporting any further increase in the loads, provided any change in geometry is negligible. Such a critical system of loads is called a *collapse- or limit-load* system, and the distribution of the unlimited plastic flow is often called a *collapse mechanism*. The alternative definition is that the collapse-load system is the system of loads at which the structure of rigid–plastic material begins to undergo deformation.

In limit or collapse analysis, the loads are assumed to be applied *statically*, *monotonically*, *proportionally*, or *simultaneously*. A structure is often subjected to several loads, each of which can be applied repeatedly and can vary independently of the others within certain prescribed limits. Such a load system, referred to as *variable repeated loading*, can lead a structure to a state of plastic failure, even if no simultaneous application of any load combination causes plastic collapse or *instantaneous collapse*. The failure by cyclic plastic deformation can take place in one of two ways. If the loads are essentially alternative in character, one or more of the members might be subject to yielding in tension and compression alternately or bent back and forth repeatedly, indicating a possibility of rupture due to fatigue. Such behavior is termed *alternating plasticity*, or *yield*. In the other type of failure, plastic deformation is built up incrementally in each cycle, eventually leading to an unacceptably large deformation by a number of critical combinations of loads following one another in fairly definite cycles. The structure is then said to have failed by *incremental collapse*, *deflection instability*, or *progressive plastic deformation*.

Terms such as *ratchetting* and *cyclic creep* are also used to indicate cumulative strains.†

In a statically indeterminate structure, which is accompanied by some plastic deformation, a state of *residual stress* or of *self-stress* or *self-equilibrated state of stress* can exist, which keeps the structure in equilibrium without any external loads. It is possible for the structure to be free from failure by cyclic plastic deformation, if, at an early stage of the loading program, such a state of residual stress is built up so that the stress due to elastic response of the structure to subsequent loads within the prescribed limits can be superimposed on these residual stresses without exceeding the yield limit at any point of the structure. The structure is then said to *shake down* to a state of residual stress and the corresponding state of permanent deformation. This phenomenon is also referred to as *adaptation*. It is recognized that the plastic collapse and shakedown loads are determined independently of the past history or path of the loading within the restrictions of perfect plasticity and the absence of change in geometry. Excessive loads cause *failure by continued plastic deformation*.

2.2. Simple Model and Shakedown Theorems

In order to show how these phenomena occur, a simple model of a planar truss is considered, as shown in Figure 1. A load P acts in the plane of the truss in a fixed orientation with slowly varying intensity, so that any inertia effects can be neglected. Since two bars are sufficient to support the load, this truss is statically indeterminate with a single indeterminacy. Each bar is assumed to be elastic–perfectly plastic. In the elastic range, the axial elongations Δ_i are related linearly to the tensile forces N_i through $\Delta_i = c_i N_i$, with the compliances c_i as the proportional factors, where the subscripts refer to particular members ($i = 1, 2,$ and 3). The yield limit is specified by $Y_i > 0$ for tension and $Y_i' < 0$ for

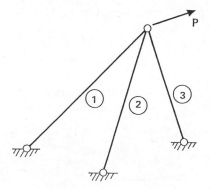

Figure 1.   Truss example.

†It is possible for the two types of plastic failure to take place simultaneously at a cross section. An example is that of a column cross section in a rigid frame subjected to alternate plastic bending and cumulative plastic axial contraction [1].

compression. Infinitesimal deformation is assumed with change in geometry neglected, so that no instability phenomena take place.

In order to determine the general behavior of the truss, the axial forces N_i are regarded as the sums of the elastic components N_i^e, which would be caused if the truss were perfectly elastic without limit and the residual components N_i^r, so that

$$N_i = N_i^e + N_i^r \tag{1}$$

The elastic response is determined by the instantaneous load as

$$N_i^e = p_i P \tag{2}$$

where the proportionality factors p_i are given from the distribution of the compliances c_i and overall geometry. The residual forces N_i^r, on the other hand, depend on the entire history of loading because of the history-dependent or irreversible nature of plasticity, and are equal to the axial forces that are attained by the complete removal of the load if the unloading process is elastic. The set of the components N_i^r forms a state of residual stress. The fictitious strain energy E corresponding to the axial forces N_i is introduced as

$$E = \tfrac{1}{2} \sum_{i=1}^{3} c_i N_i^2 = \tfrac{1}{2} \sum_{i=1}^{3} c_i (N_i^e)^2 + \sum_{i=1}^{3} c_i N_i^e N_i^r + \tfrac{1}{2} \sum_{i=1}^{3} c_i (N_i^r)^2 \tag{3}$$

The second term in the right-hand side of Equation (3) can be interpreted as the sum of the work done by the internal forces N_i^r on the elastic elongations $c_i N_i^e$. Since N_i^r are in equilibrium with $P = 0$ and since $c_i N_i^e$ are compatible with a displacement at the joint of load application, the principle of virtual work gives the sum as equal to the work done by the zero load on the joint displacement. The second term, therefore, vanishes. The first and third terms are rewritten by noting that N_i^e are related to P through Equation (2) and that the two independent equations of equilibrium at the joint are combined to express the three forces N_i^r in terms of a parameter R in the form

$$N_i^r = r_i R \tag{4}$$

where r_i are given from the geometry of the truss, whereas R depends on the loading history. With x and y proportional to P and R, respectively, and defined from

$$x = P \cdot \left(\tfrac{1}{2} \sum_{i=1}^{3} c_i p_i^2 \right)^{1/2} \qquad y = R \cdot \left(\tfrac{1}{2} \sum_{i=1}^{3} c_i r_i^2 \right)^{1/2} \tag{5}$$

it follows that

$$E = \frac{P^2}{2} \sum_{i=1}^{3} c_i p_i^2 + \frac{R^2}{2} \sum_{i=1}^{3} c_i r_i^2 = x^2 + y^2 \tag{6}$$

Considering that

$$N_i = p_i P + r_i R \tag{7}$$

and taking note of Equation (5), it is seen that the state of stress of the truss is represented by a point in the plane with x and y taken as the horizontal and vertical coordinate axes, as shown in Figure 2. It is also seen from Equation (6) that the energy E equals the square of the distance between the point and the origin of the coordinate axes. The perfect plasticity fixes the limits for the stress point by

$$Y_i' \leqq \left\{ p_i \middle/ \left[\tfrac{1}{2} \sum_{i=1}^{3} (c_i p_i^2) \right]^{1/2} \right\} x + \left\{ r_i \middle/ \left[\tfrac{1}{2} \sum_{i=1}^{3} (c_i r_i^2) \right]^{1/2} \right\} y \leqq Y_i \tag{8}$$

so that three pairs of parallel lines, $1, 1'$; $2, 2'$; $3, 3'$, enclose the convex allowable domain,† which is symmetric with respect to the origin if $Y_i = -Y_i'$. It

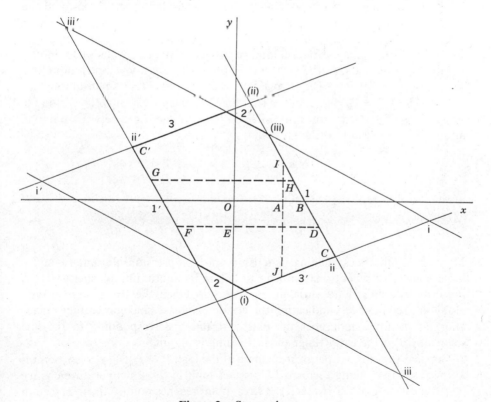

Figure 2. Stress plane.

†It can happen that some of these lines fall outside of the allowable domain, so that the corresponding members never experience yielding for large degrees of indeterminacy.

is found from Equations (5) and (7) that any states of stress in equilibrium with the load P have stress points on the equilibrium line IJ perpendicular to the x axis. The actual stress point lies on this line segment, and has an ordinate corresponding to the true residual state of stress. If the truss is free from any previous plastic straining, the stress point must be located at point A, the intersection of the line of equilibrium with the x axis, without any residual axial forces.‡ The stress point moves along the x axis for the initial elastic response corresponding to the varying load, and member 1 yields when the stress point arrives at point B, the truss becoming statically determinate. Linearity of the truss behavior terminates at this point, and a further increase in the load is accompanied by a movement of the stress point along line 1 until point C is reached, where the yield limit is also attained in member 3. There is no allowable state of equilibrium beyond this load, $P = P_c$. This is the limit state for instantaneous collapse to occur. Plastic deformation can increase indefinitely, whereas the load and axial forces remain constant.

If the load reverses its direction at point D somewhere before reaching point C, an elastic recovery takes place alone line DF parallel to the x axis, and the complete removal of the load, $P = 0$, leaves a residual state of stress as represented by point E. A further decrease in the load, $P < 0$, causes a further shift of the stress point to the left, and member 1 is stressed to its yield limit at point F, with the sign of the axial force opposite to that developed along line 1, say, in compression. The arrival at point C' indicates the instantaneous collapse, with $P = P'_c < 0$. Suppose that the load can vary arbitrarily within the limits $P_G \leqq P \leqq P_D$, where state G lies on line 1' somewhere between points F and C'. The repetition of the particular cycle

$$0 \to P_D \to P_G \to P_D \to 0$$

in the load forces the stress point to move in a cyclic manner as in

$$0 \to A \to B \to D \to E \to F \to G \to H \to B \to D \to E$$

Member 1 undergoes yielding in tension and compression alternately, and a large number of repetitions can lead to failure by alternating plasticity. If it is prescribed that the load varies in $P_F \leqq P \leqq P_D$, then the stress point travels along line segment DF, indicating the elastic response; shakedown takes place after the member undergoes the plastic straining corresponding to the line segment BD in the early range of the loading program.

Suppose that the temperature changes at the load $P = P_A$. The stress point is allowed to move along segment IJ; stresses built up by a temperature change constitute a state of self-stress. If the stress point reaches point I, the intersection of line IJ and line 1, then the truss is turned into a statically determinate one

---

‡This is also seen from the principle of minimum-strain energy for elastic behavior.
§The unloading process is not always elastic, though it is in this example.

with plastic straining in member 1 under constant axial forces on all the members. A similar situation arises at point J, the intersection of lines $3'$ and IJ, with plastic straining in the member 3. If such a cycle of temperature changes is repeated, plastic strains accumulate with the same sign in each of the members, and a large number of repetitions eventually can lead to incremental collapse. This takes place with the same collapse mechanism as at the instantaneous collapse of point C. If, on the other hand, the range of the temperature variation corresponds to a length less than that of IJ, a shakedown state is reached, possibly after some plastic straining, and subsequently the truss responds in an elastic manner. It may be of interest to note in this connection that the state of stress varies only in a restricted range, no matter how large the range of the temperature variation may be. Another interesting feature is that the state of stress depends only on the overall distribution of the temperature and not on the precise local distribution. It is not difficult to demonstrate the occurrence of the incremental collapse under load variation alone.

Since an elastic response to a temperature change causes a change in the ordinate or the state of self-stress, it is convenient to separate the components N_i^r into the components due to the elastic response to the temperature change and the remaining residual components $\overline{N_i^r}$, so that the axial forces N_i are the sums of the three components, viz, the components due to the fictitious elastic response to the load, the components due to the fictitious elastic response to the temperature change, and the components $\overline{N_i^r}$. The sums of the first two components are written as N_i^e in the new sense; N_i^e are identified with the axial forces that would be developed in the perfectly elastic truss due to the change in temperature as well as load. It is then recognized from the foregoing discussions that *shakedown occurs if and only if a state of self-stress $\overline{N_i^r}$ can be found such that superposition $\overline{N_i^r} + N_i^e = N_i$ of this state and the purely elastic response N_i^e to the given variation of loads and temperature does not at any point or instant lead to stresses at or above the yield limit* of bounds Y_i and Y_i'. This is the static or equilibrium theorem of shakedown and is known in association with the name of Melan [2–5]. This implies that the truss has an ability to withstand the loading by rearranging its internal forces to best advantage. This principle provides a means of determining a safe range for the limits of variation in loads and temperature in order for no failure by cyclic plastic deformation to occur without restrictions on the number of repetitions; a trial distribution of the history-independent residual forces $\overline{N_i^r}$ gives a safe range.

It is seen from Figure 2 that as the load increases, the allowable range of a temperature change becomes smaller. When the load is close to the collapse load P_c, a small change in temperature causes incremental plastic deformation in the two members alternately. At the limit state C, no change in temperature is required in order for the two members to yield simultaneously, and an unlimited plastic flow can take place under the constant load and temperature. Thus, the instantaneous collapse is recognized to be a limiting case of an incremental collapse occurring due to variable repeated loading. The limit load itself is independent of any change in temperature. The previous considerations suggest that a theorem for the instantaneous collapse be derived from the shakedown theorem as an application to the limiting case. The absence of variation indicates that the axial forces need not be separated into the variable and constant components. It follows that *instantaneous collapse does not or is just about to occur if and only if an equilibrated state of stress N_i (satisfying the equations of equilibrium and boundary conditions on forces) can be found that nowhere exceeds the yield limit* of bounds Y_i and Y_i'. This is nothing but the static or lower-bound theorem of limit analysis. Generalization of the static theorem to a general continuum or structural type is clear by rephrasing force and deformation quantities in proper terms.

The absence of any variation in the internal forces indicates the absence of variation in the elastic deformation Δ_i^e at the limit state; deformation increases in a purely plastic manner. A collapse mechanism is meant to describe the spatial distribution of such plastic flow. At the limit state C, where

$$P = P_c \qquad N_1 = Y_1 > 0 \qquad Y_2' < N_2 < Y_2 \qquad N_3 = Y_3' < 0 \qquad (9)$$

member 1 can extend plastically and member 3 can contract plastically, whereas member 2 undergoes no change in length. In terms of increments in elongation,

$$d\Delta_1^p \geqq 0 \qquad d\Delta_2^p = 0 \qquad d\Delta_3^p \leqq 0 \qquad (10)$$

The truss joint, therefore, can move in a direction perpendicular to the axis of member 2, at the limit state C. The collapse mechanism must be of the form shown in Figure 3. It does not include the deformation that has occurred before the limit state is reached, but shows the distribution, or proportion in space, of incremental deformations that can occur upon reaching the limit state under the constant plastic collapse load. This plastic flow can increase indefinitely, since changes in geometry and work-hardening of material are ignored. Let this mechanism be ii. It is clear that the collapse mechanism for the limit state C' is opposite to the one in Figure 3; note that a load with reversed direction does negative work on the collapse mechanism of Figure 3. The following discussion is restricted to positive loads and the corresponding collapse mechanisms, on

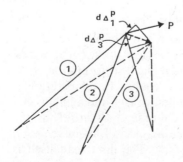

Figure 3.   Collapse mechanism ii.

which positive loads do positive work. The point denoted by (ii) in Figure 2 corresponds to the case in which member 3 yields in tension instead of compression, but is otherwise similar to ii. The corresponding deformation would be such that

$$d\Delta_1^p \geqq 0 \qquad d\Delta_2^p = 0 \qquad d\Delta_3^p \geqq 0 \qquad (11)$$

This condition, however, would not be satisfied unless the truss members were disconnected. In other words, Equation (11) does not satisfy the condition imposed on deformation and is not kinematically admissible.

Among collapse mechanisms, those in which the positive load does positive work are shown in Figures 4 and 5. In collapse mechanism iii of Figure 4, member 3 undergoes no plastic action, and

$$d\Delta_1^p > 0 \qquad d\Delta_2^p > 0 \qquad d\Delta_3^p = 0 \qquad (12)$$

From the flow rule or from the condition that the plastic work be positive, the force system must be such that

$$P = P_{iii} \qquad N_1 = Y_1 \qquad N_2 = Y_2 \qquad Y_3' \leqq N_3 \leqq Y_3 \qquad (13)$$

The corresponding stress point is denoted by iii in Figure 2, which indicates that the last condition of Equation (13) is not satisfied with $N_3 < Y_3'$ and that $P_{iii} > P_c$. The stress point (iii) corresponds to a deformation where bar 2 would be subjected to plastic action in compression, and hence is not kinematically admissible. Figure 5 shows collapse mechanism i, in which

$$d\Delta_1^p = 0 \qquad d\Delta_2^p < 0 \qquad d\Delta_3^p < 0 \qquad (14)$$

Therefore, conditions

$$P = P_i \qquad Y_1' \leqq N_1 \leqq Y_1 \qquad N_2 = Y_2' \qquad N_3 = Y_3' \qquad (15)$$

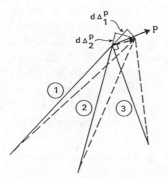

Figure 4. Collapse mechanism iii.

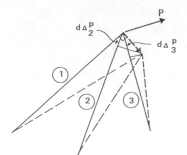

Figure 5. Collapse mechanism i.

must be satisfied. The corresponding stress point i in Figure 2 indicates, however, that $N_1 > Y_1$ and that $P_i > P_c$. The stress point (i) indicates $N_2 = Y_2$, so that the corresponding flow rule $d\Delta_2^p \geqslant 0$ would not lead to a mechanism that is kinematically admissible. Note that examination has been completed of all the points of intersection for the positive load.

From this scrutiny, it is seen that, subject to the condition that the work done by the load is positive, there are three conceivable collapse mechanisms satisfying kinematical conditions. Among these, *false mechanisms* (i and iii) *correspond to loads not less than the collapse load and correspond to the states of internal forces that violate the yield condition somewhere* in the truss. This establishes the kinematical or upper-bound theorem of limit analysis: *If a collapse mechanism exists for a given load, then the structure must collapse plastically under this load.* This implies that a structure of perfectly plastic material must fail plastically whenever a collapse mechanism can be formed or whenever a path to failure exists. In cases where loads act simultaneously in different directions, so they cannot be distinguished as large or small, then a scalar load factor should be properly defined in order to express the magnitude of loading in terms of large or small. When this is done, the upper-bound theorem can be rephrased to state that *any loads corresponding to the assumed collapse mechanisms cannot be smaller than the correct collapse load.*

Since a collapse mechanism constitutes a kinematically admissible system satisfying the condition of compatibility, it is conveniently used in the virtual-work equation to determine the corresponding load. The internal forces that do work on the collapse mechanism are determined from the condition of yielding.† In the example under consideration, N_i are equal to $Y_i > 0$ or $Y_i' < 0$, according to positive or negative $d\Delta_i^p$ in the assumed collapse mechanism. Note that those internal-work components are always positive, being dissipated in plastic deformation. In truss members where $d\Delta_i^p = 0$, the condition of yielding does not determine the corresponding forces N_i. These values do not appear, however, in the virtual-work equation to be used to determine the collapse load.

---

†There exist such cases in a general state of stress where the condition of yielding does not determine the state of internal forces uniquely, but the corresponding internal work is determined uniquely.

Some of these are the internal forces that can violate the yield condition in a false collapse mechanism. The upper-bound theorem is rephrased by the use of the virtual-work equation: *a structure collapses plastically under the load system that is determined by equating the external work to the internal work for an assumed collapse mechanism.* The safety against plastic collapse is gained by examining all the possible collapse mechanisms.

Figure 2 provides further information about the instantaneous collapse load. Since the collapse load is governed by the pairs of parallel lines of yielding, it is not decreased by expanding the allowable domains that are prescribed by the yield limits Y_i and Y_i'; it is not increased by reducing the yield limits in magnitude. The fact that the limit state is not affected by the position of the initial or current stress point indicates that *thermal stress, initial stress, initial deformation, loading history, or elastic property does not affect the limit load,* unless it alters the geometry.

Another instructive consideration is made of energy dissipation for the case of variable repeated loads without temperature variation. A characteristic feature of plastic failure under repeated loads lies in the cyclic changes in the states of the internal as well as external forces. In Figure 2, the state of internal forces and that of internal residual forces change in a cyclic manner as in the variation along the parallelogram $DHGF$. The cyclic change in the internal residual forces is accompanied by plastic deformation, and hence by energy dissipation. The elastic strain energy being recovered after a complete cyclic change, the amount of the energy dissipation in each cycle, is given from the area of the parallelogram. An unlimited number of repetitions of such a cycle may require an unlimited amount of energy to be dissipated in plastic deformation. Similarly, at the state of instantaneous collapse, an indefinitely large energy dissipation may be accompanied by an indefinitely large plastic deformation. Failure by continued plastic deformation can thus be characterized by the limitless property of energy dissipation or plastic work; it is seen as the counterpart of this statement that shakedown is defined properly by the limitedness in the amount of the total sum of the energy dissipation under limitless repetitions of loading.

Cyclic changes in force and deformation systems serve to establish a kinematical theorem for shakedown against repeated loads in an isothermal condition. When failure by cyclic plastic deformation occurs, residual forces at the end of a cycle return to their values at the beginning of the cycle. Increments in plastic deformation vary cyclically during this cycle of loading. Accumulation of these increments over the cycle amounts to form a collapse mechanism in the case of incremental collapse, and vanish in the case of alternating plasticity. Such accumulation constitutes a kinematically admissible distribution of plastic deformation in the both cases. This distribution does not occur instantaneously, but occurs as a result of the cyclic change. A set of increments $d\Delta_i^p$ in the plastic deformation at any instant is accompanied by a change in the residual-force system and, hence, by a set of increments $d\Delta_i^r$ of the corresponding elastic deformation. The combination of these increments, $d\Delta_i^p + d\Delta_i^r$, constitutes a kinematically admissible distribution at any instant of the cycle, although each is not itself admissible. Whereas the accumulation of the increments $d\Delta_i^p$ of plastic deformation during the cycle amounts to an incremental collapse mechanism, or a vanishing distribution of plastic deformation in the case of alternating plasticity, the accumulation of the increments $d\Delta_i^r$ of the elastic deformation vanishes upon completion of the cycle. It is thus understood that the truss does *not shake down but fails ultimately by cyclic plastic deformation if any kinematically admissible cycle of the increments $d\Delta_i^p$ in plastic deformation and external loads within the prescribed limits can be found for which the work done by the loads equals the plastic work or the energy dissipated over the cycle.* Shakedown is guaranteed if no such admissible cycles exist. This is the kinematic theorem of shakedown, and formulated by Koiter [6, 7] for three-dimensional continua. This principle is not very convenient to apply for solving specific shakedown problems. It can be simplified when the components of plastic-strain increments have a fixed relative proportion among themselves. It takes a particularly simple form in the case of single stress and strain components such as a beam or truss. Derivation follows for the truss example, instead of this principle, along the line of Symonds and Neal [8, 9].

Suppose that in an incremental collapse mechanism, the ith member is subjected to plastic elongation $d\Delta_i^p$. Considering that the yield limit is supposed to be reached in this member at some loading stage, either the maximum or minimum elastic axial force, $(N_i^e)_{\max}$ or $(N_i^e)_{\min}$, respectively, is taken and either of equations

$$(N_i^e)_{\max} + \overline{N_i^r} = Y_i \qquad \text{if } d\Delta_i^p > 0$$

$$(N_i^e)_{\min} + \overline{N_i^r} = Y_i' \qquad \text{if } d\Delta_i^p < 0 \tag{16}$$

is multiplied by $d\Delta_i^p$. The summation over all the yielding members gives

$$\sum_i \begin{Bmatrix} (N_i^e)_{\max} \\ (N_i^e)_{\min} \end{Bmatrix} d\Delta_i^p + \sum_i \overline{N_i^r}\, d\Delta_i^p = \sum_i \begin{Bmatrix} Y_i \\ Y_i' \end{Bmatrix} d\Delta_i^p \tag{17}$$

It is clear that each term in the summation of the right-hand side of Equation (17) is positive regardless of the sign of $d\Delta_i^p$. The sum of the second term in the left-hand side vanishes, since it is interpreted as the work done by the internal forces $\overline{N_i^r}$, which are in equilibrium with the external load $P = 0$, on the elongations $d\Delta_i^p$, which are compatible with a displacement at the joint of load application, and hence is equal to the work done by the vanishing load, as given by the principle of virtual work. It is thus seen that *incremental collapse must occur if there are such collapse mechanisms and loading cycles that satisfy the equation*

$$\sum_i \begin{Bmatrix} (N_i^e)_{\max} \\ (N_i^e)_{\min} \end{Bmatrix} d\Delta_i^p = \sum_i \begin{Bmatrix} Y_i \\ Y_i' \end{Bmatrix} d\Delta_i^p \tag{18}$$

This implies that if there is a path to failure, the truss gives up. When alternating plasticity is just about to occur, Equations (16) are satisfied alternately. By noting that the elimination of $\overline{N_i^r}$ from the two equations gives the bound, *the conditions to prevent the failure by alternating plasticity* are easily understood to be

$$(N_i^e)_{\text{max}} - (N_i^e)_{\text{min}} < Y_i - Y_i' \tag{19}$$

In the case of a moment-resisting member in planar frames or beams, the axial force N is replaced by the bending moment M. If the moment-curvature relationship is elastic–perfectly plastic, the foregoing discussions made on the truss problem are also valid for beams and planar frames with the limit axial force replaced by the fully plastic moment and with plastic elongation by relative rotation across a yield hinge. However, special care has to be taken for the state of partial yielding in a cross section. The shakedown limit of Equation (19) is replaced by

$$(M^e)_{\text{max}} - (M^e)_{\text{min}} \leqq 2M_y \tag{20}$$

for alternate bending in a symmetric cross section, where M_y is the yield moment indicating the onset of partial yielding. (This is extended to cover kinematically hardening cross sections; see, e.g., Martin [10], p. 682.)

2.3. Theorematic Development and Significance of Shakedown

The presentation of classical theorems in the preceding section is based partly on a review paper [11]. It was Prager [12] who extended the statical theorem of Melan to include changes in temperature and gave a simple proof for a three-dimensional continuum. As the boundary condition, he specified that some part of the surface of the continuum was subjected to surface tractions and required displacements to vanish on the remainder of the surface. However, the present authors assert that the displacements need not vanish but can be prescribed, because nonzero displacements do not impair the validity of the proof. In fact, it has been shown in an example employing a portal frame that the existence of a time-independent residual-force system is sufficient for shakedown to occur under the condition of cyclic displacement loading [1].

The classical theorems of shakedown are based on elastic–perfectly plastic material behavior and on the original structural geometry. Effects of strain-hardening, viscosity, geometry changes, and inertia are neglected. Extended approaches have been proposed in the last several decades to include some of these effects. Numerical techniques such as finite elements and mathematical programming have been utilized for the analysis of shakedown problems. Attempts have been made for the use of generalized variables suitable for structural types to replace physical variables of stress and strain. Shakedown-based optimum design or minimum-weight design, and stochastic analysis have been reported. Experiments have been also carried out to confirm the validity of the shakedown theory. The reader is referred to review papers and monographs on these developments [10, 11, 13–19].

The fundamental theorems for an elastic–perfectly plastic structure discussed in the preceding section provide the means of determining the shakedown limit, which is not affected by the exact history of loading. Variable loads and changes in temperature are allowed to be repeated without a bound in the number of cycles, and each can vary independently of one another under prescribed limits. The shakedown analysis can determine the most critical combination of independent loads that would cause the instantaneous collapse. Therefore, it does not require separate analysis for plastic collapse. A disadvantage of the shakedown analysis is the necessity of determining the elastic behavior of structures along with the plastic.

The shakedown theorems are not concerned with exact structural deformations, which are regarded as being of the order of elastic deformation within the context of perfect plasticity and negligible changes in geometry. Inclusion of strain-hardening predicts a possibility of the shakedown limit beyond the instantaneous collapse limit. However, residual deflections can reach intolerable amounts. A crucial effect of deformation is structural instability. In fact, when steel structures are subjected to severe repeated loading, failure by local or lateral–torsional instability or P–Delta effect is not always less stringent than failure by continued plastic deformation. In some cases, the shakedown limit can be exceeded if deflections are sufficiently small. A most realistic mode of failure can be caused by the interaction of continued plastic deformation and destabilizing effects; the number of repetitions of loading is limited, but changes in geometry due to deformation play an important role. A most elemental example of such interactive behavior is presented in some detail in the following section.

3.   HYSTERETIC INSTABILITY

3.1.   Combined Nonlinearity of Geometry and Material Behavior

So far, infinitesimal deformation has been assumed, so that deformation does not alter the geometry of a structure appreciably. Equilibrium considerations have been based on the original structural geometry before loading, and nonlinearity has been taken into account only in material behavior. Under certain circumstances, however, geometrical change plays an important role. It is well known, for example, that the phenomenon of buckling of a compressed bar cannot be analyzed without resorting to its deflection in the equation of equilibrium. In general, tensile axial forces have a favorable structural effect in conjunction with large deflections, whereas compressive axial forces tend to destabilize structural behavior. Since the theory of plasticity, based on material ductility, applies in deformation ranges where the elastic limit is exceeded, geometrical changes must be duly accounted for. The term *finite deformation* is often used as opposed to *infinitesimal deformation*, and so is *geometrical nonlinearity* against *material nonlinearity*. An example of combined nonlinearity is discussed in the following section in reference to the behavior of an axially loaded elastic–perfectly plastic bar. Repeated action of tension and compression is allowed in order to observe favorable and unfavorable effects of the axial force taking place through finite deflection.

3.2. Formulation of an Axially Loaded Bar

Suppose an initially straight bar has an effective length, so that its ends can be regarded as having no constraints against rotation. External loads act through the ends. The loads must be a pair of equal and opposite forces in order to maintain equilibrium. Acting through the centroid of the cross section, the loads cause tension or compression in the bar. The bar can be bent at some stage due to buckling and subject to plastic deformation. A contractible yield hinge appears at the center of the bar, where the deflection is a maximum. A subsequent tension tends to decrease the deflection, and the yield hinge becomes extensible. An example of such hysteretic behavior is shown in Figure 6. The ordinate represents in all cases the variation of the load, whereas the abscissa indicates the displacement of one end relative to the other, Figure 6(*a*); the central deflection, Figure 6(*b*); the hinge rotation, Figure 6(*c*); and the bending moment at the bar center, Figure 6(*d*). Inertia is neglected, so the load must vary slowly or the inertia of the bar must be negligible. A symmetric uniform cross section is assumed, so that the bar is deflected in a definite plane without twisting.

When the bar is subjected to tension from its virgin state at point ⓪, the elastic–perfectly plastic load-displacement relationship is expected as in ⓪ → ① → ② in Figure 6(*a*). The load is subsequently decreased and becomes compressive. The bar remains straight until it buckles at state ③. Lateral deflection is initiated abruptly and so is the bending moment, as shown in Figures 6(*b*) and 6(*d*). As the deflection keeps increasing, the yield limit is reached at state ④, and a further increase in the deflection is accompanied by plastic deformation at the yield hinge. The load magnitude decreases as in ④ → ⑤. Hinge rotation keeps increasing during this process, as seen in Figure 6(*c*), and the resultant stresses of the bar center vary, as represented by a point tracing along the yield curve, which is shown as an octagon in Figures 6(*d*) and 7. The shapes of the deflection curve are also shown in Figure 7, as well as the directions of the applied loads.

If the bar ends stop approaching each other and begin to separate at a state such as ⑤, then the yield hinge disappears, leaving a kink at the center of the deflection curve. The variations ⑤ → ⑥ is characterized by elastic recovery with decreasing deflection. A particular state is reached during this process such that the load is completely removed. The resultant stresses vanish, as represented by ⓪ in Figure 6(*d*), but the deformation quantities do not. The deflection

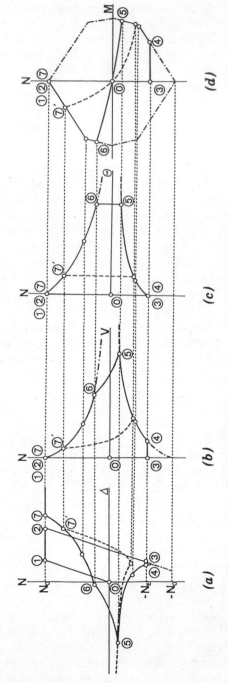

Figure 6. Behavioral diagrams: (*a*) Load vs. axial displacement. (*b*) Load vs. lateral deflection. (*c*) Load vs. half-hinge rotation. (*d*) Load vs. bending moment.

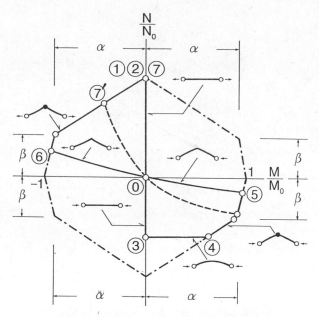

Figure 7. Yield octagon and deflection curves.

curve is composed of two straight segments, which are inclined relative to each other by an amount equal to the hinge rotation acquired at state ⑤.

At state ⑥, the yield hinge again forms, with a tensile axial force, and the kink angle decreases in the process ⑥ → ⑦. The bar returns to its straight configuration at ⑦, where the load becomes the yield limit in pure tension. Unloading from a state such as ⑦′ before the complete recovery at ⑦, gives rise to elastic recovery. The difference between ② and ⑦ in the relative displacement is the net plastic elongation, which is acquired at the yield hinge, during the complete load cycle.

Analytical determination of such behavior is accomplished by deriving the basic equations that relate the relative displacement Δ between the bar ends to the load N. The deflection curve $y(x)$ must also be found for each specified value of N. The origin of cartesian coordinates (x, y) is taken at the left end of the bar, as is the arc length parameter s, as shown in Figure 8. Let L and L_0 denote the current and original lengths of the bar, respectively, and Φ, the slope angle at a generic point of the deflection curve.

It follows from the definition of Δ that

$$L_0 + \Delta = \int_0^{L_0 + \Delta} dx$$

This is combined with the relation

$$dx = \cos \Phi \, ds \tag{21}$$

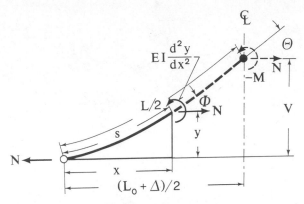

Figure 8. Notation.

to give

$$\Delta = \int_0^L \cos \Phi \; ds - L_0 = \left(\int_0^L \cos \Phi \; ds - L \right) + (L - L_0) \qquad (22)$$

It is seen that the displacement Δ of one end relative to the other, taking place along the original bar axis x, is composed of two parts. The first of these is induced when the bar of length L is deflected laterally, which results in one end approaching the other. This must be a negative component of Δ, and is written as Δ^g, when Δ is taken as positive for relative separation. Thus,

$$\Delta^g = \int_0^L \cos \Phi \; ds - L \leq 0 \qquad (23)$$

The second part, $L - L_0$, is clearly due to a change in the length of the bar. This is expressed as the sum of the elastic elongation Δ^e, plastic axial deformation Δ^p at the yield hinge, and plastic elongation Δ^t acquired in a straight configuration. Allowance can be made for thermal loading by simply adding the term of axial deformation due to a change in temperature [20], but this effect is neglected here. It follows that

$$\Delta = \Delta^e + \Delta^g + \Delta^p + \Delta^t \qquad (24)$$

The last component Δ^t can increase, i.e., $d\Delta^t \geq 0$, when the bar yields under uniform tension at a load N_0. Unless the bar becomes locally unstable due to necking, it can be assumed that $\Delta^t \geq 0$ is distributed along the straight bar and is controlled by the end constraints. It is therefore treated as known.

As shown in Figure 8, equilibrium requires that the resultant force acting at a generic point is a horizontal force equal in magnitude to the load N, whose component $N \cos \Phi$ in the direction of the tangent to the deflection curve equals the axial force. If the deflection is so small that Φ^2 or $\tan^2\Phi$ is negligible in comparison with unity, then $\cos\Phi$ is approximated as unity, and the axial force is nearly equal to N. By applying the usual sign convention of positive axial force for tension to the load N, the elastic elongation Δ^e equals the length L times elastic strain, which is the ratio of the axial force N to the axial rigidity EA. Use is also made of the assumption of small deflection to approximate the component Δ^g of Equation (23). It follows that

$$\int_0^L \cos\Phi \, ds - L = \int_0^L (1 + \tan^2\Phi)^{-1/2} \, ds - L$$

$$\cong -\frac{1}{2} \int_0^L \tan^2\Phi \, ds$$

$$= -\frac{1}{2} \int_0^L \left(\frac{dy}{dx}\right)^2 ds \tag{25}$$

It is to be noted that the small-deflection assumption does not impair the consideration of finite deflection, because a change in geometry is considered; the deflection vanishes initially. Another assumption also made is that of small deformation such that the change in bar dimensions can be neglected. Change in bar length, in particular, is assumed negligible compared with the original length. The approximation $L \cong L_0$ leads to†

$$\Delta^e \cong \frac{NL_0}{EA} \tag{26}$$

$$\Delta^g \cong -\frac{1}{2} \int_0^{L_0} \left(\frac{dy}{dx}\right)^2 dx \tag{27}$$

where the original area of cross section is taken for A.

The deflection curve $y(x)$ is determined by considering a portion of the bar to be in equilibrium, as shown in Figure 8. Because of the symmetry of the deflection curve with respect to the center, it is adequate to consider the left half, $0 \leq x \leq L_0/2$. Moreover, plastic bending can take place only at the center, so that elasticity prevails throughout the half bar. By neglecting shear deformation, the conventional theory of bending relates the bending moment Ny to the product of flexural rigidity EI times curvature, which is approximated by the second derivative of deflection with respect to the axial coordinate. With consideration of sense and direction of these quantities, it follows that

$$EI \frac{d^2y}{dx^2} - Ny = 0 \tag{28}$$

†The assumption of small deflection does not permit the term $L - L_0$ in Equation (22) to be neglected, since the equation involves only small quantities of the same order.

The integration of this equation is combined with the boundary condition $y(0) = 0$ at the left end to give

$$y = C \sinh\left[\left(\frac{N}{EI}\right)^{1/2} x\right] \tag{29}$$

where the integration constant C is to be determined by the boundary condition at the center, which reads either

$$y(L_0/2) = V \quad \text{or} \quad \lim_{x \to L_0/2} dy/dx = \Theta \tag{30}$$

where $V \geq 0$ denotes the central deflection, and $\Theta \geq 0$ is the slope angle just to the left of the center. Plastic yielding determines the former, whereas the elastic process ensures that the latter remains constant, at a value that is determined through the preceding plastic action. It is found from Equations (29) and (30) that V and Θ are related by

$$\Theta = \frac{\left(\frac{N}{EI}\right)^{1/2}}{\tanh\left[\frac{L_0}{2}\left(\frac{N}{EI}\right)^{1/2}\right]} V \tag{31}$$

Let the bending moment at the center be M with positive direction corresponding to increasing slope angle Θ, as shown in Figure 8. The equation of moment equilibrium for the half bar reads

$$M + NV = 0 \tag{32}$$

It is to be noted that M and N are of opposite signs, since the deflection is positive, if not zero. Plastic action can occur when M and N satisfy the yield condition, which is here taken to be represented by the yield octagon of Figure 7. The parameters α and β are to be chosen between zero and unity, subject to the condition that their sum be not less than unity because of the convexity requirement. For a state of resultant stress lying in the second or fourth quadrant of the plane $(M/M_0, N/N_0)$, the yield condition is written as

$$\left|\frac{M}{M_0} - \frac{1-\alpha}{\beta}\frac{N}{N_0}\right| = 1 \quad \text{if } \left|\frac{N}{N_0}\right| \leq \beta \tag{33}$$

$$\left|\frac{N}{N_0} - \frac{1-\beta}{\alpha}\frac{M}{M_0}\right| = 1 \quad \text{if } \beta \leq \left|\frac{N}{N_0}\right| \leq 1 \tag{34}$$

where M_0 is the limit moment in pure bending. The hinge rotation or the relative rotation across the yield hinge equals the discontinuity 2Θ in slope angle at the center of the deflection curve, and the axial deformation Δ^p of the yield hinge is equal to the relative axial displacement across the yield hinge, their signs complying with those of M and N, respectively. The flow rule associated with Equations (33) and (34) gives the ratio of the increments $d(\Delta^p)$ and $d(2\Theta)$ occurring in a plastic process. This leads to

$$N_0 d\Delta^p = -\frac{2(1-\alpha)}{\beta} M_0 d\Theta \qquad \text{if } \left|\frac{N}{N_0}\right| < \beta \tag{35}$$

$$N_0 d\Delta^p = -\frac{2\alpha}{1-\beta} M_0 d\Theta \qquad \text{if } \beta < \left|\frac{N}{N_0}\right| < 1 \tag{36}$$

These are easily integrated to determine Δ^p from Θ in a plastic process.

Note that there are no equalities in the conditions in Equations (35) and (36) as against the corresponding conditions in Equations (33) and (34). This is because the direction of the flow vector is not determined uniquely at the corners of the yield octagon. The corners at $|N/N_0| = \beta$ should cause no trouble, since the stress point does not remain there to make any contribution to plastic deformation, but moves from one neighboring regime to the other in the process of deformation. Plastic flow at the corner $N/N_0 = 1$ is taken into account in the component Δ^t. Note that the condition $N = N_0$ is attained only in a straight configuration, $y = \Theta = 0$. Difficulty at $N/N_0 = -1$ is avoided here by assuming that a straight bar cannot remain straight but buckles laterally under the yield load in compression. This assumption excludes extremely short or stubby bars, which can contract plastically appreciably before instability begins.

The equilibrium condition of Equation (32) is combined with the yield condition, Equations (33) and (34), to give the central deflection:

$$V = \frac{M_0}{N_0}\left(\left|\frac{N_0}{N}\right| - \frac{1-\alpha}{\beta}\right) \qquad \text{if } \left|\frac{N}{N_0}\right| \leq \beta \tag{37}$$

$$V = \frac{M_0}{N_0}\frac{\alpha}{1-\beta}\left(\left|\frac{N_0}{N}\right| - 1\right) \qquad \text{if } \beta \leq \left|\frac{N}{N_0}\right| \leq 1 \tag{38}$$

These are valid during a plastic process, viz, when the yield condition is satisfied. The hinge rotation that takes place during this process is determined by substituting Equation (37) or (38) into Equation (31). Combining Equations (27), (29), and (30) gives

$$\Delta^g = -L_0\left\{\frac{\left(\dfrac{N}{EI}\right)^{1/2}}{2\sinh\left[\dfrac{L_0}{2}\left(\dfrac{N}{EI}\right)^{1/2}\right]}V\right\}^2\left\{1 + \frac{\sinh\left[L_0\left(\dfrac{N}{EI}\right)^{1/2}\right]}{L_0\left(\dfrac{N}{EI}\right)^{1/2}}\right\} \tag{39}$$

or

$$\Delta^g = -L_0 \left\{ \frac{\Theta}{2\cosh\left[\frac{L_0}{2}\left(\frac{N}{EI}\right)^{1/2}\right]} \right\}^2 \left\{ 1 + \frac{\sinh\left[L_0\left(\frac{N}{EI}\right)^{1/2}\right]}{L_0\left(\frac{N}{EI}\right)^{1/2}} \right\} \tag{40}$$

The former, Equation (39), is useful in a plastic process, and the latter, Equation (40), in an elastic process, in which Θ remains constant.

If N takes on negative values, imaginary functions are involved in Equations (29), (31), (39), and (40). Expressions only in terms of real functions are found by replacing N by the absolute value $|N|$, and the hyperbolic functions by the corresponding trigonometric functions. This can be confirmed by making use of their functional relations with inclusion of the imaginary unit $(-1)^{1/2}$. This can also be confirmed through treatment similar to that before, after setting $N = -|N|$ in the equilibrium differential equation, Equation (28). It may be worth noting that the deflection curve for a half bar is represented by a hyperbolic-sine function for $N > 0$ and a sine function for $N < 0$.

When the load is negative and takes a particular value

$$N_E = \frac{\pi^2 EI}{L_0^2} \tag{41}$$

in magnitude, Equation (40) involves a term

$$\cos\left[\frac{L_0}{2}\left(\frac{N_E}{EI}\right)^{1/2}\right] = \cos\frac{\pi}{2} = 0 \tag{42}$$

in the denominator. It means that the component Δ^g is not necessarily zero even if $\Theta = 0$. The value N_E is the Euler load for elastic buckling. For such a long bar that $N_E < N_0$, it is taken that a straight bar buckles elastically at $N = -N_E$. The deflection V and the component Δ^g are not determined uniquely from the small-deflection theory.† It is approximated here that V and Δ^g keep increasing under constant load $-N_E$ until the yield limit is reached at the center.

The particular case of vanishing load $N = 0$ is treated as the limit $N \to 0$ in the results obtained before. It is seen from Equations (31) and (40) that

$$\Theta = 2V/L_0 \qquad \Delta^g = -L_0\Theta^2/2 \tag{43}$$

The basic equations obtained previously are adequate to determine the hysteretic behavior of a bar for any specified history of axial loading within a range of finite but small deformations.

†A large-deflection theory shows that the postbuckling elastic behavior is stable, and V and Δ^g keep increasing with increasing $|N|$ with a one-to-one correspondence. This increase in $|N|$ is negligibly small for bars of structural interest [21].

3.3. Hysteretic Structural Behavior

The results presented in the preceding section have been obtained in the course of research into the clarification of the ultimate behavior of steel-framed structures [20–24]. Representing the restoring-force characteristics of axial force-carrying members in a closed analytic form, this formulation is simple enough to be applicable for determining the hysteretic behavior of a heavily and repeatedly loaded bar structure such as a truss or braced frame. Despite the drastic assumption of the elastic–perfectly plastic behavior of cross sections, it has been confirmed that this theory serves as a first-order approximation to predict the inelastic hysteresis of a steel prismatic bar, unless other types of instability, such as local buckling or a fracture, intervene. An example of comparison with experimental observation is shown in Figure 9 for the dimensionless axial force-displacement relationship. The particular specimen

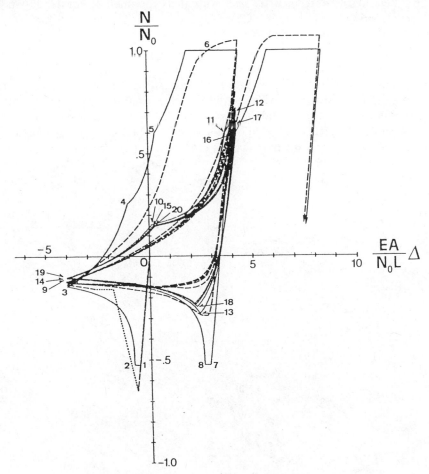

Figure 9.   Theoretical–experimental correlation for a steel bar.

cited was made of mild steel and had a square cross section of such a slenderness that $N_E = 0.523N_0$[25]. The experimental curves are drawn as dashed lines against the theoretical curves of solid lines, in which 0.9 and 0.6 are assigned to α and β, respectively. The small serial numerals indicate the order of state variation.

In view of the fact that the stable behavior of moment-resisting frame members in steel has long been a subject of investigation and is fairly well understood, the formulated restoring-force characteristics of axial-force-carrying members widen the capability of determining the hysteretic behavior of bar structures. Figure 10 shows an example of the application of the afore-described basic equations to analyze the load-displacement relation of a braced frame of wide flanges under constant gravity and varying horizontal loads. The theoretical curves are drawn as dotted lines by taking the effective length to be half the entire length of the braces, which were tightly connected at their ends [26]. Reasonable agreement is seen with the experimental curves drawn by interconnecting measurements of small hollow circles. (Isolated dots indicate the test results for monotonous loading.)

Once the restoring-force characteristics of structural constituents are known, of which the prior is an example, the response of a structure to any prescribed loading can be determined; aggravating processes due to intense repeated loading can be followed, provided ample ductility is assured. When the number of repetitions and/or amplitude in cyclic loading becomes excessive, there is a fair possibility for structural components to suffer breakage or to rupture, which makes the formulation invalid. Typical examples of rupturing phenomena are

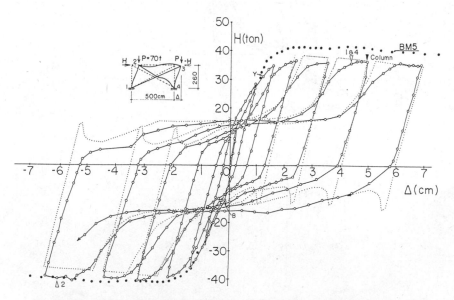

Figure 10. Theoretical–Experimental correlation for a steel-braced frame.

summarized in the following from the results of an extensive experimental study
on bracing members [24].

3.4. Experimental Observation on Rupture

It is a general experimental observation that steel braces or braced frames
undergo cracking and eventual rupture after a number of loading repetitions of
the order of 10, when subjected to repeated alternate plastic bending and
elongation. The behavior of rupture greatly depends on cross-sectional shapes,
although overall restoring-force characteristics are similar when expressed in
terms of properly nondimensionalized variables as previously described. In the
case of solid cross sections, such as round or flat bars, the ultimate failure often
takes place near joints, even if the joints themselves are proportioned strongly
enough to transmit the yield loads in tension. It seems that cracking is initiated
by the effects of heat treatment at weldments or of stress concentration due to
abrupt changes in geometrical dimensions.

Thin-walled cross sections are involved in localized deformation of plate
elements (transverse bending of plates), with alterations in the cross-sectional
shapes. Local buckling takes place in the flanges of H- or wide-flange cross
sections in an early stage of repeated loading into the plastic range, as shown in
Figure 11. The localized deformation is accumulated in the course of repeated
cycles. It does not affect the restoring-force characteristics significantly until

Figure 11.   Local deformation of a wide-flange brace.

Figure 12. Breakage of a wide-flange brace.

Figure 13. Breakage of a wide-flange brace.

414

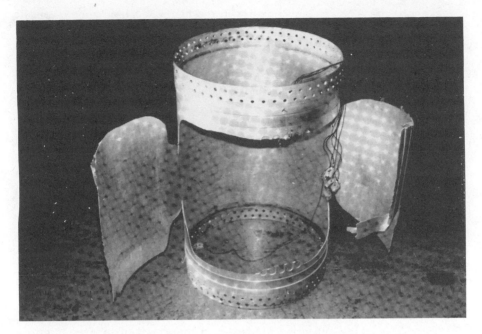

Figure 14. Local deformation of an angle brace.

Figure 15. Brekage of a double-angle brace.

415

cracking occurs. The cracking expands in the highly localized zone of plastic deformation and eventually leads to rupture. The outbreak of rupture is commonly observed while locally deformed plate elements are engaging in tension. Breakage is also detected at welds near joints, as shown in Figures 12 and 13.

Local deformation develops in the legs of an angle cross section and causes conspicuous twisting in addition to overall bending, as shown in Figure 14. Accumulation of the localized deformation is pronounced and induces cracking and eventual rupture, as shown in Figure 15. The process of cracking and rupturing is illustrated in Figure 16. When the legs of an angle are jointed to gusset plates with bolts, they suffer breakage near the bolting holes, as shown in Figures 17 and 18. Cracking follows necking at the location of the minimum effective cross-sectional area, starting either from the hole circumference, Figure 18(*a*), or from the longitudinal edge, Figure 18(*b*).

A circular tube undergoes ovaling deformation during bending. This does not alter the overall restoring-force characteristics, however. Ultimate failure often takes place in the vicinity of the weld at its juncture with the gusset plate, as shown in Figure 19. This would occur in spite of the fact that the joint might have been checked and found strong enough to carry the yield load in monotonous tensile loading. A typical rupturing process is illustrated in Figure 20. Internal cracking appearing at the tip of the gusset plate spreads into a

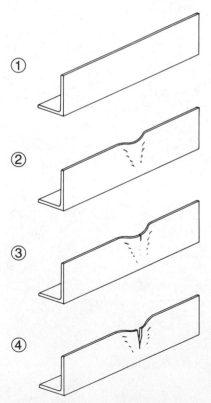

Figure 16.   Rupture of an angle due to developing local deformation.

Figure 17. Rupture of an angle brace.

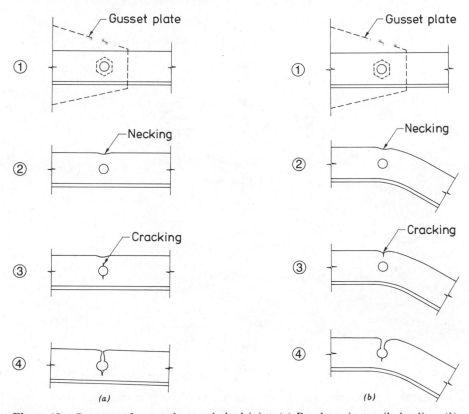

Figure 18. Rupture of an angle at a bolted joint: (*a*) Breakage in tensile loading. (*b*) Breakage in compressive loading combined with bending.

Figure 19. Breakage of a tube brace.

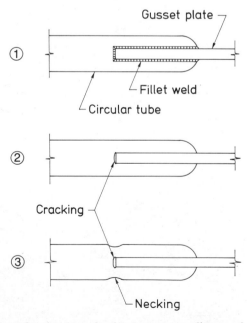

Figure 20. Rupture of a circular tube due to an expanding crack in the welding zone.

418

rupture zone, forming a slit. This accelerates necking and eventually brings about rupture.

Affected by stress concentration or heat treatment of welding, the ultimate rupture just described may have resulted from the repetition of severe plastic straining and may principally be attributed to the phenomenon of low-cycle fatigue, which is the topic of the next section.

4.   STRUCTURAL FATIGUE

4.1.   Damage Due to Fatigue

When a steel structure is subjected to intense repeated loading, as in a destructive earthquake, some constituents of the structure can undergo plastic deformation repeatedly. The number of repetitions may be limited, but the degree of plasticization may become so high that the structure may fail through low-cycle fatigue. Fatigue damage can be an important factor in structural safety. It depends not only on material properties, but on structural configuration, loading conditions, details of structural constituents, etc.

A survey from recent literature follows on the problems of structural low-cycle fatigue, with particular reference to the earthquake-resistant capability of steel-framed structures. Most of the investigations are naturally of an experimental nature. Attention is focused on repeated loading with the number of repetitions of the order of 10 to several hundreds. The survey is preceded by a brief description of basic cyclic behavior, as well as definitions of relevant terms.

4.2.   Cyclic Bending of a Steel Member

When a steel member of symmetric cross section is subjected to alternate cyclic bending in the plane of symmetry, it behaves in one of two ways, as shown in Figure 21. If the amplitude of the force or deformation quantity is fixed and sufficiently small, the restoring-force characteristics, as represented by a force-deflection or moment-curvature diagram, tend to form a fixed hysteresis loop after the first several cycles. Such a state is called *steady* or *stationary*.

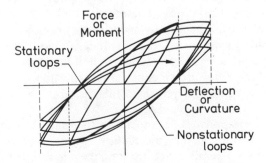

Figure 21.   Stationary and nonstationary hysteresis loops.

Application of a large number of cycles causes fatigue and eventually leads to rupture, unless the amplitude is less than the *fatigue limit*. If the amplitude is large, on the other hand, torsional, lateral–torsional, or local deformation, as caused by buckling of plate elements such as flanges, intervenes the stationary loops at some stage of loading. Such local deformation accumulates with the increasing number of cycles. In the case of a constant amplitude in the deformation quantity, both strength and stiffness decrease and the hysteresis loop becomes thinner. The loop is *nonstationary*; *deterioration* is said to take place. Continuation of such cyclic bending ends up with rupture or fatigue failure at less cycles than in the stationary case.

Tanabashi et al. [27–29] subjected SS41 rolled wide-flange beams to alternating plastic bending with constant deflection amplitudes. It has been shown that, in spite of complex local buckling and lateral–torsional deformations developed during deflection-cycling in specimens subjected to larger deflection amplitudes, in general, a load-deflection curve approaches a stationary hysteresis loop within a few cycles. The hysteretic and skeleton stress–strain relations using a modified Ramberg–Osgood model are derived from the test results of flange-plate specimens and the stationary response of the wide-flange beam is thereby confirmed theoretically. Goto et al. [30] have indicated that, in alternately reversed cyclic loading of simply supported beams with constant deflection amplitude, the critical amplitude in which lateral-torsional buckling occurs decreases to about one-fifth of the corresponding deflection in monotonic loading.

Krawinkler and Zohrei [31] conducted cyclic-loading tests of cantilever specimens consisting of structural wide-flange shapes welded to a column stub or to a base plate. The experimental work has indicated that deterioration takes place in one of two modes: those are the local buckling mode and the crack-propagation and fracture mode at weldments, as shown in Figure 22, where d and f in the abscissas designate the deflection and the number of repetitions, respectively. In the case of the local buckling mode of Figure 22(a), the deterioration threshold (number of cycles without noticeable deterioration) is small (point A) and deterioration occurs at a high rate that is associated with the continuous growth of the flange buckles (portion AB). In the next range, deterioration proceeds at a low and almost constant rate due to the stabilization in buckle size (portion BC). These are followed by a period of rapid deterioration that is caused solely by crack propagation at the welds (portion CD). In the crack-propagation and fracture mode of Figure 22(b), localized crack growth does not cause noticeable deterioration of the component strength and stiffness until the crack has grown considerably (portion EG). Once deterioration takes place, it occurs at a very high rate, leading soon to failure (portion GH).

While structural response is dependent on the exact restoring-force characteristics as well as loading history, mechanical energy absorbed or dissipated in plastic work indicates overall structural toughness in a rather reliable manner. Represented by the integration of the area of hysteresis loops, this can, therefore,

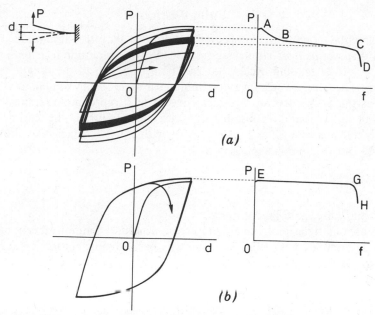

Figure 22.   Models of deterioration: (*a*) Local buckling mode. (*b*) Crack propagation and fracture mode.

be a reasonable measure of structural safety, especially when loading is random, as in earthquake excitation, and can govern the fatigue life of the structure. Popov and Pinkney [32] tested steel cantilever beams attached to short column stubs. The energy-absorption capacity per cycle increases with increasing tip deflection. They have suggested a simple linear dependence of the dissipated energy on the residual plastic deflection. They demonstrated that flange local buckling did not precipitate an immediate loss of load-carrying capacity. The ability to buckle and thus distribute damage can be of significance in prolonging the life of a member. Based on experiments on beam-to-column subassemblages, Bertero et al. [33] have concluded that for an efficient design of earthquake-resistant structures, energy-absorption capacity and especially energy dissipation of the system, rather than strength, should be used as the basic criterion.

Attention was focused on the effects of local buckling of flanges in some tests on wide flanges. A linear relation between residual plastic deflection and energy dissipated per cycle, suggested by Popov and Pinkney, is confirmed except for the case of delayed local buckling [34]. From load-controlled tests of wide-flange beams, Fukuchi and Ogura [35] have found that rotation capacity is governed by the width-to-thickness ratio of flanges. Udagawa et al. [36, 37] have shown the existence of the critical deflection amplitude that discriminates stationary and nonstationary hysteresis loops.

4.3.   Failure Criteria

A fundamental condition of repeated loading is that the amplitude of the induced displacement or strain is constant. As the amplitude a increases, the number F of the loading cycles that can be applied up to the fatigue failure decreases. The Manson–Coffin rule stipulates a relationship similar to what is known as the S–N curve or Woehler curve, which represents graphically the relation of stress (S) and the number of cycles (N) at failure of the material [38, 39]. This rule relates $\log a$ and $\log F$ linearly with a negative factor, as shown in Figure 23 [40,41], and is expressed as

$$a = bF^{\,j} \qquad\qquad (44)$$

where b and j are material constants.

The state of fatigue failure in a structure or structural component is not as yet definitive. Examples of its definition comprise the following: outbreak of fracture in the tension flange of a wide-flange beam [27]; onset of cracking in a wide-flange beam [30, 42]; a crack reaching a certain length (5 mm) [34]; strength reduction reaching 10% of the maximum strength [31] or reaching 40% (reinforced concrete) [43]; 50% reduction of dissipated energy per cycle in comparison with that in a stationary range [44]; and accumulated axial deformation arriving at a reference value (4.5 mm) in a column (reinforced concrete) [45].

It seems meaningless to compare the abundant numerical expressions for the relationship of the Manson–Coffin rule obtained from each test, because the respective definitions of a failure state and the materials of the test specimens vary. It is proposed in formulating the failure criterion that the number F should be replaced by the total energy dissipated [27, 30, 42, 44] or by an effective number equal to the total energy divided by the energy absorbed per cycle in a stationary process [46]. The evaluation of the deformation amplitude is based either on the difference between the maximum and minimum deformations or on its residual component, which is taken to be the difference in the deformations corresponding to the vanishing load.

Fatigue-failure modes are classified by Goto et al. [30] into two types: the plastic bending type, in which the resistance and dissipated energy of a beam decrease monotonically with increasing number of loading cycles; and the buckling type, in which the resistance and dissipated energy decrease suddenly

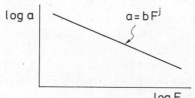

log F   **Figure 23.**   The Manson–Coffin rule.

at the onset of buckling. The relation between deflection amplitude and the number of cycles at failure is different in the two types, but the relation between the deflection amplitude and the total dissipated energy is shown in both cases to lie along the same straight line on a log-log plot.

In the tests by Kaneta and Kohzu [46] on welded joints, no significant differences in fatigue behavior between the base metal and the welded joint were observed at stationary states. In their successive tests [47], they simulated biaxial stress states in a panel zone under cyclic bending moments and column axial forces. The low-cycle fatigue strength has been well predicted by the Manson–Coffin formula in terms of the equivalent fatigue life of energy dissipation.

4.4. Effects of Axial Forces

The existence of a constant compressive axial force during cyclic bending, as in a column, accelerates the deteriorating phenomenon. The accumulation of localized deformation is more significant; the fatigue failure is more imminent. The rate of approach to the stationary state is somewhat decelerated by the compression in the small-amplitude cycling. It is possible that stationary loops later become nonstationary.

Yamada [48] carried out tests on a variety of structural members subjecting to alternately repeated plastic bending under constant axial compression. He has found a linear relation between $\log a$ and $\log F$ for the low-cycle fatigue limit, as in the Manson–Coffin rule, with consideration of the axial compression, whose ratio to the plastic squash load enters as a parameter. He proposes the use of this relationship in evaluating structural ductility as a basis of aseismic design criteria. Kato and Akiyama [49, 50] have presented the procedures for calculating the inelastic deformation and the ultimate strength of a member subjected to combined thrust and bending, and compared the theoretical load-deformation predictions with their test results. The accumulated plastic deformation capacity under repeated loading is shown to be limited and equated to the plastic deformation capacity under monotonic loading. Suzuki and Tamamatsu [44, 51] have investigated the energy-absorption capacity of H-shaped steel columns subjected to a lateral force at midheight under a constant thrust. They have found that the deflection amplitude has a logarithmically linear relationship with the total absorbed energy, provided that the amplitude is large enough to produce out-of-plane deformation in the first cycle during constant-amplitude cycling. They have discovered a larger capacity than this relation for small-amplitude cycles delaying out-of-plane deformation.

Importance of structural instability due to cyclic-strain accumulation in the plastic range has been pointed out by Neale and Schroeder [52]. From the test results of columns subjected to cycles of controlled axial displacements, it has been observed that cyclic-strain accumulation causes a gradual increase of transverse deflection, leading to a gradual reduction in the critical compressive load for instability. Their conclusion is that a structure designed against low-cycle fatigue failure may become unstable due to a gradual reduction in stability after a number of load applications in the plastic range long before the fatigue life is reached. Yokoo et al. [53] have reported that experimental behaviors of beam-columns under repeated completely reversed plastic bending and a constant axial force are classifiable into divergent and convergent ones. In the case of divergent behavior, the column does not fail by low-cycle fatigue, but by accelerated growth of antisymmetric deflections. Uetani and Nakamura [54, 55] defined new critical states, *symmetry limit* and *steady-state limit*. The symmetry limit represents the critical state at which transition from a symmetric steady state to an asymmetric steady state occurs and the steady-state limit is the state at which divergence occurs. They presented theories for predicting these limits.

4.5.   Nonstationary Loading

The fatigue life or damage in the case of varying amplitude is estimated by making reference to its equivalence to the constant-amplitude cycling. Miner [56] proposed a linear cumulative damage law. He assumed that the pheno-menon of cumulative damage under repeated loads was related to the net work absorbed by a specimen. The number of loading cycles applied expressed as a percentage of the number to failure at a given stress level would be the proportion of useful life expended. When the total damage, as defined by this concept, reached 100%, the fatigued specimen should fail. Suppose that fatigue failure takes place at a constant stress σ_i greater than the fatigue limit, with the number of cycles F_i. If f_i cycles are applied ($f_i < F_i$) under this stress level, the cumulative fatigue damage is set as the ratio f_i/F_i. Miner's law stipulates that fatigue failure be reached under the condition that

$$\sum_i f_i/F_i = 1 \tag{45}$$

where the summation is taken for different stress levels σ_i. Application of this law to a structural component is made by replacing the stress by a stress-resultant, pertinent deformation quantity, or their amplitudes.

It is pointed out from the results of some tests on a simply supported beam that the validity of Miner's law depends on the history of loading and the occurrence of local buckling and/or a yield hinge [30, 42]. Modifications of Miner's law also have been attempted. Suidan and Eubanks [57] and Kaneta et al. [58] suggest taking account of the mean stress or strain, if not zero, for cycling under varying stress or strain amplitudes. Suzuki and Tamamatsu [51] propose to make use of the ratio of energy dissipation in place of the ratio of the number of cycles. Another modification is to adopt for the total fatigue damage a linear combination of the cumulative fatigue component and the effect of the maximum deflection, which is expressed in terms of conventional ductility factor, the ratio of the maximum deflection to the yield-limit deflection (reinforced concrete) [59, 60].

Suidan and Eubanks have investigated the viability of cumulative fatigue damage as a criterion for single-degree-of-freedom steel seismic structures. The cumulative damage is found to be significant in structures having periods of vibration at the lower end of the medium period range (0.4–2.0 sec) of the design earthquake spectra and to decrease rapidly with increasing values of the natural period of vibration and the strength factor, which is defined as the ratio of the yield-limit restoring force to the weight of the structure. An alternative criterion is also proposed for the estimation of damage that depends on the hysteretic dissipated energy pertaining to the critical section of the structure. Kaneta et al. performed shaking-table tests and concluded that the safety of a beam-to-column connection could be assessed by assuming that fracture would occur when the right-hand side of Equation (45) is replaced by one-half. A recent attempt has been made to stochastically assess seismic damage (reinforced concrete) [61].

4.6. Damage Due to Earthquakes

Examples are given in photographs of the damaged structures due to the Miyagi-Ken-oki earthquake, which brought about characteristic failure to steel-framed structures in Sendai City in northern Japan on June 12, 1978. The peak horizontal ground acceleration was recorded and ranged between 0.25 to 0.40 times the acceleration of gravity. The majority of severe damage in steel structures was observed in bracing members or their connections. This indicated inadequate capability in carrying the horizontal load or in the ductility of joints causing rupture or excessive deflection.

A two-story steel-frame structure suffered complete first-floor failure, as shown in Figure 24. This structure had a rectangular plan and consisted of rigid frames in the ridgewise direction and of gabled roof frames in the spanwise direction, with bracing at each perimeter face. Failure occurred at the end of angle (L-65 × 65 × 6) braces framed in the first floor, as shown in Figure 25.

Figure 24. Failure of a two-story steel-frame warehouse: Sendai Unyu Soko. (Courtesy of Prof. M. Shibata, Setsunan University, Japan.)

Figure 25. Failure of braces—a warehouse adjacent to that of Figure 24: Sendai Unyu Soko. (Courtesy of Prof. M. Shibata, Setsunan University, Japan.)

Breaking of bolts or tearing of gusset plates must have caused a reduction in the horizontal load-carrying capacity of the structure and the frames underwent significant deformation and eventual rupture. In such storage buildings, the live loads can vary at times and can be very large. The collapsed warehouse was supposedly under excessive gravity loads on the second floor. The soft soil at this site must have amplified the ground acceleration.

A three-story steel-frame structure sustained bracing failure in the first story, as shown in Figures 26 and 27. Columns and beams of this structure were made of an H-shaped (wide-flange) steel. In the ridgewise plane corresponding to the weak-axis bending of the column sections, diagonal braces were arranged at both exterior faces using an angle section (L-50 × 50 × 6), as shown in Figure 28. All the bracing members in the first floor experienced out-of-plane deflection and were broken at their ends, which were jointed with high-strength friction bolts. The large deformation of bracing members and the rupture of connections as observed in these figures indicate the destructive effect of load repetition. When a single-angle brace is jointed with a gusset plate at one leg alone, the effective cross-sectional area to transmit the longitudinal force is much reduced by the effect of eccentricity. It turns out from the authors' investigation that the connections of this type reduce the load-carrying capacity to about 70% of the tensile yield load of the angle.

Figure 26.   Damage in a three-story steel-frame store: Kyuei Furniture Store.

5.   CONCLUDING REMARKS

Based on the elastic–perfectly plastic behavior of a bar element, explanatory discussions have first been made of the types of plastic failure of structures under variable repeated loading. Failure by continued plastic deformation is classified into instantaneous collapse and failure by cyclic plastic deformation. The latter is caused either by progressive plastic deformation or by alternating plasticity, when loads apply in a cyclic manner; whereas a single application of a load

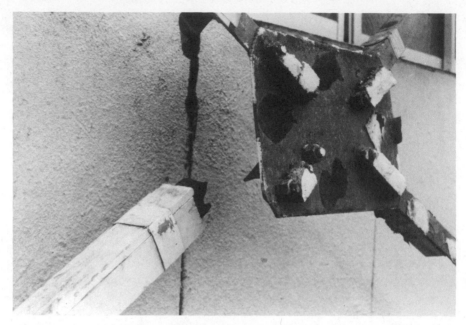

Figure 27. Failure of diagonal braces at their midheight connection: Kyuei Furniture Store.

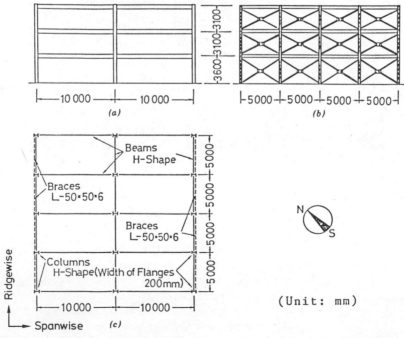

Figure 28. Damaged steel-frame structure: (*a*) Spanwise elevation. (*b*) Ridgewise elevation. (*c*) Plan.

system can give rise to the former, which is regarded as a limiting case of the latter. Classical theorems of shakedown have been outlined that provide the means of determining the bounds of the range in which loads can vary and apply repeatedly for a large number of cycles without leading to the failure by cyclic plastic deformation.

Geometrical nonlinearity has then been incorporated in the following discussions on hysteretic behavior. An axially loaded elastic–plastic bar is considered as an example. Whereas a tensile axial force brings about a favorable effect, compression destabilizes structural behavior, when accompanied by a change in geometry due to deflection. The derived basic equations are capable of determining the restoring-force characteristics onf axial-force carrying members in a closed form with first-order accuracy, provided ample ductility and local stability can be assumed.

The question as to the validity of the last assumption is raised by calling attention to the possibility of structural fatigue. Problems of low-cycle fatigue have been discussed as the final topic. A survey of the literature has been presented with particular reference to steel structures subjected to seismic loading. Cyclic bending of a constant deflection amplitude is characterized by hysteresis loops, which may either be steady or deteriorating. The latter is often associated with local deformation or instability, and leads to fatigue failure with a rate higher than the former. The existence of a compressive axial force accelerates the deteriorating behavior. A seemingly common experimental observation is that the logarithm of the amplitude has a linear relationship, with a negative factor, with the logarithm of the number of cycles at failure, the state of which, however, is not conclusively defined to date. Various proposals have been made for fatigue life and cumulative damage, but not with unanimity. This is an important practical area in which much research is required. More fundamentally, it is necessary to fully clarify the actual mechanism of fatigue failure.

REFERENCES

1. T. Nonaka, "Elastic-Perfectly Plastic Behavior of a Portal Frame with Variation in Column Axial Forces," *J. Struct. Mech.*, **6**(1), 61–84 (1978).
2. E. Melan, "Theorie statisch unbestimmter Systeme aus idealplastischem Baustoff," *Sitzungsber. Akad. Wiss. Wien, Math.-Naturwiss. Kl., Abt. 2A*, **145**(3 and 4), 195–218 (1936).
3. E. Melan, "Theory of Statically Indeterminate Systems," *Prelim. Publ. Congr., Int. Ass. Bridge. Str. Eng.*, 2nd, pp. 43–64 (1936).
4. E. Melan, "Der Spannungszustand eines Mises-Hencky'schen Kontinuums bei veraenderlicher Belastung," *Sitzungsber. Akad. Wiss. Wien, Math.-Naturwiss. Kl., Abt. 2A*, **147**(1 and 2), 73–87 (1938).
5. E. Melan, "Zur Plastizitaet des raeumlichen Kontinuums," *Ing.-Arch.*, **9**(2), 116–126 (1938).

6. W. T. Koiter, "A New General Theorem on Shake-Down of Elastic–Plastic Structures," *Proc. K. Ned. Akad. van Wet., Ser. B: Phys. Sci.*, **59**, 24–34 (1956).

7. W. T. Koiter, "Chapter IV. General Theorems for Elastic–Plastic Solids," *Prog. Solid Mech.*, **1**, 165–221 (1960).

8. P. S. Symonds and B. G. Neal, "Recent Progress in the Plastic Methods of Structural Analysis. Part I," *J. Franklin Inst.*, **252**, 383–407 (1951).

9. P. S. Symonds and B. G. Neal, "Recent Progress in the Plastic Methods of Structural Analysis. Part II," *J. Franklin Inst.*, **252**, 469–492 (1951).

10. J. B. Martin, *Plasticity: Fundamentals and General Results*, MIT Press, Cambridge, MA, 1975.

11. T. Nonaka and S. Morino, "7. Failure by Plastic Deformation in Structures under Repeated Loading," in M. Naruoka and T. Nakamura, Eds., *Survey on Analytical Methods of Frames*, Baifukan Publ. Co., Tokyo, 1976, pp. 137–158 (in Japanese).

12. W. Prager, "Shakedown in Elastic, Plastic Media Subjected to Cycles of Load and Temperature," *Symposium sulla plasticita nella scienza delle construgioni* (*Varenna, 1956*), *Bologna*, pp. 239–244 (1975).

13. A. Sawczuk, "Shakedown Analysis of Elastic–Plastic Structures," *Nucl. Eng. Des.*, **28**, 121–136 (1974).

14. G. Maier, "Chapter 6. Shakedown Analysis," in M. Z. Cohn and G. Maier, Eds., *Engineering Plasticity by Mathematical Programming*, Pergamon, New York, 1977, pp. 107–134.

15. D. A. Gokhfelt and O. F. Cherniavsky, *Limit Analysis of Structures at Thermal Cycling*, Sijthoff and Noordhoff Interantional Publishers, Alphen aan den Rijn, 1980.

16. M. A. Koenig and G. Maier, "Shakedown Analysis of Elastoplastic Structures: A Review of Recent Developments," *Nucl. Eng. Des.*, **66**, 81–95 (1981).

17. M. Zyczkowski, *Combined Loading in the Theory of Plasticity*, PWN–Polish Scientific Publishers, Warszawa, 1981.

18. G. Maier and J. Munro, "Mathematical Programming Applications to Engineering Plastic Analysis," *Appl. Mech. Rev.*, **35**(12), 1631–1643 (1982).

19. J. A. Koenig, *Shakedown of Elastic–Plastic Structures*, Am. Elsevier, New York, 1987.

20. T. Nonaka, "Formulation of Inelastic Bar under Repeated Axial and Thermal Loadings," *J. Eng. Mech., Am. Soc. Civ. Eng.*, **113**(11), 1647–1664 (1987).

21. T. Nonaka, "An Analysis for Large Deformation of an Elastic–Plastic Bar under Repeated Axial Loading. I. (Derivation of Basic Equations), and II. (Correlation with Small Deformation Theory)," *Int. J. Mech. Sci.*, **19**, 619–627(I) and 631–638(II) (1977).

22. T. Nonaka, "An Elastic–Plastic Analysis of a Bar under Repeated Axial Loading," *Int. J. Solids Struct.*, **9**(5), 569–580 (1973); erratum in Vol. 9, No. 10.

23. T. Nonaka, "Approximation of Yield Condition for the Hysteretic Behavior of a Bar under Repeated Axial Loading," *Int. J. Solids Struct.*, **13**(7), 637–643 (1977).

24. N. Yoshida, "Studies on Elastic–Plastic Behavior of Braces," Thesis submitted in partial fulfillment of the requirements for the degree of Doctor of Engineering at Kyoto University, 1984 (in Japanese).

25. M. Wakabayashi, T. Nonaka, T. Nakamura, S. Morino, and N. Yoshida, "Experimental Studies on the Behavior of Steel Bars under Repeated Axial Loading. Part 1. Rectangular Cross-Section," *Annu., Disaster Prev. Res. Inst. Kyoto Univ.*, **16B**, 113–125 (1973) (in Japanese).

26. M. Wakabayashi, C. Matsui, K. Minami, and I. Mitani, "Inelastic Behavior of Full-Scale Steel Frames With and Without Bracings," *Bull. Disaster Prev. Res. Inst., Kyoto Univ.*, **24** (216), Part 1, 1–23 (1974).

27. R. Tanabashi, Y. Yokoo, M. Wakabayashi, T. Nakamura, H. Kunieda, H. Matsunaga, and T. Kubota, "Load-Deflection Behaviors and Plastic Fatigue of Wide-Flange Beams Subjected to Alternating Plastic Bending. Part I. Experimental Investigation," *Trans. Archit. Inst. Jpn.*, **175**, 17–29 (1970).

28. R. Tanabashi, Y. Yokoo, T. Nakamura, T. Kubota, and A. Yamamoto, "Load-Deflection Behaviors and Plastic Fatigue of Wide-Flange Beams Subjected to Alternating Plastic Bending. Part II. Hysteretic and Skeleton Stress–Strain Relations and Plastic Fatigue of Flanges," *Trans. Archit. Inst. Jpn.*, **176**, 25–36 (1970).

29. R. Tanabashi, Y. Yokoo, and T. Nakamura, "Load-Deflection Behaviors and Plastic Fatigue of Wide-Flange Beams Subjected to Alternating Plastic Bending. Part III. Steady-State Theory," *Trans. Archit. Inst. Jpn.*, **177**, 35–46 (1970).

30. H. Goto, H. Kameda, T. Koike, R. Izunami, K. Wakita, and Y. Sugihara, "A Consideration of Failure Process of Structural Steel under Repeated Flexural Loads," *Annu., Disaster Prev. Res. Inst., Kyoto Univ.*, **17B**, 157–169 (1974) (in Japanese).

31. H. Krawinkler and M. Zohrei, "Cumulative Damage in Steel Structures Subjected to Earthquake Ground Motions," *Comput. Struct.*, **16**(1–4), 531–541 (1983).

32. E. P. Popov and R. B. Pinkney, "Cyclic Yield Reversal in Steel Building Connections," *J. Struct. Div., Am. Soc. Civ. Eng.*, **95**(ST3), 327–353 (1969).

33. V. V. Bertero, E. P. Popov, and H. Krawinkler, "Beam-Column Subassemblages under Repeated Loading," *J. Struct. Div., Am. Soc. Civ. Eng.*, **98**(ST5), 1137–1159 (1972).

34. Y. Mukudai, A. Matsuo, and T. Imada, "On Post-Local Buckling Properties of Flange of Beam Subjected to Cyclic Bending," *Summ. Tech. Pap. Annu. Meet., Archit. Inst. Jpn.*, pp. 1083–1086 (1976) (in Japanese).

35. Y. Fukuchi and M. Ogura, "Experimental Studies on Local Bucklings and Hysteretic Characteristics of H-Shape Beams," *Trans. Archit. Inst. Jpn.*, **228**, 65–71 (1975) (in Japanese).

36. K. Udagawa, K. Takanashi, and H. Tanaka, "Restoring Force Characteristics of H-Shaped Steel Beams under Cyclic and Reversed Loadings. Part I. Rotation Capacity of Plastic Hinge under Cyclic Loads at Constant Deflection Amplitudes," *Trans. Archit. Inst. Jpn.*, **264**, 51–59 (1978) (in Japanese).

37. K. Udagawa, K. Takanashi, and H. Tanaka, "Restoring Force Characteristics of H-Shaped Steel Beams under Cyclic and Reversed Loadings. Part II. Strength Deterioration of Beams under Random Deflections," *Trans. Archit. Inst. Jpn.*, **265**, 45–52 (1978) (in Japanese).

38. C. C. Osgood, *Fatigue Design*, 2nd ed., Pergamon, Oxford, 1982, pp. 41–100.

39. A. Puskar and S. A. Golovin, *Fatigue in Materials: Cumulative Damage Processes*, Elsevier, Amsterdam, 1985, pp. 148–178.

40. S. S. Manson and M. H. Hirschberg, "Fatigue Behavior in Strain Cycling in the Low- and Intermediate-Cycle Range," in J. J. Burke, W. T. Read, and V. Weiss, eds., *Fatigue—An Interdisciplinary Approach*, Syracuse University Press, Syracuse, NY, 1964, pp. 133–178.

41. J. F. Tavernelli, and L. F. Coffin, Jr., "Experimental Support for Generalized Equation Predicting Low Cycle Fatigue," *J. Basic Eng.*, **34**, 533–541 (1962).

42. H. Goto, H. Kameda, T. Koike, I. Aoyama, and K. Wakita, "Probabilistic Considerations on Plastic Fatigue Failure of Structural Steel," *Annu., Disaster Prev. Res. Inst., Kyoto Univ.*, **18B**, 377–393 (1975) (in Japanese).

43. H. Iemura, "Earthquake Failure Criteria of Deteriorating Hysteretic Structures," *Proc., World Conf. Earthquake Eng.*, *7th*, Vol. 5, pp. 81–88 (1980).

44. T. Suzuki and K. Tamamatsu, "Experimental Study on Energy Absorption Capacity of Columns of Low Steel Structures. Part 1. Energy Absorption Capacity of H-Shaped Steel Columns Subjected to Monotonic Loading and Cyclic Loading with Constant Deflection Amplitudes," *Trans. Archit. Inst. Jpn.*, **279**, 65–75 (1979) (in Japanese).

45. T. Nishigaki and K. Mizuhata, "Experimental Study on Low-Cycle Fatigue of Reinforced Concrete Columns," *Trans. Archit. Inst. Jpn.*, **328**, 60–70 (1983) (in Japanese).

46. K. Kaneta and I. Kohzu, "Low-Cycle Fatigue Strength of Welded Steel Structural Joints. Part 1. Low-Cycle Fatigue of Cylindrical Specimens Using Mild to High Tensile Strength Steels," *Trans. Archit. Inst. Jpn.*, **313**, 30–38 (1982) (in Japanese).

47. K. Kaneta and I. Kohzu, "Low-Cycle Fatigue Strength of Welded Steel Structural Joints. Part 2. Fatigue of Idealized Beam-To-Column Welded Joints Considering Anisotropic Properties of Mild Steel," *Trans. Archit. Inst. Jpn.*, **317**, 15–22 (1983) (in Japanese).

48. M. Yamada, "Low Cycle Fatigue Fracture Limits of Various Kinds of Structural Members Subjected to Alternately Repeated Plastic Bending under Axial Compression as an Evaluation Basis or Design Criteria for Aseismic Capacity," *Proc. World Conf. Earthquake Eng.*, *4th*, Vol. 1, B2, pp. 137–151 (1969).

49. B. Kato and H. Akiyama, "The Ultimate Strength of the Steel Beam-Columns. Part 4," *Trans. Archit. Inst. Jpn.*, **151**, 15–20 (1968) (in Japanese).

50. B. Kato and H. Akiyama, "Inelastic Bar Subjected to Thrust and Cyclic Bending," *J. Struct. Div., Am. Soc. Civ. Eng.*, **95**(ST1), 33–56 (1969).

51. T. Suzuki and K. Tamamatsu, "Experimental Study on Energy Absorption Capacity of Columns of Low Steel Structures. Part 2. Energy Absorption Capacity of H-Shaped Steel Columns Subjected to Cyclic Loading with Varying Deflection Amplitudes," *Trans. Archit. Inst. Jpn.*, **280**, 19–25 (1979) (in Japanese).

52. K. Neale and J. Schroeder, "Instability under Cycles of Plastic Deformation," in H. Leipholz, Ed., *Instability of Continuous Systems*, Springer-Verlag, Berlin, 1971, pp. 329–333.

53. Y. Yokoo, T. Nakamura, K. Uetani, and I. Takewaki, "Experimental Investigation of Convergence and Divergence of In-plane Deformations of Cantilever Steel Beam-Columns Subjected to Repeated Reversed Plastic Bending," *Trans. Archit. Inst. Jpn.*, **316**, 41–52 (1982) (in Japanese).

54. K. Uetani and T. Nakamura, "Symmetry Limit Theory for Cantilever Beam-Columns Subjected to Cyclic Reversed Bending," *J. Mech. Phys. Solids*, **31**(6), 449–484 (1983).

55. K. Uetani, "Symmetry Limit Theory and Steady-State Limit Theory for Elastic-Plastic Beam-Columns Subjected to Cyclic Reversed Bending," Thesis submitted in partial fulfillment of the requirements for the degree of Doctor of Engineering at Kyoto University, 1984 (in Japanese).

56. M. A. Miner, "Cumulative Damage in Fatigue," *J. Appl. Mech.*, **12**, A159–A164 (1945).

57. M. T. Suidan and R. A. Eubanks, "Cumulative Fatigue Damage in Seismic Structures," *J. Struct. Div., Am. Soc. Civ. Eng.*, **99**(ST5), 923–943 (1973).

58. K. Kaneta, I. Kohzu, and H. Nishizawa, "Cumulative Damage of Welded Beam-To-Column Connections in Steel Structures Subjected to Destructive Earthquakes," *Proc., World Conf. Earthquake Eng., 8th*, Vol. 6, pp. 185–192 (1984).

59. T. Nishigaki and K. Mizuhata, "Evaluation of Seismic Safety for Reinforced Concrete Structures," *Trans. Archit. Inst. Jpn.*, **332**, 19–29 (1983) (in Japanese).

60. Y.-J. Park and A. H.-S. Ang, "Mechanistic Seismic Damage Model for Reinforced Concrete," *J. Struct. Eng., Am. Soc. Civ. Eng.*, **111**(4), 722–739 (1985).

61. Y.-J. Park, A. H.-S. Ang, and Y. K. Wen, "Seismic Damage Analysis of Reinforced Concrete Buildings," *J. Struct. Eng., Am. Soc. Civ. Eng.*, **111**(4), 740–757 (1985).

CHAPTER 13

Behavior of Composite and Metallic Superstructures Under Blast Loading

C. S. Smith

Admiralty Research Establishment
Dunfermline, Scotland

ABSTRACT

A methodology is suggested for evaluation of the nonlinear elastoplastic behavior under air-blast loading of stiffened panels forming a ship's superstructure. Two- and three-dimensional numerical models are employed to examine local response of plating and stiffened panels together with overall deckhouse sidesway. Reference is made to some experimental results for metal structures. Comparisons are made between the behavior of some light stiffened panels employing steel, aluminum, glass-fiber-reinforced plastic, (GRP) and hybrid GRP/steel construction.

1. INTRODUCTION

Military and civil defense structures may be required to withstand substantial levels of air-blast loading caused by nuclear explosions. The best means of meeting this requirement in static land-based structures where weight savings are unimportant is likely to be by use of earthworks and reinforced concrete. In weight-critical mobile structures, however, which are commonly of lightweight shell construction, provision of adequate blast resistance becomes a difficult and possibly critical part of the design process.

A case of some importance is that of warship superstructures, in which weight must be limited in order to maintain ship stability and a high level of blast protection may be required for personnel and vital ship systems. Ship superstructures are commonly constructed in steel or aluminum with light plating welded to predominantly transverse framing. Although the response of such structures to air-blast loading has been extensively researched [1–5], much

435

scope remains for development of analysis and design methods that are both accurate and tractable.

Considerable interest has evolved recently in the use of fiber-reinforced plastics for ship superstructures. Use of GRP (glass-fiber-reinforced plastic) construction offers two main advantages:

1. weight savings of up to 50% relative to a conventional steel structure
2. elimination of fatigue cracking associated with hull-superstructure interaction (by virtue of the low material stiffness of GRP, which is between 5 and 10% that of steel)

This approach to superstructure design, including consideration of jointing, flammability, ballistic resistance, EM characteristics, and fabrication economics, is fully discussed in Reference 6. In the case of warships, an aspect of GRP superstructure design that requires particular attention is that of air-blast response. Unlike steel or aluminum structures, in which large ductile deformations may be allowed in design, a GRP superstructure must rely on elastic stiffness, possibly enhanced by large-displacement effects, to withstand blast loading.

The purpose of this chapter is to outline an approach to the evaluation of air-blast response in steel or aluminum superstructures, to consider the application of this approach to GRP (or hybrid GRP/metal) superstructures, and to contrast the behavior of some representative metal, GRP, and hybrid structures under typical air-blast loading.

2. AIR-BLAST LOADING

Literature on the subject of a nuclear air blast is largely classified. Useful published accounts of air-blast mechanics can, however, be found in References 7–9.

An atmospheric nuclear explosion is most likely to occur at some height, which may be optimized to produce maximum damage effects, vertically above "ground zero" on the earth or sea surface. The blast wave is reflected from the surface, and at a certain distance from ground zero, primary and reflected waves combine to form an enhanced nearly vertical "Mach" front (or "stem") that propagates outwards with diminishing intensity: the peak incident blast overpressure p_i at any distance from ground zero resulting from an explosion of specified yield and height of burst can be found from charts given in References 7–9. An object in the path of the Mach stem experiences a virtually instantaneous rise of pressure to a peak incident value p_i; the overpressure p then decays exponentially, approximately in accordance with the expression

$$p(t) = p_i(1 - \alpha_p t)e^{-\alpha_p t} \tag{1}$$

where $1/\alpha_p = t_{p+}$ is the duration of positive blast pressure (obtainable from charts in Reference 7), and t is the time after arrival of the blast front. The velocity U of the blast front is given by

$$U = c_0 \left(1 + \frac{6p_i}{7p_0} \right)^{1/2} \tag{2}$$

where c_0 is the ambient speed of sound in undisturbed air ($c_0 \approx 332 + 0.6T$ m/sec, where T is the temperature in °C) and p_0 is the ambient atmospheric pressure.

Passage of the blast front is followed immediately by a transient "blast wind" that exerts a supplementary dynamic pressure q on objects in its path. Peak dynamic pressure q_i can be estimated from the relation [7]

$$q_i = \frac{5}{2} \left(\frac{p_i^2}{7p_0 + p_i} \right) \tag{3}$$

Subsequent decay of dynamic pressure is given approximately by

$$q = q_i(1 - \alpha_q t)e^{-2\alpha_q t} \tag{4}$$

where $1/\alpha_q = t_{q+}$ is the duration of positive dynamic pressure (somewhat greater than t_{p+}).

For the purpose of air-blast analysis, a ship's superstructure can be treated usually as a rectangular box-like structure, as shown in Figure 1. Although a blast front can strike a ship from any angle, the most severe condition, which

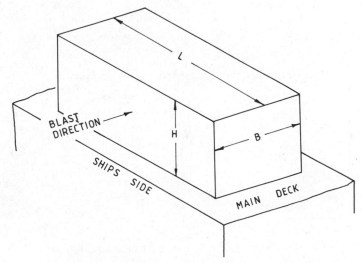

Figure 1. Ship superstructure under air-blast load.

would normally be examined for design purposes, is that of a blast direction normal to the ship's side. From Figure 1, blast loading can then be described as follows:

1. Reflection of the incident blast wave from a vertical superstructure side results in an instantaneous increase of pressure to a peak reflected level p_r given by

$$p_r = 2p_i \left(\frac{7p_0 + 4p_i}{7p_0 + p_i} \right) \tag{5}$$

which is between two and eight times the peak incident pressure p_i. As the blast wave diffracts round the structure, reflected pressure drops rapidly over a period t_s (typically between 15 and 70 msec) to the "stagnation pressure" $p_s = p_i + p_d$, where the "drag pressure" p_d is equal in this case to the dynamic pressure q. The time t_s depends on the geometry of the structure and has been found empirically to approximate

$$t_s = 3S/U \tag{6}$$

where S is the smaller of H and $L/2$ (see Figure 1). Finally, the stagnation

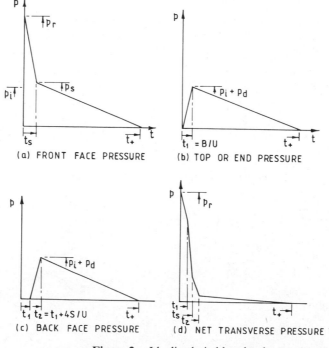

Figure 2.   Idealized air-blast load.

pressure decays exponentially over the remainder of the positive pressure phase which, in the case of a large, e.g., megaton explosion, can extend to several seconds. For practical purposes, loading can be assumed to have the trilinear form shown in Figure 2(a), defined by the five parameters: p_i, p_r, q_i, t_s, and t_+.

2. Loading on the ends and top of a superstructure, which lie in planes parallel to the direction of blast propagation, can be represented as shown in Figure 2(b): pressure increases over a short rise time $t_1 = B/U$ to a peak average value $p_i + p_d = p_i + C_d q_i$, where C_d (in the range 0.2 to 0.5) is a drag coefficient [7], and then decays in accordance with Equations (1) and (4). For moderate blast pressures ($p_i < 1$ bar), the drag pressure p_d is a small fraction of p_i and can be ignored.

3. On the back face of a superstructure, it has been found empirically [7] that the diffracted blast pressure takes time $t_2 = 4S/U$ to build up to its peak value $p_i + C_d q_i$ and can be represented as shown in Figure 2(c). The net horizontal load acting on the superstructure unit as a whole can be estimated by combining the front-face and back-face pressures, i.e., by superposing Figures 2(a) and 2(c) to obtain a net pressure–time curve as shown in Figure 2(d).

3. STRUCTURAL DESIGN REQUIREMENTS

Nuclear blast can be regarded as an exceptional load condition for which substantial relaxations can be made in normal design criteria. In the case of steel or aluminum superstructures, elastic design is generally too conservative: fairly large inelastic deformations, as illustrated in Figure 3, are acceptable provided that the superstructure maintains its protection of internal systems and personnel. Design requirements are, therefore,

1. that deformations should not exceed prescribed limits, chosen so that mechanical damage to piping, cables, electronic units, and other internal components does not occur (permanent displacements of between 50 and 200 mm may be acceptable in some stiffened panels)
2. that superstructure modules remain attached to the main hull and are not ruptured, admitting blast pressures to internal compartments: this involves use of materials with sufficient toughness and ductility to ensure that strains associated with permissible displacements do not cause brittle fracture or ductile tearing.

In the case of an all-GRP structure, assuming a laminate formed by glass-fiber-reinforcement in a thermosetting (polyester, vinylester, epoxy, or phenolic) resin matrix, the material remains effectively elastic up to its ultimate failure strain. No significant plastic deformation occurs, although the laminate may experience microscopic damage (resin cracking and fiber-debonding) under tensile stress at about 20 to 30% of ultimate strain, causing some reduction in elastic moduli and possibly some loss of wet durability. For these reasons, in

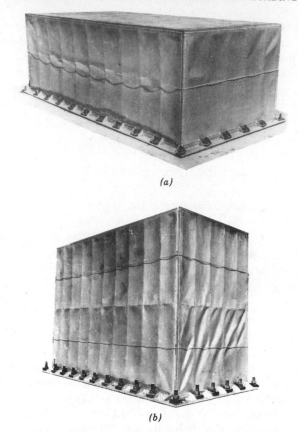

(a)

(b)

Figure 3. Forms of blast damage in metal superstructures: (*a*) Structure *A*: single-tier deckhouse model. (*b*) Structure *B*: two-tier deckhouse model.

designing against normal loads, tensile strains are usually restricted to less than 30% of ultimate. In the case of nuclear air-blast loading, however, this restriction is unduly conservative and it is appropriate to use material ultimate strains as design failure criteria, together with displacement limitations as in a metal superstructure. Similar considerations apply in the case of a hybrid super-structure employing GRP paneling attached to steel or aluminum frames. In both cases, special attention must be paid to the attachment of GRP panels to framing and to the steel main deck, probably involving a combination of bonding and bolting.

4. RESPONSE OF METAL SUPERSTRUCTURES

Evaluation of air-blast resistance requires consideration of the following:

1. Local response of plate panels between stiffeners

2. local response of stiffened panels between supporting bulkheads and decks

3. overall response (sidesway) of the complete superstructure

Permanent inelastic deformations associated with each form of response are illustrated in Figure 3. A reasonable first approach, as in static and vibration analysis, is to assume that these responses are uncoupled, although it may be desirable to check this assumption by examining coupled behavior, particularly as regards interaction between stiffeners and plating.

4.1. Strain-Rate Effects

It is well known that the ultimate and yield strengths of metals are affected by the rate at which load is applied and that yield strength, in particular, can increase substantially with the strain rate $\dot{\epsilon}$. The effect of strain rate on yield and UTS of steel has been shown [10] to be greatest in mild steel and to diminish with increasing steel strength. The relationship between static and dynamic yield stresses is sometimes estimated using the Cowper–Symonds expression

$$\sigma_{0d}/\sigma_{0s} = 1 + (\dot{\epsilon}/D)^{1/r}$$

where D and r are empirical coefficients equal to 40 sec^{-1} and 5, respectively, for mild steel [11], and σ_{0s} refers to an effectively zero strain rate ($\dot{\epsilon} = 10^{-5}$ sec^{-1}, say). It appears from Reference 10 that this expression relates quite closely to upper yield, but exaggerates the influence of strain rate on the more signigicant value of lower yield stress.

The strain rates experienced by a stiffened shell under dynamic load obviously vary throughout the volume of the structure and with respect to time. As part of a time-stepping numerical analysis of nonlinear structural response, which necessarily refers to the current state of strain in every part of the structure at each time step, it is straightforward in principle to identify the instantaneous strain rate at any point and hence to adjust the yield stress locally. This correct treatment of visco–plastic effects does not however appear to be included in most nonlinear analysis codes. An alternative, simpler but approximate approach is to estimate an average strain rate and adopt a corresponding fixed dynamic yield stress: this approach has been followed in the analysis to be described, with σ_{0d}/σ_{0s} taken equal to 1.3 for mild steel, 1.2 for medium steel ($\sigma_0 \approx 300$ MPa), and 1.1 for high-strength steel ($\sigma_0 > 500$ MPa) in accordance with data from Reference 10 and assuming an average strain rate of between 0.1 and 10 sec^{-1}.

4.2. Local Response of Plating

Since the length a of plate elements between stiffeners is commonly large compared with the width b, local response of the plating can usually be examined using a two-dimensional model, i.e., a strip of plating of unit width and span b. As adjacent plate elements are identically loaded, it is reasonable to assume clamped boundary conditions, i.e., $w = \partial w/\partial x = 0$ at $x = 0, b$, where w is the lateral displacement and x is directed along the span b. In a superstructure containing many such plate elements it is reasonable to assume full axial restraint ($u = 0$ at $x = 0, b$) at the ends of the plate strip, except near the superstructure ends, where zero restraint may be a more appropriate assumption.

Plating response in some representative structures has been examined using a special-purpose finite-element program (FABSTRAN) for large-displacement elastoplastic analysis of frame and beam structures under static and transient dynamic loads [12]. A two-dimensional model was employed, as previously described, with each plate strip assumed to satisfy conditions of symmetry at midspan and to be fully clamped at its ends (i.e., at stiffener positions). Each strip was divided into 10 elements over its half-length (plane-frame elements being employed in order to allow for large displacements) and each element was subdivided into 10 "fibers" over its depth to account for progressive development of plasticity. Equations of motion were set up in the standard form:

$$(K + K_G)\delta + M\ddot{\delta} = F \tag{7}$$

where K and K_G are conventional and geometric (tangent) stiffness matrices, M is a diagonal matrix of lumped masses, F is a column matrix of time-dependent nodal forces, and δ and $\ddot{\delta}$ are nodal displacements and accelerations. Shear deformation, rotatory inertia, and damping were ignored. Equation (7) was solved using a Newmark–β time-stepping integration routine [13], assuming constant acceleration in each time step. Large displacements were accounted for using an updated Lagrangian formulation, involving redefinition of structural geometry following each time step, and an iterative equilibrium correction was applied to the solution for each time step. Effective values of Young's modulus $E' = E/(1 + \mu^2)$ and yield stress $\sigma_0' = 2\sigma_0/3^{1/2}$, where E, μ, and σ_0 are, respectively, material Young's modulus, Poisson ratio, and uniaxial yield stress, were adopted to represent bending and stretching under conditions of plane strain.

Results were computed for the scale model of a steel superstructure shown in Figure 3(a), with geometry and static material properties as indicated in Figure 4, which was subjected to blast loading from a conventional explosion [14]. The average recorded incident pressure–time history was as shown in Figure 5, where $p_i = 0.86$ bar and $t_+ = 0.19$ sec: by using Equations (3), (5), and (6), parameters of the front-face loading, assumed to have the trilinear form of Figure 2(a), were found to be $q_i = 0.23$ bar, $p_r = 2.28$ bar, and $t_s = 0.0048$ sec. The computed central displacement–time history was as shown in Figure 6, indicating a permanent displacement of amplitude 3.8 mm. This can be compared with the permanent plating deformation measured across the mid-height of the front face following the test (Figure 7): displacement amplitudes have a maximum value of 6.6 mm, with a mean of 4.1 mm, which compare well with the calculated value.

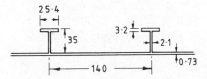

DIMENSIONS IN mm

PLATING AND STIFFENER GEOMETRY
IN SIDES, ENDS AND DECKS
(VERTICAL STIFFENERS IN SIDES AND ENDS,
TRANSVERSE STIFFENERS IN DECKS)

L = 1397, B = 699 IN STRUCTURES 'A' & 'B'
H = 521 IN STRUCTURE 'A', H = 1041 IN
STRUCTURE 'B' WITH INTERMEDIATE DECK
AT MID - DEPTH.

MATERIAL PROPERTIES:-

	E (GPa)	μ	σ_o (MPa)
PLATING	207	0·3	240
STIFFENER WEB	207	0·3	212
STIFFENER TABLE	207	0·3	359

Figure 4. Structures A and B. Geometry and material properties.

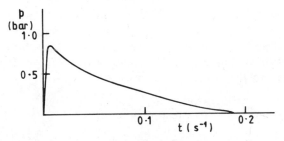

Figure 5. Structures A and B: recorded incident blast pressure.

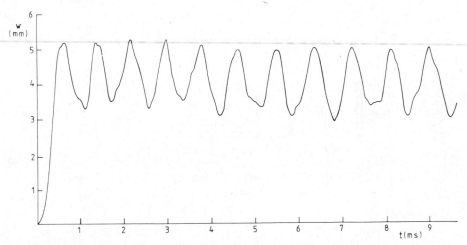

Figure 6. Structures A and B: computed plating displacement.

443

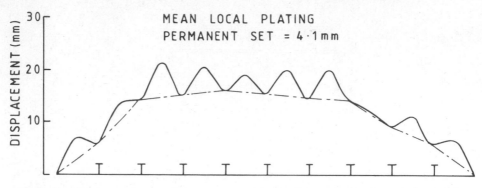

Figure 7. Structure A: recorded permanent deformation across the front face at mid-depth.

Plating response has also been computed for a large-scale stiffened steel panel (structure C) tested under air-blast loading from a simulated nuclear (4000 tonne TNT) explosion [5]. Geometry and static material properties were as shown in Figure 8. This panel was mounted horizontally on a rigid reinforced-concrete test bed in order to provide fully fixed boundary conditions. The recorded pressure–time history, which may be taken as the effective blast load on the horizontal panel, is shown in Figure 9. Displacements recorded at position P in the middle of one of the plate panels and position Q at midspan on an adjoining stiffener (see Figure 8) are indicated in Figure 10: distortion of the plating relative to the stiffeners, obtained by subtracting the Q- from the P-

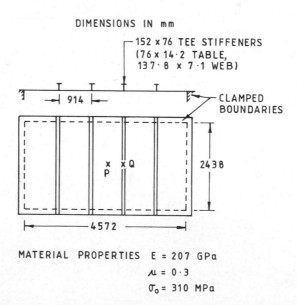

Figure 8. Structure C: geometry and material properties.

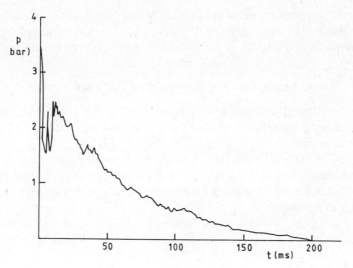

Figure 9. Structure *C*: recorded incident blast pressure.

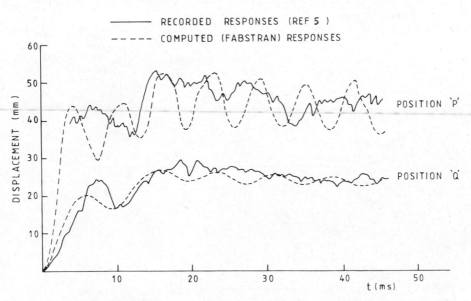

Figure 10. Structure *C*: recorded and computed displacements.

displacement records, reveals a peak local plating displacement of about 22 mm. This compares with a peak plating displacement of 27 mm indicated by analysis of a 10-element strip model of the plating as previously described.

4.3. Local Bending Response of Stiffened Panels

The sides, ends, and decks of a light superstructure are commonly formed by panels of plating with unidirectional transverse framing, corresponding to the form of construction in structures A and B (Figures 3 and 4). Bending deformation of such stiffened panels between supporting decks and bulkheads can be examined by evaluating the response to blast pressure of a single stiffener with an attached strip of plating, treated as a beam with appropriate support conditions at its ends.

The plating of a light superstructure is usually very slender, with breadth/thickness ratios (b/h) typically greater than 100. It is well known that the stiffness and strength of such plating are much reduced by local buckling under compressive stress, as can be induced by stiffener bending. Compressive buckling of the deckhouse-side plating caused by a blast load is clearly evident in Figure 3(a). As a result of initial weld-induced distortions, the effectiveness (i.e., initial stiffness) of slender plating can be low even in the stress-free condition. Some data curves defining effective average (static) stress–strain relationships for rectangular plates under tension and compression acting in the longer direction are shown in Figure 11 for a range of values of the plate slenderness parameter $\beta = \sigma_0/E)^{1/2}b/h$: these curves, derived by a combination of nonlinear finite-element analysis and regression analysis of a large body of test data [15], account for local buckling and yielding and for the effects of average imperfections.

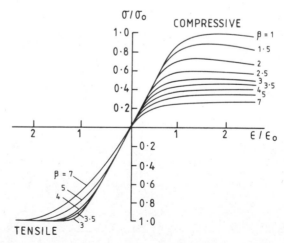

Figure 11. Effective stress–strain curves for plates under longitudinal compression and tension.

Deformation of the front face of structure A (Figure 7) under blast loading as indicated in Figure 5 was evaluated by examining the response of a representative stiffener with an attached strip of plating using the program FABSTRAN. The structure was divided into 20 frame elements over its 521 mm height; the stiffener was subdivided into 20 fibres over its depth and the plating was represented as a single fiber with effective (tangent) modulus derived from the slope of the stress–strain curves shown in Figure 11, which are represented numerically in the program. Peak midspan displacements, computed assuming the following boundary conditions, are listed in Table 1:

Case 1: rotationally and axially clamped
$(w = u = dw/dx = 0$ at $x = 0, H)$

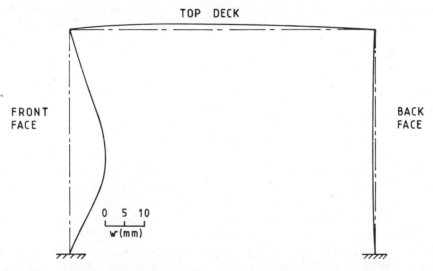

Figure 12. Structure A: computed deformation using the portal frame model (case 6).

TABLE 1 Peak Displacements of Stiffened Panel (Structure A) for Various Boundary Conditions

Case	Model	Peak Displacement, w_{max} (mm)
1	Beam	1.9
2	Beam	10
3	Beam	36
4	Beam	6.5
5	Portal frame	3.0
6	Portal frame	9.5

Case 2: simply supported with axial fixity
($w = u = d^2w/dx^2 = 0$ at $x = 0, H$)

Case 3: simply supported and axially unrestrained

Case 4: clamped at base, simply supported, and axially unrestrained at top

Results demonstrate the dependence of response on both axial and rotational restraint, the former becoming important where displacements are large The effective boundary conditions in an actual superstructure can be difficult to determine. A clamped condition might reasonably be assumed at the base of a superstructure where attachment is to a main deck with heavy transverse frames. At the top edge ($x = H$) of a deckhouse side, a condition of zero lateral displacement can be adopted ($w = 0$), since sidesway is inhibited by shear rigidity of the deckhouse top, with zero axial (vertical) restraint; a condition of rotational restraint is exerted by the adjoining deckhouse top. An improved assessment of front-face response might be expected from a portal frame model representing the front, top, and back faces of a deckhouse, with sidesway suppressed (Fig. 12). Peak displacements at mid-height on the front face, computed for a portal frame model assuming the following support conditions at the base, are included in Table 1:

Case 5: fully clamped

Case 6: simply supported with axial fixity

Results listed in Table 1 highlight the sensitivity of stiffened-panel response to both rotational and axial boundary conditions. Computed displacements evidently bracket the experimental permanent set of 15 mm (Figure 7).

Displacements have also been computed for a stiffener with attached strip of plating representing structure C (Figure 8) under blast loading as indicated in Figure 9: in this case, fully clamped boundary conditions can reasonably be assumed. A similar, plane-frame (FABSTRAN) finite-element idealization was employed, with plating stiffness derived from an appropriate stress–strain curve (Figure 11, $\beta = 7$). A comparison of midspan displacement–time curves, shown in Figure 10, indicates reasonable agreement between analysis and experiment.

4.4. Coupled Response of Plating and Stiffeners

An alternative finite-element model for analysis of stiffened-panel response is shown in Figure 13. Plating is represented by shell elements, whereas a tee-section stiffener can be represented either by beam elements or by subdivision into shell elements. The front-face stiffened panel of structure A was analyzed in this way using the general-purpose nonlinear finite-element program ADINA [16], which accounts for large-displacement elastoplastic response, including local buckling and postbuckling behavior of the plating. Plating was represented by eight-node isoparametric quadrilateral shell elements and the stiffener by

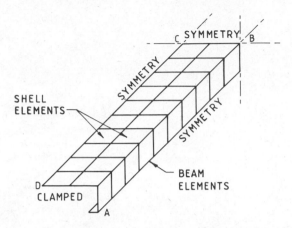

Figure 13. Finite-element model of structure A, front face.

three-node isoparametric beam elements, accounting accurately for interaction between stiffener and plate bending. Planes of symmetry were assumed along the centerline of the stiffener (AB) and midway between stiffeners (DC), with a transverse plane of symmetry at mid span (BC) and conditions of axial, lateral, and rotational fixity at the support position (AD). Analysis was carried out on a VAX computer at ARE, Dunfermline.

Computed displacement–time histories at midspan on the stiffener and at the center of a plate panel under the blast loading of Figure 5 are shown in Figure 14. Peak stiffener displacements and peak displacements of the plating relative to stiffeners evidently correspond fairly well with uncoupled stiffener and plating displacements (Figure 6 and Table 1). Displacements under static pressure were found to correspond similarly. Linear elastic natural frequencies indicated by the ADINA model were 244 and 1030 Hz for the lowest local (plating) and overall (stiffener) modes: corresponding beam-theory frequencies for the un-coupled plate strip and stiffener were 206 and 1021 Hz respectively.

A similar ADINA analysis has been described in Reference 5 for structure C. Computed responses reported in Reference 5 correlate satisfactorily with both experimental responses and uncoupled computed responses. These, together with results previously described for structure A, support the assumption that plating and stiffener responses can be treated as being uncoupled and suggest that this will usually be a satisfactory basis for design calculations. In cases where the lowest (uncoupled) natural frequencies of plating and stiffener vibration are approximately equal, it may be desirable to check this approach by examining the response of a finite-element model, as illustrated in Figure 13. It should be noted that finite-element analysis of this type using a program such as ADINA accounts for the influence of inertial forces on plate buckling: in this respect use of the static plate-stiffness curves (Figure 11) as previously described is likely to be slightly conservative.

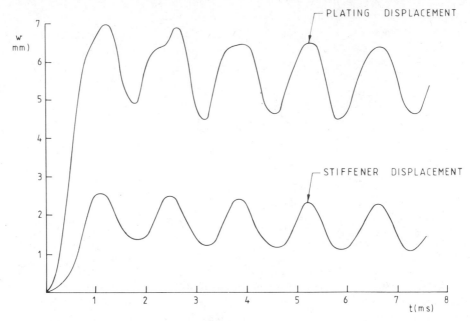

Figure 14. ADINA analysis of coupled plating and stiffener response.

4.5. Rigid–Plastic Analysis of Plating and Stiffener Response

Where displacements of a stiffened panel are large so that most of the strain energy is associated with plastic deformation, elastic strains can be ignored and structural response examined using rigid–plastic analysis. Application of this method to plates and beams under static and impulsive loads, including the geometric effects of large displacements, has been described by Jones, Symonds, and others [17–19]. Expressions have been developed for the large permanent plastic displacements of clamped and simply supported beams and plates under uniform impulsive line loads or pressures with rectangular pulse shape [17–19]. A method (applicable strictly only in the case of small displacements) has also been suggested for converting impulses of any shape to equivalent rectangular form [20]. This approach is, however, only applicable where the impulse duration t_i is small relative to the fundamental period of vibration T, where, for a clamped beam, $T = 1/f$, and frequency f is given by

$$f = \frac{11.2}{\pi b^2} \left(\frac{EI}{m} \right)^{1/2} \tag{8}$$

in which b is the beam length, EI is the elastic flexural rigidity, and m is the mass per unit length. The lowest frequency of a long rectangular plate is given by the same expression if the flexural rigidity per unit width $D = Eh^3/[12(1 - \mu^2)]$ is substituted for EI and m is taken as the mass per unit area.

The periods T of plate and stiffener vibration derived from Equation (8) for the experimental structures described above are

(i) for structure A, 4.9 and 0.98 msec respectively
(ii) for structure C, 24 and 7.6 msec respectively

It is evident that the effective duration of blast loading is in each case of the same order as or greater than T, so that the dynamic rigid–plastic solutions of References 18 and 19 are not applicable. In the case of plating response, peak displacements w_{max} ($13h$ in structure A, $2.8h$ in structure B) are larger than given by the rigid–plastic expressions for static application of peak pressures, i.e., $w_{max}/h = \frac{1}{2}p/\bar{p}_c$, where $\bar{p}_c = 4\sigma_0 h^2/b^2$ is the small-displacement plastic collapse pressure for a clamped strip [17]. Peak displacements are, however, large enough, particularly in structure C, for rigid–plastic assumptions to be effective: this suggests that there is scope for development of dynamic rigid–plastic solutions for nonrectangular load profiles with durations intermediate between the impulsive and effectively static cases.

4.6. Overall Superstructure Response

Overall deformation of a superstructure, i.e., sidesway under blast pressure acting on a deckhouse side, is resisted primarily by shear rigidity of decks, transverse bulkheads and deckhouse ends. Large deformations are likely to arise mainly as a result of shear buckling of slender plating, as illustrated in Figure 3(b). This form of response has been examined previously [21] using a lumped-parameter model in which the shear rigidity of decks and bulkheads was based on an assumed bilinear or trilinear shear stress–strain curve, with discontinuities corresponding to elastic buckling and yield. Recent research, including extensive parametric application of nonlinear analysis, has led to a much improved understanding of the behavior of rectangular plates and stiffened panels under shear loads [22]. A set of data curves defining the effective shear stress–strain relationship for rectangular steel plates with average imperfections is shown in Figure 15: these curves, derived from Reference 22, refer to plates of any slenderness β and can be applied with sufficient accuracy to plates with any aspect ratio in the practical range ($1.5 < a/b < 5$).

A simple nonlinear finite-element method, employing the curves of Figure 15, has been developed for the analysis of overall sidesway of superstructures of general geometry. Stiffened panels forming decks, transverse bulkheads and deckhouse ends are represented by rectangular shear elements, as shown in Figure 16, where the width B of the element corresponds to the width of the deckhouse (Figure 1). Element stiffness coefficients k_{ij} relating incremental shear

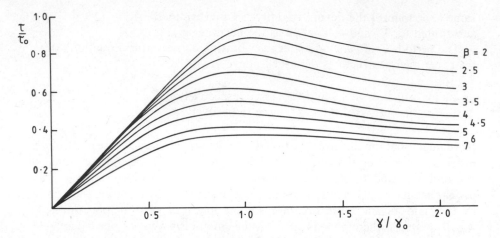

Figure 15. Effective shear stress–strain curves for rectangular plates [22].

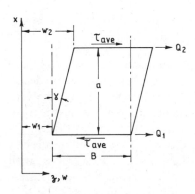

(a) SIMPLIFIED SHEAR DEFORMATION

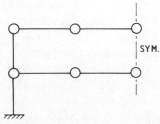

(b) IDEALIZATION OF STRUCTURE 'B'

Figure 16. FE model for analysis of super-structure sidesway.

452

forces ΔQ_i to incremental transverse displacements Δw_i are defined by

$$\begin{bmatrix} \Delta Q_1 \\ \Delta Q_2 \end{bmatrix} = \begin{bmatrix} k_{11} & k_{12} \\ k_{12} & k_{22} \end{bmatrix} \begin{bmatrix} \Delta w_1 \\ \Delta w_2 \end{bmatrix} \tag{9}$$

where $k_{11} = k_{22} = -k_{12} = \phi GA_s/a$, $A_s = Bh$ is the total shear area, G is the shear modulus, and $\phi = d(\tau_{\text{ave}}/\tau_0)/d(\gamma_{\text{ave}}/\gamma_0)$ is the slope of an appropriate stress–strain curve (Figure 15) defining the effective (tangent) rigidity. Shear stress $\tau_{\text{ave}} = Q/A_s$ and hence shear strain $\gamma = (w_2 - w_1)/a$ are assumed to be constant in the element. An assembly of elements representing a symmetrical half of the two-tier superstructure model of Figure 3(b) is shown in Figure 16(b): assuming fixity at the base, the idealization contains only six degrees of freedom (one per node), with masses and blast forces assumed to be concentrated at node points. The resulting equations of motion [Equation (7)] are solved using a Newmark−β procedure.

In order to test some of the assumptions in the foregoing approximate analysis, particularly as regards shear-stress distribution, linear elastic dynamic

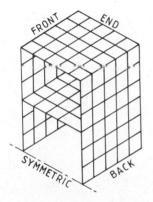

(a) FINITE ELEMENT MODEL

0 5 10 mm

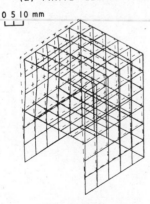

(b) PEAK DEFORMATION

Figure 17. Structure B: linear analysis of overall response.

response analysis was carried out using ADINA for a finite-element idealization of the two-tier superstructure, as shown in Figure 17(a) [23]. Plating was represented by eight-node isoparametric shell elements and stiffeners by three-node beam elements. Loading, assumed to act on the front face only, was as indicated in Figure 2(a). Computed peak elastic displacements are shown in Figure 17(b) and the distributions of peak shear stresses in the deckhouse end and decks (normalized with respect to τ_0) are indicated in Figure 18. Results suggest that the shear-stress distribution assumed in the nonlinear treatment is reasonable. Shear-stress levels indicated by linear analysis evidently exceed τ_0 in the lower part of the deckhouse end. Bending stresses in the front-face stiffeners were also found to exceed yield, in accordance with experimental observations of plastic deformations.

Nonlinear response of the two-tier deckhouse has also been computed by

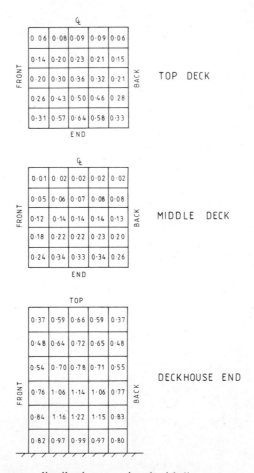

Figure 18. Shear-stress distribution associated with linear response.

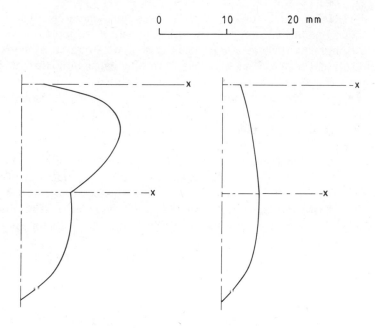

Figure 19. Structure *B*: overall sidesway.

applying the approximate method to the idealization shown in Figure 16(*b*). A net-load condition (front-face pressure minus back-face pressure) was assumed, as indicated in Figure 2(*d*). Computed permanent sidesway at the ends and center of the deckhouse, represented by crosses, is compared in Figure 19 with experimental permanent sets. Numerical and experimental results evidently do not correspond very closely, but are of the same order, suggesting that with some refinement the approximate model can provide a worthwhile assessment of overall response.

5. RESPONSE OF GRP AND HYBRID SUPERSTRUCTURES

The essential difference between response of GRP structures and that of steel or aluminum structures under air-blast loading is the absence of plastic deformation in the former case. Because of low material stiffness, response of a GRP structure is likely to involve large elastic displacements; failure can reasonably be equated with first-ply failure of the laminate or can alternatively occur as a result of failure of connections, either between stiffeners and plating or at the boundaries of stiffened panels. A reasonable design approach is to ensure, on the basis of calculations (including consideration of joint loads derived from air-blast analysis) and tests, that panel strength can be developed fully without premature joint failure.

An all-GRP superstructure might in some circumstances be considered too brittle under air-blast loading, or insufficiently rigid to support displacement-sensitive equipment such as components of weapon or radar systems under normal loading. In such cases, a hybrid form of construction can be adopted, employing GRP cladding bonded and bolted to steel framing. The inherent advantages of GRP (weight savings, effective decoupling of the superstructure from the main hull and effective fire containment) are thus retained and combined with the advantages of steel (high stiffness, ductility and strength retention under fire conditions).

As in the case of metal superstructures, local deformation of plating between stiffeners, local deformation of stiffened panels, and overall sideway of a GRP or hybrid superstructure can be examined using the two-dimensional models previously described. It is again reasonable to assume in the first instance that these modes of response are uncoupled, with evaluation of coupled response as a possible refinement.

A difficulty that arises in evaluating the response of stiffened panels is that of estimating the effective width of GRP plating, which has been less thoroughly investigated than that of steel plating. Some guidance can be obtained from studies of the compressive postbuckling behavior of stiffened GRP panels [24, 25] and from approximate formulas for elastic postbuckling effective widths such as those of Marguerre [26] and Koiter [27]. As imperfections are normally small, the initial effective width of a GRP plate is usually high; when buckling occurs, at a compressive stress that can be estimated conservatively as that of a long orthotropic strip, i.e.,

$$\sigma_{cr} = \frac{2\pi^2}{b^2 h} [H + (D_1 D_2)^{1/2}] \tag{10}$$

where D_1 and D_2, respectively, are longitudinal and transverse flexural rigidities and $H = G_{12}h^3/6 + \mu_{12}D_1$, the effective width drops rapidly to a level that is typically between 40 and 70% of the full plate width. For the purpose of blast-response analysis, the effective breadth of GRP plating acting as a flange to a stiffener can be derived approximately from the stress–strain curves for slender steel plates (Figure 11) or can be taken conservatively as a constant value between $0.4b$ and $0.7b$, estimated from Figure 11 or References 26 and 27, depending on the plate slenderness b/t.

A similar difficulty arises in estimating the effective postbuckling rigidity of stiffened GRP panels under shear associated with overall sideway of a superstructure. Approximate evaluation of ϕ can be made by reference to the curves for steel plates (Figure 15). Restriction of shear stresses to less than the initial buckling level τ_{cr}, which may again be taken as that for a long simply supported orthotropic strip [28], is usually unacceptably conservative. The postbuckling stiffness and ultimate strength of GRP panels under shear are currently under active numerical and experimental investigation as a basis for more accurate treatment of this problem.

6. COMPARISON OF SOME EQUIVALENT SUPERSTRUCTURE PANELS

In order to contrast the behavior of metal, GRP and hybrid structures under blast loading, a comparison has been made between the response of four panels representing alternative designs for a vertically stiffened deckhouse side of height 2.3 m. Section geometries and material properties are summarized in Figure 20. Panels were designed with the usual margins against yield and excessive deformation under a static design pressure of 0.25 bar [6]. Analysis of local plating response and overall stiffener bending under blast loading was carried out for two-dimensional (beam) models, as previously described, using the ARE program FABSTRAN. Incident overpressures of 0.25 and 0.5 bar were considered, giving reflected pressure–time histories of the form shown in Figure 2(a) with parameters as follows:

$$p_i = 0.25 \text{ bar} \qquad p_r = 0.55 \text{ bar} \qquad q_i = 0.022 \text{ bar} \qquad t_s = 0.018 \text{ sec}$$

$$p_i = 0.5 \text{ bar} \qquad p_r = 1.2 \text{ bar} \qquad q_i = 0.083 \text{ bar} \qquad t_s = 0.017 \text{ sec}$$

A positive phase duration t_+ of 2.0 sec was assumed in each case.

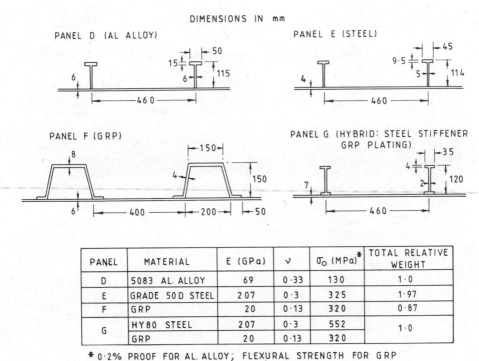

PANEL	MATERIAL	E (GPa)	ν	σ_0 (MPa)*	TOTAL RELATIVE WEIGHT
D	5083 AL. ALLOY	69	0·33	130	1·0
E	GRADE 50D STEEL	207	0·3	325	1·97
F	GRP	20	0·13	320	0·87
G	HY80 STEEL	207	0·3	552	1·0
	GRP	20	0·13	320	

* 0·2% PROOF FOR AL. ALLOY; FLEXURAL STRENGTH FOR GRP

Figure 20. Metal, GRP, and hybrid superstructure panel designs.

In the case of plating response, fully clamped boundary conditions were assumed at stiffener locations for panels D, E, and G, whereas for panel F a continuous-strip model was adopted with simple support at the base of the hat-section stiffener and planes of symmetry at the stiffener centerline and midway between stiffeners. In the case of stiffened-panel response, conditions of simple support with full axial fixity were generally assumed at the top and bottom edges. Some alternative boundary conditions were also examined for panel G. For panels D and E, plating effectiveness was determined using the curves of Figure 11 (it has been shown [29] that the stress–strain curves for steel plates can be applied with reasonable accuracy to aluminum plates if the 0.2% proof stress is substituted for σ_0); fixed effective breadths of $0.6b$ and $0.5b$ were assumed for panels F and G respectively.

Results are summarized in Table 2, which lists maximum plating and stiffener displacements, together, in the case of panels F and G, with peak strains that may be related to a maximum permissible (laminate failure) strain of 0.016. As might be expected, the steel structure (panel E), which is over twice the weight of the GRP structure (panel F), shows the smallest response amplitudes. Panel F performs satisfactorily at the lower incident pressure of 0.25 bar, but under 0.5 bar, outer-fiber strains in the stiffener exceed the material failure level by about 20%. Computed midspan displacements of plating and stiffeners are shown in Figure 21. Deformations of the hybrid panel G are large, but under the

TABLE 2   Comparison of Air-Blast Responses in Aluminum, Steel, GRP, and Hybrid GRP/Steel Panels

		Stiffened Panel Response				Local Plating Response	
	Load p_i	Boundary Conditions		w_{max}	ϵ_{max} (GRP)	w_{max}	ϵ_{max}
Panel	(bar)	Rotational[a]	Axial	(mm)	Stiffener	(mm)	(GRP)
D	0.25	SS	Fixed	60	–	6.5	–
D	0.5	SS	Fixed	126	–	11.6	–
E	0.25	SS	Fixed	16	–	5.8	–
E	0.5	SS	Fixed	73	–	8.5	–
F	0.25	SS	Fixed	44	0.0094	11.7	0.0045
F	0.5	SS	Fixed	81	0.019	15.4	0.0086
G	0.25	SS	Fixed	74	–	11.0	0.0084
G	0.5	SS	Fixed	171	–	15.5	0.016
G	0.25	C	Fixed	10	–	11.0	0.0084
G	0.25	SS	Free	80	–	11.0	0.0084
G	0.5	C	Fixed	44	–	15.5	0.016
G	0.5	SS	Free	580	–	15.5	0.016

[a]SS = simply supported, C = clamped.

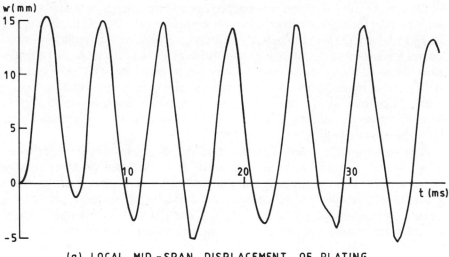

(a) LOCAL MID-SPAN DISPLACEMENT OF PLATING

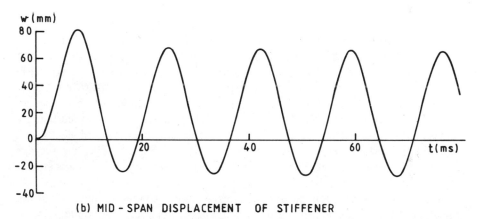

(b) MID-SPAN DISPLACEMENT OF STIFFENER

Figure 21. Elastic responses of GRP panel.

more severe load, strains in the GRP panel only just reach the ultimate value. Results for panel *G* again demonstrate the sensitivity of displacement amplitudes to axial and rotational boundary conditions.

7. CONCLUSIONS

The response to air-blast loading of light stiffened steel or aluminum super-structure panels can be examined by numerical analysis of simple two-dimensional beam or frame models that account for elastoplastic deformation and large displacements. Similar methods can be used to evaluate the response of GRP or hybrid GRP/steel superstructures. In carrying out such analysis, it is important to include the effects of local compressive buckling of slender plating: this can be done approximately by reference to effective (static) stress–strain curves for imperfect rectangular plates. Account should also be taken of strain-rate effects, either by including a variable strain-rate-dependent yield criterion within the incremental-analysis procedure or, more approximately, by adopting a fixed "dynamic" yield stress that for typical blast loading is higher than static yield by a factor of about 1.3 in the case of mild steel or 1.1 in the case of high-strength (HY80) steel.

The response of stiffened superstructure panels can be very sensitive to the conditions of rotational restraint and, in the case of larger displacements, axial restraint at the boundaries. Such boundary conditions are difficult to estimate in a simple-beam model of stiffened panel response: better results are likely to be obtained by adopting a portal frame model, possibly with a condition of elastic restraint at the base representative of supporting deck stiffness.

The assumption that local plating response and stiffened-panel response are uncoupled should usually give adequate results. A more accurate (but much more expensive) numerical treatment, accounting for coupling between plating and stiffeners and for dynamic plate buckling, can be achieved by nonlinear finite-element analysis in which plating is modeled using shell elements.

Nuclear-blast pressure is likely to have a duration of the same order as or longer than the period of the first mode of vibration in typical superstructure panels. Existing expressions for rigid–plastic response of plates and beams under impulse load therefore do not appear to be applicable in most cases.

Overall sidesway of a deckhouse under blast load can be examined using a simple finite-element model in which nonlinear shear stiffness of decks and bulkheads, accounting for buckling and yield, is represented by effective average shear stress–strain curves. Further experimental evaluation and perhaps empirical adjustment of this model are needed.

The behavior of GRP superstructures is likely to be satisfactory under moderate blast loads, exhibiting a damage-free response under conditions in which a metal structure would experience slight permanent deformation. Under more severe blast loads, however, where an equivalent metal superstructure would undergo large plastic deformations but might be able to maintain protection of internal systems, an all-GRP structure is susceptible to catastrophic failure. The most effective form of construction, which retains most of the weight savings offered by an all-GRP superstructure, together with effective decoupling of the superstructure from the main hull, while incorporating the ductility of a steel structure, appears to be use of GRP plating bonded and bolted to welded steel transverse framing.

ACKNOWLEDGMENTS

Acknowledgments are due to the author's colleagues, Messrs. W. Kirkwood and P. Murphy, and to Lyle Morgan of DTNSRDC, Bethesda, for assistance in carrying out computations.

REFERENCES

1. N. M. Newmark, "An Engineering Approach to Blast Resistance Design," *Trans. Am. Soc. Civ. Eng.*, **121**(Paper No. 2786) (1956).

2. L. C. Dye and B. W. Lankford, "A Simplified Method of Designing Ship Structure for Air Blast," *Nav. Eng. J.*, 693, August (1966).

3. J. E. Slater and R. Houlston, "Dynamic Analysis of Beams and Plates Subjected to Air Blast Loading," *Proc. 3rd, Int. Modal Anal. Conf., 1985* (1985).

4. R. Houlston, J. E. Slater, N. Pegg, and C. G. DesRochers, "On Analysis of Structural Response of Ship Panels Subjected to Air Blast Loading," *Comput. Struct.*, **21**(1/2) (1985), p. 273.

5. R. Houlston and C. G. DesRochers, "Nonlinear Structural Response of Ship Panels Subjected to Air Blast Loading," *Proc. 6th, Conf. Nonlinear Anal. ADINA, 1987* (1987).

6. C. S. Smith and D. W. Chalmers, "Design of Ship Superstructures in Fiber-Reinforced Plastic," *Trans. Roy. Inst. Naval Arch.*, **129** (1987).

7. S. Glasstone, *The Effects of Nuclear Weapons*, U.S. Govt. Printing Office, Washington, DC, 1964.

8. L. W. McNaught, *Nuclear Weapons and their Effects*, Brassey's Defence Publishers, London, 1984.

9. *Fortifikasjonshandbok* (Fortification Handbook, Parts 1 and 2), Norwegian Defence Building Service, Oslo, 1973 (in Norwegian).

10. P. Soroushian and K. Choi, "Steel Mechanical Properties at Different Strain Rates," *J. Struct. Eng., Am. Soc. Civ. Eng.*, **113**(4), 663 (1987).

11. G. R. Cowper and P. S. Symonds, Tech. Rep. No. 28, ONR, Contract Nonr-562(10), NR-064-406, Div. Appl. Mech., Brown University, Providence, RI, 1957.

12. R. S. Dow and C. S. Smith, "FABSTRAN: A Computer Program for Frame and Beam Static and Transient Response Analysis (Nonlinear)," unpublished ARE report, 1985.

13. N. M. Newmark, "A Method of Computation for Structural Dynamics," *J. Eng. Mech., Div. Am. Soc. Civ. Eng.*, **85**, 67 (1959).

14. W. D. Hart, unpublished ARE report, 1968.

15. C. S. Smith, P. C. Davidson, J. C. Chapman, and P. J. Dowling, "Strength and Stiffness of Ships' Plating under In-Plane Compression and Tension," *Trans. Roy. Inst. Naval Arch.*, **130**, 1988.

16. ADINA Engineering, *Automatic Dynamic Incremental Nonlinear Analysis: Users Manual*, ADINA Engineering Inc., USA, 1984.

17. N. Jones, "Plastic Behaviour of Ship Structures," *Soc. Nav. Archit. Mar. Eng., Trans.*, **86**, 115 (1976).

18. P. S. Symonds and T. J. Mentel, "Impulsive Loading of Plastic Beams with Axial Constraints," *J. Mech. Phys. Solids*, **6**, 186 (1958).

19. N. Jones, "A Theoretical Study of the Dynamic Plastic Behavior of Beams and Plates with Finite Deflections." *Int. J. Solids Struct.*, **7**, 1007 (1971).

20. C. K. Youngdahl, "Correlation Parameters for Eliminating the Effect of Pulse Shape on Dynamic Plastic Deformation." *J. Appl. Mech.*, **37**(3), 744 (1970).

21. S. B. Kendrick and C. Ng, unpublished ARE report, 1970.

22. J. C. Chapman and P. C. Davidson, "Behaviour and Design of Full Depth Web Panels in Shear," *Struct. Eng.*, **65**(B), No. 4, 65, December (1987).

23. S. L. Morgan and C. S. Smith, unpublished ARE report, 1987.
24. C. S. Smith and R. S. Dow, "Compressive Strength of Longitudinally Stiffened GRP Panels," in I. H. Marshall, Ed., *Composite Structures–3*, Elsevier Applied Science Publishers, London, 1985.
25. C. S. Smith and R. S. Dow, "Interactive Buckling Effects in Stiffened FRP Panels," in I. H. Marshall Ed., *Composite Structures-4*, Elsevier Applied Science Publishers, London, 1987.
26. K. Marguerre, "Die Mittragende Breite der gedruckten Platte," *Luftfarhtforschung*, **14**(3) (1937); translated in *NACA Tech. Note* **833** (1937).
27. W. T. Koiter, *The Effective Width of Flat Plates for Various Longitudinal Edge Conditions at Loads far beyond Buckling Load*, Rep. S287, National Luchtvaart-Laboratorium, Netherlands, 1943.
28. P. S. Bulson, *The Stability of Flat Plates*, Chatto & Windus, London, 1970.
29. A. F. Dier and P. J. Dowling, *Aluminium Plated Structures: Plates under Uniaxial Loading*, CESLIC Rep. APS1, Imperial College, London, 1980.

CHAPTER 14

Catastrophic Failure Modes of Marine Structures

Torgeir Moan
Department of Marine Technology
The Norwegian Institute of Technology
The University of Trondheim
Trondheim, Norway

Jorgen Amdahl
SINTEF
Division of Structural Engineering
Trondheim, Norway

ABSTRACT

Catastrophic failure modes of marine structures include progressive structural failure and, in case of buoyant structures, capsizing and sinking. The probabilistic and mechanical aspects of the structural failure modes are described, followed by a review of analysis and behavior of structural components and systems. Examples of recent analyses of the behavior of damaged marine structures are presented.

1. INTRODUCTION

Typical marine structures are shown in Figure 1. Possible failure modes can be characterized according to cause, type and extent of damage, and consequences in terms of fatalities, pollution, and economic losses. Component-failure modes comprise ductile collapse, fracture due to single overloads, or incremental collapse, and low- or high-cycle fatigue due to time-varying loads such as waves and earthquakes. A component failure can develop into a progressive failure. Overturning or sinking/capsizing failures can also occur by a single overload or progressive flooding. The extent of initial damage can range

463

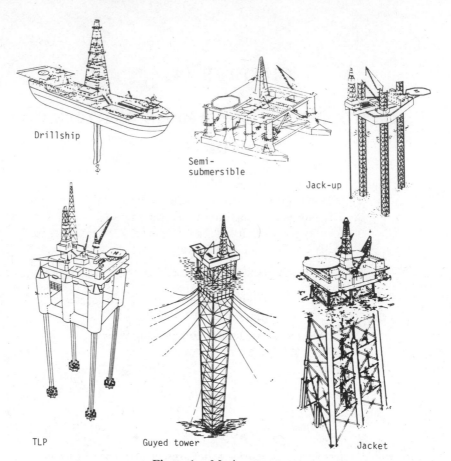

Figure 1. Marine structures.

from that of the order of fabrication tolerances to complete loss of load-carrying capacity. Fatalities induced by structural failures of marine structures are typically caused in connection with global failure modes, especially when the time available for evacuation and rescue is short. Significant pollution or economic losses can be incurred even by more moderate structural damages.

Here the focus is on catastrophic events, i.e., those associated with large final consequences such as total loss of systems strength, overturning, or capsizing. However, as these failure modes can develop progressively from a small initial damage, scenarios with a spectrum of initial conditions have to be assessed.

The failure potential depends to a large extent upon how well the various failure modes are anticipated and prevented by decisions made during design, fabrication, and operation. To illustrate the technical features, as well as the influence of human aspects on failures, a brief account of catastrophic failures is first given, followed by a discussion of the safety philosophy relating to catastrophic failure modes, with an emphasis on design requirements relating to damage tolerance. The implementation of such a philosophy requires information about the probabilistic and mechanical aspects of structural behavior. Therefore, the probabilistic aspects are briefly introduced before the main topic of catastrophic structural failure modes is addressed.

2. CATASTROPHIC FAILURE MODES

In this section, catastrophic failure modes are identified, mainly by reference to experiences. Failures and accidents are continuously reported in e.g., References 1 and 2. Catastrophic failures are moreover normally documented through inquiry reports, as will be shown.

Floating structures positioned by catenary mooring or thrusters, Figure 1, most likely suffer total collapse by sinking/capsizing induced by buoyancy loss or by progressive structural failure. Buoyancy loss can be caused by, e.g., ballast errors, subsea gas blowouts, and structural damage leading to leakage [3, 4].

Experiences with ship and mobile platforms indicate an annual frequency of total loss of 5×10^{-3} to 10^{-2} per vessel year [3-6]. Total loss of ships is typically induced by fire and explosion, grounding and collision, or structural abnormal environmental conditions, and crack-type defects. Normally, the initial event leads to flooding, and thereby capsizing and sinking. The fact that floating structures can fail not only by total structural loss, but also through capsizing or sinking, make them more vulnerable than bottom supported structures.

Notable total ship losses are the grounding of *Amoco Cadiz*, the fire of *Andros Patria*, the explosion of *Berge Vanga*, and the *Betelgeuse* case, as summarized in Reference 6. In the latter event, severe corrosion contributed to an initial structural failure followed by an explosion and the breaking of the ship's aft section.

Ship collisions have caused damages and represented near-miss situations with a catastrophic outcome. The increasing vessel sizes and traffic intensity implies concern about the associated failure potential.

The burning Ixtoc blowout heated a semisubmersible rig to the extent that it was abolished afterwards. A gas cloud released in a blowout on a jack-up in the Matagorda Island area exploded and destroyed the superstructure. In addition, buoyancy losses by nonstructural causes, such as subsea gas blowouts, ballast operation errors, etc., have led to total losses (by capsizing), as summarized in Reference 3.

Abnormal environmental conditions have frequently occurred outside South Africa, where wind-generated waves, swell, and strong currents have created steep high waves and led to permanent deflection of the hull girders [7]. References 8 and 9 report collapse of the hull girder of a container and a bulk carrier due to heavy slamming, respectively.

Brittle fractures of hulls of all-welded ships during the period 1942–1956 focused attention on this most serious failure mode. Up to then, rivet joints between ship plates had provided sufficient discontinuity in the material to stop possible transverse cracks from leading to total rupture and, hence, total casualty. The brittle fractures were initiated by bad workmanship (e.g., causing gross weld defects), maloperation (causing excessive still-water load), poor design details (e.g., square-ended bilge keels, rat holes in bilge keels, and square-cornered hatchways), and the notch sensitivity of the welded material. As brittle fractures also were experienced in other important welded structures, such as storage tanks, pressure vessels, pipelines, and bridges, an extensive research effort [10] to develop and implement material-toughness criteria and materials with adequate properties, etc. has reduced the susceptibility to brittle fracture. Service experience has shown that the special steel grades in thinner sections have been effective in restricting the extent of service failures [11]. However, *Energy Concentration* [10] and *Glomar Java Sea* [12] cases, for instance, demonstrate that concern about this failure mode is necessary, especially in monohulls like ships, where the crack can grow through fatigue and result in fracture without warning. Reduction in the plate cross-sectional area due to corrosion has contributed to some catastrophic accidents in ships [10, 13, 14]. Fatigue was a factor in the case reported in Reference 15.

Framed offshore platforms provide a better strategy for fracture control since fracture normally implies a single member in the first place. However, in the case of the semisubmersible *Alexander L Kielland*, a fatigue failure in one brace led to subsequent progressive failure in the nonredundant brace system supporting one of its columns. The column was lost and the platform capsized, resulting in 123 fatalities, the largest death toll in any offshore accident. Fatigue failures in platforms are reviewed in Reference 16.

Progressive failure of the positioning system for a platform leads to driftoff, and can become catastrophic if the platform collides with another installation, as discussed in Reference 17.

Fixed platforms, such as the jacket in Figure 1, have so far only experienced total losses due to fires and explosions. However, severe damages and near misses due to other hazards such as ship collisions have occurred fairly frequently.

For new concepts, such as the TLP (Figure 1), there is limited experience, and possible catastrophic failure modes must be inferred based on experiences with related concepts and analyses. The TLP is held in position by excessive buoyancy in the hull, causing a pretension in the mooring system. Possible catastrophic failure modes for the TLP include those to floaters, together with failure of the mooring system. Failure of all tethers or the foundation in one corner (one leg) results in large transient motions and also possibly progressive failure of the remaining tethers. Even if the TLP is stable in the freely floating mode, the severe transient motion suggests that the aforementioned failure of a tension leg is considered catastrophic.

3. SAFETY PHILOSOPHY

Failures can be judged from a technical–physical or a decision-making point of view [3–5]. If the decision-making approach is taken, the causes of failure can be categorized as follows:

- unknown phenomena, such as unpredictable freak waves outside South Africa and brittle fractures initiated by new welding procedures introduced in connection with the "Liberty Ships"
- design and fabrication errors causing deficient strength; and operational errors causing accidental loads
- inadequate safety margins in the design criteria to account for the variability and uncertainty associated with loads and strength

The second category is the dominant cause. It should be reflected in quality assurance and control (QA/QC), structural design, and contingency plans for operation. QA/QC is beyond the scope of this chapter. The two other aspects are now discussed.

The main decisions affecting safety are made during design, when layout and scantlings, as well as fabrication, operation, and inspection procedures, are determined. Up to now, design codes have focused on component-failure modes. Minor and moderate damages associated with components contribute to the costs, but systems failure is more important because potential fatalities are primarily associated with total failure [5].

Catastrophes such as those with the *Titanic* and the *Alexander L. Kielland* occurred as a result of a relatively small initial damage. They led progressively to total loss because the systems were not sufficiently damage-tolerant. As this is a typical feature of catastrophic accidents, requirements as to systems robustness are crucial.

Criteria directed toward avoiding progressive failure of a system due to accidental damage have been used in connection with the floating stability of marine structures, in terms of so-called damage-stability criteria. They are suggested in model codes for structures [18] and introduced in codes of offshore structures [19].

The basis for design decisions is predicted behavior. The actual phenomena are inherently uncertain. Consider, for instance, the planning of an offshore facility in relation to the risk of ship collision. The outcome of a subjective assessment of the impact likelihood obviously varies from one individual to another. Hence, a more rational decision process requires the use of probabilistic measures of the likelihood. This aspect is further discussed in the next section.

During fabrication or, especially, operation, the situation for safety management differs from the design situation, in that more information about loads and strength are available. In particular, decisions during operation are often made in connection with intentional change of function or the occurrence of damage. Since these situations are more well-defined than the scenarios envisaged at the design stage, the focus in the following paragraphs is on the design situation.

4. PROBABILISTIC MEASURES

4.1. Introduction

Total failure of a system is clearly the most interesting catastrophic failure model. The probability of systems failure, p_{fSYS} due to natural hazards, H_i (wind, waves, earthquake, ice), and man-made hazards, H_i' (ship collisions, explosions, abnormal strength), can formally be expressed as suggested in Reference 20 as Reference 20 as

$$p_{fSYS} = \sum_i p_{fSYS|H_i} \cdot p_{H_i} + \sum_i \sum_j \sum_k p_{fSYS|D_{ijk}} \cdot p_{D_{ijk}} \qquad (1)$$

where $p_{fSYS|H_i}$ denotes p_{fSYS} conditional upon the event H_i. The first sum accounts for natural events, while man-made hazards, properly combined with natural hazards, are implicit in the damage state D_{ijk} of the second sum. The infrequent occurrence and short duration of the accidental action often justifies considering an extreme accidental hazard H_i' in conjunction with an expected value of an actual set of the environmental condition $\{H_i\}$.

Natural hazards H_i can lead to systems failure through overloads or cumulative damage. The probabilities $p_{fSYS|H_i}$ in the first sum of Equation (1) are determined by structural-reliability analysis. In case of failure by overload, the hazard H_i can refer to a return period of, e.g., 100 years, and p_{H_i}, the corresponding lifetime exceedance probability of the hazard. In case of fatigue, the long-term distribution of wave cycles is used and p_{H_i} is 1.0.

$p_{fSYS|D_{ijk}}$ represents the failure probability of the system given a damage due to hazard H_i' at location j with a conditional probability p_k. This implies that the damage for hazard H_i' is discretized in terms of the location on the platform and of the extent of damage corresponding to the probability p_k.

$p_{fSYS|D_{ijk}}$ is also determined by the structural reliability theory. However, as illustrated in References 20–22, uncertainties beyond those normally inherent in load effects and resistances contribute to the probability content in this case.

The damage relating to structural failure and sinking/capsizing is measured by strength reduction and buoyancy loss, respectively, and is characterized by its type (dent, crack, flooding), extent, and location. For a given hazard H_i' and location of the damage, its extent can be related to the magnitude (IH_{ij}) of the hazard. If, in addition, the relation between exceedance probability $Q_{IH_{ij}}(IH)$ and the magnitude of the hazard is known, a similar relationship between damage exceedance probability and damage can be established. From this relationship, a set of discrete events D_{ijk} with a conditional probability p_k can be determined.

While systems failures due to extreme natural hazards H_i' normally are assumed to occur at the time ($t = 0$) when the hazard acts, systems failures due to a man-made hazard H_i' can take place when H_i' acts, or later, in a time period T when the intensity of the natural hazard is higher than at $t = 0$. The period T itself can be a random variable, and depends upon the likelihood of detecting the actual damage, delay of repair after detection due to weather conditions, etc., and can last for a storm period and up to months or even years for smaller damages. Damage detection depends, for instance, on the nature of the damage, i.e., whether there is evidence of its occurrence such as given by accidental loads, or not; and the damage extent itself. For these reasons, the second term in Equation (1) should be divided into contributions from the time period when the hazard acts, a period with structural damages, possibly a longer period with partial repair of the damage, etc.

4.2. Progressive Limit State (PLS) Criterion

A measure of the sensitivity for systems failure probability with respect to the extent of damage (d) caused by a given type of man-made hazard (i) at a given location (j) was suggested in Reference 20 in terms of the probability of systems failure [$p_{\mathrm{fSYS}_{ij}}(d)$] associated with damage (d) and associated environmental conditions. This probability is expressed by

$$p_{\mathrm{fSYS}_{ij}}(d) = p_{\mathrm{fSYS}|D_{ij}}(d)Q_{D_{ij}}(d) \tag{2}$$

where $p_{\mathrm{fSYS}|D_{ij}}(d)$ is the conditional systems failure probability given $D_{ij}(d)$, and $Q_{D_{ij}}(d)$ is the exceedance probability of damage d. Possible variations of $p_{\mathrm{fSYS}_{ij}}$ with d are shown in Figure 2. It is typical that the largest contribution to systems failure probability comes from moderate damages and not necessarily from scenarios with extensive damage, around $d = d^*$. Here, d^* represents the damage that implies a sure total collapse under the acting permanent loads.

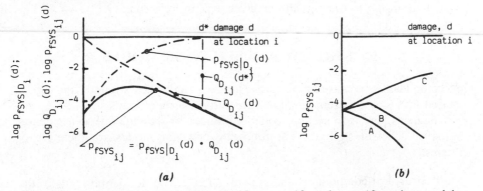

Figure 2. (a) Exceedance probability $Q_{D_{ij}}(d)$, $p_{\mathrm{fSYS}|D_{ij}}(d)$, and $p_{\mathrm{fSYS}_{ij}}(d)$ vs. damage d due to man-made hazard H_i' and an associated set of natural hazards $\{H_i\}$. (b) Typical variation of $p_{\mathrm{fSYS}_{ij}}(d)$ vs. d.

Risk control measures can be assessed with reference to the systems failure probability vs. the damage level d [Figure 2(a)] as follows:

- excessive $p_{fSYS_{ij}}$ relating to $d = 0$ is dealt with by conventional component limit state criteria

- excessive $p_{fSYS_{ij}}$ relating to $0 < d < \bar{d}$, where $\bar{d}$ is a small quantity, is handled by progressive limit state (PLS) criteria to ensure that small damages do not yield disproportionately large consequences

- excessive $p_{fSYS_{ij}}$ relating to intermediate damage levels, $d \geqslant \bar{d}$ are dealt with by combined measures

- excessive $p_{fSYS_{ij}}$ relating to d of the order d^* requires event-control measures, including siting and orientation of the structure, specification of equipment, operational procedures, distance between the hazardous area and the main load-carrying structure, surveillance of ship traffic, etc.

Curve A in Figure 2(b) shows a case of inherent hazard robustness, and no additional safety measures are necessary. Safety improvement can be obtained by PLS requirements in case B. For case C, the most efficient risk-control measure may be event control.

The hazard robustness/fragility check should be accomplished for all hazards and for various damage locations to determine whether PLS criteria are necessary, and, if so, to establish the damage condition and associated natural load conditions and partial factors that yield the target safety level.

The probabilistic approach previously outlined is illustrated by examples relating to jacket, semisubmersible, and TLP under various hazards in References 20–22.

As mentioned, accidental effects are guarded against by event control or direct design against accidental loads. Event-control measures are discussed in References 3, 4, and 23 and are not further treated herein. The remainder of this chapter is devoted to prediction of structural behavior as a basis for direct design according to the regulations given, e.g., in Reference 19.

5.   STRUCTURAL BEHAVIOR MODES

Structural failure is caused by excessive loads or inadequate strength, and can be caused by a single extreme load or cumulative effects of loads.

The focus here is to systems collapse induced by damage due to accidental loads or deficient strength. The following behavior modes are therefore of interest:

(i) structural components and systems subjected to accidental loads of the following type:

- abnormal weight or pressure

- ship collision
- dropped objects
- fire
- explosions

(ii) damaged structural components and systems subjected to functional and natural loads. The damage can be in terms of effects of accidental loads or cracks and other structural defects.

Accidental loads are often assumed to act together with certain functional loads only, i.e., with no environmental loads due to the small correlation between accidental and environmental loads. A damaged structure is subjected to functional and environmental loads.

Abnormal weight and pressure are, e.g., due to leakage, erroneous operation, or malfunction of the ballast system. The determination of response in this case can be easily achieved.

Ship collisions and dropped objects are normally described in terms of kinetic energy. The damage of interest here is perforation, local denting, and overall permanent deflection. The damage is normally determined by assuming that the kinetic impact energy is transformed into strain energy and other types of energy. Reference 24 provides a general introduction to impact analysis. A review of methods for damage assessment of offshore structures is given in Reference 25. Recent work on ship collisions is summarized in References 26 and 27. General numerical methods are reviewed in Chapters 10–11 of Reference 28.

The major fire effects are reduction of yield stress and E-modulus, thermal expansion, and creep. Unprotected steel subjected to significant fire loads attains critical temperatures of a few hundred °C, where its strength is lost, within few minutes. This makes fire load-effects analysis of components often superfluous. The systems effect can then be evaluated by removing from the system the components affected. In practical designs, fire loads are normally considered by requiring sufficient protection to ensure time for safe evacuation. In Reference 29, it is claimed that if important structural members are protected by thermal insulation, the fires are usually extinguished before the temperature rises to 400°C.

In the building sector, there seems to be a trend toward full systems analysis of the structure subjected to thermal loading. The development of FE-based programs are verified by tests of two- and three-dimensional steel frames in References 30 and 31. An overview of structural fire design methods is given in Reference 32. The development of a simplified FE method for systems analysis of offshore structures under fire loading is reported in Reference 33.

Explosion blast waves are characterized by a pressure–time relationship. Two types of explosions are often envisaged: detonations and deflagrations. Detonations can give significant local overpressure and the duration is of the order of milliseconds. Deflagrations have a lower attainable overpressure, but the duration is of the order of a second, and represent the normal explosion type.

The most important factors with respect to structural response are the rise time of the pressure pulse, maximum pressure, and the duration relative to the fundamental period of vibration. The natural period of elastic vibration for the panels of offshore platforms is in the range of 5–30 msec (for the plating) and 0.02–0.5 sec for the stiffened panels. Thus, the major dynamic effect stems from the rise time of the pressure pulse [34].

It is often convenient to study the static response first, applying plastic theory, and then evaluate the dynamic magnification on the basis of a one degree of freedom (d.o.f.) model.

Various methods for design of offshore structures against gas explosions are discussed in Reference 35. A general description of explosion hazards and evaluation is given in Reference 36.

Deficient strength can especially be due to corrosion and wear, and crack-type damage. The corrosion and wear problems can most often be easily controlled by adequate maintenance. The focus here is on cracks, which also can be controlled to a certain extent through inspection and maintenance. Abnormal cracks occur due to fabrication or design faults, and can propagate due to a large number of stress cycles with relatively small amplitude (high-cycle fatigue). In certain circumstances, especially in connection with a sequence of a few wave cycles, i.e., a wave group [7] of large amplitude, they can lead to failure (low-cycle or incremental collapse).

The number of possible accident scenarios is often quite large. A systematic evaluation requires that extensive analyses of component and systems behavior be carried out. The significant nonlinear effects involved make the structural analysis complex. Use of general numerical methods such as FEM can in principle represent all failure modes to a high level of accuracy. On the other hand, the level of sophistication should be balanced against the inherent uncertainty in the description of the accidental loads. Hence, simple response-analysis methods become attractive for design purposes and are emphasized in the following sections.

6.   DUCTILE BEHAVIOR OF STEEL COMPONENTS

6.1.   Introduction

This chapter is devoted to the determination of damage due to impact type loads and residual strength of damaged steel components subjected to single extreme loads. The damage considered is permanent overall deflection and local denting. Both the ultimate and postultimate behavior of components are of interest in a systems-collapse analysis, and are discussed.

Structural behavior on a component level depends on the type of loading and boundary conditions. This is illustrated for beam column in Figure 3. Figure 3(a) shows the behavior of a beam subjected to lateral accidental loads, such as ship

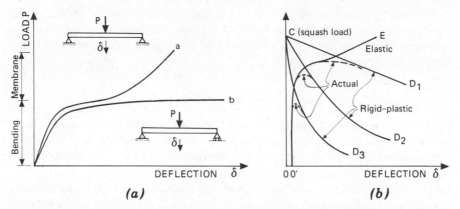

Figure 3. Load-deflection curves for a beam column: (*a*) Lateral load. (*b*) Axial load.

oollioiono, falling objects, and explosions, where the response is a matter of energy dissipation rather than load resistance.

Because these loads demand quite a large energy-dissipation capability compared to the elastic energy available in the elements, inelastic effects dominate the response. Hence, rigid–plastic methods of analysis as outlined, e.g., in Reference 37 are often applicable.

The curves in Figure 3(*b*) describe the behavior under predominantly axial compressive loads representative for members that are not directly subjected to the accidental load or where the load is distributed over multiple structural members (environmental loads and fire loads). The behavior is generally very complex, involving significant elastoplastic and nonlinear geometry effects. The ultimate capacity can often be determined approximately by the simplified technique proposed in Reference 38 and shown in Figure 3(*b*). The bifurcation load is easily determined from the intersection between the elastic and rigid–plastic solution. In the vicinity of the ultimate capacity, the "true" behavior departs from the ideal solution due to elastoplastic effects. However, the actual curve can often be approximated by engineering judgment, as indicated in Reference 38. The postcollapse behavior is often adequately described by a rigid-plastic model.

From these considerations, it is evident that rigid–plastic methods play an important role in the analysis of component behavior. Simple, analytical formulas are often available for perfectly ductile beams and beam columns under idealized conditions. However, real members do not meet these perfect conditions. Sources of imperfections are, for example, degradation of cross-sectional shape caused by denting, ovalization or local buckling or cracks, fracture due to excessive strains that are possibly accelerated by cracks, low temperatures or high strain rates, insufficient capacity of the supporting structure, etc.

The static or quasi-static behavior of various steel components under actual load conditions is reviewed in Sections 6.2–6.4, followed by a treatment of dynamics effects in Section 6.5. Sources of nonductile behavior are further described in Section 7.

6.2. Unstiffened Tubulars

Unstiffened tubulars are widely used in supporting structures, e.g., jackets, jack-ups, and braces of semisubmersibles, as shown in Figure 1. Functional, environmental, as well as impact loads are relevant for these components.

Tubular Beam Columns   There is a significant amount of work relating to beam columns with different load conditions. References 25 and 37 provide reviews of rigid–plastic solutions to laterally loaded beams. The strength of beam-columns with axial loads can be predicted by means of beam-column curves [39, 40]. The postcollapse behavior can be estimated by means of rigid–plastic methods, paying attention to possible local buckling, as will be discussed.

Local Denting Under Lateral Loads   Except for very thick-walled tubes, local denting of the cross section takes place. This reduces the plastic collapse load in bending (but not the capacity in the membrane phase). On the other hand, the denting process contributes also to the energy dissipation. A comparative study of various methods for calculating the denting load is performed in Reference 41.

The local denting process interacts with the global behavior of the beam. An attempt was made in Reference 42 to incorporate the beam forces in an analytical denting model. This problem is very difficult to analyze in itself, and is further complicated by the fact that the beam forces can alternate between compression and tension during impact. The accuracy of simplified methods can, in this respect, only be checked by an integrated nonlinear dynamic analysis, where the impacted element is modeled by shell elements. Such models are applied in studies of pipe-on-pipe impact problems in the reactor industry, as exemplified in Reference 43.

A practical strategy is to assume that the denting process continues until the load reaches the limit-collapse load in bending, whereby the beam mode is initiated, as shown in Figure 4. Dent growth in the postcollapse region is not considered.

For typically sized brace members in jackets, the denting energy is small and can be neglected. Thus, for significant energy dissipation to take place, it has to be ascertained that the deformations can proceed well into the membrane mode. For impacts on jacket legs and braces in semisubmersibles the denting energy can be significant, especially for legs with a ground insert pile. In Reference 44, it is proposed to add the energy dissipations from the leg and the insert pile.

Local Buckling/Ovalization   Depending on the wall slenderness, tubular beams can undergo a limited amount of rotation before local buckling is

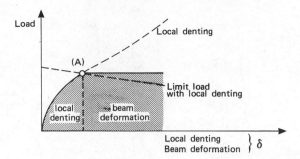

Figure 4. Simplified procedure for calculating plastic beam deformation, including local denting [44].

initiated on the compression side. This reduces the load-carrying capacity in bending while the membrane behavior, in principle, is not influenced. This effect is particularly important in connection with semisubmersible braces, which often have a very high slenderness (diameter/thickness $\sim 100-120$).

The limits to plastic rotation of tubes have been subjected to extensive studies, in, e.g., Reference 45. This work constitutes the basis for the code requirements to allowable wall slenderness for "plastic compactness" to be assumed in the analysis of plastic mechanisms for ultimate limit state check. These studies refer to cylindrical tubulars. Further investigations are required on the local buckling adjacent to the tubular joints.

The load-carrying capacity is not exhausted by initiation of local buckling. The postbuckling behavior is accordingly of interest in a overall collapse analysis. A semiempirical model is given in References 46 and 47, and a theoretical model considering buckle growth is presented in Reference 48.

Tube with Dents and Permanent Lateral Deflection Subjected to Combined Loads Extensive investigations have been devoted to the effect of damage (in the form of local dent and permanent lateral deflection) on the behavior of tubes in compression and bending [46, 47, 49, 50]. Figure 5 shows the model used in Reference 46. It accounts for premature local buckling and includes an empirical relationship for the growth of dent/buckle in the postultimate regime. The importance of including such effects is indicated in the sample load-end shortening curve in Figure 6.

The computer program DENTA has been validated against 107 tests covering a wide range of diameter/thickness ratios, damage types (dent depth and overall deflection), boundary conditions (flexurally simply supported and fixed), and load combinations (axial force/bending moment). The test results are compared with predictions by DENTA in Figure 6(b). A very small bias and coefficient of variation are demonstrated.

Tubular Joint Behavior Tubular joints contribute to the stiffness and strength of the system, and hence affect the energy-absorption capacity. Structural design

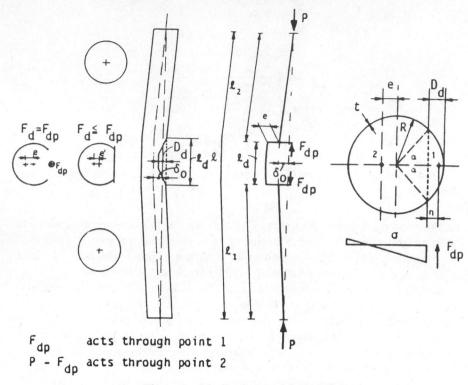

F_{dp} acts through point 1

$P - F_{dp}$ acts through point 2

Figure 5. Idealized geometrical model.

codes provide much information about unstiffened tubular joint strength. However, these formulas apply to ultimate strength design. They should be used with care in evaluation of energy absorption because the failure criterion adopted in the underlying experiments varies considerably, e.g., crack initiation, ultimate strength, and defined excessive deformation. Moreover, the tests have often been carried out for a single load component. Recent reviews of tubular joint behavior are presented in References 51 and 52.

Recently, results have become available for interaction between in-plane/out-of-plane bending and axial force [53, 54]. As a practical check, it is proposed to verify that the interaction surface for the joint circumscribes the interaction surface for the brace member, as illustrated in Figure 7. In this way, adequate capacity is obtained in the bending and membrane modes as well as in the combined modes.

6.3. Stiffened Cylindrical Shells

Stiffened shells typically constitute columns or large legs, with a diameter exceeding, say, $3m$, in floating and fixed platforms.

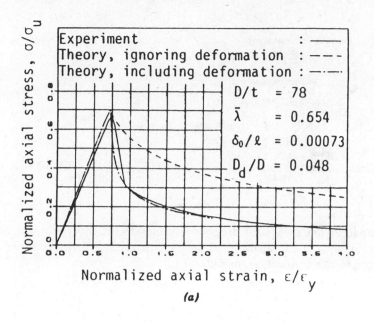

(a)

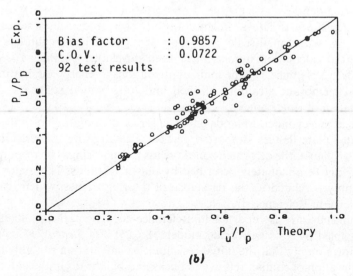

(b)

Figure 6. Behavior of dented tubes: (*a*) Load-end shortening curve. (*b*) Test results compared to theoretical predictions.

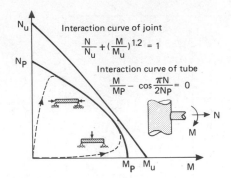

Figure 7. Plastic capacity of a brace vs. ultimate strength of a joint.

Impacts from ships and dropped objects are the most relevant accidental loads for cylindrical shells in offshore structures. Two aspects are of main concern in this connection, namely,

(i) rupture of the shell followed by flooding of buoyant volumes
(ii) local and global deformation of the platform during impact and its strength in damaged condition

Shell rupture can either be due to local penetration of a sharp object or by excessive straining in an overall deformation mode. Theoretical calculations of both types of rupture are subject to considerable uncertainty, because these modes depend on local properties of strength and impact geometry. An assumption often applied is that local penetration causes puncture to the indentation depth.

The resistance to lateral indentation has been studied in a series of quasi-static tests with small-scale ring- and stringer-stiffened cylinders [55–57]. Theoretical calculation models have been developed on the basis of plastic methods in Reference 57. The main contributions to energy dissipation are due to axial membrane straining of shell and longitudinal stiffeners and radial deformation of ring frames.

Hence, the cylinder is modeled as a series of longitudinal stiffeners with associated plate flanges supported vertically by the ring frames. Once a ring starts to collapse, the deformed zone spreads to a new ring. The response of each stiffener can be adequately described by plastic methods. The major source of uncertainty is related to the resistance of the ring frames, which is studied in more detail in Reference 57.

A similar study is reported in Reference 58. Theoretical calculations are accompanied by tests with scale models (1:6.25) of a *Zedco 711* column. The tests show notably that the initial resistance to indentation is highly dependent on the location of contact relative to decks and bulkheads in the column, which have been ignored so far, as they increase the complexity of the calculation models considerably.

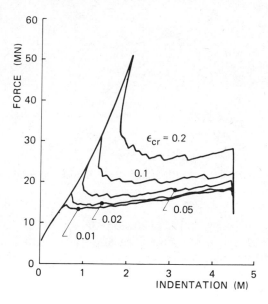

Figure 8. Load-lateral indentation relationships for an Aker H3 column.

The theoretical and experimental results indicate that collision energies in the order of 14 MJ (which is a typical design energy used) produce lateral indentation of approximately 0.4–1.2 m, depending on the fraction of collision energy to be absorbed by the column. This damage level is normally much too small to place the structural integrity of the column in jeopardy.

In a recent investigation [59], the response up to ultimate collapse is assessed for the column of an *Aker H3-2* platform. The principles of Reference 57 are adopted in a somewhat simplified model. Material ductility is considered. A nominal strain is calculated for each longitudinal plate/stiffener element and compared with a fracture criterion.

Load-indentation relationships for the column are shown in Figure 8. It is observed that the maximum load, which is decisive for the amount of energy to be dissipated by the ship bow, is very dependent on the critical strain in the shell plating. For large indentations, the column energy is less influenced by the critical strain, because the response of the column in the postultimate regime is largely governed by the ring frames in direct contact with ship bow. The energy dissipated by the column at an indentation of one radius amounts to approximately 75 MN.

Residual Strength   The tests of dented cylinders reported in Reference 55 also comprised an investigation of the residual strength. For the parameter range studied, it is concluded that the strength of the damaged cylinder can be assessed approximately by assuming the damaged part of the cross section ineffective using the critical stress of an undamaged cylinder as the failure criterion. Further

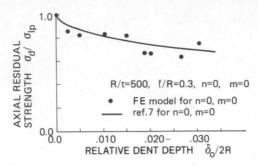

Figure 9. Axial compressive strength of damaged cylinders. Reproduced from Reference 55.

tests are needed to verify this conclusion on a general basis. For slender cylinders (diameter/thickness > 500), an additional reduction of 10% of the residual strength is recommended. Figure 9 shows the degradation of strength vs. dent depth.

6.4. Plated Structures

Stiffened plates are typically found in ship hulls, platform decks, and in pontoons of floating platforms. Lateral accidental loads to deck structures are due to explosions and dropped objects. For pontoons, they are caused by ship collisions and dropped objects.

The collision-load effects can be analyzed by means of a ship-to-ship collision analogy, due to the similarity between a pontoon and a ship hull girder.

Dropped objects seldom have the potential of directly impairing the global structural integrity of platforms. However, they can puncture buoyancy members and lead to stability loss. Furthermore, their consideration is important for the protection of personnel and equipment.

Test experience shows that in the context of dropped objects, it is useful to use static plastic displacement models combined with momentum and energy considerations.

Panels Subjected to Static Lateral Loads   The extensive work on panels with various boundary conditions and under lateral loads is reviewed in References 26, 27, 37, 60, and 61. The plastic large-deflection behavior of rectangular plates subjected to uniform lateral load is conveniently assessed by assuming a rooftop yield-line mechanism as discussed in Reference 37. This can be used for explosion loading. For dropped objects, the corresponding solution in Reference 62 to patch-type loading is more relevant.

Panels Subjected to Axial Loads   Experimental and numerical analyses of plates with different geometries and under various load conditions are summarized in Reference 60. Most of the work concerns plates with deflections of the order of fabrication tolerances. References 63 and 64 consider panels with damage.

If the stiffener as well as the plate are subjected to permanent deformations, the plate–stiffener is conveniently modeled as a beam column. The ultimate strength can be determined on the basis of a first yield criterion or from the bifurcation point between the elastic and rigid–plastic solution.

For unsymmetric profiles, such as plate–stiffener sections, collapse does not necessarily take place at the first yield hinge. This reserve strength is due to the difference between the plastic interaction curves for stiffener-induced and plate-induced failure.

Plate Girders For plate girders, the effect of permanent deformations in stiffened and unstiffened flanges is often calculated by means of the methods described for plated members. The monograph by Dubas and Gehri [60] gives a comprehensive review of methods for determining the ultimate strength of as-fabricated plate and box girders under different load conditions. Virtually no information is available on the behavior of girders with excessive permanent deflections.

It is anticipated that the strength of tension flanges in a static analysis is not influenced by initial imperfections, including that due to impacts. However, the load effect might as well be compressive due to the cyclic nature of the wave forces.

For transversely stiffened webs, the ultimate load is governed by the ability of the tension field to develop in the web plating. Conceivably, this is not impaired for moderate interframe plate deformations, provided that the plate flanges remain undamaged. (Indeed, the elastic stiffness is reduced.) Hence, the residual strength can be assumed equal to the ultimate strength of the undamaged cross section.

Longitudinal stiffeners tend to increase the ultimate strength of the web. However, beyond ultimate load, there is a rapid decrease to a postfailure plateau corresponding to the strength with no longitudinal stiffeners because the tension field develops across the stiffeners. Hence, if the web distortions include the longitudinal stiffeners, the residual strength is virtually equal to the ultimate strength of a transversely stiffened girder.

If also the transverse stiffeners and/or the plate flanges have suffered plastic deformations, the residual strength of the girder is significantly reduced. However, little information is available for quantitative assessment of the detrimental effect of such damages.

Local Buckling The previous considerations are based on ideal plastic behavior. For thin-walled sections undergoing large plastic rotations, local buckling can become important. Various codes specify limitations on the slenderness of the web and flanges so as to obtain "plastic properties." However, these requirements aim primarily at ensuring sufficient rotation capacity to maintain the moment of resistance during the deformations necessary to develop the collapse mechanism. In analyses of overall collapse, it is often necessary to consider finite deformations of the fully developed mechanism, in which cases information about the "true" rotational capacity as well as the capacity during local buckling is desired.

This has been studied in, e.g., References 65 and 66. The deformation pattern caused by local buckling can conveniently be modeled by means of yield lines. It is found in Reference 65 that the different mechanisms observed in tests can be considered to consist of an assembly of a few basic mechanisms. The total response can be found by summing over all the basic mechanisms.

Theoretically, local buckling does not affect the axial tension capacity. Thus, even if the load-carrying capacity is reduced by local buckling in the bending phase, it is regained in the membrane phase for members with axial restraint.

6.5. Dynamic Effects

Inertia Effects   The dynamic plastic behavior of beams and plates exposed to impulse loads has been subjected to extensive experimental and theoretical studies, as reviewed in, e.g., Reference 67. The results are often given in terms of the initial impulse velocity in the plate. This can be determined from the pressure impulse by means of the momentum principle, assuming a dynamic deformation mode equal to the static pattern, or by the mode-approximation technique.

The dynamic magnification factor can also be obtained by studying the response of a linear single d.o.f. system, as shown in Figure 10(*a*). For plates and beams undergoing large displacements, the plastic load-displacement relationship is linear or asymptotically linear. Thus, the "effective" natural period of the system should be based on the plastic membrane stiffness of the system.

In the case of little axial restraint, the elastic–plastic resistance can be described approximately by a bilinear function. Response charts are available for a single d.o.f. system for triangular and rectangular impulses with zero and finite rise time, as shown in Figure 10(*b*).

Strain-Rate Effects   For strain-rate-sensitive materials, rigid–plastic theory overestimates the permanent deflections. The Cowper–Symmonds equation discussed, e.g., in Reference 61 is often used to estimate the dynamic increase of yield stress. It should be noted that it is strictly valid for the lower yield stress only. It is also dependent on the actual steel material composition.

For the large-displacement range, various formulas with a reduced strain-rate effect have been proposed.

Perforation of Plates   The perforation of unstiffened and stiffened plates by dropped objects is studied in References 62 and 68–70. On the basis of tests and the yield-line model in Reference 62, a design criterion is suggested in Reference 68. Figure 11 shows results of recent experiments [71]. The required thickness to avoid perforation is plotted vs. the kinetic energy. Excellent agreement is

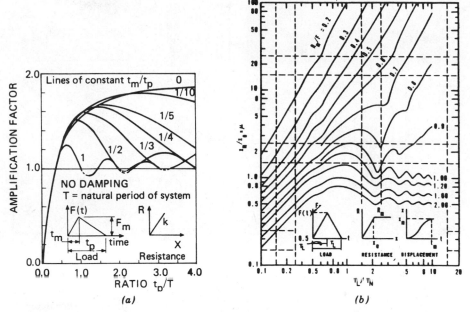

Figure 10. Response charts for dynamic single d.o.f. system [34]. (*a*) Dynamic magnification for linear resistance (plastic membrane response). (*b*) Maximum deflection for elastic–plastic bending response.

reported between test response histories and FE predictions with DYNA3D [71]. A simple design procedure is also developed. The tests show also that the dynamic deformation mode in the initial stages of impact differs significantly from the static solution. This causes a larger energy loss than that resulting from the mode-approximation technique. The principles of conservation of momentum and energy can, however, be used provided that an increased effective plate mass is assumed.

7.   NONDUCTILE BEHAVIOR OF STEEL COMPONENTS

7.1.   Introduction

Structural systems can fail locally due to fracture associated with a single extreme load or crack propagation. A single crack can lead to failure of individual members in trussworks or frames, and in overall failure of steel-plated structures such as ship hulls.

The possible fatigue mechanisms in marine structures are low-cycle and high-cycle phenomena. Both mechanisms can be experienced in intact ships, in which low-cycle effects are due to still-water loads. In offshore platforms, the last mechanism is normally experienced by intact structures.

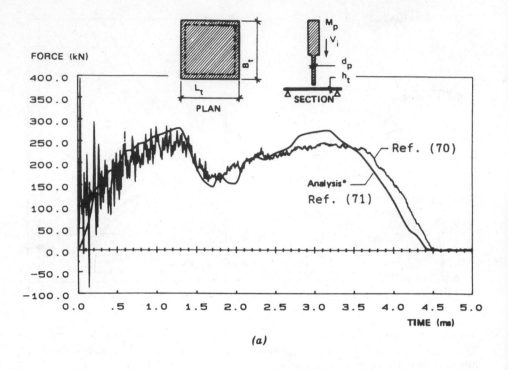

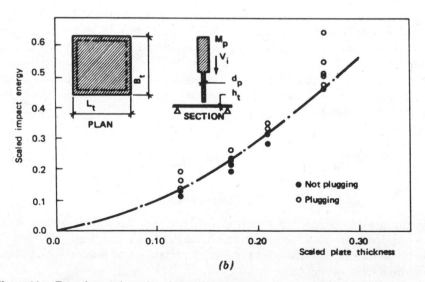

Figure 11. Experimental results of pipe dropped directly on a plate (*a*) Interface force vs. time. (*b*) Scaled impact energy vs. scaled plate thickness.

The slow (high-cycle) propagation of cracks facilitates the control of the crack before it results in a fracture. However, this is contingent on sufficient time between the instance when the crack is large enough to be reliably detected and final fracture. Inspection methods and their reliability are reviewed in, e.g., References 72–74.

In the following sections, fracture criteria and high- and low-cycle fatigue analyses are discussed.

7.2. Fracture Criteria

Introduction   Two types of criteria are required in the assessment of structural integrity:

(i) Critical strain for a member developing plastic mechanisms under large strains in the presence of fabrication tolerances. Particularly large strains are encountered in substructures subjected to impact loads (ductility criteria).

(ii) critical crack size for stable-crack propagation under moderate nominal stress levels (strength criteria).

Four flaw-related failure mechanisms can be considered in structural applications:

- rapid nonductile crack extension (cleavage)
- unstable ductile tearing prior to the limit load (under load-control conditions)
- plastic collapse, when the remaining ligament of the cracked section becomes fully plastic (limit load)
- unstable ductile tearing subsequent to the limit load (under displacement-control conditions)

The actual failure mechanism is dictated by the material toughness, strength properties, flaw size and shape, and stress distribution in the component. Toughness depends primarily on temperature, possible triaxility of stress, plate thickness, and strain rate [75].

The first category of fracture by cleavage (e.g., in ferritic steel) occurs in engineering structures from preexisting notches or cracks of a few millimeters, and at a nominal stress level less than yield, and is normally referred to as *brittle fracture*. The second type of unstable ductile crack extension occurs in even the most tough materials in the presence of sufficiently large combinations of crack size and applied stress. The third failure mechanism, plastic collapse, is particularly easy to handle and forms the basis for the systems analysis presented. The fourth mechanism resembles the second.

The principal fracture-control philosophy presently applied for marine structures is based on selecting steels of high toughness and having proper consumable and welding procedures in critical areas [76, 77]. In welded ships, where a transverse crack can cause total casualty, tough steel plates are used as crack arrestors in selected locations. For tankers, special grade D and E steels are specified for keel, sheer and bilge strake, and for strakes in the deck, bottom, and longitudinal bulkheads, and is also dependent upon plate thickness. Charpy–V criterion is the principal (dynamic) notch-sensitivity criterion applied in codes. The required average Charpy–V values are 20 and 27 J for the direction transverse and longitudinal to the rolling, respectively, obtained at a test temperature of $-20°$ and $-40°$C for grade special D and E steel, respectively. Similar criteria apply to offshore platforms for primary and special quality steels with thicknesses in the range of 25 to 50 mm. The required energy absorption capacity for high-strength steels and larger plate thicknesses is higher. For smaller plate thicknesses, the toughness requirements are relaxed. For instance, some smaller ships are built entirely of grade A steel, with unspecified toughness properties.

The reference strain for the ductility criterion in prior case (i) is the relative elongation at rupture as determined in materials testing of specimens. In laboratory tests of plated structures with thicknesses of up to a few millimeters, this limit of 20–30% has often been attained [78]. This criterion can be relevant for the base material in actual structures, but not for material adjacent to welds. In this connection, criteria based on fracture mechanics, which will be discussed, can be applied.

In the strength criterion of case (ii), the net section yield can conveniently be taken as the reference value. A simple Charpy–V criterion for the necessary ductility to attain the plastic capacity of a cracked member is given in Reference 79 based on well-documented data for test specimens, and pipes with thicknesses and diameters in the range of 2 to 150 mm and 50 to 1100 mm, respectively, and with bending and pressure loads. Pipelines and pressure vessels, as well as conventional steel materials, were included. A Charpy–V value of 45 J was found to be sufficient to develop the plastic capacity determined by a lower-bound method and the yield strength in simple tension rather than the flow stress, i.e., by neglecting strain hardening effects. A C_V-value of 20 J at $-40°$C for a mild steel is normally sufficient to exceed 45 J at $-20°$C.

The criterion in Reference 79 does not give any direct information about the true rotational capacity in a cracked pipe subjected to bending. For a more detailed assessment of the capacity of cracked members, criteria such as CTOD, J-integral/tearing modulus, or the CEGB/R6 failure-assessment diagram, which will be discussed, should be used. A general background of those methods can be found in Reference 75.

Crack Tip Opening Displacement (CTOD)   In the semiempirical approach that is used in PD 6493 [80], the crack tip opening displacement (CTOD) criterion is expressed as

$$\delta_c/2\pi\epsilon_y a_{max} = \begin{cases} (\epsilon/\epsilon_y)^2 & \epsilon/\epsilon_y < 0.5 \\ \epsilon/\epsilon_y - 0.25 & \epsilon/\epsilon_y \geqslant 0.5 \end{cases} \tag{3}$$

where δ_c is the critical CTOD, which depends on the temperature, plate thickness, and strain rate; a_{max} is the tolerable crack size of a through-thickness crack, ϵ is the actual total strain including the effect of residual (welding) strains, and ϵ_y is the yield strain. The criterion of Equation (3) is based on an upper bound fit of the right-hand side to experimental data. A procedure for dealing with surface and internal cracks is also given in the document PD 6463 [80].

The accuracy of the fit varies, so that the inherent safety margin in this approach varies from 1–2 to 20. The fact that residual strains are additive to applied strains makes the approach conservative in a number of situations.

It is also observed that the CTOD criterion is established mainly for data with ϵ/ϵ_y up to about 4–5. Extrapolation beyond this range is uncertain.

The CTOD criterion is used in Reference 81 to estimate a mean rupture criterion for as-fabricated plates subjected to large strains, and the validity of the approach is discussed. In particular, the scale effect of rupture criteria between model tests and the in-service situation is indicated. Data on crack sizes in welded joints are crucial for the application of this criterion, recognizing also possible abnormal fabrication defects as well as the effect of fatigue-crack propagation. Application of the CTOD criterion to estimate the strength of tubular members with fairly large through-thickness cracks with a δ_c corresponding to a Charpy-V value around the mentioned limit value of 15 J indicates that the full plastic capacity will not be reached. A project [83] is currently underway to further clarify the capacity of cracked large-scale members.

J-Integral Methods   An alternative to the CTOD criterion is the tearing-modulus criterion. Unstable crack extension occurs if

$$T_{app} > T_{mat} \tag{4a}$$

where T_{app} and T_{mat} are the applied and material (resistance) tearing modulus, respectively. The tearing modulus is defined by

$$T = \frac{dJ}{da}\frac{E}{\sigma_0^2} \tag{4b}$$

where J is the J-integral, σ_0 is the flow stress, and E is the elasticity modulus.

This approach was first described as a part of an NRC Pipe Crack Study Group effort [84]. Zahoor and Kanninen [85] derived J-integral values for pipes with circumferential cracks. Shibata et al. [86] compared the tearing instability and net-section collapse criteria with pipe test results and demonstrated good correlation. Further data are, however, required.

Failure Assessment Diagram Still another elastoplastic fracture criterion can be obtained by using an interaction formula for brittle fracture, characterized by the critical stress-intensity factor K_{IC} and the plastic capacity of the net cross section. The interaction formula is [87].

$$K_r^2 = \frac{8}{\pi^2 S_r^2} \ln \sec \left(\frac{\pi}{2} S_r \right) \tag{5a}$$

where

$$K_r = K_I/K_{IC} \quad \text{and} \quad S_r = \sigma/\sigma_1 \tag{5b-c}$$

where K_I and K_{IC} are the actual and critical mode-I stress intensity factors, respectively, and σ and σ_1 are the applied and stress at collapse, respectively. A modification of the failure assessment diagram was suggested in Reference 88 based on the J-integral method, and needs further justification.

The failure assessment diagram is convenient to use in connection with estimating the residual strength in crack propagation calculations, because crack propagation calculations are based on the stress intensity factor. This is demonstrated in References 88 and 89.

7.3. High-Cycle Fatigue

Traditional prediction of high-cycle fatigue life is based on information about the long-term distribution of stress ranges in hot spots due to waves; experimentally determined S/N data for welded joints under constant stress range; and the assumption of the Miner–Palmgren hypothesis [11, 90, 91]. The fatigue life, therefore, refers to the failure condition inherent in the S/N data, that is, the time from when the crack is of the size of a fabrication flaw, say, of the order 0.5 mm deep, until final fracture, when it is usually a through-thickness crack. For this reason, the S/N approach cannot provide any information about the time when the crack goes from an abnormal size to final fracture. Neither can the S/N approach provide estimates of the possible crack-growth phase after the crack has penetrated the plate thickness. The fracture-mechanics approach, however, provides a tool for a more accurate assessment of crack growth.

In practice, crack-growth analysis is based on the Paris–Erdogan equation [92] of a one-dimensional model; this equation is:

$$da/dN = C \cdot (\Delta K)^m \tag{6}$$

where da is the crack increment in dN cycles, C and m are material constants, and ΔK is the stress-intensity-factor range, which can be written as

$$\Delta K = \Delta\sigma(\pi a)^{1/2} g(a) \tag{7}$$

in which $\Delta\sigma$ is the nominal stress range, and $g(a)$ is a function of the crack length.

Based on the long-term distribution of $\Delta\sigma$, an equivalent constant stress range can be determined, as shown, e.g., in Reference 91. The main task becomes the calculation of $g(a)$, as illustrated, e.g., in Reference 92. The determination of $g(a)$ is fairly well developed for steel-plated structures [92]. The analysis for unstiffened tubular joints is less advanced, as the behavior is more complex. Some progress is reported in Reference 93.

The relation between the number of cycles N required from a crack length a_i to $\bar{a}$ can then be obtained from

$$N = \int_{a_i}^{\bar{a}} \frac{da}{C(\Delta K)^m} \tag{8}$$

The number of cycles N can be related to real time T by invoking the average period $\bar{T}$ of a cycle.

In using Eq. (8) for determining the time to final fracture, e.g., $\bar{a} = a_f$, the critical crack length a_f has to be estimated as previously discussed.

The fracture mechanics calculations have been compared with experimental results of large-scale structures [94] and used to trace in-service failures [95, 96] and to support decisions regarding design criteria/inspection planning [89, 97, 98].

The example presented in Reference 95 shows the importance of adequate design/inspection criteria in relation to fatigue.

Figure 12 displays results presented in Reference 89 regarding possible crack

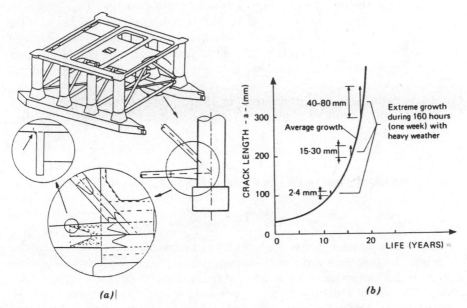

(a)| *(b)*

Figure 12. Fracture mechanics analysis of crack growth in a brace of a semisubmersible rig: (a) geometry. (b) Crack growth of through-thickness crack.

growth in a brace of a semisubmersible platform. It is shown that the crack can spend a significant period of time as a through-thickness crack without fracture of the member. This fact demonstrates adequate time is available for detecting the crack by the leak-before-break principle and for repairing the crack in this case.

7.4. Low-Cycle Fatigue and Incremental Collapse

Intact bottom-supported platforms can be subjected to large cyclic strains during earthquakes. Ultimate load design criteria and the nature of wave loads normally effectively limit the cyclic strains due to waves. However, in damaged platforms, cyclic plastic behavior can be relevant. This is because large single overloads can be carried by redundant structures by redistribution of loads by elastoplastic mechanisms, implying large strains.

Fatigue of structures subjected to high tensile strains causing failure in less than 10,000 cycles is briefly reviewed in Chapter 12 of Reference 90, where other references, especially the fundamental works of Orowan, Manson, and Coffin, are reviewed.

Test results for plates and tubes subjected to large cyclic compressive strains and out-of-plane displacements are presented in References 99–101.

There is a significant difference in the behavior of members with an as-fabricated geometry and members with dents or cracks. This is, for instance, illustrated by tests of tubular members under cyclic loading in Reference 100. The behavior is characterized by the maximum compressive strain $\epsilon_{c,max}$ in the first cycle and the total strain range. The test results show that if $\epsilon_{c,max}$ exceeds the critical local buckling strain $\epsilon_{cr} \cong 0.25t/R$ (where R is the radius, and t is the thickness of the tube), ripples occur in the first cycles and develop into a dent and failure before the low-cycle fatigue life under tensile strains can be attained. If $\epsilon_{c,max}$ is less than ϵ_{cr}, low-cycle failure occurs according to a modified Manson–Coffin law.

Clearly, the strain concentration in dented tubes creates large strains, especially at the wrinkles of the dent for loads that cause changes in the dent depth.

8. ANALYSIS AND BEHAVIOR OF STRUCTURAL SYSTEMS

8.1. Introduction

Analyses to determine the strength of intact or damaged systems can be carried out in a very simplified way. Initial damage is determined by a local analysis of a single or few components by means of plastic methods or more sophisticated tools. The global integrity check is performed with a linear analysis and the criterion that systems collapse occurs when the first member collapses. Components with local damage are omitted from the model. However, the effect of accidental and other loads acting on the damaged region is represented by forces at the boundary of the damaged region.

Generally, this approach is conservative, as the capability of redistributing forces in redundant systems is accounted for to a very limited extent. For this purpose, methods that account for geometric and material nonlinearities associated with load redistribution should be applied.

General finite element codes have been developed to cope with nonlinear effects [71, 102, 103]. In principle, such analyses yield the most accurate predictions. To achieve a reasonable accuracy, several finite elements must be used per structural element. To capture local failure modes, very refined meshes have to be used. In addition, numerical integration must be carried out over the cross section as well as along the length of the member.

These recognitions have inspired the development of various simplified procedures, based on the following alternative element formulations:

- phenomenological models, based mainly on experimental data
- simple finite element models, based on assumed displacements, stresses, or a combination of the two
- hybrid approaches

So far, the development has been devoted mostly to systems with static behavior. In the following sections, emphasis is placed on such simplified methods.

8.2.   Frame and Trusswork Platforms

Introduction   Methods for nonlinear analysis of trussworks and frames are reviewed in Reference 104. Finite element methods are also applied in systematic studies of the ultimate behavior of trussworks.

Phenomenological brace models based on a given relationship between axial force and displacement for a truss element in earthquake analyses of jackets are presented and applied in References 105 and 106. The idealized structural unit method (ISUM) [107] was extended in Reference 108 to analyze tubular frame structures. This method is later refined in Reference 109. The practical study in Reference 110, applying the INTRA system [105], also provides valuable information.

Other methods are based upon explicit derivation of elastic large-deflection stiffness matrices and the yield-hinge collapse mechanism. They have been applied to offshore structures [33, 111–114] as well as building structures [115, 116].

A simplified model of the elastoplastic behavior of unstiffened tubular joints is presented in Reference 52.

The Computer Program USFOS   The basic idea in USFOS is to use one model element per physical element in the structure, as shown in Figure 13. Nonlinear elastic displacement characteristics are established on the basis of the exact solution of the equation of a beam subjected to axial force and lateral bending. In this way, closed-form solutions are obtained for all stiffness terms. Numerical integration is no longer required neither over the cross section nor along the element length. Material nonlinearities are modeled by means of plastic hinges, where all plastic displacements are concentrated, whereas the beam remains elastic between hinges.

The corresponding element stiffness matrices are modified according to plastic flow theory.

Automatic incrementation of loads is performed, including load step scaling due to occurrence of new plastic hinges or global instability. Possible unloading of plastic hinges into the elastic region is checked at each step. Equilibrium iterations can be performed, but excellent accuracy is often obtained with pure incrementation. An important feature is that the method allows for the same finite element discretization as used in elastic analysis.

Plastic interaction surfaces are incorporated, e.g., for rectangular, circular and I-shape cross sections. Other cross-sectional shapes can be easily implemented provided that the plastic interaction surface can be formulated.

Several important physical phenomena have been incorporated in the model, such as

- initiation of local buckling due to plastic rotation and growth of buckle in the postultimate regime
- behavior of dented tubes analogous to the model adopted in DENTA described earlier
- elastic shear failure in deep girder web plates

The major advantage of the plastic hinge concept is that all these effects can be

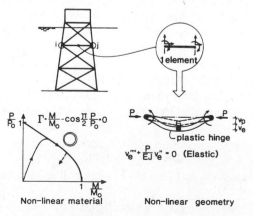

Figure 13.   Basic concepts of USFOS.

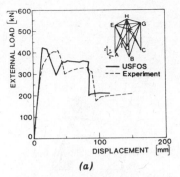
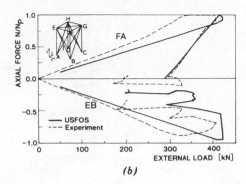

(a)   (b)

Figure 14. Ultimate behavior of a three-dimensional frame [112]: (a) Load-displacement. (b) Axial-force histories.

accounted for by performing simple modifications of the plastic interaction curves.

The program is intended for progressive collapse analysis for a broad class of accidental loads and damage conditions, both in design and during operation. The program is validated by comparison with conventional non-linear finite element programs in Reference 111 and experimental results in Reference 112. The predicted and experimental behavior of a three-dimensional tubular frame subjected to a quasi-static load is given in Figure 14.

USFOS has been used in several design studies of jackets and semisub-mersibles for the North Sea. Figure 15(c) shows response curves for a jacket platform at a water depth of 110 m. The loads are given in multiples of the design environmental load effects (corresponding to a return period of 100 years). Intact condition (case 1) as well as three damage cases are studied. The calculations show that the intact platform can sustain loads that are 2.3 times the design value. In damaged condition, the strength is reduced to a load factor of approximately 2. The platform is more sensitive to damage in the lower part because only the four corner legs are piled to the soil.

The nonlinear analyses show that the available force redistribution capability is sufficient to survive with a large safety margin in the damaged conditions studied. Thereby, the PLS criterion of NPD [19] is satisfied. By contrast, linear analysis and a first-yield criterion predict platform strength in the range of 0.68–0.99 for the cases 2–4. This fact emphasizes the importance of including nonlinear effects in PLS analyses. The ultimate strength levels in this case study are rather low. Normally, they are in the range of 2.5–5.0 for North Sea jackets. These values may seem high. However, in view of the large uncertainties and variabilities related to the load and strength parameters, they are often not unreasonable. Such aspects should be accounted for in a probabilistic framework, as previously discussed.

Global Dynamic Behavior Under Ship Impact In certain situations, global dynamic inertia effects can be significant in impact situations.

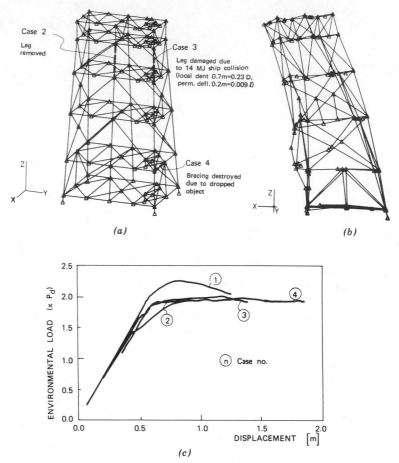

Figure 15. Ultimate behavior of a jacket at a water depth of 110 m: (*a*) Finite-element model of the platform. (*b*) Collapse mode of the intact platform. (*c*) Load-displacement relationship at top of the platform center.

In Reference 117, a simple procedure is presented for the analysis of ship impacts on jack-up platforms, considering the overall structure as an elastic single d.o.f. system.

In Reference 118, a method that accounts for large deflections and plastic behavior in concentrated hinges is applied in the impact analysis of a jacket at a 110 m water depth. It is shown that dynamic effects reduce the load-carrying capacity significantly when the impact duration is about half the fundamental eigenperiod. However, when the response during the impact period is governed by the strength of members adjacent to the contact location, the static solution normally applies.

The collapse pattern in Figure 16 illustrates the important role played by the large inertia of the topside structure.

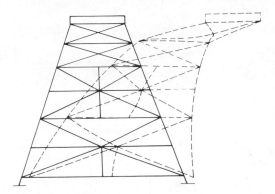

Figure 16. Global collapse of jacket under ship impact [118]. Amplification factor for displacements is 30:1.

Global Behavior After Fracture of a Member   The behavior of a system after fracture of a member is very much like the situation after an initial ductile failure. However, the sudden redistribution of force from the fractured member can be accompanied by dynamic inertia effects. The magnitude of the dynamic load factor depends on the time of force redistribution relative to the eigenperiod of the associated behavior mode. The unloading period by fracture in a steel tube, etc., can be estimated by assuming a fracture velocity of 100 m/sec. The maximum transient response occurs up to a few seconds after the fracture. This implies that the condition of external loads on the structure can be assumed to be that at the instance of fracture.

The dynamic magnification associated with the load redistribution after the failure of the brace in the *Alexander L. Kielland* was found in Reference 95 to vary between 1.4–2.6 in the various members, whereas the dynamic load factor for a single d.o.f. system is 2.0 for instantaneous unloading.

A special dynamic problem is encountered when a tether in a TLP (Figure 1) fails due to fatigue or fracture and results in progressive failure in a tension leg due to the load distribution or impact of the debris. A method for predicting the behavior of collapsing submerged tethers after a sudden loss of tension is presented in Reference 119. Large deflections and yield hinges are considered. An interesting aspect of the collapse is that buckling occurs at a load much higher than the static buckling load. Analytical predictions are compared with test results for model tethers in air in Reference 119. Systematic studies were accomplished to determine collapse mechanisms, as shown in Figure 17(*b*) and impact energies associated with different impact scenarios. The impact energy at the bottom is only 10–20% of the initial potential energy, due to the energy dissipation, especially in water.

The collapsing tether can impact other tethers, risers, or subsea templates, all being tubular structures. Studies reported in Reference 43 on the so-called pipe-whip problem relating to pipe failures in nuclear reactor or chemical process plants provide useful information in this respect.

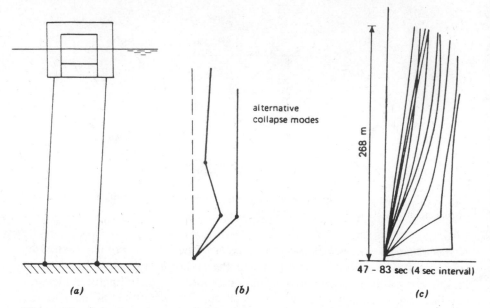

Figure 17. Behavior of a tether after sudden loss of tension [119]. Tether diameter $D = 1000$ mm and thickness $t = 30$ mm. (*a*) Layout. (*b*) Possible collapse modes. (*c*) Calculated collapse behavior.

Systems Under Cyclic Loads   Most previous work on extreme cyclic loads refers to seismic loads.

An extensive experimental and analytical study on systems behavior is found in Reference 120. Two types of two-bay X-braced frames at 1:6 scale were subjected to imposed cyclic displacements causing buckling, postbuckling strength deterioration, tensile restraightening, and tensile stretching. The tests aimed particularly at modeling the cyclic behavior due to earthquake loading, but provide important information as to the response under a sequence of extreme waves.

Several test series are reported in References 121 and 122. Usually, simple beam models with assumed flexural boundary conditions are applied to predict the behavior.

In Reference 122, the cyclic behavior of a trusswork/frame system with a cracked member is studied. It is demonstrated that incremental collapse can occur for loads below ultimate strength under a single load.

8.3.   Steel-plated structures

A variety of behavior modes can be envisaged for steel-plated structures, depending upon geometry and load conditions.

Box Girders Subjected to Bending   A ship hull girder subjected to bending is an important example of a box girder. A hierarchy of methods can, in principle, be used to estimate the ultimate bending capacity of the hull, ranging from simple elastic bending considerations (elastic section modulus times an ultimate stress) to finite-element methods. A simple procedure for estimating the bending capacity of experimental results for model structures is found in References 124–127. This approach accounts for the progressive loss of stiffness caused by buckling and postcollapse load reduction in longitudinal elements of the cross section. The capacity is determined by incrementing the vertical and horizontal curvature of the hull girder, on the assumption that plane sections remain plane.

This method is rather straightforward. The only difficulty is how to derive the stress–strain relationships of structural elements taking into account the effect of possible (1) local collapse of panels between stiffeners, (2) local collapse of stiffeners, e.g., tripping, (3) collapse of longitudinally stiffened plates between transverse girders, and (4) overall collapse of orthogonally stiffened plates between transverse bulkheads. The element formulation can be a phenomenological approach, by using load-end shortening relationships from experiments, or a finite element approach, including elastoplastic behavior and large deflections. In the latter case, one interframe element is used. Appropriate flexural boundary conditions have to be assumed.

The effects of transverse and lateral stresses have to be accounted for in a simplified manner, e.g., by using reduced capacities through interaction formulas. The longitudinal strength of elements carrying predominantly transverse loads or local lateral loads can, for simplicity, be neglected [128].

The effect of idealized damage conditions relating to slamming and explosion or crack damage in a special production ship of tanker type [128] is studied with element properties of the type shown in Figure 18(*b*). σ_{cr} is taken as the code value for the ultimate strength of stiffened plates. The strength of the corners are based on the effective breadth of adjacent plates. Table 1 demonstrates the sensitivity found.

General Steel-Plated Structures   A more general approach is achieved by using two-dimensional discretization of the panels that constitute the hull. In the structural unit method [78, 107, 129, 130], the stiffened plates between girders and frames, as well as girders between supports on bulkheads or transverse frames, are considered as elements.

Such elements have been reported in References 107 and 129–134.

Plate and panel elements are derived on the basis that the in-plane displacements of the boundaries are linear functions determined by two nodal parameters in each corner and that the flexural boundary condition is simply supported. For the plate element, normal and shear stresses are considered. For the panel element, only in-plane normal stress is considered. Lateral loads have not yet been included. The nonlinear behavior is established based on analytical or numerical solutions of the governing equations and experimental work. The incremental form of the stiffness equation is obtained for the following

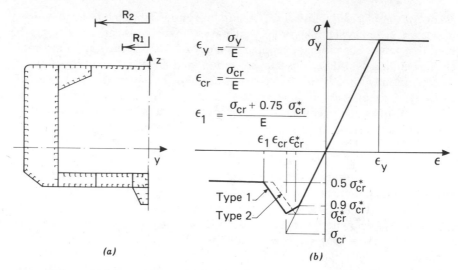

(a) (b)

Figure 18. Hull girder model: (a) Cross section of example hull. (b) Element characteristics applied in example.

TABLE 1 Ultimate Strength of Section [Figure 18(a)]

System	Normalized Moment Capacity M_u/M_{pl}[a]	
	Hogging	Sagging
Intact	$0.50\ (0.54)$[b]	$0.61\ (0.69)$[b]
Moderate slamming damage	0.45	0.60
Compression strength in outer bottom is 65% of intact value		
Tensile stiffness is 50% of intact value		
Severe slamming damage	0.40	0.59
Compression strength in outer bottom is 35% of intact value		
Tensile stiffness if 50% of intact value		
Moderate explosion or crack damage	0.49	0.59
Material in region R_1 removed [Fig. 18(a)]		
Severe explosion or crack damage	0.49	0.56
Material in region R_2 removed [Fig. 18(a)]		

[a]M_{pl} is the full plastic moment capacity. Buckling effects are neglected.
[b]Number within the parentheses refers to element characteristic of type 1 [Figure 18(b)]. Otherwise, type 2 is applied.

498

conditions of the panel: (a) linear elastic prebuckling, (b) postbuckling, and, depending on whether buckling occurs or not, (c) ultimate or (d) fully plastic condition.

The behavior of the panel element is treated differently, depending on the rigidity of the stiffeners. For weak stiffeners, overall panel collapse occurs and the behavior is modeled by orthotropic plate theory. For stiffeners with intermediate and high rigidity, local buckling of the plate is assumed to occur, and the postbuckling properties of the plate (between stiffeners) is applied. Unstiffened deep girder elements, which have been developed, account for the behavior modes shown in Figure 19, including the effect of manholes on the bending/shear behavior of the girder web.

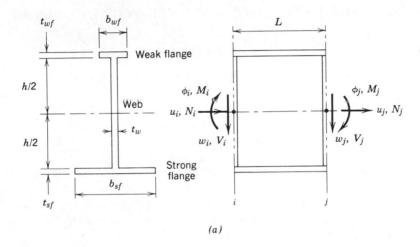

(a)

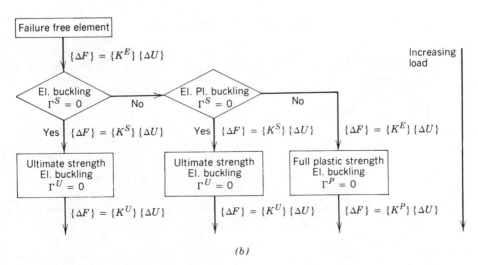

(b)

Figure 19. Deep-girder model: *(a)* Deep-girder element. *(b)* Behavior modes.

The method has been used to predict the response of a double-bottom model subjected to a concentrated load, simulating a grounding incident. As the girder elements comprised effective flanges in the top and bottom plate, these plates are modeled only to account for the interaction between normal stresses. Rupture of the bottom plate is estimated to occur at a strain of 30%. The calculated overall behavior is in good agreement with test results.

Interesting calculations of an actual double bottom is also accomplished in Reference 52. The effect of various damage conditions, including corrosion, on the hull girder strength is illustrated by application of this kind of method in Reference 135.

9.   RISK OF COLLISION-INDUCED TOTAL COLLAPSE

9.1.   Introduction

The present regulations for mobile platforms normally require that they have sufficient reserve buoyancy to remain in a stable condition after flooding of one to two compartments. After the *Alexander L. Kielland* accident, the Norwegian Maritime Directorate introduced a survival requirement associated with the loss of a major buoyancy member (for example, with a column). These damage stability requirements do not relate to a unified risk level.

In a recent project [59], the inherent safety level with respect to accidental loss of buoyancy reflected by various stability criteria was estimated for several platforms, considering all significant hazards. In the following section, the probabilistic aspects of the total collapse induced by ship collisions in an example platform are briefly summarized.

9.2.   Probability of Total Collapse

Figure 20 shows the model used for ship–column collision. The collision duration is assumed to be short relative to the fundamental period of the governing motion. By applying the principles of conservation of momentum and energy, a closed-form solution for the fraction of the impact energy that remains as kinetic energy after the first impact period is found. For a ship and platform with displacements of 6.500 and 25.000 tons, respectively, the residual kinetic energy for the sample case is in the range of 14–27% of the collision energy. For larger vessels, the percentage is larger. The difference energy must be dissipated as strain energy in the platform and ship. The response analysis shows that neither the global strength nor the bending capacity is likely to be exceeded under the maximum collision load. Thus, the initial response is governed by local indentation of the column. The model used for column deformation is described in Section 6.3.

When the indentation is equal to the radius of the column, the load caused by local indentation is likely to exceed the limit-collapse load in bending for the

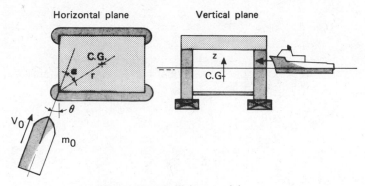

Figure 20. Collision model.

column (taking into account the reduced bending capacity in the collision zone). This is taken as the collapse criterion, disregarding possible energy dissipation in global bending. It is also reasonable to assume that the deck girders connected to the column are subjected to inelastic deformations to such an extent that cracks occur. This leads in turn to rapid flooding of the reserve buoyancy volume in the deck within a short period. Hence, the collapse criterion of the column serves also as a capsizing criterion for the platform.

The dissipation of energy in ship bows is assessed by means of plastic methods [136]. The major parameters with respect to bow stiffness are ship beam and plate thickness, which can be expressed approximately as functions of ship displacement.

Due to the uncertainties involved, the probability of collapse is calculated by means of a standard probabilistic algorithm. The parameters considered as basic random variables are mass and speed of the colliding ship, angle of collision, model uncertainty with respect to dissipation of energy in the ship platform, and the rupture-strain criterion.

The annual collision probabilities are assessed for different size categories at various sites in Reference 137. These figures along with the conditional and total probabilities of collapse are shown in Table 2 for the example platform at a location on the Troll field.

For this specific case, the highest risk of collision is associated with merchant vessels with displacements of 6500 tons. It is noted, however, that the conditional probability of survival is 0.33 in spite of the high demand (characteristic collision energy of 165 MJ) as compared to the resistance (characteristic energy dissipation in ship and platform of ~ 100 MJ). It is also observed that the contribution to the total failure probability depends strongly on the relative distribution of the collision frequencies for each category. A shift in this distribution can cause a shift in the "governing category." This points to the importance of considering different magnitudes of collision energy in such evaluations. Finally, it is worth noting that the failure probabilities are somewhat higher than the target values (given, e.g., in Reference 138: as 10^{-4}/year). Hence, both event-control and direct design measures should be

TABLE 2   Annual Probability of Total Collapse Due to a Ship Collision with a Column[a]

Ship Category	Displacement (tons)	Charact. Speed (knots)	Charact. Energy (MJ)	Collision Prob., P_1	Collapse Prob., P_2	Total Prob., $P_1 \times P_2$
Merchant vessel	1.100	11.1	20	1.2×10^{-2}	3.7×10^{-4}	4.1×10^{-6}
	2.200	11.6	40	1.4×10^{-2}	1.7×10^{-2}	1.9×10^{-4}
	6.500	13.2	165	2.8×10^{-3}	0.67	1.9×10^{-3}
	18.000	15.3	560	1.3×10^{-4}	~ 1	1.3×10^{-4}
	100.000	15.3	3430	6.7×10^{-5}	~ 1	6.7×10^{-5}
Fishing vessel	1.100	11.1	20	2.7×10^{-3}	3.7×10^{-4}	1.0×10^{-6}
Supply vessel, exp.	2.200	14.6	70	2.8×10^{-3}	0.12	3.4×10^{-4}
Supply vessel, develop.	2.200	14.6	70	2.8×10^{-3}	0.12	6.0×10^{-4}

[a]The platform location is a cautiously selected site on the Troll field.

considered in possible risk-reduction efforts. It is emphasized that the probability of impacts by merchant vessels is very sensitive to the distance to major shipping lanes. For other locations on the Troll field, the results can be significantly different.

10. CONCLUSIONS

The probabilistic and mechanical aspects of catastrophic failure modes of marine structures are reviewed. The emphasis is on methods for predicting structural failure modes, such as buckling and collapse under accidental impact loads, and crack growth and fracture in relation to deficient strength. Examples relating especially to systems collapse are presented. While there is significant information available on ductile behavior under single extreme loads, more information is needed on damage systems subjected to cyclic loads and on systems with crack-type damage.

REFERENCES

1. Lloyd's Register of Shipping, *Casualty Returns. Annual Summary and Quarterly Returns*, LRS, London, Annual editions.

2. *Worldwide Offshore Accident Databank*, Annu. Stat. Rep., Veritec, Oslo, Annual editions.

3. T. Moan, *Safety Appraisal of Floating Production Systems*, Semin. Float. Prod. Syst., IBC, London, 1985.

4. J. E. Vinnem, *Risk Assessment of Buoyancy Loss-Project Synthesis*, Siktec Rep. No. ST-87-RR-013-01, Siktec, Trondheim, 1987.

5. International Ship Structures Congress, *Report of Committee IV.1. Design Philosophy*, Genoa, 1985; Copenhagen, 1988.

6. C. G. Soares and T. Moan, *Risk Analysis and Safety of Ship Structures*, Congresso Ordem dos Engenheiros, Lisbon, 1981.

7. International Ship Structures Congress, *Report of Committee I.1. Environmental Conditions*, Paris, 1979.

8. Y. Yamamoto et al., "Analysis of Disastrous Structural Damage of a Bulk Carrier," *Proc. Int. Symp. PRADS, 2nd*, Tokyo (1983).

9. Y. Yamamoto et al., "Analysis of a Container Ship due to Slamming," *J. Soc. Nav. Archit. Jpn.*, **155** (1984) (in Japanese).

10. *Report of Commissioner of Maritime Affairs in Matter of Major Hull Fracture of "Energy Concentration": In Europort, Netherlands on 22nd July 1980*, Ministry of Finance, Republic of Liberia, Monrovia, 1981.

11. International Ship Structures Congress, *Report of Committee III.1. Ferrous Materials*, Paris, 1979; Gdansk, 1982; Genoa, 1985.

12. *Marine Accident Report—Capsizing and Sinking of the US Drillship GLOMAR JAVA SEA in the South China Sea, 65 Nautical Miles SW of Hainan Island, China, October 1983*, NTSB Rep. MAR 84/08, 1984.

13. Y. Akita, *Lessons Learned from Failure and Damage of Ships*, Jt. Sess. 1, Eight ISSC, Gdansk, 1982.
14. International Ship Structures Congress, *Report of Committee III.3. Fabrication and Service Factors*, Genoa, 1985.
15. Y. Masuda et al., "Fracture Mechanics Analysis of Failure of Ship Structural Members," *Proc. Int. Symp. PRADS, 3rd* (1987).
16. T. Moan, "Overview of Steel Structures," in A. Almar-Næss, ed., *Fatigue Handbook*, Tapir Publishers, Trondheim, 1985, Chapter 1.
17. T. Moan and J. Amdahl, "On the Risk of Flotel-Platform Collision," *Am. Soc. Civ. Eng., Spec. Conf. Prob. Mech. Struct. Reliab.* (1984).
18. JCSS, "General Principles on Reliability for Structural Design," *Jt. Comm. Struct. Saf.*, **35** Part II (1981).
19. Norwegian Petroleum Directorate, *Regulations for Load-Carrying Structures for Extraction or Exploitation of Petroleum*, NPD, Stavanger, 1985.
20. T. Moan, "On Hazard-Tolerance Criteria for Marine Structures," in N. Lind, Ed., *Reliability and Risk Analysis*, Inst. Risk Res., University of Waterloo, 1987.
21. J. Amdahl et al., *Risk Assessment of Buoyancy Loss. Structural Damage*, Rep. No. 6, Siktec, Trondheim, 1986.
22. T. Moan, "Structural Risk Assessment for Design and Operation," *Offshore Saf. Conf., IBC*, London (1987).
23. *Design Guidance for Offshore Structures Exposed to Accidental Loads*, Veritec/SINTEF/NTH, 1988.
24. W. Johnson, *Impact Strength of Materials*, Edward Arnold, London, 1972.
25. C. O. Ellinas and S. Valsgård, "Collisions and Damage of Offshore Structure: A State-of-the Art," *Proc. Int. Symp. OMAE, 4th* (1985).
26. International Ship Structure Congress, *Report of Committee II.2. Non-linear Structural Response*, Genoa, 1985.
27. International Ship Structures Congress, *Report of Committee II.3. Transient Dynamic Loadings and Response*, Genoa, 1985.
28. J. Zukas, T. Nicholas, H. F. Swift, L. B. Greszczuk, and D. R. Curran, *Impact Dynamics*, Wiley, New York, 1982.
29. O. Bach-Gansmo et al., *Design against Accidental Loads in Mobile Platforms. Project Summary Report*, Rep. No. 15, Veritec Rep. 85-3094, Veritec/Otter, 1985.
30. A. Rubert, *Experimentelle Untersuchungen zum Brandverhalten Kompletter, ebener Rahmentragwerke aus Baustahl* (in German), Krupp Forschunginstitut, Essen, 1984.
31. K. Nakamura, K. Shinoda, M. Hirota, and K. Kawagve, "Structural Behaviour of Steel Frame in Building Fire," *Proc. Int. Symp. Fire Saf. Sci., 1st* (1985).
32. O. Pettersen, "Structural Fire Behaviour Development Trends," *Proc. Int. Symp. Fire Saf. Sci., 1st* (1985).
33. J. Amdahl et al., "Progressive Collapse Analysis of Mobile Platforms," *Proc. Int. Symp. PRADS, 3rd* (1987).
34. Departments of the Army, Navy, Air Force (US), *Structures to Resist the Effect of Accidental Explosions*, Washington, D.C., 1969.
35. S. Fjeld, "Design for Explosion Pressure/Structural Response," Semin., Gas Explosions—Consequences and Measures, Norw. Soc. Char. Eng., Geilo, 1983.

36. W. E. Baker, P. A. Cox, P. S. Westine, J. J. Kulesz, and A. Strehlow, *Explosion Hazards and Evaluation*, Fund. Stud. Eng., Elsevier, Amsterdam, 1983.

37. N. Jones, "Review of the Plastic Behaviour of Beams and Plates," *Int. Shipbuild. Prog.*, **19**(218), 313–327 (1972).

38. N. W. Murray, "Recent Research into the Behaviour of Thin-Walled Steel Structures," Steel Struct.—Recent Res. Adv. Appl. Des., Elsevier Applied Science Publishers, London, 1986.

39. ECCS (Eds.), *Introductory Report*, 2nd Int. Colloq. Stab., 1976–1977.

40. W. F. Chen and D. J. Han, *Tubular Members in Offshore Structures*, Pitman, Boston, 1985.

41. Lloyds Register of Shipping, *Boat Impact Study*, Offshore Technol. Rep., OTH 85224, LRS, London, 1985.

42. J. G. de Oliveira, T. Wierzbicki, and W. Abramowicz, *Plastic Behaviour of Tubular Members under Lateral Concentrated Loading*, Det Norske Veritas, Rep. No. 82-0708, Høvik, 1982.

43. *Structural Mechanics in Reactor Technology. Structural Mechanics*, biannual events since 1981.

44. J. Amdahl, *Impact Capacity of Steel Platforms and Tests on Large Deformations of Tubes under Transverse Loading*, Det Norske Veritas, Prog. Rep. No. 10, Rep. No. 80-0036, 1980.

45. D. R. Sherman, *Inelastic Flexural Buckling of Cylinders*, Steel Struct., Recent Res. Adv. Appl. Des., Elsevier, Applied Science Publishers, London, 1986.

46. J. Taby, "Ultimate and Postultimate Strength of Dented Tubular Members," Dr.ing. Thesis Report, UR-86-50, Div. Mar. Struct., with Appendices, NTH, Trondheim, 1986.

47. J. Taby and T. Moan, "Ultimate Behaviour of Circular Tubular Members with Large Initial Imperfections," *Proc. Ann. SSRC Conf.* (1987).

48. I. S. Sohal and W. F. Chen, "Local Buckling in Tubular Braces," *Proc. Int. Symp. OMAE, 6th* (1987).

49. C. S. Smith, W. Kirkwood, and J. M. Swan, "Buckling Strength and Post-Collapse Behaviour of Tubular Bracing Members Including Damage Effect," *Proc. BOSS '79*, pp. 303–326 (1979).

50. C. S. Smith, W. L. Sommerville, and J. M. Swan, "Residual Strength and Stiffness of Damaged Steel Bracing Members," *Proc. OTC*, Pap. No. 2981 (1981).

51. P. W. Marshall, "Connections for Welded Tubular Structures," in *Welding of Tubular Structures*, Int. Institute of Weld., Pergamon, Oxford, 1984.

52. Y. Ueda et al., "Flexibility and Yield Strength of Joints in Analysis of Tubular Offshore Structures," *Proc. Int. Symp. OMAE, 6th* (1987).

53. P. W. Hoadley, U. Clarkson and J. A. Yura, "Ultimate Strength of Tubular Joints Subjected to Combined Loads," *Proc. OTC*, Pap. No. 4854 (1985).

54. Y. Mahinom, Y. Kurobane, S. Takizawa, and N. Yamamoto, "Behaviour of Tubular T- and K-Joints Under Combined Loads," *Proc. OTC*, Pap. No. 5133 (1986).

55. P. J. Dowling, B. F. Ronalds, A. Onoufriou, and J. E. Harding, "Resistance of Buoyant Columns to Vessel Impact," *Proc. Int. Symp. PRADS, 3rd* (1987).

56. A. Onoufriou, A. S. Elnashai, J. E. Harding, and P. J. Dowling, "Numerical Modelling of Damage to Ring Stiffened Cylinders," *Proc. Int. Symp. OMAE-ASME*, *6th* (1987).

57. B. F. Ronalds and P. J. Dowling, "A Denting Mechanism for Orthogonally Stiffened Cylinders," *Int. J. Mech. Sci.* (to be published).

58. V. A. Zayas and B. V. Dao, "Experimental and Analytical Comparisons of Semisubmersible Offshore Rig Damage Resulting From a Ship Collision," *Proc. OTC*, Pap. No. 4888 (1985).

59. *Risk Assessment of Buoyancy Loss—Case Study 1—Deep Sea Bergen and Treasure Scout*," Siktec/SINTEF, Trondheim, 1987.

60. P. Dubas and E. Gehri, *Behaviour and Design of Steel-plated Structures*, Publ. No. 44, ECCS, 1986.

61. N. Jones, "Structural Aspects of Ship Collisions," in N. Jones and T. Wierzbicki, Eds., *Structural Crashworthiness*, Butterworth, London, 1983.

62. J. G. de Oliveira, *Design of Steel Offshore Structures against Impact Loads due to Dropped Objects*, Rep. No. 81-6, Dep. Ocean Eng., Massachusetts Institute of Technology, Cambridge, MA, 1981.

63. C. S. Smith and R. S. Dow, "Residual Strength of Damaged Steel Ships and Offshore Platforms," *J. Construct. Steel Res.*, **1**, 4 (1981).

64. J. Czujko, *Strength of Plates with Large Initial Deflections Subjected to Biaxial Compression and Lateral Loads*, Det Norske Veritas, Rep. No. 83-0071, 1983.

65. N. W. Murray, "The Static Approach to Plastic Collapse and Energy Dissipation in Some Thin-Walled Steel Structures," in N. Jones and T. Wierzbicki, Eds., *Structural Crashworthiness*, Butterworth, London, 1983.

66. K. Roik and U. Kuhlmann, "Rotation Capacity of I-profiles considering the Effects of Plastic Plate Buckling," *Proc. Int. Colloq. Stab. Plate Shell Struct.*, Rijksuniversiteit, Gent, 1987.

67. N. Jones, "A Literature Review on the Dynamic Plastic Response of Structures," *Shock Vib. Dig.*, August, pp. 89–105 (1975); see also September, pp. 21–33 (1978); October, pp. 13–19 (1978); October, pp. 3–16 (1981); pp. 35–47 (1984).

68. A. Wenger et al., "Design for Impacts of Dropped Objects," *Proc. OTC*, Pap. No. 4471 (1983).

69. G. R. Ellis and K. R. Perret, "The Design of an Impact Resistant Roof for Platform Wellhead Modules," *Proc. OTC*, Pap. No. 3907 (1980).

70. M. Langseth, "Dropped Objects—Plugging Capacity of Steel Plates. An Experimental Investigation," Dr.ing. Thesis, Div. Struct. Eng., Norw. Inst. Technol., Trondheim (to be published).

71. J. O. Hallquist, "DYNA 3D User's Manual. Non-linear Dynamic of Solids in Three Dimensions," *Lawrence Livermore Lab.* Rep. *UCRL*, **UCRL-19592**, Rev. 1 (1984).

72. P. Dunn, "Offshore Platform Inspection," *Proc. Int. Symp. Role Des., Inspect. Redundancy Mar. Struct. Reliab.*, Natl. Acad. Press, Washington, DC, 1984.

73. F. Dyhrkopp, "Inspection of Floating Offshore Platforms", *Proc. Int. Symp. Role Des., Inspect. Redundancy Mar. Struct. Reliabil.*, Natl. Acad. Press, Washington, DC, 1984.

74. Tanker Structure Cooperative Forum (Eds.), *Manual for the Inspection and Condition Assessment of Tanker Structures*, 1986.

75. S. T. Rolfe and J. M. Barsom, *Fracture and Fatigue Control in Structures: Applications of Fracture Mechanics*, Prentice-Hall, Englewood Cliffs, NJ, 1977.

76. Det norske Veritas, *Rules for Classification of Steel Ships*, Oslo, 1987.

77. Norweigan Petroleum Directorate, *Guidelines for the Structural Design of Steel Structures*, NPD, Stavanger (to be published).

78. Y. Ueda et al., "Ultimate Strength Analysis of Double Bottom Structures in Stranding Conditions," *Proc. Int. Symp. PRADS, 3rd, 1987*, pp. 1043–1059 (1987).

79. H.-J. Golembiewski and G. Vasoukis, "Ductility Minimum for Application of Plastic Limit Load Concept to Failure Analysis of Structures with Imperfections," *Int. J. Pressure Vessels Piping*, **24**, 27–36 (1987).

80. British Standards Institution, *Guidance on some Methods for the Derivation of Acceptance Levels for Defects in Fusion Welded Joints*, PD 6493, BSI, London, 1980.

81. E. Pettersen and S. Valsgård, "Collision Resistance of Marine Structures," in N. Jones and T. Wierzbicki, Eds., *Structural Crashworthiness*, Butterworth, London, 1983, Chapter 12.

82. J. Lereim, "The Influence of Cracks on Energy Absorption Capacity and Structural Damage During Collisions," *Proc. Int. Symp. OMAE, 6th* (1987).

83. Norwegian Institute of Technology, *Research Program for Marine Structures 1986– 88. Partproject 2.3 Residual Strength*, Div. Mar. Struct., NTH, Trondheim, 1988.

84. Pipe Crack Study Group, *Investigation and Evaluation of Stress Corrosion Cracking in Piping of Light Water Reactor Plants*, NUREG-0531, U.S. Natl. Research Council, Washington, DC, 1979.

85. A. Zahoor and M. F. Kanninen, "A Plastic Fracture Instability in a Circumferentially Cracked Pipe in Bending. Part I. J-Integral Analysis," *J. Pressure Vessel Technol.*, **103**, 352–358 (1981).

86. K. Shibata et al. "Ductile Fracture Behaviour of Circumferentially Cracked Type 304 Stainless Steel Piping under Bending Load," *Nucl. Eng. Des.*, **94**, 221–231 (1986).

87. R. P. Harrison et al., *Assessment of the Integrity of Structures Containing Defects*, CEGB Rep. R/H/R6-Rev. 2, 1980.

88. B. G. Wade, "Fatigue of Circumferential Cracks in a Semisubmersible," *Proc. OTC*, Pap. No. 5353 (1986).

89. G. Hessen and T. Moan, "Fracture Mechanics Analysis of Stiffened Tubular Members," *Proc. Int. Symp. PRADS, 3rd* (1987).

90. T. R. Gurney, *Fatigue of Welded Structures*, Cambridge Univ. Press, London and New York, 1979.

91. A. Almar-Næss, (Ed.), *Fatigue Handbook*, Tapir Publishers, Trondheim, 1985.

92. K. Engesvik, "Fracture Mechanics as a Tool in Fatigue Analysis," in A. Almar-Næss, Ed., *Fatigue Handbook*, Tapir Publishers, Trondheim, 1985, Chapter 3.

93. D. J. Burns et al., "Crack Growth Behaviour and Fracture Mechanics Approach," in C. Noordhoek and J. de Back, Eds., *Steel in Marine Structures*, Elsevier, Amsterdam, 1987.

94. H. J. Wessel and T. Moan, "Fracture Mechanics Analysis of Crack Growth in Plate Girders," *Proc. IABSE Annu. Conf.* (1988).

95. T. Moan et al., "Analysis of the Fatigue Failure of the 'Alexander L. Kielland'," *Am. Soc. Mech. Eng. Winter Annu. Meet.* (1981).

96. Y. Masuda et al., "Fracture Mechanics Analysis on Failure of Ship Structural Members," *Proc. Int. Symp. PRADS, 3rd* (1987).

97. W. C. Chen, "Fracture Control Strategy for TLP Tethers," *Proc. Int. Symp. OMAE, 6th* (1987).

98. O. Bach-Gansmo et al., "Fatigue Assessment of Hull Girder for Ship Type Floating Production Vessels," *Proc. Conf. Mobile Offshore Units*, City University, London, 1987.

99. M. Kaminski, "The Behaviour of Imperfect Plates subjected to Cyclic Compressive Loads, *Proc. Int. Colloq. Stabil. Plate Shell Struct.*, Rijksuniversiteit, Gent, 1987.

100. T. Nomoto and M. Enosawa, "On the Cyclic Inelastic Buckling of Tubular Columns," *Proc. Int. Meet. Saf. Criteria Des. Tubular Struct.* (1986).

101. Y. Fukumoto and H. Kusama, "Cyclic Bending Tests of Thin-walled Box Beams," *Proc. JSCE, Strength Eng./Earthquake Eng.*, **2**(1), 117–127 (1985).

102. H. D. Hibbitt et al., *ABAQUS-EPGEN: A General Purpose Finite Element Code*, Rep. NP-2709-CCM, Electr. Power Res. Inst., Palo Alto, 1982.

103. P. Bergan and A. Arnesen, "FENRIS—A General Purpose Finite Element Program," *Proc. Int. Conf. Finite Elem. Syst., 4th* (1983).

104. A. Engseth, "Finite Element Collapse Analysis of Tubular Steel Offshore Structures," Doctoral Thesis, Rep. UR-85-46, Div. Mar. Struct., Norw. Inst. Technol., Trondheim, 1984.

105. P. W. Marshall, W. E. Gates, and S. Anagnostopoulos, "Inelastic Dynamic Analysis of Tubular Offshore Structures," *Proc. OTC*, Pap. No. 2908 (1977).

106. V. A. Zayas, S. A. Mahin, and E. P. Popov, "Ultimate Strength of Steel Offshore Structures," *Proc. Int. Conf. BOSS, 3rd* (1982).

107. Y. Ueda and S. M. H. Rashed, "An Ultimate Transverse Strength Analysis of Ship Structures," *J. Soc. Nav. Archit. Jpn.*, **136** (1974).

108. S. M. H. Rashed, *Behaviour to Ultimate Strength of Tubular Offshore Structure by the Idealized Structural Unit Method*, Rep. No. SK/R51, Div. Mar. Struct., Norw. Inst. Technol., Trondheim, 1980.

109. Y. Ueda and S. M. H. Rashed, "Behaviour of Damaged Tubular Structural Members," *Proc. Int. Symp. OMAE, 4th* (1985).

110. J. R. Lloyd and W. C. Clawson, "Reserve and Residual Strength of Pile Founded Offshore Platforms," *Proc. Int. Symp. Role Des., Inspec. Redundancy Mar. Struct. Reliab.* (1983).

111. T. Moan et al., "Collapse Behaviour of Trusswork Steel Platforms," *Proc. Int. Conf. BOSS, 4th* (1985).

112. T. Søreide et al., "The Idealized Structural Unit Method on Space Tubular Frames," *Proc. Int. Conf. Steel Alum. Struct.* (1987).

113. H. Nedergaard and P. T. Pedersen, "Analysis Procedure for Space Frames with Material and Geometrical Non-linearities," *Eur.–US Symp. Finite Elem. Methods Non-linear Probl.*, Norw. Inst. Technol., Trondheim, 1985.

114. G. Shi, C. T. Yang, and S. N. Atluri, "Plastic Hinge Analysis of Flexible-Jointed Frames Using Explicitly Derived Tangent Stiffness Matrices," *Proc. Int. Symp. OMAE, 6th* (1987).

115. J. G. Orbison, W. McGuire and J. F. Abel, "Yield Surface Application in Non-linear Steel Frame Analysis Computer Methods," *Appl. Mech. Eng.*, **33**, 557–573 (1982).

116. S. I. Hilmy and J. F. Abel, "Material and Geometric Non-linear Dynamic Analysis of Steel Frames Using Computer Graphics," *Comput. Struct.*, **21**(4), 825–840 (1985).

117. E. Pettersen and K. R. Johnsen, "New Non-linear Method for Estimation of Collision Resistance of Mobile Offshore Units," *Proc. OTC*, Pap. No. 4135 (1981).

118. J. Xu, "Nonlinear Static and Dynamic Analysis of Tubular Framed Structures," Dr. Thesis Div. Mar. Struct., Norw. Inst. Technol., Trondheim (to appear).

119. T. Moan and W. C. Webster, "Consequence Analysis of Tether Failure," *Proc. Int. Conf. BOSS, 5th* (1988).

120. V. Zayas et al., "Inelastic Structural Analysis of Braced Platforms for Seismic Loads," *Proc. OTC*, Pap. No. 3979 (1981).

121. K. Inoue et al., "Buckling Strength and Post-Buckling Behaviours of Tubular Truss Towers," in *Welding of Tubular Structures*, Int. Inst. Weld., Pergamon, Oxford, 1984.

122. K. Ogawa et al., "Buckling and Post-Buckling Behaviour of Complete Tubular Trusses under Cyclic Loading," *Proc. OTC*, Pap. No. 5439 (1987).

123. T. Yao and T. Moan, "Elastic–Plastic Behaviour of Structural Members and Systems with Crack Damage," *J. Soc. Nav. Archit. Jpn.*, **161** (1987).

124. C. S. Smith, "Influence of Local Compressive Failure on Ultimate Longitudinal Strength of a Ship's Hull," *Proc. PRADS Conf., 1st* (1977).

125. R. S. Dow et al., "Evaluation of Ultimate Ship Hull Strength, Symp." *Proc. Extreme Load Response*, SNAME, Arlington, VA, 1981.

126. C. S. Smith, "Structural Redundancy and Damage Tolerance in Relation to Ultimate Ship-hull Strength," *Proc. Int. Symp. Role Des., Inspect. Redundancy Mar. Struct. Rehab.*, Natl. Acad. Press, Washington, DC, 1984.

127. J. Adamchak, "An Approximate Method for Estimating the Collapse of a Ships' Hull in Preliminary Design," *Proc. Ship Struct. Symp.*, SNAME, Arlington, VA, 1984.

128. H. Brekke et al., *PTS 'PETROJARL.' Ultimate Strength of the Midship Section*, SINTEF Rep. STF71 F86047, 1986.

129. S. M. H. Rashed, "An Ultimate Transverse Strength Analysis of Ship Structures (The Idealized Structural Unit Method)," Dr. Eng. Dissertation, Osaka University, 1975.

130. Y. Ueda and S. M. H. Rashed, "The Idealized Structural Unit Method and its Application to Deep Girder Structures." *Comput. Struct.* **18**(2), 277–293 (1984).

131. E. Pettersen, "Analysis and Design of Cellular Structures," Doctoral Thesis, Div. Mar. Struct., Rep. UR 79-02, Norw. Inst. Technol., Trondheim, 1979.

132. M. Katayama, "Static Strength and Damage Analysis Method of Double Bottom Structures" (in Japanese), Dr. Eng. Dissertation, Osaka University, 1979.

133. Y. Ueda et al., "Plate and Stiffened Plate Units of the Idealized Structural Unit Method" (1st Report), *J. Soc. Nav. Archit. Jpn.*, **156**, 389–400 (1984) (in Japanese).

134. J. Paik, "An Ultimate Strength Analysis of Ship Structures by the Idealized Structural Unit Method" (in Japanese), Dr. Eng. Dissertation, Osaka University, 1987.

135. K. Skaar et al., "How Low can Steel Weight go with Safety and Economy," *Proc. Inst. Symp. PRADS, 3rd* (1987).

136. J. Amdahl, "Energy Absorption in Ship-Platform Impacts," Doctor Thesis, Div. Mar. Struct., Rep. UR-83-34, Norw. Inst. Technol., Trondheim, 1983.

137. *Risk Assessment of Buoyancy Loss—Ship Model Collision Frequency*, Technica, London, 1987.

138. Norwegian Petroleum Directorate, *Guidelines for Safety Evaluation of Platform Conceptual Design*, NPD, Stavanger, 1981.

CHAPTER 15

Industrial Experience with Structural Failure

J. M. Thomas

Failure Analysis Associates
Engineering and Scientific Services
Palo Alto, California

ABSTRACT

Four factors continually arise for consideration in structural-failure investigations: stress, environment, defects, and material properties. Each is discussed separately in a generic context, followed by some discussion of how they frequently interact. A number of examples of industrial failures and a summary of how each was analyzed are presented.

1. INTRODUCTION

Structural failure in the world's industrial complex is not a rare occurrence. In spite of increased engineering knowledge, failures still occur, as structures and machines are pushed toward new thresholds of performance and complexity. The aging process also takes its toll because it is generally more economical, at least in the short run, to keep an old system in service than to build a new one. These and other factors result in a steady stream of failures that need to be analyzed and understood so that corrective measures can be applied and recurrences avoided. Exchanges of experience between analysis practitioners is one way to maximize the positive benefits that can come from failure analysis.

Other chapters in this book have addressed numerous special techniques for analyzing particular types of circumstances of structural failures. In this chapter, we attempt to synthesize the process of structural-failure analysis in such a way as to complement the other chapters, which mainly deal with failure due to extreme or catastrophic loads. Only limited success can be expected in such an attempt; the variety of problems and approaches to their solution are too great to be encompassed by a single general approach. Nevertheless, it is both challenging and informative to search for such a framework.

There are a number of ways that a general structural-failure-analysis framework can be constructed. For example, one attractive alternative is to examine known modes of failure one by one, which is in keeping with the structure of this book. To these might be added fatigue, creep, corrosion, wear, and several others. However, proceeding in this fashion would likely duplicate, in summary fashion, many excellent textbooks and other publications that are organized by failure mode. Most of these publications address the failure mode in a broader context than that of failure analysis.

An examination of the chronology of a number of structural-failure investigations revealed that four factors continually arise for consideration. These are stress, environment, defects, and material properties. These factors cannot be claimed as an exhaustive, or all-encompassing, list, but they suffice as a springboard for discussing structural failures in a generic context, and then for examining several specific failures as examples. The chapter is organized along these lines.

2. STRESS

Stress arises more often than any other factor as a consideration in structural failures. It represents the simultaneous consideration of the loads (including thermal) and constraints applied to a system and the geometry of the system. Computerized numerical methods, particularly finite-element analyses [1, 2], are typically used to calculate stresses in a structural system. However, closed-form solutions and handbook approaches [3] also have an important role for getting a rapid solution for simpler systems and as a scoping or bounding solution that precedes the numerical analysis for more complex systems.

The degree of refinement needed in the stress analysis depends greatly on how the results are to be used. For example, it is well known that for most ductile materials under a monotonically increasing load, failure is more closely related to the average stress over the critical cross section than to peak stresses at notches, holes, and other discontinuities. The opposite is true for fatigue failure, which tends to depend on the peak stresses. Thus, the refinement needed in the stress analysis depends upon (among other considerations) the failure mode under analysis. Further complicating this issue is the fact that some published fatigue allowable stresses include the effects of particular types of discontinuities and others do not.

In very few structural failures does a uniaxial state of stress exist at the location of interest, yet the vast majority of material test data is for uniaxial stress. It is quite often very important to determine the multiaxial state of stress, which is readily accomplished by the available numerical methods. Of significantly greater difficulty is relating this stress state to failure of the material, particularly for cyclic stresses.

The temporal behavior of stress is significant in virtually every failure analysis. Monotonically increasing stress at a moderate loading rate is by far the simplest to treat, because failure depends on a direct comparison of the maximum stress in the part with the commonly published yield and ultimate strength of the material. Sustained constant or variable loading can sometimes be important, especially when coupled with a thermal or chemical environment.

In the majority of structural failures, cyclic stress is an important consideration. This is seldom so simple as to have one amplitude of stress that is repeated until failure, which would be typical of laboratory fatigue testing. Instead, the system is often exposed to a complex spectrum of cyclic stress. If so, a major task in the failure analysis can be simplifying the actual spectrum into one that is useful for analysis. Even with the simplified spectrum, a direct comparison of the spectrum with (preexisting) laboratory fatigue data is unlikely to be possible. Instead, it is often necessary to resort to (sometimes questionable) damage theories that permit the combination of fatigue damage for variable-amplitude stresses [4, 5]. Laboratory tests for the actual or simplified spectrum are often advisable.

Separate technical approaches have evolved for predicting crack initiation and crack growth under cyclic loads [6, 7]. One or both approaches may be applicable to a particular investigation, depending on the serviceability of a part with an existing growing crack. In both cases, there are circumstances where the sequence of applied stresses, as well as the stress amplitudes and number of cycles, can be important.

The stress spectrum can be further complicated by the superposition of sustained constant stress with the cyclic spectrum. These can arise from thermal stresses, interferences, bolt-clamping forces, residual stresses, and many other effects. Such sustained stresses typically influence the capability of the material to withstand cyclic stresses.

One of the most uncertain aspects of stress analysis is residual stress, which can have a significant effect on part performance, particularly with respect to fatigue and crack growth [7]. Residual stresses are typically difficult to calculate, difficult to measure, and highly variable between nominally identical parts. In fatigue-prone parts, it is generally advisable to take steps in the fabrication process to eliminate or control the state of residual stress. Since this is often not done, the possible presence of residual stresses frequently complicates the analysis of structural failures.

3. ENVIRONMENT

The environment that a system must endure is frequently a factor in structural failures [8]. In the present discussion, loads (forces) applied to the system, although technically a part of the environment, have been treated under the section on stress and are not discussed again here. Instead, this discussion focuses on environmental factors that tend to affect performance of the material.

Of special importance are temperature, chemical, and irradiation environments. For these and other environments, time is usually an important factor. These environments can act alone, in combination with each other, or in combination with other factors to adversely affect material performance.

High temperatures can cause scaling that results in loss of material cross section. It can also alter the material atomic structure and the alloying phases. High temperature can alter the material grain structure and cause alloying elements to migrate to grain boundaries. Material strength, toughness, and stiffness properties can be strong functions of operating temperature.

Chemicals can interact with structural materials, causing corrosion, pitting, grain boundary dissolution, and other adverse effects. Some chemicals can embrittle certain structural materials.

Interaction of temperature and chemical environments with other factors is discussed in a separate section.

4. DEFECTS

Defects arise all too often as contributing factors in structural failures. These can include geometric errors such as size and configuration anomalies. Material errors such as improper design selection, substitution of incorrect material, and processing anomalies are not uncommon.

Probably the most prevalent type of defect is material flaws. These can take the form of voids, inclusions, or cracks. Welds are a likely location for material flaws. In fact, it is generally accepted that all welds contain flaws, and that the question is not whether flaws are permissible, but how large a flaw is acceptable. Flaws are also common in castings, where the question is similar to that for welds. Forgings are also prone to defects, though not nearly as much so as welds and castings.

The use of composite materials introduces an entirely new realm of defects and their disposition. Much more so than metals, each nominally identical part is unique. Significant variations in density are not uncommon, between parts and within the same part. Delaminations between plies can easily occur and go undetected. Fortunately, composite materials are inherently more "forgiving" than metals because defects do not propagate readily from one fiber strand to another or from one layer to another.

Most of the activity surrounding defects in a failure analysis addresses two issues:

1. the effect of the defect on part performance
2. the detectability of the defect

The first line of inquiry regarding effect on part performance is usually a reference to standards that generally define the limitations on types and sizes of defects that will not degrade performance of the material below some specified level. However, such standards are known to be inexact and will certainly be called into question if a failure is traced to an "acceptable" defect.

Fracture-mechanics approaches have become the methods of choice in assessing the effect of crack-like defects on part performance. These methods serve two main purposes:

1. assessment of the crack size that can be tolerated without unstable crack propagation, possibly under a single high-intensity load
2. calculation of the rate of crack growth under cyclic loading

The uncertainties in defect detectability are frequently an issue in any failure analysis where defects are suspected of having a role. Nondestructive methods using eddy currents, ultrasonics, die penetrants, and X-rays search for defects in critical parts. Two appropriate questions are

1. How reliably can the method detect a defect of concern?
2. How accurately can the method size the defect once it is found?

5. MATERIAL PROPERTIES

Material properties represent the other side of the coin as compared to stress, environment, and defects. That is, stress, environment, and defects might be regarded as the "demand" placed on the structural system, and material properties as the "capacity" to meet the demand.

The most fundamental material properties as related to performance in a structural system are strength and stiffness. These are the first properties considered by the designer in selecting the material, and they are usually the first properties addressed in a failure analysis. Also important, but not as frequently considered, is the material fracture toughness. Roughly speaking, this property is a measure of the strength of a material when it contains a defect. Tests to determine this property are considerably more difficult than strength and stiffness tests. In these and other properties, welds must always be considered separately from the parent material.

Resistance of the material to fatigue and crack growth is important when the structure is subjected to cyclic loading. Tests for these properties are more difficult than those for strength and stiffness, but the properties are generally available for commonly used parent materials. (Care must be exercised not to use parent material properties for welds.) Fatigue resistance is generally measured in terms of the number of cycles to failure at a given stress amplitude. Crack-growth rate is usually related to the cyclic stress-intensity factor as calculated by fracture-mechanics principles.

Finally, the importance of the material's resistance to environmental factors is entirely dependent on what constitutes the environment. Adverse environments can result in creep, corrosion, stress corrosion cracking, and other failure modes. All too often, this is either overlooked, its importance is not understood, or data to establish the resistance (or lack of resistance) is not available. The innumerable combinations of chemical concentration, temperature, stress (and stress-related variables such as strain rate), time, and other factors make a priori laboratory testing a difficult task. Usually, such tests must be accelerated in some fashion to produce results within a useful time frame. This compromises the use of such results for design or for failure-analysis purposes.

6. INTERACTIONS

So far in this chapter, the interaction effects of stress, environment, defects, and material properties have been occasionally mentioned but have not been emphasized. Most failures result from such interactions. Some important examples are discussed in this section.

A common failure mechanism for high-temperature structures is creep, which is caused by a combination of stress, temperature, and time. Very high-temperature metallic parts must be operated at low stresses to avoid gradual deformation (creep) of the part that can lead to dimensional instability or rupture. Creep problems are common in aircraft engines, boiler tubes and headers, and high-temperature piping in power plants.

Stress corrosion cracking (SCC) occurs when a part made of a susceptible material is exposed to a critical combination of sustained tensile stress, chemical environment, and time. If any of these three factors is absent, SCC does not occur. This problem is prevalent in steam turbine discs, nuclear-power-plant piping, and steam-generator tubes.

Corrosion fatigue can occur when cyclic loads are applied to a susceptible material in a corrosive media. The effect is that the fatigue mechanism is aggravated by chemical attack, so that lower stress produces failure for a given number of cycles. The effect can be such that the endurance limit (stress that produces no fatigue damage) that exists in air does not exist in the corrosive environment.

In many structural failures, only one or a few of a number of nominally identical systems experiences a failure. Finding out why this is so is perhaps the most valuable contribution that a failure investigation can make to the industrial complex. If the reasons can be understood, decisions can be made regarding continued operation or corrective actions to the remaining systems. Rarely does the investigation pinpoint "the" cause of a failure or failures. Instead, the investigation is more likely to produce an explanation or "model" of how the various factors interacted to result in failure. Application of the same model to surviving units gives some perspective for continued successful operation. Properly treated, the perspective is not likely to provide an absolute statement about, for example, the amount of time the surviving unit can operate without failure. A complete analysis is more likely to produce probability statements about continued successful operation [8].

7. INDUSTRIAL EXAMPLES

Several examples of industrial structural-failure investigations are briefly reviewed to illustrate how the stress, environment, defect, and material property factors are addressed. Not every factor is present in every example. Some attempt is made to progress from simpler to more complex examples.

7.1. Scale-Model Nuclear Containment

The first example involves an intentional test failure of a metallic scale model of a nuclear-power-plant containment vessel [9]. The structure is a cylindrical shell with a spherical dome on top and a reinforced circular penetration through the cylinder to simulate a personnel hatch. The loading is internal pressure that produces a simple state of stress away from the boundaries and the penetration. The objective was to analyze the effect of the penetration on the failure pressure of the structure.

A finite-element model was used for this purpose. Figure 1 illustrates the idealized structure and the displacement behavior of a point on the shell at the penetration as compared to a closed-form solution for a similar point on a shell without a penetration.

As illustrated on the figure, two different meshes of different refinement were used to ensure that the solution had approximately converged. Numerical difficulties that correspond to plastic strains in the range 0.13–1.9% (the Luder-band region) were encountered at about 102.5 psi pressure, but these were

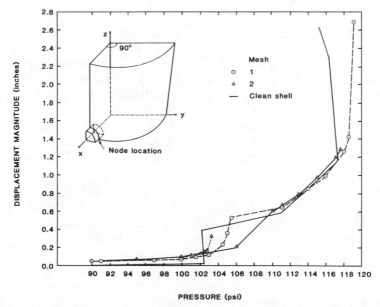

Figure 1. Displacement vs. pressure of a node adjacent to the penetration sleeve.

eventually overcome. Beyond the pressure of 119.1 psi, the last point plotted for mesh 1 on Figure 1, the solution would not converge, which was interpreted as an unstable large displacement. Failure pressure predicted by this analysis corresponded very favorably to catastrophic rupture test results for this structure.

7.2. Spray Cooler

Several spray coolers designed to operate in a canal as part of the cooling system of a nuclear power plant had experienced severe cracking problems very early in their intended operating life [10]. A circular-shaped pontoon provided flotation for a propeller, nozzle, motor, and a superstructure that served as a support for a fan and shaft, as illustrated in Figure 2. The purpose of the fan was to draw cooling air through the warm water sprayed into the air through the annular nozzle. Cracking was primarily in the superstructure.

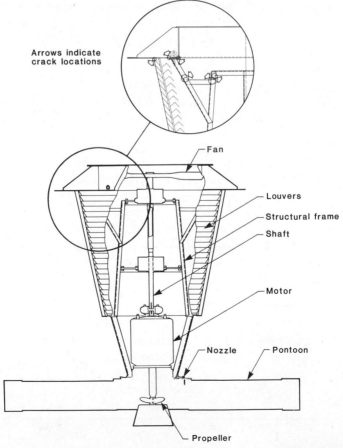

Figure 2.   A superstructure of the spray water cooler indicating location of cracks.

Calculations were performed to estimate the lower natural frequencies of vibration and the thermal stresses in the superstructure. A test unit was thoroughly instrumented with strain gauges and accelerometers and run for several hours. Results from the test unit were confirmed with a limited test program on a field unit. The analysis and testing revealed extremely high steady stresses (up to yield strength) and high vibratory stresses in the superstructure. A laboratory examination of crack surfaces from a failed unit revealed no significant initial defects.

It was concluded that the high steady stresses resulted from temperature differences between the tower structure and the ring surrounding the fan at the top of the unit. No allowance for thermal expansion was built into the design. The high vibration stresses resulted from low natural frequencies of the superstructure that were excited by the motor and propeller. The superstructure was completely lacking in lateral bracing between vertical members, which resulted in a very flexible structure in bending. Failures (cracking of major structural members) had occurred because of fatigue that was initiated by the combination of high steady and high cyclic stresses.

7.3. Salad-Oil-Plant Vessel

An extensive rupture occurred in the walls of a winterizer column in a 20-year-old salad oil plant [11]. The winterizer column was an eight-chamber baffled steel vessel, 54 inches in diameter by 25.5 feet tall. A motor-driven agitator shaft ran from top to bottom through the center of the vessel and drove paddles above and below each baffle. During operation, miscella, a mixture of cottonseed oil and hexane, flowed from top to bottom through the vessel and was gradually collected by refrigeration coils.

The unit had been out of service for approximately a month prior to the explosion; however, it was full of miscella. On the day preceding the explosion, the winterizer column was drained in preparation for maintenance. Normally, as the tank is drained, the miscella in the tank is displaced with nitrogen gas introduced at the top of the vessel. In the normal course of events, the empty tank would contain only a noncombustible gaseous mixture of nitrogen and hexane vapors, as well as surface deposits of miscella. Winterizers have reportedly been repeatedly drained without incident.

Normal operating pressures in the unit are far below that required to rupture the steel vessel. Metallurgical examination of the fracture surfaces revealed no defects or other abnormal material conditions.

The occurrence of the explosion was the result of deviation from normal conditions. The nature and extent of the damage indicated that a powerful blast wave propagated from top to bottom of the vessel. Two factors were necessary for the explosion to occur: the presence of an explosive mixture within the vessel, and a source of ignition to trigger an explosion. Neither of the factors could be precisely identified. The explosive mixture may have consisted of large quantities of oxygen with either hexane or hydrogen or both. Unless initially compressed well above atmospheric pressure, it is questionable that hexane alone would have the necessary explosive energy to cause the observed damage. Hydrogen may have been generated by decomposition of ammonia leaking from the refrigeration system. Inspection of all the significant system components did not reveal any defects capable of causing either condition. Operationally, however, there were numerous opportunities for error. For example, valves, switches, gauges, and hose connections were not distinct and were, generally, unlabeled. Air, rather than nitrogen, could have been drawn into the tank during draining operations by venting the tank to the atmosphere, or the tank could have been inadvertently purged with compressed air rather than nitrogen. The source of ignition may have been from sparking between mechanical components, or it may have traveled through the open vent at the top of the vessel. Although other possibilities were not completely ruled out, operator error was considered the most likely cause of the explosion.

7.4. Williamsburg Bridge Main Suspension Cables

The Williamsburg Bridge spans the East River in New York City, connecting the boroughs of Manhattan and Brooklyn. The main span of the bridge, opened in 1903, is some 1600 feet long and carries heavy automobile and subway rail traffic. The roadway is suspended from four main support steel-wire-cable bundles that pass over tower columns 300 feet high and are embedded in concrete anchorages at the shorelines [12].

The four main cables actually consist of 7696 No. 6 gauge (0.192 in.) bright steel wires spun together using a process pioneered by John Roebling. The individual wires are a low-alloy medium carbon steel with an initial ultimate tensile strength of 225 ksi. The force carried by each cable is currently estimated to be approximately 12,225 kips, representing the mean live load due to the traffic, the static load due to the weight of the bridge, and the wind loading. Assuming that none of the wires are broken or exhibit a reduced cross section anywhere along their lengths, then the safety factor should be about 4.

This, however, is not the case. The individual wires were not galvanized during initial installation, even though the process was well known at that time. Instead, they were coated with a combination of graphite and slushing oil, which did not provide adequate protection. This lack of protection, coupled with the humid, chloride- and sulfate-containing air, and the use of deicing salts in the winter, initiated corrosion damage to the main cables almost immediately. Periodic examinations throughout the 80-year life of the bridge has revealed a steadily worsening situation. A detailed examination between 1980 and 1982 of 300 feet of one of the cables in the back span region revealed some eight broken wires on the surface of the wire bundle alone, with many more wires nearly corroded through or otherwise extensively damaged. It proved impossible to assess the degree of damage to the wires in the interior of the bundle.

In 1982, an engineering analysis effort was initiated to determine the controlling corrosion mechanisms, to estimate the rate at which these corrosion processes are occurring, to estimate the residual life of the cables, assuming the corrosion damage continues at this rate, and to determine whether the corrosion processes could somehow be mitigated by the addition of properly formulated inhibitor compounds. Detailed microscopic examination of wire samples removed from the bridge showed the principal corrosion process to be pitting/localized corrosion, greatly accelerated by the presence of the graphite in the original protective coating. Laboratory experiments were conducted in which bundles of wires were alternately immersed in a simulated rain-water environment and dried to recreate the wetting and drying cycles to which the cables are exposed. The wire bundles were disassembled periodically and the individual wires weighed to determine weight loss due to pitting as a function of time. The results led to an estimated material removal rate of between 0.6 and 2.5 mil per year. Inhibitor additions to the simulated rain water substantially reduced this corrosion rate. Tensile tests were also performed on samples of wire, and the current ultimate strength of an individual wire was estimated from these data to be 198 ksi.

This information was used to construct a computer model of the cable-strength degradation process. The model allows pits to initiate randomly along the lengths of the wires at some preselected initiation rate. The pits then grow at a rate selected between the experimentally measured values. The model then sums the amount of damage at a given time, and computes the mean load-bearing capacity of an individual wire. This is compared to the measured current value, and the initiation rate is adjusted until the computed and measured values match. Using this initiation rate, the model is rerun and the total damage to the cable is computed by summing over all the wires in the cable. The model identifies broken wires when the cross-sectional area is reduced to the point where the applied load exceeds the load-bearing capacity. The effect of interwire friction in these clamped bundles is simulated by allowing a broken wire to continue to carry load outside of a specified clamping distance. The breaking force of the cable at any given time is then computed from the net cross-sectional area of the unbroken wires in the cable.

Figure 3 shows the calculated breaking force as a function of time for various combinations of corrosion rate and clamping distance. Clearly, the model predictions are sensitive to these parameters. Selecting a corrosion rate of 1.0 mil per year and a clamping distance of 400 feet (curve F in Figure 3), both realistic values, then the current safety factor is about 2.5 and decreasing rapidly. The safety factor is predicted to drop below 1.0 around 2005. Irrespective of the values of the parameters selected, however, the overall result is still the same: the safety factor has been degraded, and is continuing to dangerously degrade at an ever increasing rate.

The effect of inhibitor additions on cable degradation were simulated in the model by modifying the pit-initiation and growth rates. This is illustrated in Figure 4. Clearly, inhibitor additions can have a profound effect on the residual life of the cable. Unfortunately, laboratory experiments on cable models showed that inhibitor penetration through the entire cable could not be guaranteed.

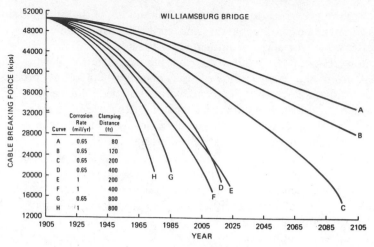

Figure 3. Effect of corrosion rate and friction on the cable-strength history.

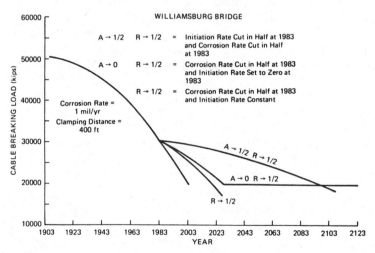

Figure 4. Effect of corrosion inhibitor applied in 1983 on the cable-breaking load.

Since partial inhibitor coverage can be worse for pitting corrosion than no inhibitor additions at all, this method of cable rehabilitation could not be recommended. As a result, the New York City and State Departments of Transportation have elected to replace the main suspension cables of the Williamsburg Bridge as soon as possible. Further studies are currently underway to determine whether the situation is sufficiently dangerous to warrant closing the bridge to traffic.

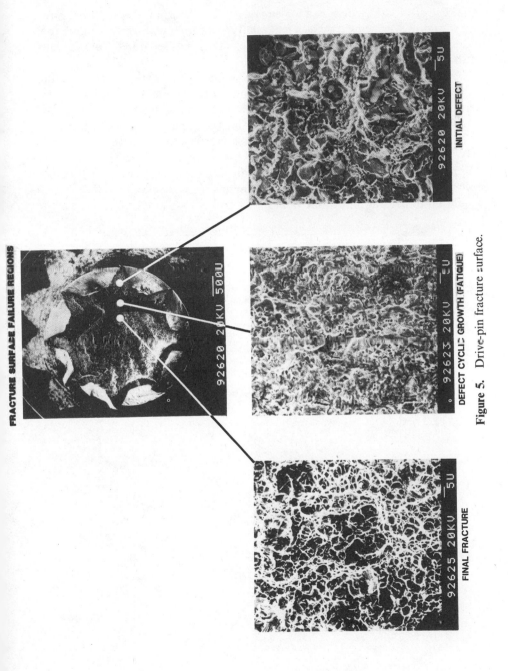

Figure 5. Drive-pin fracture surface.

7.5. Ladder Fastener

A folding steel ladder designed for emergency escape from a second- or third-story window was being used as a makeshift means for a performer to climb to a tightrope act. During a practice session, one of the rungs separated from the rails of the ladder, causing the performer to fall.

Visual examination revealed that the ladder rungs were attached to the ladder rails by drive pins, which passed through drilled holes in the rail and were driven into undersized holes, one in each end of the rungs. Both of the drive pins attaching the failed rung to the rails were broken. The rung was not bent.

A simple stress analysis revealed that the cross-sectional dimensions of the drive pins were sufficient to support several times a man's weight. However, tests on several rungs on an exemplar ladder exhibited a wide scatter in the drive-pin failure load among individual rungs, ranging from approximately a man's weight up to several times that value.

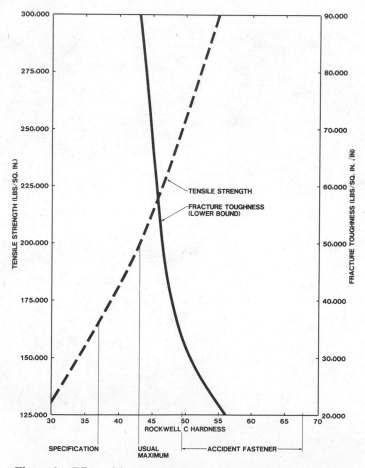

Figure 6.   Effect of fastener hardness on strength and toughness.

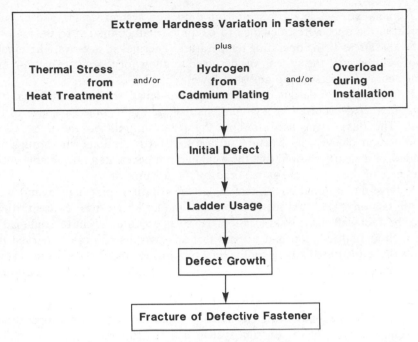

Figure 7. Ladder-failure sequence.

Microscopic examination of one of the accident drive-pin fracture surfaces revealed an initial defect, a fatigue zone growing from the defect, and a fast-fracture zone, as illustrated in Figure 5. Microhardness traverses revealed an extremely hard surface on the fastener, far beyond the hardness consistent with design specifications and the maximum hardness expected for the particular alloy involved. The material could be expected to be extremely brittle in this hardness condition. The hardness of the fastener and the effect of hardness on the strength and toughness of the fastener are illustrated in Figure 6. The drive pin was cadmium plated, which admitted the possibility of hydrogen or other environmental attack associated with the plating operation.

It was concluded that the drive pin had failed due to the defect and the extremely low tolerance of such a brittle material to the defect. The defect may have been induced during the service life of the ladder, but much more likely it was caused by the same extreme quenching condition that probably caused the high hardness. The sequence leading to the failure is illustrated in Figure 7.

7.6. Process-Plant Duct

A large duct in a processing plant suddenly ruptured during start-up operations after years of continuous successful service. The separation was circumferential, adjacent to a rotation joint that was part of the thermal-expansion design for the ducting system.

Nominal stresses in the system from pressures, system weight, and product weight in the duct were calculated to be quite small compared to the material allowable stress at the operating temperature. Thermal stresses were nominally also small because the system was designed to allow for thermal expansion and contraction during start-up and shutdown.

Examination of design drawings and the failed duct revealed that the separation had occurred at a circumferential weld that joined a flange to the duct. The flange was used to bolt the rotation joint to the duct. Close examination of the weld fracture surface and cross sections cut through the welded joint clearly showed that the weld had not penetrated through the entire thickness of the duct wall, as required by the design drawings.

A three-dimensional stress analysis of the rotation joint, flange, and weld details demonstrated that the rotation joint induced high stress concentrations into the duct wall at the weld location. This was because the entire axial load in the duct at that location had to pass through two diametrically opposed pin joints that constituted a hinge in the duct to allow for thermal expansion. Thus, the stresses in the duct wall from this stress concentration was three to four times the nominal stress.

It was also determined by analysis that pressure stresses and product-weight stresses could be higher during start-up than during normal operations. Dynamic behavior of the ducting system due to stoppages and unstoppages during start-up was also characterized and found to be of some significance. The calculated stresses from all combined effects exceeded the ultimate tensile strength of defect-containing welds that were cut from the unit and tested in the laboratory.

From a quantitative standpoint, the defective weld was by far the most significant factor in the failure. The stress concentration from the pin joint was also a very important factor. Of much less significance were the loading conditions caused by the start-up transient.

7.7. Power-Line Clevis

A 1045 steel zinc-coated clevis supporting a utility power line failed after only a few days in service [13]. No unusual loading conditions were known to have occurred. This was at least the third instance of such failure in the same utility within a period of a few months. The failure was in the "eye" region of the clevis.

A metallurgical examination included optical and scanning electron microscopy, metallography, hardness and microhardness tests, and determinations of composition. No large defects were found on the fracture surfaces. However, one of the four "legs" of the failed eye region (see Figure 8) showed a very flat fracture surface, with only small shear lips. In the scanning electron microscope, this surface exhibited a fringe of intergranular failure at most points around the fracture surface, varying from a few thousandths to about 0.02 inches deep.

Hardness tests revealed that the bulky region of the clevis, where quality-control tests had been made, had a hardness only slightly above the maximum

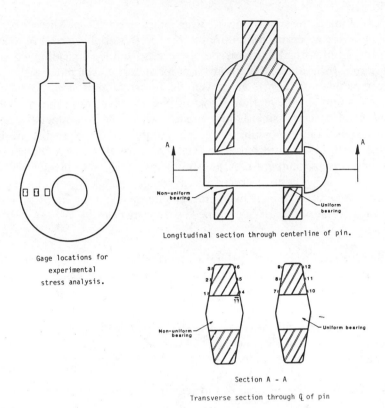

Gage locations for
experimental
stress analysis.

Longitudinal section through centerline of pin.

Section A - A

Transverse section through ₵ of pin

Figure 8. Power-line clevis characteristics.

specified level of Rockwell C32. However, the eye region where failure occurred had a maximum hardness of Rockwell C44. These hardness determinations were consistent with the microstructural characteristics.

Visual and low-power microscopic examination revealed that the eye at leg *D* had been bearing nonuniformly on the pin, as illustrated schematically in Figure 8. The effect of the nonuniform pin bearing, as well as the elastic-stress distribution for a uniform bearing condition, were determined experimentally with a tensile load test of a strain-gaged exemplar clevis. Peak strains for the nonuniform condition were approximately 2.5 times the peak strain for the uniform bearing condition.

Based on visits to the clevis manufacturer, heat treater, and galvanizer, it was deduced that hydrogen could have been induced into the material surface as a result of a pickling operation to remove an unacceptable zinc coating during galvanizing of one batch of clevises.

As a result of these and other activities, it was determined that clevis failures likely resulted from the combination of a hard hydrogen-embrittled surface and nonuniform loading on the pin that caused very localized stresses to exceed the material ultimate tensile strength. It is likely that all of these abnormal conditions were necessary for failure to occur in this application. For example, for a material of normal hardness, deformations from local yielding would like result in redistribution of the nonuniform loading from the pin.

Recommendations aimed at preventing future failures need to take into account not only the prevention of hydrogen embrittlement problems, but also prevention of other possible failure modes, where the recommended actions might increase the probability of such failure modes. Possible problems identified were inadequate tensile strength at normal ambient temperatures, brittle failure at low temperatures, and problems arising from inconsistent heat-treatment or quality-control procedures as a consequence of framing specifications that are difficult to meet using normal commercial practices. Consequently, recommendations were developed that satisfied all of the following objectives:

1. Prevent hydrogen embrittlement.
2. Ensure adequate tensile strength at normal ambient temperatures.
3. Ensure adequate toughness at minimum service temperature and maximum service loading rates.
4. Formulate specifications on heat-treatment and quality-control procedures that can be met using normal commercial heat-treatment practices, with a low percentage of rejects.

These recommendations were supported and confirmed through appropriate testing.

7.8. Coal-Gasification-Plant Bellows

A 304 stainless-steel toroidal bellows (Figure 9) was used to allow motion of a valve seat relative to the valve housing in a coal gasification plant [14]. The toroid had a 1 in. outside diameter with 0.065 in. wall thickness and was TIG welded to a 304 stainless-steel supporting ring. Cracks that penetrated entirely through the bellows thickness had developed parallel to the TIG weld after several days in service and only 150 cycles of operation.

Bellows service conditions were 650 psig internal pressure and temperatures up to 500°F. Fluid flowing through the valve was a mixture of water, steam, coal ash, and by-products of coal gasification. The toroid must withstand a flex of 0.032 inch.

Microscopic examination revealed that cracks were initiating at the base metal and weld metal interface on the outer surface of the bellows and propagating radially to the inner surface. General surfaces and crack surfaces were so corroded that no detailed metallurgical analysis of the fracture or the failure mode could be made. Corrosion pits were observed in several locations. Weld quality varied along the weld, and in some locations, undercutting was observed. Although branching at crack tips suggested an environmental

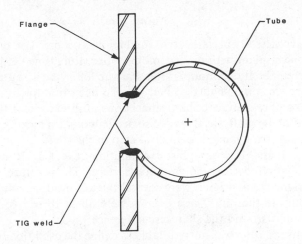

Figure 9. Bellows geometry.

contribution to crack growth, no aggressive species were found in several crack tip branches. However, potential attacking species were identified in an ash sample.

A stress analysis was performed using an axisymmetric finite-element model (Figure 10). Axial (z) displacements, internal pressures, and a lateral displacement of one flange relative to the other were applied to the model. Although the cyclic strains from axial load were significant, they were not large enough, by themselves, to induce failure (fatigue cracking) in the observed operating cycles. Cyclic strains from lateral motion that could possibly result from misalignments were higher and in the range that might cause fatigue-cracking.

Although the failure cause could not be conclusively established, strains from lateral misalignment, possibly assisted by environmental attack, appeared to be the most likely cause.

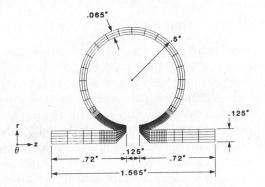

Figure 10. Model geometry and finite-element mesh.

7.9. Pump Shaft

After years of successful operation, two shaft failures occurred on one of the twin pumps in a power plant within a few months' operating time [15]. Each of the failures occurred about two months after start-up following an outage; however, the first failure was on a shaft that had been in operation for over 15 years.

Both breaks were complete shaft separations in threads near the end of the shaft. This end of the shaft was threaded to accommodate a locking nut that held a thrust disc in place against a diametral step in the shaft.

Metallurgical and fractographic examination revealed clear evidence of fatigue propagation across the fracture surface. The fracture, and several additional fatigue-crack origins, were several threads inside the nut rather than on threads outside the nut. These origins were all at thread roots and were evenly distributed around the shaft circumference.

Sufficient information was not available to define the cyclic loading experienced by the shaft. A review of limited vibration records did not reveal any clear indications of vibration changes over time and were not in locations from which loading data could be directly derived. Therefore, loading was studied parametrically in the stress analysis.

An axisymmetric finite model was developed of the shaft region near the failure location. This model included the thrust disc and the locking nut. With this model, a study was made of the effect of locking-nut preload, an external load (applied through the shaft and reacted by the thrust disc), and combinations of the two on cyclic stresses that could initiate fatigue cracks. Using stresses from this model, a study was also made of the size of the initial or initiated defect that could be expected to propagate to failure.

With the finite-element model, the effect of a possible loose locking nut was thoroughly studied. The amount of preload required to prevent separation, hence, high cyclic stresses, due to cyclic thrust-disc loading was also evaluated. In this manner, the combinations of preload and cyclic load that would result in the observed fatigue failures were ascertained.

7.10. Steam-Generator-Tube Dents

In some nuclear-power-plant steam generators, an interaction among the tube-hole crevice environment, tube, and support plate has caused a corrosion product to form on the carbon-steel support plate [16]. The corrosion product occupies more volume than the original metal; the tube-to-support plate crevice volume is, thus, consumed with corrosion product, and further corrosive action results in a radially inward force on the tube and a radially outward force on the corroding support plate. This has resulted in indentation (denting) of the tube, accompanied by occasional cracking. In very rare instances, the cracking has progressed through the tube wall. Large in-plane deformation and cracking of the support plate have also been observed in the most severely affected plants.

Some serious side effects, such as deformation and cracking of tube U-bends caused by support-plate deflections, have been observed.

Examination of tubes extracted from operating plants has revealed that stress corrosion cracking is the failure mechanism for cracked tubes. Thus, the combination of stress (strain), environment, and time to cause cracking is of interest. A considerable effort was directed toward determination of stresses (or strains) in dented tubes. The interest was first to determine the strains in dented cracked tubes, then to be able to estimate strains from in situ measurements of tubes in the field. Figure 11 illustrates idealized shapes of some dents that had been observed and were investigated. Strains in these dents were far into the plastic range.

Observed cracks at dents were axial, indicating that circumferential strains were of primary importance. Furthermore, changes in dent shape with axial station along the tube were generally gradual. For these reasons, circumferential strains were calculated for shapes such as those in Figure 11 as though the entire length of the tube had the same deformed cross section.

Three strain estimation approaches were used:

1. experimental methods
2. finite-element methods
3. approximate methods for field measurements

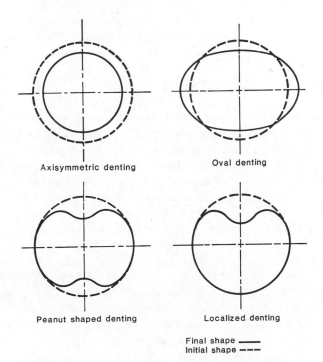

Final shape ──
Initial shape ─ ─ ─

Figure 11. Denting shapes under investigation.

The essential idea from this collection of methods was to establish the validity of the approximate methods for field strain measurements by comparing them with experimental and finite-element results. The finite-element method was used to establish this validity for a wider range of dent depths and shapes than was practical for laboratory testing.

Laboratory tests for strains at dents were accomplished by etching a finely spaced grid into the tube wall, then microscopically measuring the grid spacing before and after denting the tube. The finite-element method was used to analyze these same dent shapes, and demonstrated that the laboratory strain results could be reproduced analytically.

Approximate methods were then applied to the same dents. Probes to measure dents in the field consisted of eight radial measurements, equally spaced around the circumference, at numerous axial stations along the tube. Thus, the necessity was to estimate circumferential strains, given only the nominal tube radius and these eight radius measurements. To accomplish this, two types of functions were fit through the eight points to approximate the shape of the entire tube cross section: a Fourier series approximation and a cubic spline approximation. Once the approximate tube shape was established in functional form, tube circumferential strains would be calculated from kinematics. Tubes for which measured strains exceeded the strain threshold for stress corrosion cracking could be removed from service.

Both approximations gave reasonable results, especially for tubes with gradually changing dent shapes. Bounds were established on the accuracy of the approximate methods for several shapes and depths of dents.

7.11. Booster Tank Reentry

Introduction   Some multi-stage rocket boosters use liquid fuel stored in tanks. The boosters are separated from the upper stages and returned to Earth's atmosphere. Normally, these structures burn up during reentry, but to protect populated areas from impact by residual debris that may survive reentry, it is important to understand the mechanics of fracture and the break-up mode of the booster. The following discussion gives a treatment of the subject of booster break-up. A hypothetical liquid hydrogen tank is assumed for the study. After separation from an upper stage, a typical booster tank normally reenters the atmosphere with residual internal pressure in the tanks and oxygen feedline. Due to friction heating and the flow patterns induced by tumbling and proturbances, the thermal protection can burn off and the tank can develop hot spots that reach the melting temperature of aluminum ($\sim 1000°F$). The aluminum loses strength rapidly with increasing temperature, with an ultimate strength at 500°F that is less than half of the strength at room temperature. Consequently, at some altitude, the tensile strength of the tank may no longer be sufficient in the local hot-spot areas to contain the residual internal pressure, and the tank may rupture.

To maintain the desired footprint of tank debris, the tank must either break up relatively gently (pieces separating from the center of mass at low relative velocity) or break up at low enough altitude so that atmospheric-friction drag is sufficient to prevent a wide dispersion of pieces from the booster.

To characterize tank breakup during reentry, an investigation involving several tasks was conducted. These included:

1. a limited test program to confirm some analysis assumptions and enhance analytical models
2. stress and fracture-mechanics analyses to study crack initiation, and unstable propagation from hot spots

An important aspect of the study was to estimate the pressure to which the booster tank would need to be vented to prevent violent rupture. The investigation included a typical hydrogen cylindrical tank and dome, an oxygen tank, and an oxygen feedline. Since the basic approach for each structure was the same, discussion in this chapter is limited to the hydrogen cylindrical tank as an example.

The first task was intended to provide insight on the initiation of benign and violent rupture. For this purpose, a series of small-scale cylinders were pressurized and heated locally to produce failure. The cylinders had some features similar to the hydrogen-tank barrel, and the results of a previous study were used to design the tests. Both violent and benign failure modes were produced as planned, and the results led to a refined initiation model that gives good agreement with the tests.

The second task included stress analysis and fracture mechanics. The stress analysis used axisymmetric finite-element analysis, with orthotropic "smearing" of longitudinal stiffeners and thicker weld lands where appropriate. The stresses provided a theoretical basis for the prediction of the most probable crack-growth direction. They were also used in conjunction with the spatial distribution of temperature vs. time to predict initiation location, time, and orientation.

The fracture-mechanics analysis included both initiation and crack-arrest considerations. The initiation model considers the stable growth of a rupture zone, which occurs when the applied stress (including thermal effects) exceeds the ultimate strength of the material at the given temperature. The temperature distribution is found by interpolation in time and space from empirical data. The stress relaxation due to material "softening" and compressive thermal stresses is taken into account by a semiempirical equation derived from test results. Reduction of the gas pressure due to venting is considered, and the rupture size as a function of time is obtained as the solution of a coupled elastothermodynamic analysis. The rupture zone behaves like an unstable crack when its stress intensity exceeds a critical value. In the arrest model, there is assumed to be a running crack that tends to arrest at a stiffening member or weld land. The critical pressure is the pressure below which the tank would need to be vented in order for the crack to remain subcritical in the initiation model or become subcritical in the arrest model. In most cases, the critical pressure is predicted to be higher for the initiation model than the arrest model.

Cylinder Tests Three aluminum alloy 2219-T87 cylinders were used for testing crack initiation from a hot spot when subject to internal pressure. The cylinders were nominally 21 in. in diameter, 33 in. long, and 0.08 in. thick, with a ribbed test section reduced to 0.015 in. thickness. They were pressurized with helium and heated locally with a propane burner. Although no attempt at accurate scale-modeling was made, the fracture mode and test data provided vital information and verification of the assumptions used in the fracture-mechanics analysis.

Two cylinders were tested under low internal pressure (15 psi), with the tank skin heated locally up to about 880°F. The heated area was small enough so that the resulting rupture was smaller than the critical crack size. A stable self-arresting crack (benign rupture mode) with venting through the crack was observed. One cylinder was tested under high internal pressure (70 psi) with the tank skin locally heated to about 500°F. The heated area was large enough so that the initial rupture was bigger than the critical crack size. A fast unstable fracture (violent rupture mode) was observed.

The temperature of the cylinder skin was monitored by eight thermocouples placed on the inside in the heated area. The hoop strains were recorded by four strain gauges. The test articles were placed vertically in a cross-shaped trench, with video camera coverage via mirrors. One video camera was used to record a closeup of the crack-initiation area at a filming speed of 12,000 frames per second. A second camera was used for an overall shot of the ribbed area at a filming speed of 2000 frames per second.

Cylinder 1 was tested first, at 15 psi. Air in the propane line prevented the burner from producing a peak temperature over 500°F during the test. At that temperature, there was not any crack initiation, only local bulging of the skin. The propane plumbing was subsequently revised to prevent recurrence of the problem by adding a solenoid valve at the burner.

Cylinder 2 was tested under the same 15-psi gas pressure. Heating continued up to a peak temperature of 880°F. At this instant, a crack was formed and, as planned, it was shorter than the critical crack size. The opening was unfortunately oriented such that the escaping gas quenched or deflected the flame. Thus, there was no increasing hot-spot size, and, hence, no heat-rate-controlled crack growth. The venting area remained small, resulting in a very slow reduction of gas pressure from 15 to 12 psi over about two minutes.

The final crack size was measured after the end of the test. The total crack length was 0.56 in., with a crack opening of 0.035 in. over a length of 0.308 in., and prominent bulging in the vicinity of the crack. Comparison with the critical length obtained from Figure 12 for fracture toughness between 70 and 100 ksi(in.)$^{1/2}$ and pressure equal to 15 psi shows that the final crack length is about 20% of the critical length. Figure 12 was developed using a K solution for cylindrical shells with pressure stresses only (no thermal relief stress).

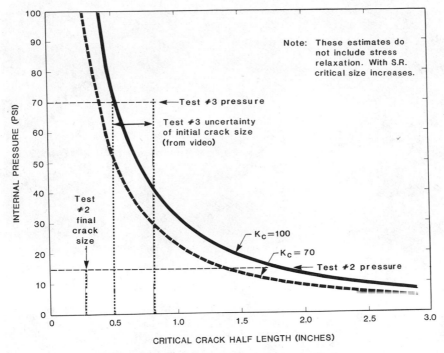

Figure 12. Critical crack length for test cylinders.

Cylinder 3 was tested under internal pressure of 70 psi, rupturing at peak temperatures on the order of 510°F. In this test, the goal was to create an initial crack with length greater than critical and simulate a fast-fracture phenomenon. Two propane burners were used instead of one to be sure of having a large enough hot spot. The total length of the two burners was 3.8 in., and the flame was slightly larger.

High-speed video images of the failure site and overall cylinder are shown in Figure 13 for the rupture event in test 3. These images are interpreted as follows. The upper right corner contains the elapsed time after starting the camera (23.7125 sec for Figure 13), increasing by 0.5 msec per frame. The right third of each frame shows an overall view of the ribbed portion of the test article. The black spots are the strain relief compound put on the strain-gauge leads. The bright area at the top is just above the location where the crack initiates. The left two-thirds of the frame shows six split images of the central channel between ribs, rotated 90°. Each image is 83 μsec later than the one above it, an effective speed of 12,000/sec. Only the top 10 in. are shown.

A crack of initial length of about 1.6 in. appeared just before the images in Figure 13. Its length is greater than the critical size of 0.7–1 in. predicted from the curve of Figure 12 at a pressure of 70 psi. Consequently, the crack would be predicted to grow in an unstable (dynamic fracture) mode, as is seen to be the

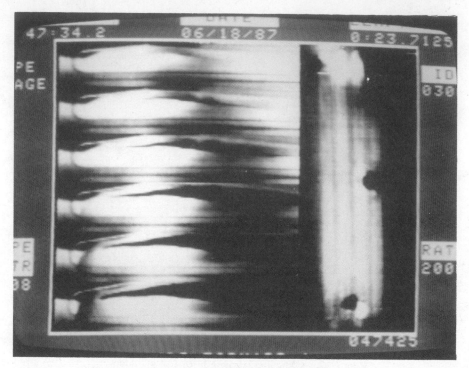

Figure 13.   Video of cylinder 3 at $t = 23.7125$ sec.

case in subsequent frames. The final state after the test is shown in Figure 14. Note that the crack branched and ran circumferentially just before reaching the end of the thinned section opposite the initiation point.

The crack-propagation velocity, as estimated from the video, is different for each of the two crack tips. The lower crack tip (right side of the six-way split image) propagates at a higher velocity, which is a function of time. On the basis of total crack running time of 1.16 msec and crack travel of about 16.3 in., an average velocity for the lower crack tip during the entire event is 1166 ft/sec. This high speed justifies the assumption that venting can be neglected once fast fracture begins.

Correlation Between Initiation Model and Cylinder Tests   This section describes the rupture criterion, shows how it was calibrated, and compares predicted vs. actual rupture temperatures for the tests. The rupture criterion is a refinement of the simple model that rupture occurs when and where

$$pR/h \geqslant \sigma_u(T) \tag{1}$$

This model underestimated the temperatures required to rupture the test cylinders by a significant amount, predicting rupture at about 700°F for test 2

Figure 14. Cylinder 3 after rupture.

(vs. 880°F actual), and about 400°F for test 3 (vs. 510°F actual). Consequently, an initial model with a thermal-stress-relief term was developed, i.e.,

$$\sigma_I - \sigma_r(T) \geqslant \sigma_u(T) \tag{2}$$

where σ_I is the principal stress perpendicular to the rupture, i.e., pR/h for a cylinder.

The thermal-stress distribution for a hot spot in an infinite plate is studied in Reference 17, with the result:

$$\sigma_r(T) = fE(T)\alpha(T - T_\infty) \tag{3}$$

where $f = 0.5$, E is Young's modulus at temperature T, $(T - T_\infty)$ is the temperature rise with respect to a uniform skin temperature away from the hot spot, and α is the coefficient of linear thermal expansion of aluminum. The solution is based on the hot spot remaining in the same plane as the cooler surrounding plate. This solution can be extended to a shell by adjusting the factor f to account for curvature and other geometry effects and stress redistribution. However, this approach is valid only for small hot spots, where the local heating effect is constrained by elastic shell behavior around the hot spot. If applied with $f = 0.5$, it overpredicts the thermal stresses in a shell, because it does not consider bulging.

Combining Equations (2) and (3) with the pressures and temperatures at rupture in tests 2 and 3 provides two equations with one unknown quantity, the factor f, which is considered to be independent of temperature. Since there are two estimates, a value of f of 0.46 is found, so that the error in predicting rupture temperature is approximately equal and opposite in sign for tests 2 and 3. This value of f yields failure-initiation temperatures equal to 895°F for test 2 and 498°F for test 3. The percentage error with respect to the maximum measured temperatures of 880 and 510°F are approximately equal, opposite in sign, and less than 3%. Having determined the value of $f = 0.46$ for the cylinder tests, the same value is utilized in the analysis of the tank.

Stress Analysis   Figure 15 illustrates the hydrogen-tank geometry. The purpose of the stress analysis is to determine the initial stresses in the last barrel (adjacent to the aft dome) of the tank and the subsequent redistribution of stress that occurs when an axial crack propagates from the aft end of the barrel to the front end. Similar but separate analyses are performed for the cracked and uncracked tank.

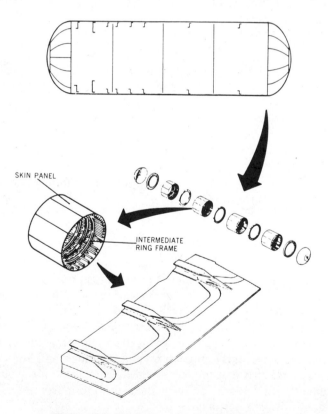

SKIN PANEL

INTERMEDIATE
RING FRAME

Figure 15.   Geometry of a hydrogen tank.

An axisymmetric finite-element shell model was used to perform the stress analysis for this portion of the tank. To eliminate end effects, the model included the entire aft dome, the last barrel section, and half of the adjacent barrel. The major ring frames at stations 4 and 8 and the intermediate ring stiffeners within each barrel were also modeled.

At station 1, axial displacement and rotation were constrained. At the apex of the ellipsoidal dome, radial displacement and rotation were constrained. Internal pressure loading was applied to the shell elements representing the tank skin. Other loads present but not applied to the model (thermal gradients, vibration, aerodynamic and inertial effects) are either assumed to be negligible or neglected in order to simplify the analysis.

Axisymmetric shell elements were used to model the tank skin, major ring stiffeners, and intermediate ring stiffeners using the MARC code. Special axisymmetric beam elements modeled the longitudinal stiffeners within the barrel sections. These axisymmetric shell and beam elements, when used in conjunction with each other, are well suited for modeling the behavior of axially stiffened cylindrical shells. These two elements basically have identical isoparametric formulations, modified to satisfy thin-shell and curved-beam theory, respectively, for the shell and beam elements.

An elastic analysis was performed for the uncracked tank. For the cracked tank, three nonlinear elastic–plastic analyses were performed to determine the redistribution of stress that occurs when an axial crack propagates in the last barrel of the tank. Ideally, three dimensional finite-element modeling and analyses would be used for this problem, since an axially cracked tank is not axisymmetric. However, with the use of orthotropic properties, the presence of a crack can be approximately simulated using the axisymmetric finite-element model, saving considerable time and expense. The orthotropic properties have the full elastic modulus in the axial direction, but a greatly reduced elastic modulus in the circumferential direction to simulate the lack of hoop strength across the crack. The remainder of the model is assigned elastic–plastic material properties.

The finite-element stress-redistribution analyses were performed using the MARC general-purpose nonlinear code. An elastic–plastic solution was obtained by incrementally applying the pressure to the inside of the tank. Incremental plasticity and isotropic hardening rules were employed. The updated Lagrangian formulation was used to account for large-displacement and finite deformation effects. The nonlinear analysis is essential; an elastic analysis gives meaningless results in the cracked configuration.

The axisymmetric shell and beam element in this analysis use cubic displacement fields, which provide for strain variation within each element. This feature makes these elements well suited for the nonlinear large-displacement analysis required to determine stress redistribution in the cracked tank.

The first analysis was performed for a tank with a crack extending the full length of the last barrel section. The purpose of this analysis was to calculate the redistribution of hoop stress from the axially cracked skin to the intermediate ring stiffeners to determine the pressure at which these intermediate rings fail under net tensile overload. A second stress-redistribution analysis was performed, identical to the first, except that the intermediate ring stiffeners in the cracked barrel were removed. The objective of this analysis was to determine the stress redistribution and critical pressure for the major ring frames at the ends of the cracked barrel *after* the intermediate rings have failed. A third stress-redistribution analysis was performed for a tank with a partially cracked last barrel section. This analysis determined the stress in the intermediate ring at station 6 as the crack runs, to investigate whether the ring will fail as the crack passes it. Figure 16 is an example of the type of results obtained from these analyses.

The results from the first analysis (full-length crack with intermediate rings intact) indicate that, at an internal pressure exceeding 24 psi, the intermediate ring at station 6 is unable to hold the split barrel together, and it will fail due to tensile overload. Subsequent transfer of this load to the smaller intermediate rings at stations 5 and 7 will cause them to also fail in tension. At pressures above 24 psi, the hoop stress redistributes entirely to the major ring frames at stations 4 and 8, since the intermediate rings have failed. Consequently, the first analysis is only valid to 24 psi; above that, the second analysis is appropriate.

The results from the second analysis (full-length crack with intermediate rings removed) give the stress redistribution that occurs on the major ring frames at pressures above 24 psi. First, the major ring frame at station 4 is more heavily

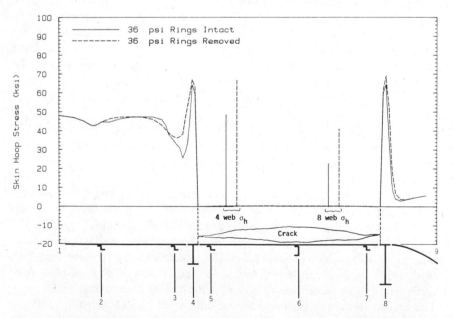

Figure 16. Comparison of hoop-stress plots for a tank with a full-length simulated crack, with and without intermediate rings ($p = 36$ psi).

loaded than the one at station 8. This is due to the restraining effect of the aft dome, which is also welded to the ring at station 8. Second, at 36-psi internal pressure, the hoop stress in the web at station 4, although past initial yield (61 ksi), has not yet reached ultimate. However, this analysis, like the other two stress-redistribution analyses, has the restriction that the simulated crack size cannot be changed during the course of the analysis.

It is possible to approximate the stress redistribution from an additional increment of crack growth by applying the load in the top flange of station 4 (between the web and assumed crack tip) to the web. This assumes that the crack progresses through the top flange to a point where the tip is lined up with the web. With this small additional crack extension at 36 psi, the stress in the web at station 4 exceeds the ultimate stress (83 ksi). It is concluded then that at 36 psi, a crack propagating axially in the last barrel section of the tank will not only fail the intermediate rings by tensile overload, but the major ring frame at station 4 will also fail. Furthermore, since this last barrel section is the strongest of the four barrels that make up the tank, an axially running crack in the hydrogen tank at 36 psi would be predicted to cause a catastrophic failure of the entire tank.

The third analysis (partially cracked tank) shows that the intermediate ring at station 6 does not reach yield for tank pressures up to 36 psi. One of the considerations of this last analysis was to determine the force in the intermediate ring (station 6) as the crack approaches. Since the ring does not fail until the crack passes, it can contribute to crack arrest. This information is necessary for the subsequent fracture-mechanics analyses of the cracked tank.

The stress-redistribution analyses have determined the pressures when various parts of the tank fail under tensile overload. It is important to note here that failures can occur due to crack propagation at lower stresses, as analyzed in the next section.

Crack-Initiation, -Arrest, and -Branching Analysis An initiation model was developed in this work with the following primary steps:

1. determine spatial and temporal temperature distribution
2. estimate strength and net stress perpendicular to the assumed crack path at all locations along the crack, based on temperatures from step 1
3. apply a rupture criterion to determine the length of rupture along the crack path
4. reduce the initial pressure as a function of time by blowdown through the time-varying rupture orifice
5. calculate stress intensity from net stress, assuming the rupture behaves as a crack of the same length
6. compare stress-intensity factor with fracture toughness to determine whether the crack will run unstably

Note that in this model, the rupture area initially grows slowly with time, limited by the rate of heating. Only when it exceeds the critical length for a sharp crack does it run at dynamic-fracture speeds.

At each point along the assumed crack path, the pressure-related stresses and thermal relief stresses are calculated and compared to the ultimate strength, as in the rupture criterion of Equations (2) and (3). The finite-element pressure stresses were used, linearly adjusted to account for the current pressure P from the venting calculations. The entire length for which the inequality of Equation (2) is satisfied is considered ruptured. The shell factor f in Equation (3) was assumed to be 0.46, as calibrated in the cylinder tests.

The time at which blowdown begins is t_i, at which time, theoretically, only one point has reached a temperature where rupture will occur. Fast fracture begins at t_c, after which complete tank rupture is assumed to occur within milliseconds, so no further blowdown calculations are required.

The stress-intensity factor is calculated from the net stress, that is, $\sigma_I - \sigma_r$. This is equivalent to the simplifying assumption that the thermal relief stress is constant over the crack length. It is also assumed that the rupture zone behaves like a sharp crack of equivalent length, going unstable when $K_I \geqslant K_c$. The applied stress intensity K_I was calculated by a recently developed approach [18] that gives the results illustrated in Figure 17.

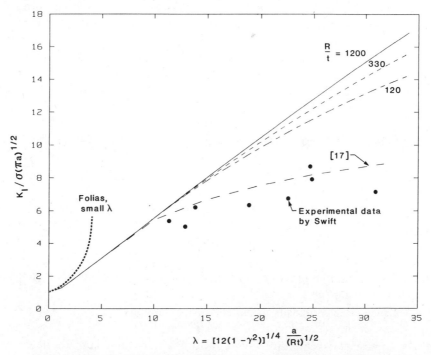

Figure 17. Stress-intensity coefficient for a cylindrical geometry.

The following basic assumptions were made in the crack-arrest model. First, a running crack exists, moving so fast that depressurization can be neglected. Second, there is a change in the thickness or stiffening rib or ring in the path of the crack that will substantially reduce the stresses and stress intensity K_I for the running crack. In the case of a stiffening rib or ring, there are also pinching forces equal to the material yield strength multiplied by an appropriate area, which will tend to hold the crack closed. Third, the crack arrests if $K_I \leqslant K_c$, where K_i is calculated with the stress field over the entire length, including the change in section, reduced by K_R arising from the pinching forces, if any. An examination of the merits of including dynamic effects (material inertia) in the calculations was conducted. It was decided that quasi-static analysis was sufficient for analysis of a typical booster.

There is some difficulty determining the appropriate value of K_c to use for determining when the critical stress intensity is reached, both in the initiation and arrest models. Examination of the sparse data that were available resulted in selection of 70 to 100 ksi(in.)$^{1/2}$ as a credible range for the thin sections of tank.

The crack-branching model used the simple criterion that whenever the applied strain-energy release rate (K^2/E) exceeds twice the critical strain-energy release rate (K_c^2/E), there can be two active crack tips, hence, crack branching. Thus, for $K \geqslant 2^{1/2}K_c$, there can be branched cracks. The higher the pressure, the more branches and greater likelihood of fragment separation.

Crack-curving cannot be predicted analytically, but it is likely to occur in dynamic fracture. Consequently, all crack paths are highly uncertain, particularly when the principal stresses are similar in magnitude.

The fracture-mechanics model was applied to the hydrogen-barrel cracking problem for a particular trajectory and thermal conditions. At full pressure, 36 psi, initiation of an unstable (dynamic) crack in the barrel near station 7 is expected at 4862 sec for the conditions studied. The critical hot spot or crack size is less than 6 in. The corresponding altitude is 262,000 ft. Crack-branching and fragmentation are expected.

Venting of the booster tanks is one of several alternatives for preventing violent rupture. The initiation model was applied to determine the required venting pressure to prevent initiation of an unstable crack in the barrel. If the pressure is less than 2.4 psi by $t_i = 4880$ sec, a rupture will initiate but safely vent the tank before reaching $K_c = 100$ ksi(in.)$^{1/2}$.

Barrel crack-arrest calculations were used to find the critical pressures for arrest of a longitudinal crack. These were approximately 2.7 psi at the ring frame at station 6 and 4.5 psi at the major ring frame at station 4. These are comparable to the required pressure to prevent initiation of an unstable crack.

Significant Uncertainties in Results The initiation model used in this study includes a thermal-stress relief term that was calibrated to the cylinder test results. The approach is simple, but it does not model the bulging typical of heated pressurized shells. In the cylinder test, the hot spot was small relative to the size of the cylinder and could be considered well-constrained by elastic material around it. It is unknown what the coefficient f in the thermal-relief term should be when the hot spots are relatively large. In the limiting case, when a cylinder is hot along an axial line, the thermal-relief stress would be zero.

In the cylinder tests, it was demonstrated that fracture mechanics is relevant for predicting the critical size of rupture at which unstable crack propagation would occur. However, the problem was only bounded by those tests, with one test at a subcritical crack size and another with a critical initial rupture. Consequently, the point of transition between stable and unstable growth from a thermally induced rupture is not well established. If the initiation mode is to be used for design predictions, three key parameters in the model must be better defined. These are strength properties in the high-temperature regime (e.g., $E(T)$, $\sigma_u(T)$ for $T > 600°F$), fracture toughness in thin material at temperatures of $T > 600°F$, and parameter f as a function of hot-spot size or degree of elastic constraint.

The spatial-temperature distribution vs. time is the key input to the initiation model, since that model assumes that the size of the rupture area does not grow any faster than the material heats up. Since the critical flaw sizes are on the order of 5–20 in., whereas the grid spacing for temperature calculations is several feet, a significant improvement in predictive capability could be obtained if the reentry analysis had a finer grid.

8.   CONCLUDING REMARKS

A variety of analytical and laboratory approaches are available for analyzing structural failures. A number of excellent approaches and their applications are presented in the foregoing chapters of this book. However, selection of which tools to apply to a particular failure is likely to remain an experience-based art for the foreseeable future. The attempt to synthesize the failure-analysis process and illustrate it by examples in this chapter is a start.

REFERENCES

1. P. R. Johnston and W. Weaver, Jr., *Finite Elements for Structural Analysis*, Prentice-Hall, Englewood Cliffs, NJ, 1984.
2. P. R. Johnston and W. Weaver, Jr., *Structural Dynamics by Finite Elements*, Prentice-Hall, Englewood Cliffs, NJ, 1987.
3. R. J. Roark, *Formulas for Stress and Strain*, 5th ed., McGraw-Hill, New York, 1976.
4. H. O. Fuchs, D. V. Nelson, M. A. Burke, and T. L. Toomay, *Shortcuts in Cumulative Damage Analysis*, SAE Publ. No. 730565, Soc. Automot. Eng., Warrendale, PA, 1973.
5. N. E. Dowling, "Fatigue Failure Predictions for Complicated Stress-Strain Histories," *J. Mater.*, **7**(1), 71–89 (1972).

6. R. A. Sire, P. M. Besuner, C. C. Schoof, and J. M. Thomas, *Techniques for Fatigue Life Predictions from Measured Strains*, Failure Anal. Assoc. Rep. FaAA-I-84-6-1, June 1984.

7. J. M. Thomas, D. O. Harris, and P. M. Besuner, "Fracture Mechanics Life Technology," *AIAA/ASME/ASCE/AHS Struct., Struct. Dyn. Mater. Conf. 27th, 1986* (1986).

8. J. M. Thomas, "Probabilistic Analysis of Contributions to OTSG Tube Cracking," *1985 Am. Soc. Mech. Eng. Tech. Conf.* (1985).

9. G. Derbalian, G. Fowler, and J. Thomas, "Three-Dimensional Finite Element Analysis of a Scale Model Nuclear Containment Vessel," *Am. Soc. Mech. Eng.* [*Pap.*] **84-PVP-55** (1984).

10. C. A. Rau, Jr., J. M. Thomas, and S. W. Hopkins, *Evaluation of the Structural Integrity of the Modified Kool Flow Design with Fan*, Failure Anal. Assoc. Rep. FaAA-77-3-3, FAA, 1977.

11. Private communication, J. D. Osteraas to Gil Malmgren, 1984.

12. L. E. Eiselstein and R. D. Caligiuri, "Atmospheric Corrosion of the Suspension Cables on the Williamsburg Bridge," *ASTM Spec. Tech. Publ.* **STP 965** (1987).

13. J. N. Robinson, J. M. Thomas, and J. D. Osteraas, *Analysis of Clevis Failure and Recommendations to Reduce the Probability of Future Failures*, Failure Anal. Assoc. Rep. FaAA-84-5-1, May 1984.

14. J. M. Thomas, H. F. Wachob, and J. D. Osteraas, *Investigation of Bellows Failure in Coal Gasification Plant*, Failure Anal. Assoc. Rep. FaAA-84-3-7, March 1984.

15. S. Hopkins et al., *Feedwater Pump Shaft Failure Investigation*, Failure Anal. Assoc. Report FaAA-PA-R-86-10-17, January 1987.

16. S. A. Rau, G. K. Derbalian, and J. M. Thomas, *Steam Generator Tube-Plugging and Tube-Sleeving Criteria: Assessment of Current Practices*, NP-29214, Electr. Power Res. Inst., Palo Alto, CA, 1983.

17. J. F. Harvey, *Theory and Design of Pressure Vessels*, Van Nostrand-Reinhold, New York, 1985.

18. D. O. Harris and R. A. Sire, "J-Integral Estimates for Through-Wall Axial Cracks in Cylindrical Shells with Power-Law Hardening Materials" (to be submitted for publication).

Index